Convexity and Its Applications

Edited by

Peter M. Gruber, Vienna
Jörg M. Wills, Siegen

1983

Birkhäuser Verlag
Basel · Boston · Stuttgart

Editors' addresses

Prof. Dr. Peter M. Gruber
Technische Universität Wien
Institut für Analysis, Technische
Mathematik und Versicherungsmathematik
Gusshausstrasse 25–27
A-1040 Wien (Austria)

Prof. Dr. Jörg M. Wills
Universität Siegen
Lehrstuhl für Mathematik II
Hölderlinstrasse 3
D-5900 Siegen 21 (FRG)

Library of Congress Cataloging in Publication Data

Convexity and its applications.
1. Convex sets Addresses, essays, lectures. 2. Convex bodies Addresses, essays, lectures. 3. Convex functions Addresses, essays, lectures. I. Gruber, Peter M., 1941- . II. Wills, Jörg M., 1937- .
QA640.C66 1983 516'.08 83-11859
ISBN 3-7643-1384-6

CIP-Kurztitelaufnahme der Deutschen Bibliothek

Convexity and its applications / ed. by Peter M. Gruber ; Jörg M. Wills. Basel ; Boston ; Stuttgart : Birkhäuser, 1983.
ISBN 3-7643-1384-6

NE: Gruber, Peter M. [Hrsg.]

Printed in Switzerland
ISBN 3-7643-1384-6

Contents

Preface

This collection of surveys consists in part of extensions of papers presented at the conferences on convexity at the Technische Universität Wien (July 1981) and at the Universität Siegen (July 1982) and in part of articles written at the invitation of the editors. This volume together with the earlier volume ≪ Contributions to Geometry ≫ edited by Tölke and Wills and published by Birkhäuser in 1979 should give a fairly good account of many of the more important facets of convexity and its applications. Besides being an up to date reference work this volume can be used as an advanced treatise on convexity and related fields. We sincerely hope that it will inspire future research.

Fenchel, in his paper, gives an historical account of convexity showing many important but not so well known facets. The articles of *Papini* and *Phelps* relate convexity to problems of functional analysis on nearest points, nonexpansive maps and the extremal structure of convex sets. A bridge to mathematical physics in the sense of Polya and Szegö is provided by the survey of *Bandle* on isoperimetric inequalities, and *Bachem's* paper illustrates the importance of convexity for optimization. The contribution of *Coxeter* deals with a classical topic in geometry, the lines on the cubic surface whereas *Leichtweiss* shows the close connections between convexity and differential geometry. The exhaustive survey of *Chalk* on point lattices is related to algebraic number theory. A topic important for applications in biology, geology etc. is covered by *Weil's* contribution on stereology. Modern surveys on classical topics in convexity are due to *Chakerian* and *Groemer* who consider bodies of constant width, *Fejes Tóth* whose article is on discrete geometry, *Petty* who exhibits many interesting results on ellipsoids and *Gruber*. The latter considers the problem of approximation to convex bodies. In a joint paper *Schneider* and *Weil* deal with the theory that has grown around the notion of zonoid and *McMullen* and *Schneider* wrote a long survey concerning valuations on convex bodies, a rapidly expanding new area. Polyhedral manifolds related to the Platonic solids are dealt with by *Wills*.

We are obliged to the authors who willingly accepted our invitation and who made every effort to make the surveys as informative as possible. Collaboration with them was a very pleasant experience for both of us. Many thanks are due to the Institut für Mathematische Analysis of the Technische Universität Wien, the Universität Siegen and the Research Council of Canada for providing financial support. We should also like to thank Mr. Bender from Birkhäuser for his most agreeable assistance during the publication of this volume.

Vienna and Siegen, 1983 *Peter M. Gruber* and *Jörg M. Wills*

Achim Bachem*

Convexity and Optimization in Discrete Structures

July 1982

Abstract

In this paper we survey some very recent trends in combinatorial optimization with special emphasis on matroid and oriented matroid theory. We give a brief overview of the theory of ℕℙ-completeness, discuss the interdependence of duality theorems and "good" algorithms, report on duality type theorems in combinatorial optimization and introduce oriented matroid face lattices. We also discuss a discrete version of convexity for oriented matroids and show how Linear Programming can be generalized to oriented matroids.

1. Introduction

Combinatorial optimization with all its branches has experienced a tremendous growth during the last 10 years. Moreover, recent directions of research led to an enormous progress in the area of discrete optimization. These include Cook's (1971) result in computational complexity theory-showing that all combinatorial programming problems are equally hard so that as a consequence investigations of combinatorial and algorithmic structures are the only keys for new efficient approximative algorithms-, the development of (oriented) matroid theory uncovering the combinatorial structure of discrete and linear programming problems, the fastly developing theory of submodular functions and more recently Khachian's polynomial algorithm for linear programming problems.

The purpose of this paper is to provide a summary of parts of the theory of combinatorial optimization as it is developed today. Clearly choosing topics of such an exposition the author's personal taste has played an important role. The emphasis lies on matroid and oriented matroid theory. For other topics of combinatorial optimization the reader is referred to the following recent surveys: Bachem (1982) (Concepts of algorithmic computation), Bachem and Grötschel (1982) (New aspects of polyhedral theory), Burkard and Zimmermann (1982) (Combinatorial Optimization in Ordered algebraic structures, see also Zimmermann (1982)), Grötschel (1982) (Approaches to hard combinatorial optimization problems), Korte (1982) (Matroids and independence systems) and Schrader (1982) (Ellipsoidal methods). A further source of good surveys is the book of Hammer, Johnson and Korte (1979) (these are the proceedings of the Discrete Optimization Symposium in Vancouver and Banff 1977 containing 24 surveys papers on integer programming). Some recent textbooks may also give an excellent overview of combinatorial optimization (cf. Garey and Johnson (1979), Lawler (1976), Hu (1982) and Papademitriou and Steiglitz (1982)). For a collection of combinatorial problems and exercises cf. Lovász (1979).

*Supported by Sonderforschungsbereich 21 (DFG), Institut für Operations Research, Universität Bonn

A comprehensive overview of the current literature in combinatorial programming gives the book of Randow (1982) (this is a classified bibliography on integer programming and related areas containing also a list of 119 survey papers which appeared between 1978 and 1981 in the field of integer programming).

The paper is organized as follows. We start in section 2 with an introduction of combinatorial optimization and give a short survey of complexity theory. In particular we describe precisely what the notions "easy", "hard" and "intractable" mean.

Section 3 introduces independence systems and matroids and in section 4 we show how duality theorems for a particular type of problems lead to "good" algorithms for these problems. Moreover we show how duality theorems and stopping rules for algorithms relate to each other. Because of this interrelation it is very interesting and important to characterize those discrete structures which allow a duality type theorem. It turns out that matroids, oriented matroids and special independence systems are the most general structures for this kind of duality theorems. We finally present a (asymmetric) duality type theorem for general discrete optimization problems involving subadditive functions and relate it to duality theorems involving only convex functions.

Convexity is one of the most important concepts in continuous optimization. Since combinatorial optimization deals with discrete structures it is not clear how this useful concept may translate to discrete optimization. In section 5 we discuss a version of a convexity definition for discrete structures. We first show how polyhedra theory generalize to a theory of oriented matroid polyhedra allowing a proof of a generalized Upper Bound Conjecture and a shellability and constructability theorem. We also show that polyhedra polarity results do not generalize to oriented matroid polyhedra and report that Hahn-Banach type theorems can not hold in oriented matroids. Thus, the discussions for an appropriate convexity concept in discrete structures is still open. Finally we show how Linear Programming can be generalized to oriented matroids.

2. Theory of $\mathbb{NP}$-completeness

Let E be a nonempty finite set, $I \subseteq 2^E$ and $f: I \to \mathbb{R}$. A *combinatorial optimization problem* (*CO*) is the task to determine a "feasible solution" $S \in I$ which maximize or minimize f over I or to conclude that no such $S \in I$ exists. We denote such a problem as

$$\text{CO:} \quad \min_{S \in I} f(S)$$

Computational complexity theory has developed a mathematical precise frame to classify problems according to their "inherent complexity". Here a *problem* is meant as a general question allowing several parameters left unspecified. Setting these parameters to special values gives then an *instance* of this problem. In combinatorial optimization "worst case time" is usually the only efficiency measure which is considered. The *worst case complexity* of a problem P is a function f which assigns to each length L the minimum time which the "best" algorithm (verified on a RAM or a Turing machine) needs to solve problem instances of P which have input length L (the input considered as a binary encoded string). Problems with a polynomial time complexity function are sometimes called "*easy*", "*efficient*" or "*polynomial*" solvable. The class of all "easy" problems is denoted by $\mathbb{P}$. Two combinatorial problems whose time complexity function agree up to a polynomial transformation are called (algorithmically) *polynomial equivalent*. See Aho, Hopcroft and Ullmann (1974) or Garey and Johnson (1979) for text books and Bachem (1982) or Tarjan (1978) for surveys of the theory of computational complexity. Well-known examples of polynomial solvable problems are:

(2.1) *The spanning tree problem*

Let $G = (V,E)$ *be a connected graph with vertices* V *and edge set* E. *Let* c_e $(e \in E)$ *be weights for the edges of* G. *With* $I = \{T \subseteq E | (V,T)$ *is a spanning tree of* $G\}$ *and* $f(T) = \sum_{e \in T} c_e$, *the spanning tree problem is max* $\{f(T) | T \in I\}$.

See Prim (1957), Dijkstra (1959) and Cheriton and Tarjan (1976) as examples for efficient algorithms solving the spanning tree problem.

Linear Programming has recently been discovered to be polynomial solvable. As a combinatorial optimization problem a linear programming problem can be stated as follows.

(2.2) *Linear programming*

For $E = \{1,\ldots,n\}$, A *an* (m,n) *real matrix,* $c \in \mathbb{R}^n$ *and* $b \in \mathbb{R}^m$ *let* $I :=$ $\{B \subseteq E | (A_{\cdot B})^{-1}$ *exists and* $(A_{\cdot B})^{-1}\, b \geq 0\}$. *Determine* $B \in I$ for which $c_B (A_{\cdot B})^{-1} b$ *is maximal.*

The well-known simplex algorithm for solving linear programming problems was shown to have exponential running time on some artificial examples (see Klee and Minty (1972) or Jeroslow (1973)). But recently Borgwardt (1982) and Smale (1982) showed that on the average the simplex algorithm needs only a polynomial number of pivot steps assuming identically and independently distributed data.

Khachian (1979) proved that linear programming has polynomial time complexity. He used the so called ellipsoid algorithm of Shor (1970) which unfortunately turned out to be numerically very unstable. See Schrader (1982) or Goldfarb and Todd (1982) for surveys on modifications and implementations of the ellipsoid method as well as on some of its historical antecedents. The ellipsoid method has many applications in combinatorial optimization. First studies in this direction are done by Grötschel, Lovász and Schrijver (1981).

Another interesting example of a polynomial solvable combinatorial optimization problems is:

(2.3) *The bipartite matching problem*

Let $G = (V,E)$ *be a bipartite graph with a partition* $V = V_1 \cup V_2$ *into stable sets of nodes (i.e. all nodes are pairwise non adjacent). Let* $c : E \to \mathbb{R}$ *be a function assigning weights to the edges. The bipartite maximum matching problem is then to determine a set of edges such that no two edges have a common endnode and whose sum of edge weights is as large as possible.*

The special case $|V_1| = |V_2|$ is well-known under the name assignment problem. Here we have $n = |V_1|$ persons and $n = |V_2|$ jobs. The weight c_{ij} is a qualification measure for person i to successfully do job j. The assignment problem then is to find an assignment of persons to jobs such that the total qualification measure is as high as possible.

Many interesting solution methods for matching problems are discussed in Burkard and Derigs (1980) and Lawler (1976).

For most combinatorial optimization problems there is no polynomial algorithm known. But for some of these problems the proof that a given solution is indeed a solution can be done in polynomial time. This very interesting class of problems is called $\mathbb{NP}$ (nondeterministically solvable in polynomial time) and the best known example is:

(2.4) *The asymmetric travelling salesman problem*

Let $D = (V,A)$ *be a complete digraph with vertices* V *and arcs* A. *Let* $c: A \to \mathbb{Q}_+$ *be a function assigning distances to all arcs. The asymmetric travelling salesman problem then is to determine a hamiltonian dicycle of minimum total distance, i.e. find a tour of minimum length which visits every node exactly ones.*

A combinatorial optimization problem min $\{f(S)|S \in I\}$ whose optimal function values are integral and bounded by a polynomial (for all instances in the respective input lengths) is polynomial equivalent to the decision problem: "Given a $k \in \mathbb{R}$. Does there exist an $S \in I$ with $f(S) < k$?". Clearly, if the optimal function value lies in the interval $[-a,a]$ we can use the well-known method of bisection to determine the optimal value by solving the decision problem log(2a) times.

Hence the asymmetric travelling salesman optimization problem is polynomial equivalent to the decision problem: "Given a complete digraph $D = (V,A)$, a distance function $c: A \to \mathbb{Q}_+$ and a number $k \in \mathbb{R}$. Does there exist a hamiltonian cycle H of D with total length less than k?". This problem belongs to the class $\mathbb{NP}$. If somebody presents a hamiltonian cycle claiming its total length being smaller that k we can easily (in polynomial time) proof this claim by just adding up all edge lengths. But note, the complement of the class $\mathbb{NP}$ (*the class co-*$\mathbb{NP}$) must not necessarily have this property, i.e. is not necessarily a subset of $\mathbb{NP}$. The complement of the asymmetric travelling salesman decision problem would be: "Given a complete digraph $D = (V,A)$ a distance function $c: A \to \mathbb{Q}_+$ and a number $k \in \mathbb{R}$. Does there exist no hamiltonian cycle H of D with total length less than $\mathbb{K}$?". Proving this no-question seems to be much harder than the corresponding yes-question. It is still unsettled whether $\mathbb{NP}$ = co-$\mathbb{NP}$ or not (cf. Garey-Johnson (1979)). As we shall discuss in section 4 the problems of the class

$\mathbb{NP} \cap$ co-$\mathbb{NP}$ are interesting candidates for an algorithmically powerful duality theory.

R.M. Karp (1972) developed the theory of $\mathbb{NP}$-completeness in 1972. A problem $P \in \mathbb{NP}$ is called $\mathbb{NP}$-*complete* if the polynomial time solvability of P would conclude the polynomial time solvability of any problem in $\mathbb{NP}$. The class of $\mathbb{NP}$-complete problems is denoted by $\mathbb{NPC}$. $\mathbb{NP}$-complete problems may be considered as the hardest problems within the class $\mathbb{NP}$. Karp started in 1972 with a few problems showing their $\mathbb{NP}$-completeness. Today some thousands of combinatorial problems are known to be $\mathbb{NP}$-complete (see Garey-Johnson (1979) for a list of about thousand $\mathbb{NP}$-complete problems). Among them are the travelling salesman problem and the general integer programming problem. $\mathbb{NP}$-complete problems are known from various areas such as graph theory, network design, storage and retrieval, sequencing and scheduling, mathematical programming, algebra and number theory, games and puzzles, logic, automata and languages, etc. It is very unlikely that all these problems are polynomial solvable. Hence the well-known hypothesis $\mathbb{P} \neq \mathbb{NP}$ is most probably true. There are however some results known which make it likely that the statement "$\mathbb{P} \neq \mathbb{NP}$" or "$\mathbb{P} = \mathbb{NP}$" is undecidable.

Informally $\mathbb{NP}$-complete problems are often called "*hard*" problems. Although these hard problems can not be solved (up to day) in polynomial time many (sometimes very efficient) solution techniques are well-known. Good surveys of those techniques can be found in: Balas and Padberg (1975), Burkard (1980), Fisher (1981), Grötschel (1982) and Klee (1980).

The class of problems which have exponential time complexity functions is denoted by $\mathbb{EXP}$. Clearly, those problems can never be solved efficiently regardless how fast computers ever will be. Informally they are called intractable. See chapter 7 of Garey and Johnson (1979) for examples of such problems. The class $\mathbb{EXP}$ does not play a very important role in combinatorial optimization since only very few combinatorial optimization problems are known to belong to $\mathbb{EXP}$.

Indubitable the theory of $\mathbb{NP}$-completeness is a very important tool in combinatorial optimization. But there is also some criticism. Linear programming problems and the simplex algorithm have shown that the worst case analysis may not be the best efficient measure. The simplex algorithm has an exponential worst case behaviour but works pretty well in practice. On the other hand an average time analysis may also be useless because people in practice do not know distribution functions of "real world" data. Another point of criticism regards the input length as a criterion for the size of the problem instance. Consider for instance the knapsack problem which is known to be $\mathbb{NP}$-complete: $\min\{cx \mid ax \leq b, x \in \{0,1\}^n\}$. There are algorithms for the knapsack problem whose time functions are a polynomial in b. Clearly $b \geq \exp$ (encoding of b), hence those algorithms do not solve the problem in polynomial time. But as far as b is not too big those algorithms work very fast and today real world knapsack problems with up to 5000 variables can be solved efficiently.

The knapsack problem is one example of a class of problems whose $\mathbb{NP}$-completeness is based only on the fact that numbers may become very big when the instance size increases. Garey and Johnson (1978) called such problems *number problems*. There are also so called *non-number problems* where all involved numbers do not increase when the size of the problem instance becomes bigger. The problem of determining whether a given graph contains a hamiltonian cycle is an example of an $\mathbb{NP}$-complete non-number problem.

3. Independent systems and matroids

Let E be a finite set and $I \subseteq 2^E$ be a nonempty family of subsets of E. Then (E,I) is called an *independence system* if

(3.1) $S \subseteq T \in I$ implies $S \in I$.

A subset $S \subseteq E$ is called *independent* if $S \in I$ and *dependent* otherwise. If $B \subseteq T \subseteq E$ and $B \in I$, then B is called a *base of T* if $B \cup \{x\}$ is dependent for all $x \in T \backslash B$. A minimal dependent subset $C \subseteq E$ is called a *circuit*, i.e. $C \subseteq E$ is a circuit if $C \backslash \{x\}$ is independent for all $x \in C$. Most combinatorial optimization problems can be reformulated as optimization problems over independence systems.

An independence system (E,I) is called a *matroid* if the Steinitz basic exchange principle holds for the set B of bases of (E,I), i.e. if $B_1, B_2 \in B$, then

(3.2) $\forall x \in B_1 \backslash B_2 \quad \exists y \in B_2 \backslash B_1 \quad (B_1 \backslash \{x\} \cup \{y\}) \in B.$

Matroids or combinatorial geometries have been introduced by Whitney (1935) and van der Waerden (1937) as an abstraction of linear dependence in vector spaces. Later Edmonds showed in a series of beautiful papers (cf. e.g. Edmonds (1965), (1971)) that matroids and independence systems are extremely useful in studying combinatorial optimization problems. See Iri and Fujishige (1981) for a survey of applications of matroid theory in operations research.

From an algorithmic point of view the most interesting property of a matroid is that they can be characterized by an algorithmic principle, namely the Greedy algorithm which will be introduced now.

Let (E,I) be an independence system and $c: E \to \mathbb{R}$ be an objective function. For $S \subseteq E$ we shall use the abbreviation $c(S)$ for the sum $\sum_{e \in S} c(e)$. The independence system problem is then to determine an independent set $F \in I$ with $c(F) = \max\{c(s) | S \in I\}$. In general an independence system is not given explicitly by the members of I, because the cardinality of I is usually an exponential function of $|E|$. Thus the list of all members of I as the input for an algorithm to solve the independence system problem makes it easy to let this algorithm run in linear time (with respect to the input length). Instead the independence system is given by the ground set $E = \{1,\ldots,n\}$ the function values $c_1,\ldots,c_n$ and a subroutine *IO* called *independence test oracle* which tests for a given $S \subseteq E$ whether $S \in I$ or not. We do not care how this subroutine performs this test but only count how often it is called. In most applications independence test is an easy task but there are also examples where the independence test problem itself is $\mathbb{NP}$-complete. The Greedy algorithm can now be formulated as follows:

(3.3) *Greedy algorithm*

Input : $E = \{1,\ldots,n\}, c_1,\ldots,c_n \in \mathbb{R}$, independence test oracle IO.
Output : An independent set $F \in I$ with $c(F) = \max\{c(S) | S \in I\}$

Method: *step 1*: Order the positive elements of E such that $c_{i_1} \geq c_{i_2} \geq \cdots c_{i_k} > 0$ holds.
step 2: set $F = \emptyset$
step 3: Do $j = 1$ TO i_k:
IF $F \cup \{i_j\} \in I$ THEN SET $F := F \cup \{i_j\}$.
END.

Clearly, the Greedy algorithm calls at most n times the independence test oracle. Step 1 can be performed in time $0(n \log n)$. Hence it is a very efficient procedure. Unfortunately the Greedy algorithm does not solve every independence system problem optimally.

Theorem 1. *An independence system* (E,I) *is a matroid if and only if the Greedy algorithm finds a maximal weighted independent set for all nonnegative objective functions* $c:E \to \mathbb{R}_+$

The name "Greedy algorithm" and the above theorem is due to Edmonds (1971). The theorem seemed to be known (although in a different context) already to Boruvka (1926) and was independently be discovered by Rado (1957) and Welsh (1968). See Korte (1982) for a more detailed historical account. The spanning tree problem (1.1) is an optimization problem over a matroid and the Greedy algorithm could be applied to it, although the other mentioned algorithms in section 2 are more efficient.

Let (E,I) be an independence system. Then $r^*(S) := \max\{|B| | B \subseteq S, B \in I\}$ is called the *upper rank function and* $r_*(S) := \min\{|B| | B \subseteq S, B$ a base of $S\}$ is called the *lower rank function* of (E,I). The quotient

$$q(I) := \min_{\phi \neq S \subseteq E} \frac{r_*(S)}{r^*(S)}$$

is called the *rank quotient* of (E,I). It can be used to give a performance guarantee if the Greedy algorithm is applied to general independence systems.

Theorem 2. (Jenkyns (1976), Korte and Hausmann (1978)). *Let* (E,I) *be an independence system,* $c:E \to \mathbb{R}$ *an objective function,* $c(I_g)$ *the objective function value of a Greedy solution and* $c(I_o)$ *the objective function value of an optimal solution to the independence system problem. Then we have* $c(I_o) \geq c(I_g) \geq q(I)c(I_o)$.

Clearly by the above theorem we have $q(I) = 1$ if and only if (E,I) is a matroid. As is shown in Korte and Hausmann (1978) the rank quotient of the independence system for the matching problem is at least 1/2 and for the symmetric travelling salesman problem at least 1/2 and at most $1/2 + 3/(2|V|)$. But there are also examples (vertex packing problem, acyclic subgraph problem) where the rank quotient tends to zero when $|E|$ tends to infinity.

Matroids can be defined in various ways using so called bases axioms (as we did) or circuit axioms or rank axioms etc. (cf. e.g. Welsh (1976)). All these axiom systems are equivalent in the sense that one can be deduced from the other. From an computational point of view it is also interesting whether they are also algorithmically equal powerful. For instance consider a basis test oracle (is $B \subseteq E$ a basis of (E,I)?). Can it simulate (in polynomial time) an independence test oracle? Questions of these types are investigated by Hausmann and Korte (1981) and Jensen and Korte (1981). See Korte (1982) and Korte and Schrader (1981) for surveys on those oracle techniques.

4. Duality theorems for independence systems and matroids

If we consider a combinatorial optimization problem max $\{f(S) | S \in I\}$ it is not very hard to design an algorithm A which constructs a sequence $S_1, S_2, S_3, \ldots$ of feasible sets with monoton increasing objective values $f(S_i)$. In fact any implicit enumeration algorithm does the job and at step n such an algorithm verifies the

"primal question": $\exists S \in I \; f(S) > f(S_{n-1})$?

Since we do not want to enumerate all exponentially many possible candidates $S_i \in I$ in an efficient algorithm we do need a *stopping criterion* which will verify the

"dual question": $\forall S \in I \; f(S) \leq f(S_{n-1})$?

Clearly such a stopping criterion should be algorithmically efficiently checkable, i.e. verifying the stopping criterion should be at least in the same complexity class as verifying the primal question. Hence we need a reformulation of the dual question. The inherent principal of a duality theory is to find a logical equivalent reformulation of the dual question which replaces the all quantifier by the existence quantifier.

As an example we may take the linear programming duality theory which could be viewed as a reformulation of the Farkas Lemma. Here the primal question is

$$\exists x \in \mathbb{R}^n \quad Ax \leq b?,$$

and the dual question

$$\forall x \in \mathbb{R}^n \quad Ax \nleq b?$$

can be reformulated via the Farkas Lemma as

$$\exists u \in \mathbb{R}^m \quad u \geq 0 \quad u^t A = 0 \quad u^t b < 0.$$

It is exactly this easily checkable stopping criterion which makes the simplex algorithm an efficient algorithm. More general every duality theory is useful in designing a stopping criterion for the class of algorithms to which the duality theory applies.

Hence it is quite natural to ask for the most general structure of a combinatorial optimization problem allowing a duality theory. Before we shall report on duality type theorems for independence systems, matroids and oriented matroids let us revisite Farkas' Lemma:

(4.1) Let A be an (m,n) matrix and $b \in \mathbb{R}^m$. Then Farkas' Lemma has the form

either $\exists x \in \mathbb{R}^n \quad Ax \leq b$

or $\exists u \in \mathbb{R}^m_+ \quad A^t u = 0, \quad u^t b < 0$
but not both.

Using the technique of homogenization, i.e. setting $B: = \begin{pmatrix} A & -b \\ 0 & -1 \end{pmatrix}$ (4.1) is equivalent to

(4.2) either $\exists \binom{x}{1} \in \mathbb{R}^{n+1} \; B\binom{x}{1} \leq 0$
or $\exists \binom{u}{1} \in \mathbb{R}^{m+1}_+ \; B^t\binom{u}{1} = 0$
but not both.

Again we can take a further step by using the linear subspaces $L = \{y \in \mathbb{R}^{m+1} | y = Bz\}$ and $L^\perp = \{v \in \mathbb{R}^{m+1} | B^t v = 0\}$. Denoting by *supp*$(y)$: $= \{i \in \{1,\ldots,m+1\} | y_i \neq 0\}$ the *support* of y we obtain another formulation of (4.1) by

(4.3) either $\exists y \in L \quad y \leq 0$ and $(m+1) \in \operatorname{supp}(y)$
or $\exists v \in L^\perp \quad v \geq 0$ and $(m+1) \in \operatorname{supp}(v)$
but not both.

Note that (4.3) is a theorem of the alternative about nonpositive (nonnegative) vectors of an "oriented" linear subspace L respec. $L^\perp$. To exhibit the combinatorial content of the above statements (4.1)–(4.3) we have to reduce the above existence statements over an infinite set L to an existence statement over a finite set *elem(L)*, the elementary vectors of L. Recall that $y \in L$ is an *elementary vector* if supp(y) is minimal with respect to set inclusion in L. The following well-known proposition gives an idea of how to interprete the elementary vectors $y \in \operatorname{elem}(L)$.

Proposition 3. *Let* A *be an* (m,n) *matrix of rank n and consider the polyhedron* $P(A,b) = \{x \in \mathbb{R}^n | Ax \leq b\}$, *the polyhedral cone* P(B,0) *and the linear subspace* L *defined above. Then the following statements are equivalent:*

(4.4) x is a vertex of P(A,b)

(4.5) $\binom{x}{1} = : z$ is an extremal vector of the cone P(B,0)

(4.6) $y: = Bz \leq 0$ is an elementary vector of L and $y_{m+1} \neq 0$.

It is easy to show (cf. Rockafellar (1969)) that every vector $y \in L$ has a conformal decomposition $y = y^1 + \cdots + y^k$ where $y^i \in \operatorname{elem}(L)$ for $i = 1,\ldots,k$, i.e. for all components $j \in \operatorname{supp}(y^i)$ we have $y_j y_j^i > 0$ $(i = 1,\ldots,k)$. Thus (4.3) is equivalent to

(4.7) either $\exists y \in \operatorname{elem}(L) \quad y \leq 0$ and $(m+1) \in \operatorname{supp}(y)$
or $\exists v \in \operatorname{elem}(L^\perp) \quad v \geq 0$ and $(m+1) \in \operatorname{supp}(v)$
but not both.

For $y \in L$ we can orient the support supp(y) in a natural way: $\operatorname{supp}^+(y) := \{i \in \operatorname{supp}(y) | y_i > 0\}$ and $\operatorname{supp}^-(y) := \{i \in \operatorname{supp}(y) | y_i < 0\}$. We call the set of supports $C: = \{\operatorname{supp}(y) | y \; \operatorname{elem}(L)\}$ the circuits of L and $C^\perp: = \{\operatorname{supp}(y) | y \in \operatorname{elem}(L^\perp)\}$ the cocircuits of L (circuits of $L^\perp$). Then the Farkas Lemma has the following interpretation.

Theorem 4. (Farkas Lemma). *Either there exists a circuit* $C \in \mathcal{C}$ *such that* $(m+1) \in C$ *and all elements of* C *have the same orientation or there exist a cocircuit* $C^\perp \in \mathcal{C}^\perp$ *such that* $(m+1) \in C^\perp$ *and all elements of* $C^\perp$ *have the same orientation.*

It is easy to see that $\mathcal{C}$ and $\mathcal{C}^\perp$ form the circuits and cocircuits of a dual pair of independence systems. Hence it is natural to ask whether the Farkas Lemma is valid also for a dual pair of independence systems.

Let (E,I) be an independence system. Then $\tau: 2^E \to 2^E$ defined by $\tau(S) := S \cup \{y \in E\backslash S \mid r^*(S) \cup \{y\})\}$ is called the *τ-closure* of (E,I). The map $\sigma: 2^E \to 2^E$ defined by $\sigma(S) := S \cup \{y \in E\backslash S \mid \exists C \in \mathcal{C}\; y \in C \subseteq S \cup \{y\}\}$ is called the *σ-closure* of (E,I) (r^* is the upper rank function and $\mathcal{C}$ are the circuits of (E,I)).

Theorem 5. (Farkas Lemma for matroids). *Let* (E,I) *and* $(E,I^\perp)$ *be a pair of dual independence systems with σ-closure operations* σ, $\sigma^\perp$ *respec. Then the following statements are equivalent:*

(4.8) (E,I) and $(E,I^\perp)$ are matroids.

(4.9) For every partition (3-colouring) $E = R \cup Y \cup G$ where $|R| = 1$ we have:
either $\exists$ circuit $C \in \mathcal{C}$ with $R \subseteq C \subseteq Y \cup R$
or $\exists$ cocircuit $C \in \mathcal{C}^\perp$ with $R \subseteq C^\perp \subseteq G \cup R$
but not both.

(4.10) For every partition (3-colouring) $E = R \cup Y \cup G$ where $|R| = 1$ we have: either $R \subseteq \sigma(Y)$ or $R \subseteq \sigma^\perp(G)$ but not both.

Theorem 5 was first proved in a more special setting by Minty (1966), generalizing his well known colouring lemma for graphs.

Theorem 6. (Farkas lemma for independence systems). *Let* (E,I) *and* $(E,I^\perp)$ *be a pair of dual independence systems. Then the following statements are equivalent:*

(4.11) All bases B of E are of the same cardinality.

(4.12) For every partition (3-colouring) of $E = R \cup Y \cup G$ where $|R| = 1$ we have either $R \subseteq \tau(Y)$ or $R \subseteq \tau^\perp(G)$ but not both.

This theorem is due to Matthews (1978) who also proved (in the same paper):

Theorem 7. *Let* (E,I) *be an independence system. Then* (E,I) *is a matroid if and only if* $\sigma = \tau$.

So far we discussed only non oriented versions of the Farkas lemma. To prove an analogon of (4.1) for matroids we first have to introduce orientable matroids. Let $M = (E,I)$ and $M^\perp = (E,I^\perp)$ be a pair of dual matroids. If we can partition every circuit $C \in \mathcal{C}$ and every cocircuit $C^\perp \in \mathcal{C}^\perp$ into $C = C_+ \cup C_-$ and $C^\perp = C_+^\perp \cup C_-^\perp$ such that

(4.13) $(C_+ \cap C_+^\perp) \cup (C_- \cap C_-^\perp) \neq \emptyset \neq (C_- \cap C_+^\perp) \cup (C_+ \cap C_-^\perp)$

holds (provided $C \cap C^{\perp} \neq \emptyset$) we call $(M,M^{\perp})$ a pair of *orientable matroids.* (4.13) is a combinatorial abstraction of the orthogonality property of a dual pair of orthogonal linear subspaces of $\mathbb{R}^n$. We shall discuss oriented matroids more detailed later on in section 5 where we also give some more references of related work. The signed supports of elementary vectors of a dual pair of linear subspaces of $\mathbb{R}^n$ are one class of examples of orientable matroids. The use of orientations of circuits of matroids to generalize Farkas Lemma was suggested by Rockafellar (1969). Bland and Las Vergnas independently discovered in 1975 the following theorem (cf. Bland (1977), Bland and Las Vergnas (1978), Bland and Las Vergnas (1979)).

Theorem 8. (Farkas Lemma for oriented matroids). *Let* (E,I) *and* $(E,I^{\perp})$ *be a dual pair of oriented independence systems (i.e. independence systems where the circuits and cocircuits are oriented such that (4.13) holds. Then the following statements are equivalent:*

(4.14) (E,I) and $(E,I^{\perp})$ is a dual pair of orientable matroids.

(4.15) For all partitions (3-colouring) $E = R \cup Y \cup G$ and all $x \in R$ either $\exists$ circuit $C \in \mathcal{C}$ $x \in C \subseteq Y \cup R$ and all elements in $C \cap R$ have the same orientation or
$\exists$ cocircuit $C^{\perp} \in \mathcal{C}^{\perp}$ $x \in C^{\perp} \subseteq G \cup R$ and all elements in $C^{\perp} \cap R$ have the same orientation.

Clearly if we choose $R = \{m+1\}$ and $Y = \emptyset$ we obtain Theorem 4. The general integer programming problem

(4.16)
$$\begin{aligned} &\max cx \\ &Ax = b \\ &x \geq 0 \text{ and integer} \end{aligned}$$

does not have the structure of a matroid. Hence we can not expect a duality type theorem such as Theorem 8 to hold.

Theorem 9. (Farkas Lemma for integer programs). *Let* $\mathrm{RIP}(A) := \{b \in \mathbb{R}^m \mid \exists x \in Z^n_+ \; Ax = b\}$ *and assume that all entries of* A *are rational. Then either there exists a nonnegative integer* x *with* $Ax = b$ *and* $cx < p$ *or there exists a subadditive function* $f{:}\mathrm{RIP}(A) \to \mathbb{R} \cup \{-\infty\}$ *with* $f(0) \geq 0$, $f(A_{.j}) \leq c_j (j = 1,\ldots,n)$ *and* $f(b) \geq p$ *but not both.*

Theorem 9 is due to Jeroslow (1979) and Johnson (1974); see Bachem and Schrader (1980) and Bachem, Schrader and Johnson (1982) for some generalizations. Note that if we relax the integrality of x and restrict f to be a linear function, Theorem 9 turns out to be the well-known Farkas Lemma for linear programs. Additionally the two "sides" of the alternative in Theorem 9 belong most likely to different complexity classes unless co-$\mathbb{NP} \subseteq \mathbb{NP}$. Hence Theorem 9 is not a duality type theorem as we discussed in the beginning of section 4. The class of subadditive function involved in Theorem 9 can not be restricted to the class of convex function as the following theorem of Jeroslow (1979) shows.

Theorem 10. *The word "subadditive" in Theorem 9 can be replaced by "convex" if and only if* $\mathrm{conv}(\{x \in \mathbb{Z}^n_+ | Ax = b\}) = \{x \in \mathbb{R}^n_+ | Ax = b\}$

Let D be a digraph with a distinguished arc f, and a nonnegative capacity function c defined on the arcs of D, satisfying $c_f = +\infty$. The triple (D,c,f) is called a *network*. The *maximum flow problem* for (D,c,f) is the linear program:

(4.14) $\max x_f$

$Nx = 0$

$0 \leq x \leq c$

where N is a vertex-arc incidence matrix of D. Network Flow theory was developed by Ford and Fulkerson (1962) and is today one of the corner stones of mathematical programming. The well-known max-flow min-cut theorem of Ford and Fulkerson is a duality theorem of Network Flow theory and can be shown to be a corollary of Theorem 9. Network Flow theory can also be generalized to flows in matroids. Seymour (1977, 1980) has contributed some beautiful theorems in this field characterizing those matroids having the "max-flow min-cut property". See Bixby (1981) for an excellent survey of Seymours results.

5. Oriented matroids

The theory of oriented matroids can be considered as a combinatorial abstraction of linear algebra or linear geometry of ordered fields. It was forseen by Rockafellar (1969) and independently developed by Bland (1974), Las Vergnas (1975) and Lawrence (1975) respec. Folkman and Lawrence (1978). Recent years have seen increased interest in this area because of the many parallels between the properties of polytopes and those of the lattices arising from oriented matroids. Moreover, it has been shown that the simplex method can be generalized to a pivoting algorithm solving Linear Programming problems over oriented matroids. In this section we shall mainly concentrate on very recent results of Billera, Edmonds, Fukuda, Mandel and Munson (cf. Fukuda (1981), Mandel (1981), Munson (1981)).

Let E be a finite set. A map $X: E \to \{0,+,-\}$ is called a *signed vector* and $\underline{X}$ denotes its support, i.e. $\underline{X} := \{e \in E | X_e \neq 0\}$. For two signed vectors X,Y we defined the *composition* $Z := X . Y$ by $Z_e = X_e$ for $e \in \underline{X}$ and $Z_e = Y_e$ otherwise. The set $sep(X,Y) := \{e \in E | X_e = -Y_e\}$ is called the set of *separators* of X and Y.

A collection $\mathcal{O} \subseteq \{0,+,-\}^E$ of signed vectors is called an *oriented matroid* if it fulfills the following three axioms:

(5.1) $0 \in \mathcal{O}, X \in \mathcal{O} \Rightarrow -X \in \mathcal{O}$

(5.2) $X,Y \in \mathcal{O} \Rightarrow X.Y \in \mathcal{O}$

(5.3) $X,Y \in \mathcal{O}, f \in \mathrm{sep}(X,Y) \Rightarrow \exists Z \in \mathcal{O}$ with $Z_f = 0$ and $Z_e = (X.Y)_e$ for all $e \notin \mathrm{sep}(X,Y)$

Let $X^+ := \{e \in \underline{X} | X_e > 0\}$ and $X^- := \{e \in \underline{X} | X_e < 0\}$. It is not difficult to prove that the signed vectors of an oriented matroid which have minimal support form the circuits of an orientable matroid as defined in section 4. Moreover we could also use the following axioms to define oriented matroids.

Theorem 11. $O \subseteq \{0,+,-\}^E$ *is an oriented matroid if and only if*

(5.4) $\emptyset \notin O, X \in O \Rightarrow -X \in O$

(5.5) $X,Y \in O, \underline{X} = \underline{Y} \Rightarrow X = \pm Y$

(5.6) $X,Y \in O, X \neq -Y, x \in (X^+ \cap Y^-) \cup (X^- \cap Y^+) \Rightarrow$
$\exists Z \in O\, Z^+ \subseteq (X^+ \cup Y^+)\backslash\{x\}, Z^- \subseteq (X^- \cup Y^-)\backslash\{x\}$.

If O is an oriented matroid then the supports of the minimal elements of O define the circuits of the (non oriented) matroid $M(O)$ underlying O. Clearly $M(O)$ is orientable (as defined in section 4) but there may be many orientations leading to the same oriented matroid O. Let $\delta: \mathbb{R}^E \to \{0,+,-\}^E$ be a sign function defined by

$$\delta(x)_e := \begin{cases} 0 & \text{if } x_e = 0 \\ + & \text{if } x_e > 0 \\ - & \text{if } x_e < 0 \end{cases}$$

Then it is easily verified that $\delta(\mathbb{R}^E)$ is an oriented matroid. An oriented matroid O is called *linear* if $O = \delta(V)$ for some linear subspace $V \subseteq \mathbb{R}^E$. Another example of a linear oriented matroid are the supersets of circuits (or cocircuits) of a digraph $G = (V,A)$ together with the (signed) zero vector 0.

An oriented matroid O is called *acyclic* if it has no positive circuit and *totally cyclic* if every element $e \in E$ is contained in some positive circuit. A simple corollary of Theorem 8 is the following theorem:

Theorem 12. *An oriented matroid O is either acyclic or the dual oriented matroid* $O^\perp$ *is totally cyclic but not both.*

Let O be an oriented matroid on E and $S \subseteq E$. For a signed vector X we denote by $X\backslash S$ a signed vector on $E\backslash S$ defined by $(X\backslash S)_e = X_e$ for all $e \in E\backslash S$. The set $O\backslash S := \{X\backslash S \mid X \in O\}$ is called the oriented matroid derived from I by *deleting* S and the set $O/S = \{X\backslash S \mid X \in O \underline{X} \cap S = \emptyset\}$ is called the oriented matroid derived from O by *contracting* S. Any oriented matroid derived from O by deleting or contracting S is called a *minor* of O (cf. Bland and Las Vergnas (1978)). If O' is a minor of O, then O is called an *extension* of O'.

We shall now introduce face lattices of oriented matroids. Let us motivate the following definitions by the face lattice of a polyhedral cone in $\mathbb{R}^n$. Consider the polyhedron $P = P(A,b) = \{x \in \mathbb{R}^n \mid Ax \leq b\}$ in its homogenized form $H = P(B,0)$ with $B = \binom{A, -b}{0, -1}$. The face lattice $F(H)$ of the polyhedral cone H is called the extended face lattice of P, i.e. $F(H)$ is the composition of the face lattice $F(P)$ (of the polyhedron P) and $F(\text{rec}(P))$ (of the recession cone of P) (cf. Bachem and Grötschel (1982)).

Now let $L = \{y \in \mathbb{R}^{m+1} \mid y = Bz\}$ and denote the row index set of the matrix B by E. Then $O = \delta(L)$ defines an oriented matroid and the negative circuits C of O with $m + 1 \in C$ correspond to vertices of the polyhedron P (as stated in Proposition 3). We may also consider the dual oriented matroid $O^\perp = \delta(L^\perp)$. Note that $O^\perp$ is acyclic if and only if H is a pointed cone, i.e. assuming $O^\perp$ acyclic makes life a little bit easier and we shall therefore consider only this case. Today there exists two different definitions of face lattices of oriented matroids.

Definition 13. (Las Vergnas (1980)). $F \subseteq E$ *is a* facet *of* $O^{\perp}$ *if* $E \backslash F$ *supports a positive cocircuit* $C^{\perp} \in O^{\perp}$.

Definition 14. (Edmonds, Mandel, Fukuda). $F \subseteq E$ *is a* facet *of* $O^{\perp}$ *if* F *supports a positive cocircuit* $C^{\perp} \in O^{\perp}$.

$F \subseteq E$ is a *face* of $O^{\perp}$ if there exist facets $F_1,\ldots,F_k$ with $F = \bigcap_{i=1}^{k} F_i$. By $F(O^{\perp})$ we denote the *Las Vergnas face lattice* and by $F^*(O^{\perp})$ the *Edmonds-Mandel-Fukuda face lattice* of $O^{\perp}$. Taking the above example, one easily recognizes that $F^*(O^{\perp})$ is isomorphic to the extended face lattice $F(H) = F(P) \times F(\text{rec}\ (P))$ and clearly $F(0^{\perp})$ is antiisomorphic to $F^*(O^{\perp})$ hence isomorphic to the equality face lattice of H. (An equality set $I \subseteq E$ is a maximal set such that $\{x \in P(B,0) | B_I . x = 0\}$ is a face of H).

As we have shown above every polytopal face lattice F correspond to an oriented matroid face lattice $F(O^{\perp})$ respec. $F^*(O^{\perp})$. In 1980 Lawrence found a construction "λ" how to extend an oriented matroid O to an oriented matroid $\lambda(O)$ with the following property: If the face lattice of the oriented matroid $\lambda(O)$ is polytopal (i.e. isomorphic to a face lattice of a polytope) then the oriented matroid O is linear. Since there exist many nonlinear oriented matroids this proves:

Theorem 15. *In general face lattices of oriented matroids are non polytopal.*

See Munson (1981) for a description of the Lawrence construction. Munson (1981) used this Lawrence construction to prove even more, namely he showed that the set of all Las Vergnas face lattice and the set of all Edmonds-Mandel-Fukuda face lattices differ.

A flat of a matroid M is a maximal closed set and we denote the geometric lattice of all flats of a matroid M by $L(M)$. An *adjoint* of a matroid M is a matroid M' of the same rank such that there exists an injective order reversing map $f: L(M) \to L(M')$ which maps the copoints of M onto the points of M'. An oriented matroid O^* is called a *polar* of an oriented matroid O if the face lattice $F(O)$ is antiisomorphic to $F(O^*)$. Clearly the class of all Las Vergnas lattices of acyclic oriented matroids is the class of polars of all Edmonds-Mandel-Fukuda lattices. Thus the question whether these two classes of face lattices agree reduces to the problem whether every oriented matroid face lattice has a polar. The key for proving that indeed not every oriented matroid has a polar is a result of Cheung (1974) who proved:

Theorem 16. *The Vámos matroid has no adjoint.*

Munson (1981) proved the following theorem.

Theorem 17. *Let O be an oriented matroid, $\lambda(O)$ be the Lawrence extension of* M *and $\lambda(O)^*$ the polar of $\lambda(O)$. Then there exists a subset* $D \subseteq E$ *such that the minor $\lambda(O)^*/D$ is an adjoint of O.*

Let $\mathbb{F}$ be the set of all Las Vergnas and $\mathbb{F}^*$ the set of all Edmonds-Mandel-Fukuda face lattices. Then Theorem 16 and Theorem 17 prove:

Theorem 18. $\mathbb{F} \neq \mathbb{F}^*$

Clearly, if V is the Vámos matroid and if we apply the Lawrence construction to V, then by Theorem 17 and Theorem 16 $\lambda(V)$ cannot have a polar.

As a consequence of Theorem 18 there are no analogues of the theorem of Weyl or of the theorem of Minkowski for face lattices of oriented matroids because both would imply the existence of polars. On the other hand a lot of interesting properties of face lattices of oriented matroids are known.

Theorem 19. (Las Vergnas (1980), Mandel (1981)). *All face lattices $F \in \mathbb{F}$ and $F^* \in \mathbb{F}^*$ satisfy the Jordan-Dedekind chain condition.*

A polytope P of dimension d is *shellable* if there exists an order of the facets $F_1,\ldots,F_r$ of P such that for all $2 \leq i \leq r$ $F_i \cap (\bigcup_{j=1}^{i-1} F_j)$ is a topological d-2 ball.

Edmonds and Mandel have extend the definition of shellability to oriented matroid cell complexes and Mandel (1981) proved:

Theorem 20. *The face lattice of every* d*-dimensional oriented matroid cell is a shellable* d*-ball and its boundary is a shellable* $(d-1)$ *sphere.*

This theorem of Mandel extends considerable the result of Bruguesser and Mani (1972) who proved that polytopes are shellable. As a corollary of Theorem 20 Mandel also proves:

Theorem 21. *Every oriented matroid cell* t *of dimension* d *satisfies the Euler-Poincare relation* $\sum_{i=1}^{d} (-1)^i f_i(t) = 1$. ($f_i(t)$ = *number of faces of dimension* i).

Shellability seems to be one of the most restrictive properties of polytopal face lattices. But as Theorem 15 shows it does not suffice to characterize polytopal face lattices.

All these very deep results of Edmonds and Mandel were obtain by using a topological representation of oriented matroids which were initiated by the late Folkman and later on developed by Lawrence (1975). We shall discuss this topological representation in the following. Consider an oriented matroid $O = \delta(L)$ arising from a linear subspace $L = \{y \in \mathbb{R}^m | y = Ax\}$ of $\mathbb{R}^m$. If we oriente the sides of every hyperplane $H_e := \{z | A_e . z = 0\}$ then a signed vector $X = \delta(Az)$ determines the relative position of a point $z \in \mathbb{R}^n$ with respect to the hyperplanes $H_e (e \in E)$. Clearly this position is uniquely determined up to scalar multiples. Thus we may concentrate only on intersections of the hyperplanes H_e with the n-sphere S^n, i.e. on positions of a point $z \in S^n$ relative to oriented hyperspheres. This motivates the following definition of a sphere system. Let $S^n := \{x \in \mathbb{R}^{n+1} | \|x\| = 1\}$ and denote by $S^n_-(S^n_0, S^n_+)$ the sets $\{x \in S^n | x_{n+1} < 0 (=0, >0)\}$. A set S homeomorphic to S^n is called a *sphere.* A subset $h \subseteq S$ is a *hypersphere* if $h = f(S^n_0)$ for some homeomorphism $f: S^n \to S$. We call $f(S^n_+)$ and $f(S^n_-)$ the *sides* of h.

Definition 22. Mandel (1981). *A triple* (S,E,H) *is called a* sphere system *if the following conditions are satisfied.*

(5.4) S is a sphere, E is finite and $H = \{S_e^j | e \in E, j \in \{0,+,-\}\}$ with

(5.5) *For all* $e \in E$ $(S_e^0, S_e^+, S_e^-) = (S, \emptyset, \emptyset)$ or S_e^0 is a hypersphere of S with sides S_e^+ and S_e^-.

(5.6) For all $\emptyset \neq A \subseteq E$ $F := \cap\{S_e^0 | e \in A\}$ is a sphere called a *flat* of the system.

(5.7) For all flats F and all $e \in E$ $F \subseteq S_e^0$ or $S_e^0 \cap F$ is a hypersphere of F with sides $S_e^+ \cap F$ and $S_e^- \cap F$.

(5.8) Let $S_e^{-j} := S_e^j \cup S_e^0$ and let $B \subseteq \{S_e^{-j} | e \in E, j \in \{0,+,-\}\}$. Then either $\cap B = \cap\{S_e^0 | S_e^{-j} \in B$ for a $j \in \{0,+,-\}\}$ or $\cap B$ is a ball.

Again we define a *sign* function $\delta : S \to \{0,-,+\}^E$ by

$$\delta(x)_e = \begin{cases} 0 & \text{if } x \in S_e^0 \\ + & \text{if } x \in S_e^+ \\ - & \text{if } x \in S_e^- \end{cases} \qquad (e \in E)$$

if (S,E,H) is a sphere system. The number $1 + \delta^{-1}(0)$ is called the *degeneracy* of the system. The following characterization was first proved by Folkman and Lawrence (1978).

Theorem 23. *The map assigning* S *to* $\delta(S) \cup \{0\}$ *is a bijection between equivalence classes of sphere systems of fixed degeneracy* $k \geq 0$ *and oriented matroids on the same index set.*

Mandel (1981) gave a new proof of Theorem 23 and also showed that every oriented matroid is a piecewise linear sphere (A *piecewise linear* (*LP*) *n-sphere* is a complex order isomorphic to a PL ball complex whose space is the boundary of an (n + 1)-polytope. A *PL ball-complex* is a complex K whose cells are either polyhedral balls or empty and such that for each $p \in K$, its boundary is the union of the cells of K it properly contains.)

Mandel (1981) could also extend McMullen's (1970) proof of the Upper Bound Conjecture to piecewise linear spheres proving:

Theorem 25. *Every piecewise linear sphere semilattice satisfies the Upper Bound Conjecture.*

Las Vergnas (1980) has introduced a convexity definition to oriented matroids.

Definition 25. (Convexity). *Let O be an oriented matroid and C be the circuits of O. For* $S \subseteq E$ *we define* $conv(S) = S \cup \{e \in E \setminus S | \exists C \in \mathcal{C}\, C^- = \{e\}, C^+ \subseteq S\}$

It is easily shown that for $O = \delta(L)$ and L is a linear subspace of $\mathbb{R}^n$ conv(S) corresponds to the usual convex hull in $\mathbb{R}^n$. If the oriented matroid O arises from circuits of a digraph $D = (V,A)$, then conv (S) corresponds to the transitive closure in D.

Let O be an oriented matroid on E. A subset $S \subseteq E$ is called an *open halfspace* of O if $S = S^+$ for some cocircuits C of O. A subset $H \subseteq E$ is a *closed halfspace* if $H = (E \setminus C) \cup C^+$ for some cocircuit C of O.

Theorem 26. (Las Vergnas (1980)). *Let O be an oriented matroid on* E *and* $S \subseteq E$. *Then* conv(S) *equals the intersection of all closed halfspaces containing* S.

Let F be the Las Vergnas lattice of an oriented matroid O. A face $v \in F$ is called a *vectex* if the rank of v equals 1. The following theorem may be considered as an analogue of the Krein-Milman theorem in $\mathbb{R}^n$.

Theorem 27. (Las Vergnas (1980)). *Let O be an oriented matroid on* E, $S \subseteq E$ *and* T *the set of vertices of the minor O/S. Then* conv(T) = conv(S).

Unfortunately, this convexity definition does not allow a separation type theorem for oriented matroids since Mandel (1981) has found an example of an oriented matroid O on a ground set E with the following property. For subsets $S,T \subseteq E$ $\text{conv}_{O'}(S) \cap \text{conv}_{O'}(T) = \emptyset$ (the convex hull taken with respect to all extensions O' of O) but no hyperplane of O separates S and T.

We now come to a brief discussion of how to generalize Linear Programming to oriented matroids. This work was initiated by Tucker (1963) and Rockafellar (1969). Later Bland (1974, 1977, 1980) and Fukuda (1981) developed a simplex type algorithm for Linear Programs in oriented matroids. The Linear Programming problem

(5.9) $$\begin{aligned} &\min cx \\ &Ax \leq b \end{aligned}$$

may also be written as

(5.10) $$\begin{aligned} &\min t \\ &\begin{pmatrix} c^T \\ A \end{pmatrix} x \leq \begin{pmatrix} t \\ b \end{pmatrix} \end{aligned}$$

or in a homogenized form as

(5.11) $$\begin{aligned} &\min t \\ &\begin{pmatrix} c^t & -t \\ A & -b \\ 0 & -1 \end{pmatrix} \begin{pmatrix} x \\ z \end{pmatrix} \leq 0 \\ &z > 0 \end{aligned}$$

Let us denote the coefficient matrix of the problem (5.11) by B with row index set E. The first row index (corresponding to the objective function) let be f and the last row index let be g. It is well-known that a vector $x \in \{x \in \mathbb{R}^n | Ax \leq b\}$ is optimal if and only if there is no "direction" $z \in \mathbb{R}^n$ with $Az \leq 0$ and $c(x + \varepsilon z) < cx$ for some positive ε. Now let $O = \delta(L) = \delta(\{y|y = B\binom{x}{z}\})$ be an oriented matroid derived from the linear space L. Let $A := \{X \in O|X_g > 0\}$ and $A_\infty := \{X \in O|X_g = 0\}$ and define for a subset $S \subseteq E$ the *matroid polyhedron* $P(S) := \{X \in A|X_S \leq 0\}$. Clearly for $S = E\backslash\{g,f\}$ P(S) corresponds to the polyhedron $P(A,b) = \{x \in \mathbb{R}^n|Ax \leq b\}$. Let $X \in P(S)$ and $Z \in A_\infty$. Z is called a *feasible direction* at X if $X.Z \in P(S)$.

A linear programming problem with respect to an oriented matroid can now be defined as follows. Let O be an oriented matroid on E and let $f,g \in E$ with $f \neq g$. Wit $E_0 := E\backslash\{g,f\}$ we say *X minimizes f over* $P(E_0)$ if $X \in P(E_0)$ and there is no feasible direction $Z \in A_\infty$ at X with $Z_g < 0$. Thus the *oriented matroid programming problem* is to determine an X which minimizes f or in tableau form

$$\text{(P)} \quad \left|\min_{X \in P(E_0)} f\right.$$

We call the problem (P) feasible if $P(E_0) \neq \emptyset$ and unbounded if $Z_{E_0} \geq 0$ and $Z_g > 0$ for some $Z \in A_\infty$.

Theorem 28. (Fukuda (1981), Bland (1974)). *Every feasible oriented matroid programming problem is either unbounded or has an optimal solution.*

Bland (1974) developed a finite pivot algorithm for oriented matroid programming problems generalizing the well-known simplex method and giving a constructive proof of the duality theorem (Theorem 8). This pivoting algorithm also led to a remarkable simple new finite pivot rule for the usual simplex method (cf. Bland (1977)). Fukuda (1981) found a new finite pivot method for oriented matroid programs. Moreover he found some examples of oriented matroids for which the simplex procedure produces nondegenerate cycling. The class of such oriented matroids he called non-BOM's. The complement of this class (the BOM's) include linear oriented matroids and the dual $O^\perp$ of an oriented matroid $O \in$ BOM is again within the class of BOM's.

References

Aho, A.v., J.E. Hopcroft and J.D. Ullmann, "The Design and Analysis of Computer Algorithms", Reading, London, Massachusetts: Addison-Wesley (1974).

Bachem, A. "Concepts of algorithmic computation", in: B. Korte (1982a), 3–49.

Bachem, A., and M. Grötschel, "New aspects of polyhedral theory", in: B. Korte (1982a), 51–106.

Bachem, A., E.L. Johnson and R. Schrader, "A characterization of Minimal Valid Inequalities For Mixed Integer Programs", Operations Research Letters Vol. 1 (1982), 63–66.

Bachem, A., and R. Schrader, "Minimal Inequalities and Subadditive Duality", Siam J. Control and Optimization, Vol. 18 (1980), 437–443.

Balas E., and M.W. Padberg, "Set partitioning", in: B. Roy (ed.), Combinatorial Programming: Methods and Applications Reidel, Dordrecht 1975, 205–258.

Bixby, R.E. "Matroids and Operations Research", Research Report, Department of Industrial Engineering and Management Sciences, Northwestern University Evanston, Illinois 60201, 1981.

Bland, R.G. "Complementary Orthogonal Subspaces of $\mathbb{R}^n$ and Orientability of Matroids", Ph.D. thesis, Cornell University (1974) also available as technical report No. 219 Cornell University, Ithaca (1974).

Bland, R.G. "A combinatorial abstraction of linear programming", J. Combin. Theory Ser. **B23** (1977), 33–57.
Bland, R.G. "New finite pivot rules for the simplex method", Math, Operations Research Vol. 7 (1977), 103–107.
Bland, R.G. "Linear Programming duality and Minty's lemma, Technical Report No. 449, School of Operations Research and Industrial Engl., Cornell University (1980).
Bland, R.G., and M. Las Vergnas, "Orientability of matroids", J. Combin. Theory Ser. **B24** (1978), 94–123.
Bland, R.G., and M. Las Vergnas, "Minty Coloring and Orientations of Matroids", in: Proceedings of the Second International Conference on Combinatorial Mathematics, A. Gewirtz, L.v. Quintas (eds.), Annals of the New York Academy of Sciences, Vol. 319 (1979).
Borgwardt, K.-H. "The Average of Pivot Steps Required by the Simplex-Method is Polynomial", Zeitschrift für Operations Research Vol. 26 (1982), 157–177.
Boruvka, O. "On jistem problemu minimalnim", Prace Moravske Priodovedecke Spolecnosti **3** (1926), 37–53.
Bruguesser, H., and P. Mani, "Shellable decompositions of cells and spheres", Math. Scand. **29** (1972), 197–205.
Burkard, R.E. "Subgradient methods in combinatorial optimization", in: U Pape (ed.), Graphs, Data Structures, Algorithms (Hauser, München, 1980), 141–151.
Burkard, R.E., and U. Derigs, "Assignment and Matching Problems: Solution Methods with FORTRAN Programs", Lecture Notes in Economics and Mathematical Systems Vol. 184 (Springer, Berlin-New York, 1980).
Burkard, R.E., and U. Zimmermann, "Combinatorial optimization in linearly ordered semimodules—a survey, in: B. Korte (1982a).
Cheriton, D., and R.E. Tarjan, "Finding minimum spanning trees", SIAM J. Comput. **5** (1976), 724–742.
Cheung, A.L. "Adjoints of a Geometry", Canad. Math. Bull. Vol. 17 (1974), 363–365.
Cook, S.A. "The complexity of theorem proving procedures", in: Proceedings of the Third ACM Symposium on the Theory of Computing, (ACM, New York, 1971), 151–158.
Dijkstra, E.W. "A note on two problems in connection with graphs", Numer. Mathematik **1** (1959), 269–271.
Edmonds, J. "Minimum partition of a matroid into independent subsets", J. Res. Nat. Bur. Standards **69B** (1965), 67–72.
Edmonds, J. "Matroids and the greedy algorithm", Math. Programming **1** (1971), 127–136.
Fisher, M.L. "Lagrangean relaxation methods for combinatorial optimization", Management Sciences **27** (1981), 1–18.
Folkman, J., and J. Lawrence, "Oriented matroids", J. Combin. Theory Ser. **B25** (1978), 199–236.
Ford, L.R., and D.R. Fulkerson, "Flows in Networks", Princeton University Press (1962).
Fukuda, K. "Oriented Matroid Programming", Ph.D. thesis, University of Waterloo, Waterloo (1981).
Garey, M.R., and D.S. Johnson, "Strong NP-completeness results: motivation, examples and implications", J. Assoc. Comput. Mach. (1978), Vol. 25, 499–508.
Garey, M.R., and D.S. Johnson, Computers and Intractability" (A Guide to the Theory of NP-Completeness), W.H. Freeman and Company, San Francisco, 1979.
Goldfarb, D., and M.J. Todd, "Modifications and implementation of the ellipsoid algorithm for linear programming", Math. Prog. Vol. 23 (1982) No. 1, 1–19.
Grötschel, M. "Approaches to Hard Combinatorial Optimization Problems", in: Modern Applied Mathematics, Optimization and Operations Research (B. Korte, ed.), North-Holland Amsterdam (1982), 437–515.
Grötschel, M., L. Lovász and A. Schrijver, "The Ellipsoid Method and its Consequences in Combinatorial Optimization", Combinatorica **1** (1981), 169–197.
Hammer, P.L., E.L. Johnson and B.H. Korte (eds.), "Discrete Optimization I", Annals of Discrete Mathematics Vol. 4 (1979).
Hammer, P.L., E.L. Johnson and B.H. Korte (eds.), "Discrete Optimization II", Annals of Discrete Mathematics Vol. 5 (1979).
Hartmanis, J., and J.E. Hopcroft, "An overview of the theory of computational complexity", Journal of the Association for Computing Machinery **18** (1971), 444–475.
Hausmann, D., and B.H. Korte, "Algorithmic versus axiomatic definitions of matroids", Math. Programming Study **14** (1981), 98–111

Hu, T.C. "Combinatorial Algorithms", Addison-Wesley (1982) Reading.
Iri, M., and S. Fujishige, "Use of matroid theory in operations research, circuits, and systems theory", Internat, J. Systems Sci. **12** (1981), 27–54.
Jenkyns, H.A. "The efficiency of the "Greedy" algorithm", in: Proc. 7th. S.-E. Conf. Combinatorics, Graph Theory, and Computing (1976), 341–350.
Jensen, P.M., and B.H. Korte, "Complexity of matroid property algorithms", SIAM J. Comput. Vol. 11 (1982), 184–190.
Jeroslow, R.G. "The Simplex Algorithm with the Pivot Rule of Maximizing Criterion Improvement", Discrete Mathematics, Vol. 4 (1973), 367–377.
Jeroslow, R.G. "Minimal inequalities", Math. Programming, Vol. 17 (1979), 1–15.
Johnson, E.L. "The group problem for mixed integer programming", Math. Programming Study, Vol. 2 (1974), 137–179.
Karp, R.M., "Reducibility among combinatorial problems", in: R.E. Miller and J.W. Thatcher (eds.), Complexity of Computer Compulations, Plenum Press (1972), New York, 85–103.
Khachian, L.G. "A polynomial algorithm in linear programming", Dokl. Akad. Nauk SSSR 244 (1979) (Abstract); Dokl. Akad. Nauk SSR **20** (1980), 51–68; translated in: Soviet Math. Dokl. **20(1)** (1979), 191–194.
Klee, V. "Combinatorial optimization: what is the state of the art?", Mathematics of Operations Research **5** (1980), 1–26.
Klee, V., and G.L. Minty, "How good is the simplex algorithm?", in: O. Shisha (ed.), Inequalities III (Academic Press, New York, 1972), 159–175.
Korte, B. "Modern Applied Mathematics (Optimization and Operations Research)", North-Holland, Amsterdam (1982a).
Korte, B. "Matroids and independence systems", in: Modern Applied Mathematics, Optimization and Operations Research, North-Holland, Amsterdam (1982b), 517–553.
Korte, B., and D. Hausmann, "An analysis of the greedy heuristic of independence systems", Ann. Discrete Math. **2** (1978), 65–74.
Korte, B., and R. Schrader, "A survey on oracle techniques", in Mathematical Foundations of Computer Sciences, Proc. Strbské Pleso, Czechoslovakia 1981, Lecture Notes in Computer Science (Springer, Berlin, 1981), 61–77.
Las Vergnas, M. "Matroides orientables", C.R. Acad. Sc. Paris, t. **280** (1975), Serie A, 61–64.
Las Vergnas, M. "Convexity in oriented matroids", J. Comb. Th. (B) **29** (1980), 231–243.
Lawler, E.L. "Combinatorial Optimization: Networks and Matroids", (Holt, Rinehart & Winston, 1976).
Lawrence, J. "Oriented Matroids", Ph.D. Dissertation, University of Washington, Seattle, 1975.
Lovász, L. "Combinatorial Problems and Exercises", North-Holland, Amsterdam, 1979.
Mandel, A. "Topology of Oriented Matroids", Ph.D. thesis, University of Waterloo, Waterloo (1981).
Matthews, L. "Closure in independence systems", Math. of Operations Research Vol. 7 (1982) 159–171
Minty, G.J. "On the axiomatic foundations of the theories of directed linear graphs, electrical networks, and network programming", J. Math. and Mech. **15** (1966), 485–520.
McMullen, P. "The maximum number of faces of a convex polytope", Mathematika **17** (1970), 179–184.
Munson, B.S. "Face Lattices of Oriented Matroids", Ph. thesis, Cornell University, Ithaca (1981).
Papadimitriou, C.H., and K. Steiglitz, "Combinatorial Optimization (Algorithms and Complexity)", Prentice-Hall, Englewood Cliffs (1982).
Prim, R.C. "Shortest connection networks and some generalizations", Bell System Techn. J. **16** (1957), 1389–1401.
Rado, R. "Note on independence functions", Proc. London Math. Soc. **7** (1957), 300–320.
Randow, R. von. "Integer Programming and Related Areas, A Classified Bibliography 1978–1981", Lectures Notes in Economics and Mathematical Systems Vol. 197 (Springer, Berlin-New York, 1982).
Rockafeller, R.T. "The elementary vectors of a subspace of R^n", in: R.C. Bose and T.A. Dowling (eds.), Combinatorial mathematics and its applications, Proc. of the Chapel Hill Conference (Univ. of North Carolina Press, 1969), 104–127.
Schrader, R. "Ellipsoid Methods", in: Modern Applied Mathematics, Optimization and Operations Research (B. Korte ed.), North-Holland, Amsterdam (1982), 266–311.
Shor, N.Z. "Utilisation of the operation of space delatation in the minimization of convex function", Kibernetika **6** (1970), 6–12; translated in: Cybernetics **6** (1970), 7–15.

Smale, S. "On the Average Speed of the Simplex Method of Linear Programming", Research Report 1982.
Tarjan, R.E. "Complexity of combinatorial algorithms", Siam Review **20(3)** (1978), 457–491
Tucker, A.W. "Combinatorial theory underlying linear programs", Recent Advances in Mathematical Programming, (R.E. Graves and P. Wolfe, eds.), McGraw-Hill, New York (1963), 1–16.
van der Waerden, B.L. "Modern Algebra (2nd edn.), Springer, Berlin, 1937.
Welsh, D.J.A. "Kruskal's theorem of matroids", Proc. Cambridge Phil. Soc. **64** (1968), 3–4.
Welsh, D.J.A. "Matroid Theory", Academic Press, London-New York, 1976.
Zimmermann, U. "Linear and Combinatorial Optimization in Ordered Algebraic Structures", Annals of Discrete Mathematics **10** (1981).

A. Bachem
Institut für Angewandle Mathematik
Universität Erlangen
Martenstrasse 3
D–8520 Erlangen (FRG)

Catherine Bandle
Isoperimetric inequalities

1. Introduction

It is well-known that *among all plane domains of given area the circle has the smallest circumference.* This statement is most succinctly expressed in the *isoperimetric inequality*:

(1) $L^2 \geq 4\pi A$,

where L is the length of the boundary curve and A is the area of the domain. The equality sign holds if and only if the domain is a circle.

Although this extremal property of the circle was already known to the Greeks, mathematical proofs were only given, however, in the last century. Steiner [St] proposed several ingenious ways for showing that the circle is the only curve of given length which encloses maximal area. The existence of such an extremal curve was established by Caratheodory and Study [Ca-St] in 1909. Since then this subject was taken up in a series of papers using different methods which led to numerous generalizations and extensions of (1). The geometers developed various types of symmetrizations [Bl,Had], whereas the analysts, also attracted by the problem, employed rather the tools of the calculus of variations.

It is quite clear that the spheres in higher dimensions should be characterized by a similar extremal property. The statement that *among all domains of given volume the sphere has the smallest surface area*, was first proved by Schwarz [Schw]. The proof turned out to be much more difficult than in two dimensions, partly because it cannot be reduced to convex bodies. The convex hull need not necessarily have a smaller surface area than the original body. The n-dimensional case was handled by Schmidt [Schml]. For further references see also [Had, p. 195, Bl]. The validity of an n-dimensional isoperimetric inequality depends very much on the notion of surface area.

Sharper versions of (1) were discovered for special classes of domains such as domains with precribed inradius or domains lying in a given strip [Bon, Bon-Fe p. 111].

More recently attempts were made to extend the isoperimetric inequality to Riemannian manifolds. Sharp results are now available for surfaces.

2. Parallel regions and parallel curves

We present here a method for proving the inequality (1) which appeared for the first time in a paper by Bol [Bol] who attributed the idea to Kaluza. At first glance it seems to require a greater technical apparatus than others, but, on the other hand, it gives a deep insight into the relation between the length of the boundary curve, the area of the domain and its geometry. Moreover, it can easily be extended to curved surfaces.

Throughout this section we shall assume $D \subset \mathbb{R}^2$ to be *simply connected* and its boundary curve to be piecewise analytic.

Definition. *The region* $D_\rho = \{x \in D : \operatorname{dist}(x, \partial D) \geq \rho\}$ *is called the interior* parallel region *at a distance* ρ *and* $\gamma_\rho = \{x \in D : \operatorname{dist}(x, \partial D) = \rho\}$ *is the corresponding* parallel curve.

If ρ_0 denotes the *radius of the incircle*, then obviously D_ρ and γ_ρ are defined for all $\rho \in [0, \rho_0]$. It turns out [Fi] that for almost all $\rho \in [0, \rho_0]$, the parallel curves γ_ρ are piecewise analytic.

Let $dx = dx_1 dx_2$ and s be the arc-length along a curve, and put

$$A(\rho) := \int_{D_\rho} dx \quad \text{and} \quad L(\rho) := \int_{\gamma_\rho} ds$$

for the *area* of D_ρ and the *length* of γ_ρ, respectively.

In order to estimate $A(\rho)$ and $L(\rho)$ let us consider the set of all points, say Γ_ρ, which lie on the interior normals of ∂D and are at a distance ρ from ∂D. For small values of ρ, Γ_ρ coincides with γ_ρ, but in general we have $\gamma_\rho \subset \Gamma_\rho$. By comparing $L(\rho)$ with the length of Γ_ρ we get

$$(2) \qquad L(\rho) \leq L - 2\pi\rho,$$

L being the length of ∂D. From a result of Sz.-Nagy [Sz-N] it follows that $A(\rho)$ is Lipschitz continuous and therefore differentiable a.e. Moreover we have $-A'(\rho) = L(\rho)$ which together with (2) implies

$$(3) \qquad A(\rho) \geq A - L\rho + \pi\rho^2.$$

Consequences

(A) For $\rho = \rho_0$, (3) yields

Bonnesen's Inequality $L \geq A/\rho_0 + \pi\rho_0$.

The equality sign is attained for the rectangle with two half-circles [Fig. 1]. Notice that this result holds only for simply connected domains.

Bonnesen [Bon, p. 60–61] derived this estimate for convex domains and observed [Bon, p. 62] that it remains true, if ρ_0 is replaced by the circumradius R.

Fig. 1

Hence $[\rho_0, R]$ must lie in the interval for which

$$A - Lx + \pi x^2 \leq 0$$

and consequently

$$L \geq A/r + \pi r \quad \text{for all} \quad r \in [\rho_0, R].$$

An immediate consequence is the estimate

$$L^2 - 4\pi A \geq \pi^2 (R - \rho_0)^2.$$

Other variants and a systematic discussion of Bonnesen-type inequalities are contained in [03]. Similar results exist for domains of given width [Ben].

Is there an analogous result in $\mathbb{R}^N$? Wills conjectured that for a convex domain D in $\mathbb{R}^N$ we should have

$$\rho_0 F \geq V + (N-1)V_i,$$

where ρ_0 is the radius of the largest inscribed sphere, F is the surface area and V the volume of D and $V_i = \rho_0^N \omega_N$ is the volume of the sphere of radius ρ_0. As Bokowski [Bok1] observed, it follows from an inequality of Hadwiger and Bieberbach [Had, p. 269] and its validity is, in contrast to the two-dimensional case, strictly restricted to convex domains.

(B) Elimination of ρ from (2) and (3) leads to

(4) $$L^2 - L^2(\rho) \geq 4\pi(A - A(\rho)).$$

In particular we get for $\rho = \rho_0$ the

Classical isoperimetric inequality $L^2 \geq 4\pi A$.

It is obviously not restricted to simply connected domains. Moreover by means of approximation methods it is possible to weaken the regularity condition on the boundary.

In a convex domain the parallel curves are convex. Thus, if two circles of inradius ρ_0 can be inscribed whose centers are at a distance d, (4) yields the sharper form

$$L^2 \geq 4\pi A + 4d^2.$$

A careful examination of the method of interior parallels shows (Bol) that for convex domains with corners the isoperimetric inequality can be improved considerably. Consider the domain around the unit circle formed by the tangent directions of the original domain and let σ be its surface area. Then

$$L^2 \geq 4\sigma A + L^2(\rho_0).$$

From this inequality we obtain for convex polygons whose outer normals form the angles ϕ_i, the estimate

$$L^2 \geq 4 \sum_i \operatorname{tg} \frac{\phi_i}{2} A.$$

Equality holds for a polygon circumscribed about a circle. This theorem goes back to Lhuilier in 1782. For further references and in particular extensions to higher dimensions see [Had, p. 274]. In this context it should be mentioned that for a n-sided polygon in the plane one has [Fe]

$$\frac{L^2}{A} \geq \frac{4}{n} \operatorname{tg} \frac{\pi}{n}.$$

(C) Let us call $r(\rho)$ with $\pi r^2(\rho) = A(\rho)$ the *area radius* of D_ρ. By means of the isoperimetric inequality we find

$$-2\pi r r' = -A'(\rho) = L(\rho) \geq 2\pi r$$

which implies

Schmidt's inequality $r(0) - r(\rho) \geq \rho$.

Equality holds only for the circle. Schmidt [Schm4] proved this inequality and its generalizations for an arbitrary domain in $\mathbb{R}^N$.

Historical remarks

As already mentioned before, the method of interior parallels goes back to Bol [Bol] and Fiala [Fi]. Both were mainly interested in its generalization to surfaces. It fell into oblivion until 1959, when Makai [Ma] and Pólya [Po] used it in order to estimate the principal frequency of a membrane and the torsional rigidity of a beam. Their work motivated Sz.-Nagy's paper [Sz-N] on parallel curves who found estimates of the type (2) and (3) for very general not necessarily simply connected domains. By using an ingenious approximation method he was able to avoid the question of rectifiability of γ_ρ which is the most delicate point of this theory.

Further extensions of the method of interior parallels are contained in [Ba 3, 7].

3. Inequalities on surfaces

The question arises whether the extremal property of the circle is also true on general surfaces?

In 1905 Bernstein [Bern] proved that *among all domains of given area on a sphere the circle has the smallest circumference.*

Schmidt [Schm2] showed that the same property is valid for circles on the hyperbolic plane. The isoperimetric inequality in the plane, on the sphere and on the hyperbolic plane can be expressed as follows:

(5) $L^2 \geq (4\pi - KA)A$,

where K denotes the *Gaussian curvature*. We have $K = 0$ for the plane, $K = 1/R^2$ for the sphere of radius R and $K = -1$ for the hyperbolic plane.

One cannot expect that on an arbitrary curved surface there is a domain of given area that has the smallest circumference. Such an extremal domain must be bounded by a curve of constant geodesic curvature. As Schmidt pointed out in [Schm3], there are surfaces which don't possess closed curves of constant geodesic curvature.

Instead of searching for an extremal domain we shall rather pose the question: under what conditions does an inequality of the type (5) hold?

For this purpose let us introduce the concept of an *abstract surface* S given locally by a domain D in the parameter plane and a Riemannian metric defined by means of the line element $d\sigma$. For simplicity we assume the surface S to be given locally in isothermic representation

$$d\sigma^2 = p(x)ds^2 \quad \text{(s: Euclidean arc-length)}$$

With this notation we have for the *circumference* of an arbitrary domain $B \subseteq D$ and the *area* of B

$$L(\partial B) = \int_{\partial B} d\sigma \quad \text{and} \quad A(B) = \int_B p\,dx.$$

Regular surfaces

If $p \in C^2(D)$, then the Gaussian curvature is well-defined and is of the form

$$K(x) = -\frac{\Delta \log p}{2p}.$$

Definition. *A surface for which* $K(x)$ *exists will be called* regular.

The geometrical quantity $K(x)$ is crucial for the characterization of the surfaces for which (5) is valid. Namely we have

Theorem 1. *If* $L^2(\partial B) \geq \{4\pi - K_0A(B)\}A(B)$ *holds for any domain* $B \in S$, *then the Gaussian curvature satisfies* $K(x) \leq K_0$.

The proof is based on local considerations and uses the Taylor expansion of p. This theorem has been discovered for $K_0 = 0$ by Beckenbach and Radò [Be-R] and then extended to arbitrary K_0 by several authors [Bol, Hul]. It turns out that the converse, which is a much deeper statement is also true. Indeed we have

Theorem 2. *Let* S *be a surface* $K(x) \leq K_0$ *and* $B \in S$ *be an arbitrary simply connected domain. Then the following inequalities hold.*

(i) $L^2(\partial B) \geq (4\pi - K_0A(B))A(B)$.

The equality sign is achieved only for geodesic circles on surfaces of constant Gaussian curvature K_0.

(ii) $L^2(\partial B) \geq \left(4\pi - 2\int_D K^+\,dx\right)A(B), \quad K^+ := \max(K(x),0)$.

Equality holds only for circles in the plane.

The first assertion is due to Bol [Bol], whereas the second goes back to Fiala [Fi]. Fiala stated this result for manifolds with a complete analytic Riemannian metric. Twenty years later Hartman [Har] showed that the analyticity assumption can be relaxed in the sense that the metric need only be of class C^2.

Let us sketch the main ideas of Bol's and Fiala's proof. As in the plane we set

$$B_\rho = \{x \in B : \text{dist}_\sigma(x,\partial B) \geq \rho\}$$

for the *interior parallel region* and

$$\gamma_\rho = \{x \in B : \mathrm{dist}_\sigma(x,\partial B) = \rho\}$$

for the *interior parallel curves.*

Moreover we denote by $A(\rho)$ the area of B_ρ and by $L(\rho)$ the length of γ_ρ both measured in the Riemannian metric. If ∂B is piecewise analytic, then the same is true for almost all γ_ρ and we have [Fi, Bol, Har, Ba7] the key formulas

(6) $A'(\rho) = -L(\rho)$ for almost all $\rho \in (0,\rho_0)$,

and

(7) $L'(\rho) \leq \int_{B_\rho} K(x)\,dx - 2\pi.$

Thus

$$L(\rho)L'(\rho) \leq -K_0 A'(\rho)A(\rho) + 2\pi A'(\rho)$$

or

$$L(\rho)L'(\rho) \leq -A'(\rho)\int_B K^+\,dx + 2\pi A'(\rho).$$

Formal integration yields the desired result. As Osserman [03] pointed out, (6) and (7) imply a

generalized Bonnesen inequality $L(\partial B)\rho_0 \geq A(B) + \left(2\pi - \int_B K^+\,dx\right)\dfrac{\rho_0^2}{2}$

where ρ_0 is the inradius of B measured in the Riemannian metric. As in the plane it is possible to derive from (6) and (7) a *Schmidt type inequality*, namely [Ba7]:

Let the assumptions of the previous theorem hold and denote by $\hat{B}$ *a geodesic circle on a surface of constant Gaussian curvature* K_0 *with the same area as* B. *Then the area of* B_ρ *is smaller or equal to the area of* $\hat{B}_\rho$.

From this result it follows immediately that

the inradius of B *is smaller than or equal to the inradius of* $\hat{B}$ [Ba7].

For $K_0 > 0$ the inradius is therefore always bounded from above by $\pi/\sqrt{K_0}$.

Bernstein [Bern] derived in his fundamental paper on the isoperimetric inequality for domains on the sphere inequalities for convex domains relating L, A and the width of the narrowest circular annulus containing the boundary curve. This result has been improved later by Bonnesen [Bon, p. 80–82].

Inequalities for circles on regular surfaces

Let the *circle* $C(\rho,P) \in S$ be the set of all points at a distance smaller than ρ from $P \in S$. In general, $C(\rho,P)$ is not simply connected and does not look at all like a circle. However, for small ρ or in the case of negative curvature, $C(\rho,P)$ is simply connected.

According to Osserman [03], $C(\rho,P)$ is called a *geodesic disk* if it has the property that no two geodesics starting at P intersect before a distance ρ.

For geodesic disks Osserman [03] proved the following result:

If S is simply connected and of bounded curvature $K \leq K_0$, *then* $A_\rho \geq A_\rho^{(K_0)}$, *where* A_ρ *is the area of* $C(\rho,P)$ *and* $A_\rho^{(K_0)}$ *is the area of a circle of radius* ρ *on a surface of constant Gaussian curvature* K_0. *Equality holds only for* $K \equiv K_0$.

For the length L_ρ of the boundary curve we have by Theorem 2 (i) and [Ba 7, p. 40] the estimate

$$4\pi A_\rho - K_0 A_\rho^2 \leq L_\rho^2 \leq 4\pi A_\rho - k_0 A_\rho^2$$

provided that $k_0 \leq K(x) \leq K_0$. Equality holds on both sides for circles on surfaces of constant Gaussian curvature.

This leads to a bound for the diameter of a closed convex surface. More precisely, the diameter d of a compact, simply connected surface with $0 < k_0 \leq K(x) \leq K_0$ satisfies [K1]

$$\frac{\pi}{\sqrt{K_0}} \leq d \leq \frac{\pi}{\sqrt{k_0}}.$$

Equality holds on both sides for the sphere. In addition Klingenberg studies its extensions to higher dimension and discusses questions of the classifications of Riemannian surfaces.

Further results on L_ρ, in particular on convex surfaces are found in [A1, p. 384]. Osserman [03] improved Theorem 2 (i) by using the estimates for the incircle.

General surfaces

So fare we have been dealing with surfaces whose Gaussian curvature was well-defined. Before we state the general results on general surfaces let us consider the circular cone. It can be represented as a sector in the plane whose boundary segments are identified

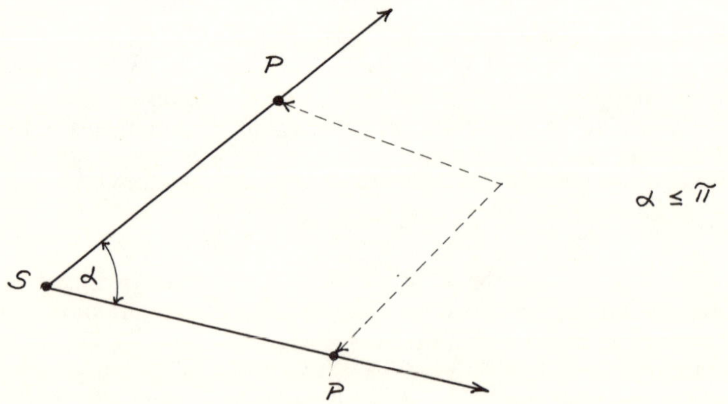

Fig. 2

It is clear that for a simply connected domain D which doesn't contain the corner S we have $L^2 \geq 4\pi A$. In general such an estimate obviously cannot be true. It has to be replaced by $L^2 \geq 2\alpha A$, equality holding for geodesic circles centered at the corner S. For $\alpha = 2\pi/n$, $n = 1,2,\ldots$ this estimate follows immediately from the isoperimetric inequality and a symmetry argument. The general cases requires a more sophisticated argument [A1, Hu1].

Notice that the isothermic representation of the circular cone is readily obtained by mapping the sector conformally onto the circle. It is of the form

$$(\mathbb{C}, |w|^{(\alpha - 2\pi)/(2\pi)}\, ds),$$

$\mathbb{C}$ being the complex plane and $w = w_1 + iw_2$.

In order to define a general surface locally represented by $(D, d\sigma)$, consider two subharmonic functions p_1 and p_2 defined on the atlas $D \in \mathbb{R}^2$. By Riesz's theorem they can be written as

$$p_i(x) = h_i(x) - \frac{1}{\pi} \int_D \log|x - y|\, d\omega_i(e_y), \quad i = 1,2$$

where $x = (x_1, x_2)$, h_i is a harmonic function and ω_i are non-negative measures such that $\omega_1(e)\omega_2(e) = 0$ for any Borel set $e \subseteq D$. Moreover we assume $v \in C^2(D)$ to be a function satisfying

$$\Delta v + 2K(x)e^{v + p_1 - p_2} = 0 \quad \text{in } D.$$

Here $K(x)$ stands for any continuous function.

We now suppose that the metric is of the form

$$d\sigma^2 = e^{v + p_1 - p_2}\, ds^2.$$

Example. *In case of the circular cone we have* $v \equiv 0$, $p_2 = 0$ *and* $p_1 = (2 - \alpha\pi^{-1})$. $\log|w|$, *that is* ω *is a point measure with* $\omega_1(\{0\}) = 2\pi - \alpha$ *and* $\omega(P) = 0$ *for* $P \neq 0$.

The quantity

$$\omega(B) := \int_B K(x)\, dx + \omega_1(B) - \omega_2(B), \quad B \subseteq D$$

is called the *total curvature* of B.

Singularities of the Gaussian curvature such as corners and edges are contained in ω_i.

Furthermore we introduce the concept of the *total curvature of* B *with respect to* K_0 in the following manner

$$\omega_{K_0}(B) := \omega(B) - K_0 A(B).$$

This notion is due to Alexandrow [A1] and it measures the deviation of the total curvature $\omega(B)$ from the total curvature of B on a surface of constant Gaussian curvature. The connection between the geometrical concept of Alexandrow and the analytical description given here has been studied by [Hu3].

Let

$$\alpha := 2\pi - \sup_{e \subseteq B} \{\omega_{K_0}(e), 0\}.$$

Under these assumptions we have for any simply connected domain B on the surface S [A1, S. 514]

Alexandrow's inequality $L^2(\partial B) \geq (2\alpha - K_0 A(B))A(B)$.

Equality holds only for geodesic circles on circular cones in spaces of constant curvature K_0, *centered at the vertex of the cones.* A model of such an extremal domain is given below

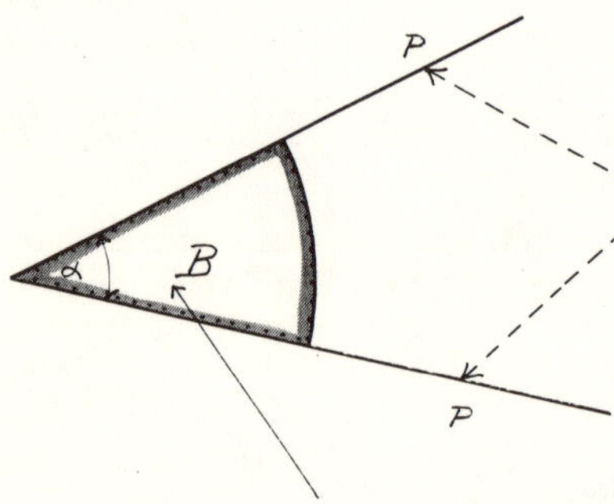

$$d\sigma^2 = b(1 + (K_0/4)br^2)^{-2} ds^2$$
$$= : e^{\hat{u}(r;b,K_0)} ds^2$$

Fig 3

Alexandrow discovered this inequality in 1945 and gave a purely geometrical proof. This result contains all previous inequalities relating L, A and the curvature. An analytical proof for the special case $v \equiv 0$ is found in [Hu1, cf. also Har]. A different method is proposed in [Ba6] in the case where $K(x) \geq 0$. For a more complete list of all recent literature on this subject we refer to [01, cf. also 03].

4. Dido type inequalities

Let $D \subset \mathbb{R}^2$ be a simply connected domain whose boundary is divided into two parts γ and Γ. The question arises *under what assumptions on* γ *is it possible to give estimates for the length of* Γ *in terms of the area of* D?

If γ is a straight segment we deduce from a reflection argument and the isoperimetric inequality (1) that

(8) $\quad L^2(\Gamma) \geq 2\pi A(D)$

with equality if and only if D is a half-circle.

It is clear by looking for example at the figure below

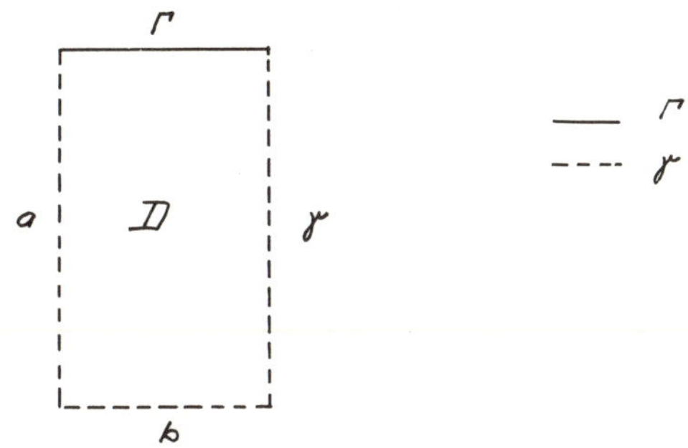

$$L(\Gamma) = b = \frac{A}{a} \to \quad \text{if} \quad a \to 0.$$

Fig. 4

that an inequality of the type (8) cannot hold without any further assumptions on γ.

The problem of maximizing the area by given length $L(\Gamma)$ appeared already in ancient time. The story tells that Queen Dido of Carthage being a refugee from Tyria asked a North African chieftain for as much land as she could encompass with the hide of a cow. She sliced the hide into small strips and lay these out along a semicircle, using the North African coast as the supplementary boundary. She obtained in this way the maximal area she could get by such a procedure.

Inequalities of this type are useful for getting bounds in elliptic equations with mixed boundary conditions. They have been investigated with this goal in mind, in [Ba2, Ba7] where also extensions to three dimensions are given. We add two examples which are obtained by means of the isoperimetric inequality on a circular cone.

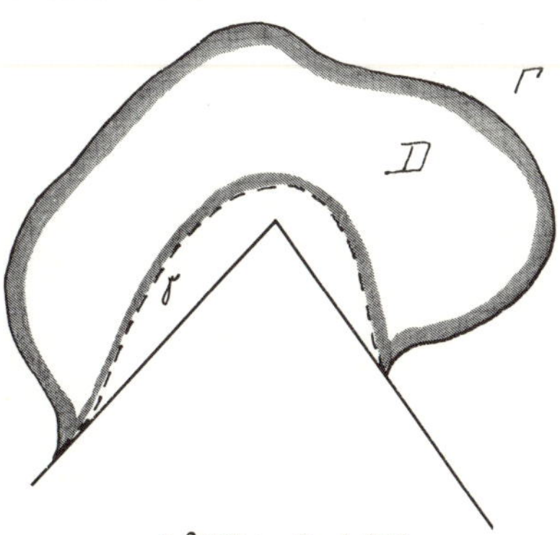

$$L^2(\Gamma) \geq 2\pi A(D)$$

Fig. 5

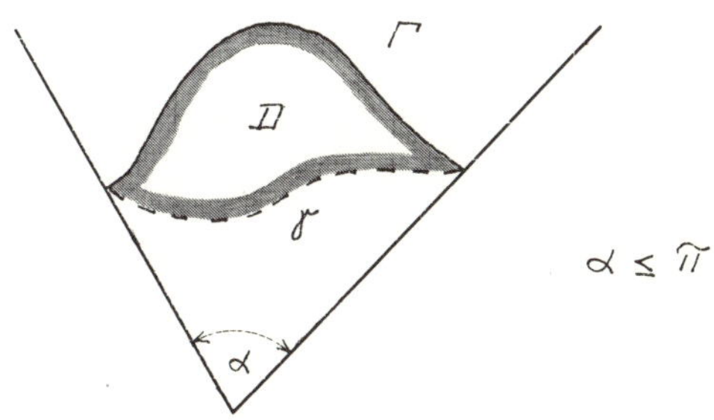

$$L^2(\Gamma) \geq 2\alpha A(D)$$

Fig. 6

The equality sign holds in the first case for the semi-circle and in the second case for a domain in a sector of angle α bounded by a circular arc.

A similar problem can be treated on surfaces. Let $D \subset S$ be a simply connected domain whose boundary consists of two arcs Γ and γ. Here the problem of estimating $L(\Gamma)$ is solved as follows. Take two copies of D and paste them along γ. Then we get a new surface for which Alexandrow's isoperimetric inequality is applicable. The result depends on the curvature of D and the geodesic curvature of γ. A more detailed study is found in [Ba4, Ba7]. As an illustration we indicate the following special case.

Let D be a domain on a regular surface with bounded Gaussian curvature $K(x) \leq K_0$ and let γ be an arc with non-positive geodesic curvature. Then

$$L^2(\Gamma) \geq (2\pi - K_0 A(D))A(D).$$

Equality is attained for semi-circles on surfaces of constant Gaussian curvature K_0.

We would also like to mention here a somewhat related question, namely: Given a convex domain D which is divided into two disjoint regions K and $D - K$ such that $A(K)/A(D - K)$ is a given fixed number. It is possible to give a lower bound for the separating curve $D \cap \partial K$? For more details on this type of problems we refer to [Bok2] where references to the previous literature are given.

5. Applications

The circle is not only the solution of geometrical isoperimetric problems, but also of various extremal problems in analysis and mathematical physics.

By use of geometrical inequalities estimates for not easily accessible quantities in function theory and in mathematical physics are obtained. Pioneering work in this area was done by Pólya and Szegö which together with earlier results is contained in their standard book on "Isoperimetric Inequalities in Mathematical Physics". A more recent survey on the development of this research field between 1951 and 1967 is due to Payne [Pa]. For further references see also [Ba7, Ba8, 01].

We outline here only a few results which are direct consequences of our inequalities on two-dimensional surfaces.

5.1 *Maximal conformal radius*

Let D be a simply-connected domain in the complex z-plane, $z_0 \in D$ an arbitrary point and $f(z) = (z - z_0) + a_2(z - z_0)^2 + \cdots$ a complex one-to-one function mapping D conformally onto the circle $\{w : |w| < R_{z_0}\}$. It follows immediately from the Riemann mapping theorem that such a function exists and that R_{z_0} is uniquely determined.

Definition. R_{z_0} *is called the* conformal radius of D with respect to z_0 *and* $\dot{R} := \sup_{z_0 \in D} R_{z_0}$ *is called the* maximal conformal radius of D.

It was observed in [Po-Sz] that the following inequality holds between the area A of D and the conformal radius:

$$\pi \dot{R}^2 \leq A,$$

the equality sign being attained if and only if D is a circle.

Consider in D a Riemannian metric $d\sigma^2 = p\,ds^2$ of bounded Gaussian curvature K_0 and denote by A_σ the total area of D with respect to this metric. The previous bound for $\dot{R}$ can be sharpened as follows [Ah, Ba7]

$$R_z^2 \leq \frac{4A_\sigma}{p(z)\{4\pi - K_0 A_\sigma\}}, \quad \text{provided that} \quad K_0 A_\sigma < 4\pi.$$

If z is the point such that $R_z = \dot{R}$, then the above estimate holds for the maximal conformal radius. Equality holds for the circle centered at the origin with the metric of constant Gaussian curvature K_0, that is

$$d\sigma^2 = \frac{b}{(1 + b(K_0/4)|z|^2)^2}\, ds^2 = :e^{\hat{u}(r;b,K_0)}\, ds^2.$$

There is a close relationship between this result and an extension of the mean value theorems for subharmonic functions [Ba7]. Namely, let v satisfy the differential inequality $\Delta v + 2K_0 e^v \geq 0$ in $\{|z| < R\}$. It has been proved in [Ba7] that under the hypothesis $K_0 \int_{\{|z|<R\}} e^v\, dx \leq 2\pi$ we have

$$\int_{\{|z|=r\}} e^v\, ds \geq \frac{2\pi r e^{v(0)}}{(1 + e^{v(0)}(K_0/4)r^2)^2} \quad \text{for all} \quad 0 < r \leq R.$$

Equality holds for the function $v = \hat{u}$ mentioned above.

Estimates for other quantities of interest in function theory such as the *conformal symmetry line* are provided in [Hu2] where references to earlier papers are found.

5.2 *Eigenvalues of the Laplacian*

The fundamental frequencies ω_i of a vibrating membrane which covers at rest a domain $D \subset \mathbb{R}^2$, are obtained by means of the eigenvalues of

$$\Delta\phi + \lambda\phi = 0 \text{ in D}, \quad \phi = 0 \text{ on } \partial D.$$

From the classical theory it follows that there exists a countable number of positive eigenvalues $0 < \lambda_1 < \lambda_2 \leq \cdots$. We have $\omega_i = \sqrt{\lambda_i}$.

In 1877 Lord Rayleigh [Ra] conjectured that *among all domains of given area the circle has the lowest principal frequency.*

This statement can again be expressed as an inequality relating the area A of D and λ_1, namely,

$$(9) \qquad \lambda_1 \geq \frac{\pi j_0^2}{A} \quad (j_0 = 2.4048\ldots, \text{first zero of the Bessel function } J_0)$$

Equality holds only for the circle.

Lord Rayleigh was led to this conjecture after having computed λ_1 for a number of special cases like the square, the equilateral triangle, the semi-circle,.... He also used a sophisticated perturbation method to approximate the value of λ_1 for a nearly circular domain.

In 1923/24 Faber [Fa] and Krahn [Kr] independently, established (9) by introducing a special system of curvilinear coordinates.

Consider now a membrane with inhomogeneous mass density p,

(10) $\Delta\phi + \lambda p\phi = 0$ in D, $\phi = 0$ on ∂D.

Nehari [Ne] extended the Rayleigh-Faber-Krahn inequality for mass densities satisfying $\Delta \log p \geq 0$. He proved that

$$\lambda_1 \geq \frac{\pi j_0^2}{\int_D p\,dx},$$

where the equality sign stands for example for the circle with constant p.

Notice that the extremal case is not uniquely determined.

If we write (10) in the form $(\Delta/p)\phi + \lambda\phi = 0$, $\Delta_S = \Delta/p$ can be interpreted as the *Laplace-Beltrami* operator on a surface with Riemannian metric $d\sigma^2 = p\,ds^2$. Problem (10) can be reformulated as

(11) $\Delta_S\phi + \lambda\phi = 0$ in $D \subseteq S$, $\phi = 0$ on ∂D.

In geometric terms Nehari's condition means that the Gaussian curvature of S is non-positive.

By means of Theorem 2(ii) of Section 3 Peetre [Pe1] showed that the first eigenvalue of Δ_S with Dirichlet boundary values satisfies

$$(12) \quad \lambda_1 \geq \frac{j_0^2}{2A_\sigma(D)}\left(2\pi - \int_D K^+\,dx\right),$$

which in view of the equivalence of (10) and (11) includes Nehari's result. Generalizations of (12) are found in [Ba1, 7; see also Ch-Fe].

The most general result states that for a simply connected domain D on a surface S satisfying the assumptions of Alexandrow's inequality we have [Ba1, 4, 7]

$$\lambda_1(D,S) \geq \hat{\lambda}_1$$

where $\hat{\lambda}_1$ stands for the first eigenvalue of the geodesic circle on a circular cone of constant curvature described in Section 4, having the same area as D.

The validity of an inequality of the type

$$\lambda_1 \geq \frac{\pi j_0^2(1-\varepsilon)}{A_\sigma}$$

was established by Peetre [Pe2] for very general surfaces. However he couldn't determine the value of ε. Similar statements were proved in [Ber-Me] for surfaces in higher dimensions.

Hersch [He1] showed that for *convex* domains with inradius ρ_0 the first eigenvalue of the homogeneous membrane satisfies

$$\lambda_1 \geq \left(\frac{\pi}{2\rho_0}\right)^2,$$

where the right-hand side is the limiting value for long thin rectangles. Other non-isoperimetric results relating λ_1 and ρ_0 are outlined in [02].

Because of the variational characterization of the eigenvalues upper bounds are much easier to construct and there are numerous isoperimetric inequalities providing such bounds. For more details we refer to [Pa2, Ba7, Sp, 01, 02]. We would nevertheless like to draw the attention to Pólya and Schiffer's inequality [Po-Sc] connecting the maximal conformal radius with the sum of the reciprocal first n eigenvalues. It is described as follows: Let $\lambda_1,\ldots,\lambda_n$ be the first n eigenvalues of the fixed membrane equation in a simply connected domain D and let $\hat{\lambda}_1,\ldots,\hat{\lambda}_n$ be the corresponding eigenvalues of the circle of radius 1. Then

$$\sum_{i=1}^{n} \lambda_i^{-1} \geq \dot{R}^2 \sum_{i=1}^{n} \hat{\lambda}_i^{-1}, \quad \dot{R} \text{ maximal conformal radius of D.}$$

This inequality has a natural extension to non-homogeneous membranes [Ba7], namely

Let D *be a simply connected domain,* $z_0 \in D$ *an arbitrary point and* p *a mass density satisfying* $\Delta \log p + 2K_0 p \geq 0$ *and* $K_0 \int_D p\,dx \leq 2\pi$. *Moreover put* $\beta := p(z_0)R_{z_0}^2$, *and* $e^{\hat{u}(r,\beta;K_0)} := \beta/(1 + (\beta K_0 r^2/4))^2$. *Observe that* β *is a conformal invariant. If* $\hat{\lambda}_i$ *is the* ith *eigenvalue of* $\Delta\hat{\phi} + \hat{\lambda} e^{\hat{u}(r,\beta;K_0)}\hat{\phi} = 0$ *in* $\{x:|x| < 1\}$ $\hat{\phi} = 0$ *on* $\{x:|x| = 1\}$, *then*

$$\sum_{i=1}^{n} \lambda_i^{-1} \geq \sum_{i=1}^{n} \hat{\lambda}_i^{-1}$$

Observe that the extremal problem is the eigenvalue problem of the Laplace-Beltrami operator of a geodesic circle on a surface of constant Gaussian curvature. This is not surprising since the proof makes use of the geometrical inequality of Theorem 2(ii).

Sharper versions of Pólya and Schiffer's result are obtained in [He2] for *symmetric regions* and an extension to *multiply connected domains* is due to Gasser and Hersch [Ga-He, cf. also [Ba7]].

Much less is known for the *free membrane* described by the eigenvalue problem

$$\Delta\psi + \nu\psi = 0 \text{ in } D \subset \mathbb{R}^2, \quad \frac{\partial\psi}{\partial n} = 0 \text{ on } \partial D$$

($(\partial/\partial n)$ outer normal derivative).

By classical results there exists a countable number of eigenvalues $0 = \nu_1 < \nu_2 \leq \cdots$. By means of conformal transplantation Szegö [Sz] proved for simply connected domains a kind of counterpart of the Rayleigh-Faber-Krahn inequality which was later on generalized by Weinberger [We2] to arbitrary not necessarily simply connected domains in higher dimensions. It expresses the following extremal property of a circle.

Among all domains of given area the circle yields the highest second eigenvalue ν_2.

This statement can again be written as an inequality

$$\nu_2 \leq \frac{\pi p_1^2}{A} \quad (p_1 = 1.841\ldots \text{ zero of the Bessel function } J_1)$$

As for the Rayleigh-Faber-Krahn inequality there exists also an extension of this inequality to free membranes on surfaces, that is to the problem

$$(13) \qquad \Delta_S \psi + \nu\psi = 0 \text{ in D}, \quad \frac{\partial\psi}{\partial n} = 0 \text{ on } \partial D.$$

As mentioned before (13) is equivalent to an inhomogeneous free membrane with mass density p(x) provided that the metric of S is of the form $d\sigma^2 = p\, ds^2$. We then have [Ba7].

Theorem 3. *Let* D *be a simply connected domain on* S *whose Gaussian curvature is bounded from above by* K_0. If the total area A_σ *of* D *satisfies* $K_0 A_\sigma \leq 2\pi$, *then the value of* $(1/\mu_{v_2}) + (1/\mu_{v_3})$ *takes its minimum for a geodesic circle on a surface of constant curvature* K_0.

Some complements for symmetric domains are found in [He2].

In his paper on inhomogeneous membranes Nehari [Ne] has also considered membranes with *mixed boundary* conditions

$$\Delta\phi + \mu\phi = 0 \text{ in D}, \quad \phi = 0 \text{ on } \Gamma, \quad \frac{\partial\phi}{\partial n} = 0 \text{ on } \gamma,$$

where $\Gamma \cup \gamma = \partial D$ and $\Gamma \cap \gamma = \phi$. He established the following theorem for the lowest eigenvalue μ_1

Theorem 4. *If* γ *is a concave arc, then*

$$\mu_1 \geq \frac{\pi j_0^2}{2A}.$$

Equality holds for semi-circles with Γ *as circular arc and* γ *as the straight segment.*

This theorem has been generalized in various ways in [Ba2, 4, 7]. First the concavity of γ has been dropped and extensions to inhomogeneous membranes are indicated. Then terms involving the curvature of γ enter into the inequalities. Without a restriction on γ an inequality of the type mentioned in the above theorem cannot be true as it is seen from the example below

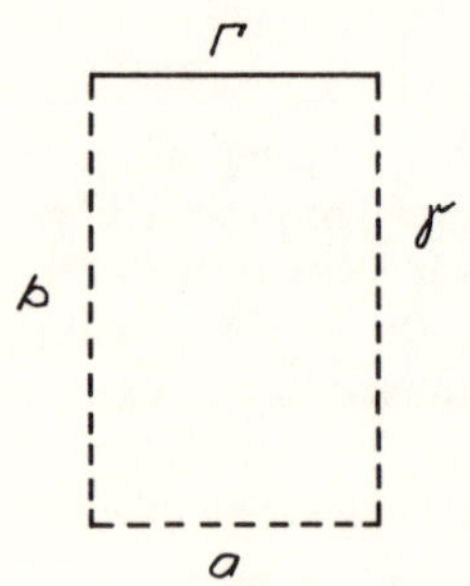

$$\mu_1 \geq \left(\frac{\pi}{2b}\right)^2 = \left(\frac{\pi a}{2A}\right)^2 \to 0 \quad \text{if} \quad a \to 0.$$

Fig. 7

The theorems are based on the isoperimetric inequalities in Section 4.

In the study of the geometrical and topological significance of the spectrum of the Laplacian on a Riemannian manifold many authors have been concerned with estimates for the first non-trivial eigenvalue μ_1. Lichnerowicz and Obata [Ber-Ga-Ma] proved *for compact 2-dimensional manifolds of positive Gaussian curvature* $K(x) \geq k_0 > 0$ *the inequality*

$$\mu_1 \geq 2k_0,$$

equality holding only for surfaces isometric to the sphere of radius $1/\sqrt{k_0}$.

Further inequalities of this type are found in [Sp, 02, Ber-Me]. Upper bounds for μ_1 in the case of a surface homeomorphic to the sphere are provided [He3] by

$$\left(\frac{1}{\mu_1} + \frac{1}{\mu_2} + \frac{1}{\mu_3}\right)\frac{1}{A} \geq \frac{3}{8\pi} \quad \text{(A area of the surface).}$$

Equality holds for the sphere.

The whole literature concerning the spectrum of the Laplacian has been put together in [Ber-Ber].

5.3 *Dirichlet problems*

In connection with investigations on the torsional rigidity of cyclindrical beams Pólya and Szegö [Po-Sz] showed that the solution of

(14) $\Delta u + 1 = 0$ in $D \subset \mathbb{R}^2$, $u = 0$ on ∂D

satisfies

$$u_{max} \leq \frac{A}{2\pi}, \quad \text{(A area of D)}$$

equality being attained for the circle.

For *simply connected* domains a lower bound for u_{max} is given [Po-Sz] by

$$\frac{\dot{R}^2}{2} \leq u_{max},$$

where the equality sign holds again only for the circle.

Both results were extended by Payne [Pa1] to multiply connected domains in higher dimensions as well as by Weinberger [We1] to more general elliptic equations. For more recent results consult also [Ta, Ba5,7, Sc-Sp].

Consider now problem (14) in a simply connected domain D on a surface surface S

$$\Delta_S u + 1 = 0 \text{ in } D \subset \mathbb{R}^2, \quad u = 0 \text{ on } \partial D.$$

If the Gaussian curvature is bounded by K_0 *and if* $K_0 A_\sigma(D) < 4\pi$, *then* [*Ba5*]

$$u_{max} \leq \frac{1}{K_0} \log \frac{4\pi}{4\pi - K_0 A_\sigma(D)}.$$

Equality holds for geodesic circles on surfaces of constant Gaussian curvature K_0.

More on this problem is found in [Ba1, 2, 5, 7]. In particular problems with mixed boundary conditions are considered.

Finally, we remark that we have not tried to give a complete up-to-date picture of the complex and diversified body of isoperimetric inequalities, but we have rather tried to collect the material related with the two-dimensional surfaces. A survey on further isoperimetric inequalities especially in mathematical physics will appear in [Ba8].

References

[Ah] Ahlfors, L.: Geodesic curvature and area, Studies in Mathematical Analysis and Related Topics, Essays in Honor of G. Pólya. Stanford University Press (1962), 1–7.

[Al] Alexandrow, A.D.: Die innere Geometrie der konvexen Flächen. Berlin (1955).

[Ba1] Bandle, C.: Konstruktion isoperimetrischer Ungleichungen der mathematischen Physik aus solchen der Geometrie, Comment. Math. Helv. **46** (1971), 182–213.

[Ba2] Bandle, C.: Extremaleigenschaften von Kreissektoren und Halbkugeln, Comment. Math. Helv. **46** (1971), 356–380.

[Ba3] Bandle, C.: A generalization of the method of interior parallels, and isoperimetric inequalities for membranes with partially free boundaries, J. Math. Anal. Appl. **39** (1972), 166–176.

[Ba4] Bandle, C.: A geometrical isoperimetric inequality and applications to problems of mathematical physics, Comment. Math. Helv. **49** (1974), 496–511.

[Ba5] Bandle, C.: Bounds for the Solutions of Poisson Problems and Applications to Nonlinear Eigenvalue Problems, SIAM J. Math. Anal. **6** (1975), 146–152.

[Ba6] Bandle, C.: On a differential inequality and its applications to geometry, Math. Z. **147** (1976), 253–261.

[Ba7] Bandle, C.: Isoperimetric inequalities and applications, Pitman Publ. London (1980).

[Ba8] Bandle, C.: Isoperimetric Inequalities, to appear in the Encyclopedia of Systems and Control, Pergamon Press.

[Bec-Ra] Beckenbach, E.F. and Radò, T.: Subharmonic functions and surfaces of negative curvature, Trans. Amer. Soc. **35** (1933), 662–674.

[Ben] Benson, D.C.: Sharpened Forms of the Plane Isoperimetric Inequality, Amer. Math. Monthly **77** (1970), 29–34.

[Ber-Me] Bérard, P. and Meyer D.: Inégalités isopérimetriques et applications, preprint C.N.R.S. 212.

[Ber-Ber] Berger, M. and Bérard, P.: Le spectre d'une variété Riemannienne en 1981. Preprint

[Ber-Ga-Ma] Berger, M., Gauduchon, P. and Mazet, E.: Le spectre d'une vatiété Riemannienne, Springer Lecture Notes 194 (1971).

[Bern] Bernstein, F.: Über die isoperimetrische Eigenschaft des Kreises auf der Kugeloberfläche und in der Ebene, Math. Ann. **60** (1905), 117–136.

[Bl] Blaschke, W.: Kreis und Kugel, Leipzig 1916.

[Bok1] Bokowski, J.: Eine verschärfte Ungleichung zwischen Volumen, Oberfläche und Inkugelradius im $\mathbb{R}^n$. El. Math. **28** (1973), 43–44.

[Bok2] Bokowski, J.: Ungleichungen für den Inhalt von Trennflächen, Archiv der Mathematik 34 (1980), 84–89

[Bol] Bol, G.: Isoperimetrische Ungleichung für Bereiche auf Flächen. Jahresber. Deutsche Math. Verein. **51** (1941).

[Bon] Bonnesen, F.: Les problemes des isopérimetres et des isépiphanes, Gauthier-Villars, Paris 1929.

[Bon-Fe] Bonnesen, F. and Fenchel, W.: Theorie der konvexen Körper, Springer 1934.

[Ca] Caratheodory, C. and Study, E.: Zwei Beweise des Satzes, daß der Kreis unter allen Figuren gleichen Umfangs den größten Inhalt hat, Math. Ann. **68** (1909), 133–140.

[Ch-Fe] Chavel, I. and Feldman, E.A.: Isoperimetric inequalities on curved surfaces, Advances Math. **37** (1980), 83–98.

[Fa] Faber, C.: Beweis, daß unter allen homogenen Membranen von gleicher Fläche und gleicher Spannung die kreisförmige den tiefsten Grundton gibt, Sitzungsberichte Bayer. Akad. der Wiss. Math. Physik (1923), 169–172.

[Fe] Fejes Toth, L.: Lagerungen in der Ebene, auf der Kugel und im Raum. Die Grundlehren der math. Wissenschaften, **65** (1972) Springer Berlin.

[Fi] Fiala, F.: Le problème des isopérimètres sur les surfaces ouvertes à courbure positive. Comment. Math. Helv. **13** (1940/41), 293–346.

[Ga-He] Gasser, T. and Hersch, J.: Über Eigenfrequenzen einer mehrfach zusammenhängenden Membran: Erweiterung von isoperimetrischen Sätzen von Pólya und Szegö, ZAMP **19** (1968), 672–675.

[Had] Hadwiger, H.: Vorlesungen über Inhalt, Oberfläche und Isoperimetrie, Springer 1957.

[Har] Hartman, P.: Geodesic parallel coordinates in the large, Amer. Math. **86** (1964), 705–727.

[He1] Hersch, J.: Sur la fréquence fondamentale d'une membrane vibrante: évaluations par défaut et principe du maximum, ZAMP **11** (1960), 387–413.

[He2] Hersch, J.: On symmetric membranes and conformal radius: some complements to Pólya's and Szegö's inequalities, Arch. Mech. Anal. **20** (1965), 378–390.

[He3] Hersch, J.: Quatre propriétés isopérimétriques de membranes sphériques homogenes, C.R. Acad. Sci. Paris A270 (1970), 1645–1648.

[Hu1] Huber, A.: On the isoperimetric inequality on surfaces of variable Gaussian curvature, Ann. Math. **60** (1954), 237–247.

[Hu2] Huber, A.: On an inequality of Fejer and Riesz, Annals Math. **63** (1956), 572–587.

[Hu3] Huber, A.: On subharmonic functions and differential geometry in the large, Comment. Math. Helv. **32** (1957), 13–72.

[K1] Klingenberg, W.: Contributions to Riemannian Geometry in the Large, Ann. Math. **69** (1959), 654–666.

[Kr] Krahn, E.: Über eine von Rayleigh formulierte Minimaleigenschaft des Kreises, Math. Ann. **94** (1925), 97–100.

[Ma] Makai, E.: Bounds for the principal frequency of a membrane and the torsional rigidity of a beam, Acta Sci. Math. (Szeged) **20** (1959), 33–35.

[Ne] Nehari, L.: On the principal frequency of a membrane, Pac. J. Math. **8** (1958), 285–293.

[01] Osserman, R.: The Isoperimetric Inequality, Bull. Amer. Math. Soc. 84 (1978), 1182–1238.

[02] Osserman, R.: Isoperimetric inequalities and eigenvalues of the Laplacian, Proceedings of the International Congress of Mathematicians, Helsinki (1978).

[03] Osserman, R.: Bonnesen-Style Isoperimetric Inequalities, Amer. Math. Monthly, 86 (1979), 1–29.

[Pa1] Payne, L.E.: Some isoperimetric inequalities in the torsion problem for multiply connected regions, Studies in Mathematical Analysis and Related Topics: Essays in honor of G. Pólya, Stanford University Press (1962), 270–280.

[Pa2] Payne, L.E.: Isoperimetric Inequalities and Their Applications, SIAM Review **9** (1967), 453–488.

[Pe1] Peetre, J.: A generalization of Courant's nodal line theorem, Math. Scand. **5** (1957), 15–20.

[Pe2] Peetre, J.: Estimates for the number of nodal domains, Proc. Thirteenth Congress Math. Scand. (1957), 198–201.

[Po] Pólya, G.: Two more inequalities between physical and geometrical quantities, J. Indian Math. Soc. **24** (1960), 413–419.

[Po-Sc] Pólya, G. and Schiffer, M.: Convexity of functionals by transplantation, J. d' Anal. Math. **3** (1954), 245–345.

[Po-Sz] Pólya, G. and Szegö, G.: Isoperimetric inequalities in mathematical physics, Princeton (1951).

[Ra] Lord Rayleigh: The Theory of Sound, London 1894/96.

[Sc-Sp1] Schaeffer, P.W. and Sperb, R.P.: Maximum principles and bounds in some inhomogeneous elliptic boundary value problems, SIAM J. Math. Anal. **8** (1977), 871–878.

[SC-Sp2] Schaeffer, P.W. and Sperb, R.P.: Corrigendum and Addendum: Maximum principles and bounds in some inhomogeneous elliptic boundary value problems, SIAM J. Math. Anal. **10** (1979), 1326.

[Schm1] Schmidt, E.: Über das isoperimetrische Problem im Raum von n Dimensionen. Math. Z. **44** (1939), 689–788.

[Schm2] Schmidt, E.: Über die isoperimetrische Aufgabe im n-dimensionalen Raum konstanter negativer Krümmung. I Die isoperimetrischen Ungleichungen in der hyperbolischen Ebene und für Rotationskörper im n-dimensionalen hyperbolischen Raum, Math. Z. **46** (1940), 204–230.

[Schm3] Schmidt, E.: Über eine neue Methode zur Behandlung einer Klasse isoperimetrischer Aufgaben im Grossen, Math. Z. **47** (1942), 489–642.

[Schm4] Schmidt, E.: Der Brunn-Minkowskische Satz und sein Spiegeltheorem sowie die isoperimetrische Eigenschaft der Kugel in der euklidischen und hyperbolischen Geometrie. Math. Nachr. **1** (1948), 81–157.

[Schw] Schwarz, H.A.: Beweis des Satzes, daß die Kugel kleinere Oberfläche besitzt als jeder andere Körper gleichen Volumens. Gesammelte Abhandlungen 2 Springer 1980, 327–340.

[Sp] Sperb, R.P.: Maximum principles and their applications, Acad. Press (1981).

[St] Steiner, J.: Gesammelte Werke, Berlin 1882.

[Sze] Szegö, G.: Inequalities for certain eigenvalues of a membrane of given area, J. Rat. Mech. Anal. **3** (1954), 343–356.

[Sz-N] B. Sz.-Nagy: Über Parallelmengen nicht-konvexer ebener Bereiche, Acta Math. Univ. Szeged **20** (1959), 36–47.

[Ta] Talenti, G.: Elliptic Equations and Rearrangements, Annali Scuola Norm. Pisa **3** (1976), 697–718.

[We1] Weinberger, H.F.: Symmetrization in uniformly elliptic problems, Studies in Mathematical Analysis and Related Topics: Essays in honor of G. Pólya, Stanford University Press (1962), 424–428.

[We2] Weinberger, H.F.: An isoperimetric inequality for the N-dimensional free membrane problem, J. Rat. Mech. Anal. **5** (1956), 533–536.

C. Bandle
Mathematisches Institut
Universität Basel
Rheinsprung 21
CH–4051 Basel (Switzerland)

G.D. Chakerian and H. Groemer*

Convex Bodies of Constant Width

1. Introduction

A spherical ball obviously has the property that it can be arbitrarily rotated between two fixed parallel planes without losing contact with either plane. It has been known for a long time, certainly since the time of Euler, that there are other convex bodies with the same property. Such bodies are called convex bodies of constant width. Other names that have also been used are 'convex bodies of constant breadth', 'equiwide convex bodies', 'orbiforms' and 'spheroforms' (in the two and three-dimensional case, respectively) and several more; the occasionally used German 'Gleichdick' being one of the most charming.

The progress of knowledge about convex bodies of constant width has been slow but steady. Most of the pertinent results that were known before 1934, together with references and some proofs, are presented in the monograph of Bonnesen-Fenchel [1934]. Consequently, our present survey concentrates on developments after the appearance of that book. There are many concepts that are related to, or may be considered extensions of the notion of constant width. Although we focus our attention on convex bodies of constant width in n-dimensional Euclidean space, we do not entirely exclude these related areas. In fact, Sections 10 and 11 deal with such matters, and frequently we discuss the situation that arises if the Euclidean metric is replaced by an arbitrary Minkowskian one.

Most books and survey papers dealing with the general area of convexity discuss at least some classical theorems about convex bodies of constant width. Besides the already mentioned standard reference Bonnesen-Fenchel [1934] we refer in particular to the books of Blaschke [1956], Eggleston [1958b], Yaglom-Boltyanskii [1961], Benson [1966] and Lay [1982], and also to the notes of Toranzos and Nanclares [1978] and the articles of Klee [1971] and Dubois [1977]. There also exist some earlier surveys of results concerning plane convex bodies of constant width. These are by Wajnstejn [1938] and Beretta-Maxia [1940a]. Some mathematicians consider a discussion of convex bodies of constant width to have substantial merit with regard to mathematical education. Papers with this objective have been written by Togliatti [1968] and Blatter [1981]. The exposition of Bontrager [1969] seems partially motivated by the same objective.

For the purpose of our survey we have subdivided the material into 11 sections. Naturally, such subdivisions are not always as clear-cut as one would desire; some topics fitting into several sections, others into none at all. In the former case we refer frequently from one section to another. In the latter case, rather than creating a section on miscellaneous results, we have put such subjects, sometimes quite forcefully, into the section that appeared in our opinion the most

* Supported by National Science Foundation research grant MCS-8001578

appropriate one. The following list gives the section headings, together with an indication of contents.

1. *Introduction*

(General remarks, previous survey literature, subdivision of material, notations and some definitions).

2. *Basic geometric properties*

(Properties involving diameters, normals, support functions and central symmetrization; generalizations to Minkowski space).

3. *Special convex bodies of constant width and approximations*

(Reuleaux polygons and analogues in three-dimensional space, very symmetrical or unsymmetrical bodies, very regular or singular bodies; approximations by Reuleaux polygons and other special classes of convex bodies of constant width).

4. *Completions of convex bodies*

(Complete sets in Euclidean and Minkowski space; preservation of symmetry and regularity when forming completions).

5. *Intersection properties*

(Representation of convex bodies of constant width as intersections of balls; intersection properties of the boundaries of plane convex bodies of constant width; a Helly-type theorem).

6. *Curvature and mixed volumes*

(Properties of mixed volumes and, in particular, mean projection measures of convex bodies of constant width; curvature properties).

7. *Inequalities*

(Inequalities between the various numbers that can be associated with convex bodies of constant width, such as width, volume, surface area, inradius, circumradius, etc.).

8. *Inscribed and circumscribed bodies*

(Particular polytopes that can be inscribed in or circumscribed about convex bodies of constant width; also other types of convex bodies that are contained in, or contain, certain bodies of constant width).

9. *Packings, coverings, lattice points*

(Translative packing problems, permeability of layers; Newton numbers; some covering and lattice point problems).

10. *Rotors in polytopes*

(Convex bodies that can rotate in a given polytope P without losing contact with the facets of P).

11. *Concepts related to constant width*

(Constant brightness, constant triangular width, constant width for space curves and in non-Euclidean spaces, polar bodies and affine images of convex bodies of constant width).

Most notational matters and definitions will be discussed whenever the need arises. However a few notations and definitions will be used consistently in several sections and we introduce them here.

E^n denotes *n-dimensional Euclidean space*, where it is always assumed that $n \geqq 2$. A *convex body* in E^n is defined as a compact convex subset of E^n with interior points. We let $\mathbf{C}^n$ denote the class of all convex bodies in E^n. For any $K \in \mathbf{C}^n$ the boundary and the interior of K will be written as bdrK and intK, respectively. B(r,p) denotes the closed ball in E^n of radius r centered at the point p, but for the unit ball B(1,o), where o is the origin of E^n, we simply write B. For any direction (unit vector) $u \in S^{n-1}$, where $S^{n-1} = \text{bdr}B$, we let $h(K,u) = h(u)$ denote the *support function* of $K \in \mathbf{C}^n$. Thus, $h(K,u) = \sup\{u \cdot x : x \in K\}$, which may also be interpreted as the signed distance from o to the supporting plane (i.e. hyperplane), say H, of K with normal vector u pointing into the half-space determined by H and not containing K. Note that we shall often refer to a hyperplane simply as a 'plane' when the meaning is clear from the context. The *width* w(K,u), or simply w(u), of $K \in \mathbf{C}^n$ in direction $u \in S^{n-1}$ is defined by $w(K,u) = h(K,u) + h(K,-u)$, which is, of course, the distance between the two supporting planes of K orthogonal to u. The convex body K is said to be of *constant width* b(K) if $w(K,u) = b(K)$ for all $u \in S^{n-1}$. If the value of b(K) is immaterial for a particular purpose we simply say that K is of constant width. The class of all $K \in \mathbf{C}^n$ of constant width will be denoted by $\mathbf{W}^n$.

When talking about continuity or approximations we assume always that $\mathbf{C}^n$ and $\mathbf{W}^n$ are equipped with the usual Hausdorff metric. There are various other metric and algebraic structures that can be imposed on $\mathbf{C}^n$ and $\mathbf{W}^n$, but these will not be used in the present survey. For more information on these matters see the papers of Ewald-Shephard [1966], Ewald [1966], and Lewis [1975].

We let d(X) denote the *diameter* of a set $X \subset E^n$. If $K \in \mathbf{C}^n$ it is well-known that $d(K) = \max\{w(K,u) : u \in S^{n-1}\}$. The *minimal width* of K is defined as $\Delta(K) = \min\{w(K,u) : u \in S^{n-1}\}$. Obviously K has constant width if and only if $\Delta(K) = d(K)$. In general v(K) denotes the *volume* and s(K) the *surface area* of $K \in \mathbf{C}^n$. However, to stay closer to the usual notations, we write in the case of plane convex bodies a(K) for the *area* and p(K) for the *perimeter* of K. Thus, if $K \in \mathbf{C}^2$ then $v(K) = a(K)$ and $s(K) = p(K)$.

As usual, if X, Y $\subset E^n$, $\lambda \in E^1$, we define $X + Y = \{x + y : x \in X, y \in Y\}$, $\lambda Y = \{\lambda y : y \in Y\}$, $-Y = \{-y : y \in Y\}$. If $K \in \mathbf{C}^n$ then the set $K^* = \frac{1}{2}(K + (-K))$ is again a convex body in E^n. It is called the convex body obtained from K by *central symmetrization.* The set $K + (-K)$ itself, also denoted by $K - K$, is the *difference body* of K.

If $K \in \mathbf{C}^n$, a point $p \in$ bdrK is called a *regular* point of K, or of bdrK, if K has only one supporting plane containing p; otherwise it is called *singular.* K or bdrK is said to be *smooth* if every point of bdrK is regular.

E^n can be viewed as a normed linear space. It is sometimes desirable to generalize the situation by considering E^n equipped with an arbitrary norm $\|\cdot\|$. This new space will be called an n-dimensional *Minkowski space.* Two-dimensional Minkowski spaces, are usually called *Minkowski planes.* If we set $Q = \{p : \|p\| \leqq 1\}$ then Q is a convex body in E^n that is centrally symmetric with respect to o. It is called the *gauge body* of the given Minkowski space. It is easily proved that any $Q \in \mathbf{C}^n$ that is symmetric in o is the gauge body of some Minkowski space. If two points p, q in a Minkowski space with norm $\|\cdot\|$ and gauge body Q are given, we call the associated distance, i.e. $\|p - q\|$, the *Q-distance* between p and q (it is also known as the 'relative distance', or the 'distance relative to Q'). Based on this concept of distance, one can introduce a support function, say $h_Q(K,u)$, of $K \in C^n$ which is the signed Q-distance from o to the appropriate supporting plane of K orthogonal to u. One finds easily $h_Q(K,u) = h(K,u)/h(Q,u)$. The width of K relative to Q, or simply the *Q-width* of K, in direction u is the Q-distance between the two parallel supporting planes of K orthogonal to u. If it is denoted by $w_Q(K,u)$ we have $w_Q(K,u) = h_Q(K,u) + h_Q(K, -u) = 2w(K,u)/w(Q,u)$. Naturally, we say that K is of *constant Q-width* b if $w_Q(K,u) = b$ for all $u \in S^{n-1}$. Instead of 'constant Q-width', the expressions 'constant relative width' or 'constant width relative to Q' have been frequently used.

For easy reference we summarize now the notations introduced in this section.

E^n	Euclidean n-dimensional space
o	origin of E^n
$\mathbf{C}^n$	class of all convex bodies in E^n
$\mathbf{W}^n$	class of all convex bodies in E^n of constant width
bdrK	boundary of $K \in \mathbf{C}^n$
intK	interior of $K \in \mathbf{C}^n$
K^*	convex body obtained from K by central symmetrization
$K - K$	difference body of K
B(r,p)	ball in E^n of radius r centered at p
B	B(1,o)
S^{n-1}	bdrB
d(X)	diameter of $X \subset E^n$
$\Delta(K)$	minimal width of $K \in \mathbf{C}^n$
$h(K,u) = h(u)$	support function of $K \in \mathbf{C}^n$ in direction u
$w(K,u) = w(u)$	width of $K \in \mathbf{C}^n$ in direction u
b(K)	width of $K \in \mathbf{W}^n$
v(K)	volume of $K \in \mathbf{C}^n$
s(K)	surface area of $K \in \mathbf{C}^n$

a(K)	area of $K \in \mathbf{C}^2$
p(K)	perimeter of $K \in \mathbf{C}^2$
Q	gauge body of a Minkowskian metric in E^n
$w_Q(K,u)$	Q-width of $K \in \mathbf{C}^n$ in direction u

2. Basic geometric properties

We start with several characteristic properties of convex bodies of constant width in E^n of arbitrary dimension and then discuss the analogous properties for Minkowski spaces. Following this we describe various other properties.

According to definition, a set $K \in \mathbf{C}^n$ is of constant width b if and only if $w(K,u) = h(K,u) + h(K,-u) = b$ for all $u \in S^{n-1}$. Since the support function of the centrally symmetrized body K^* is $h(K^*,u) = \frac{1}{2}(h(K,u) + h(K,-u))$ it follows immediately that K has constant width b exactly if $w(K^*,u) = b$ for all $u \in S^{n-1}$. We now use the often noted fact that a centrally symmetric convex body of constant width is necessarily a ball. Hence, we can formulate the following well-known characterization.

(I) A set $K \in \mathbf{C}^n$ is of constant width b if and only if $K^* = \frac{b}{2}B$.

For a given $K \in \mathbf{C}^n$ we call a closed segment $[p,q]$ a *chord* of K if $p \neq q$ and $p \in \text{bdr}K$, $q \in \text{bdr}K$. A chord $[p,q]$ of K is said to be *diametrical* if there exist two parallel supporting planes, say H_1, H_2, of K such that $p \in H_1$, $q \in H_2$. We shall say in this case that $[p,q]$ is a diametrical chord *generated* by H_1,H_2. Our second characterization is essentially taken from Eggleston [1965].

(II) A set $K \in \mathbf{C}^n$ is of constant width if and only if for each pair H_1,H_2 of parallel supporting planes of K there exists a diametrical chord of K generated by and orthogonal to H_1,H_2.

Since it is easy to see that for a given pair H_1,H_2 of parallel supporting planes of a set $K \in \mathbf{W}^n$ there is only one diametrical chord generated by H_1,H_2, it follows that (II) can also be formulated in the following way:

(III) A set $K \in \mathbf{C}^n$ is of constant width if and only if for each pair H_1,H_2 of parallel supporting planes of K every diametrical chord of K generated by H_1,H_2 is orthogonal to H_1,H_2.

The following characterization is probably more familiar.

(IV) A set $K \in \mathbf{C}^n$ is of constant width b if and only if all diametrical chords have length b.

(IV) can be summarized by saying that K is of constant width if and only if it is of ‘constant diameter’. In this form it is stated and proved (for $n = 2$) in Yaglom-Boltyanskii [1961], p. 73. A proof of (IV) in the general context of Minkowski spaces will be given later. (IV) is also used implicitly by Burke [1966] in constructing a particular plane curve of constant width. The property (for ordinary plane bodies of constant width) is explicitly mentioned by Hammer and Sobczyk [1953a] in the course of their description of outwardly simple line families in E^2 as the extended diametrical chords of bodies of constant width.

Incidentally, Hammer [1954] has shown that this characterization of outwardly simple line families is not valid in E^3.

A chord [p,q] of $K \in \mathbf{C}^n$ is called a *normal* of K at p if K has a supporting plane H at p such that [p,q] is orthogonal to H. Using this definition we can formulate two more characterizations. The first one is again from Eggleston [1965].

(V) A set $K \in \mathbf{C}^n$ is of constant width if and only if any two parallel normals coincide.

The following characterization is not only of interest in itself but also forms the basis for the usual definition of space curves of constant width and the generalization of that concept to transnormal manifolds embedded in Euclidean spaces (see Section 11).

(VI) A set $K \in \mathbf{C}^n$ is of constant width if and only if every chord [p,q] of K that is a normal of K at p is also a normal of K at q.

This result can also be expressed by saying that $K \in \mathbf{W}^n$ if and only if every normal is a *double normal.* The fact that every $K \in \mathbf{W}^n$ has this double normal property is well-known (see Eggleston [1958b] p. 126). The converse has been proved by Besicovitch [1961] for $n = 2$, and is easily extended to arbitrary dimension by noting that the orthogonal projection of a $K \in \mathbf{C}^n$ with the double normal property into a two-flat has again this property.

Since a diametrical chord is a double normal, a light ray reflecting internally at the boundary of a smooth convex body of constant width will traverse a diametrical chord repeatedly if any segment of its path lies in that chord. Sine and Kreĭnovič [1979], generalizing an earlier result of Sine [1976] in the three-dimensional case, show that a smooth convex body K in E^n is a ball if and only if each path of a light ray reflecting inside K lies in a two-flat. They also prove that the successive reflection points of a light path not on a diametrical chord inside a smooth member of $\mathbf{W}^2$ follow a clockwise order or follow a counterclockwise order around the boundary, and that this property is characteristic of smooth members of $\mathbf{W}^2$. From this they deduce that the path of a light ray reflecting inside a member of $\mathbf{W}^2$ is never ergodic.

We discuss now extensions of (I) through (VI) to Minkowski spaces. The best general reference for these matters is Eggleston [1965]. It should be remembered that Q denotes always the gauge body of the given Minkowski space.

Practically the same proof as in the Euclidean case yields:

(I′) A set $K \in \mathbf{C}^n$ is of constant Q-width if and only if $K^* = \frac{b}{2}Q$.

To transfer the other characterizations to Minkowski spaces we must first extend the definition of orthogonality. Let H be a hyperplane and $u \in S^{n-1}$ a (Euclidean) unit vector in a Minkowski space with gauge body Q. We say that u is *Q-orthogonal* to H if two supporting planes parallel to H generate a diametrical chord of Q that has direction u and passes through o. Correspondingly, we say that a line (or a line segment) is Q-orthogonal to H if it has a direction that is Q-orthogonal to H. Note that in a Minkowski space there may be infinitely many directions Q-orthogonal to H, but if Q is strictly convex there is only one such direction.

From the definition of K^* and that of Q-orthogonality, one obtains immediately the validity of the following remark:

If $D = [p,q]$ is a diametrical chord of a convex body K generated by parallel supporting planes H_1,H_2 then $D^* = [(p - q)/2,(q - p)/2]$ is a diametrical chord through o of K^* generated by supporting planes parallel to H_1,H_2. Moreover, to each such diametrical chord of K^*, say T, there exists at least one D with the above properties and $D^* = T$. As an immediate consequence we note that D and D^* are always K^*-orthogonal to H_1,H_2.

Eggleston [1965] has shown that (II) has the following extension to Minkowski spaces.

(II′) A set $K \in C^n$ is of constant Q-width if and only if to each pair H_1,H_2 of parallel supporting planes of K and each direction u Q-orthogonal to H_1,H_2 there exists a diametrical chord of K generated by H_1,H_2 of direction u.

Statement (III) has an almost identical analogue in Minkowski spaces, namely:

(III′) A set $K \in C^n$ is of constant Q-width if and only if, for each pair H_1,H_2 of parallel supporting planes of K, every diametrical chord of K generated by H_1,H_2 is Q-orthogonal to H_1,H_2.

Since no reference appears to be readily available, we include here a short proof of (III′).

Suppose first that K has constant Q-width. Let D be a diametrical chord of K generated by H_1,H_2. Then, by the above remark, D is K^*-orthogonal to H_1,H_2. But, because of (I′), $K^* = \frac{b}{2}Q$, and it follows that D is Q-orthogonal to H_1,H_2. To prove the converse let H_1,H_2 be a pair of parallel supporting planes of Q and let u be a direction K^*-orthogonal to H_1,H_2. Using again the above remark we can infer that K has a diametrical chord of direction u. Because of the pertinent assumptions in (III′) (which are assumed to be satisfied), u is Q-orthogonal to H_1,H_2 and it follows that Q has a diametrical chord of direction u generated by H_1,H_2. Hence, we may deduce with the aid of (II′) (taking K^* as gauge body and replacing K by Q) that Q has constant K^*-width. Denoting this width by c and using (I′) we find $Q = Q^* = \frac{c}{2}K^*$ and therefore $K^* = \frac{2}{c}Q$ which implies (again by (I′)) that K has constant Q-width.

The Minkowskian version of (IV) reads as follows:

(IV′) A set $K \in C^n$ is of constant Q-width b if and only if all diametrical chords have Q-length b.

Again we supply a short proof. Assume first that K has constant Q-width b and that D is a diametrical chord of K generated by parallel supporting planes H_1,H_2. If D^* is the chord of K^* as defined previously, it follows from $K^* = \frac{b}{2}Q$ that D^*, and therefore D, has Q-length b. Now suppose that each diametrical chord of K has Q-length b. Then, using again the remark preceding (II′), we find that each diametrical chord of K^* also has Q-length b. Since both K^* and Q are symmetric in o, it follows that $K^* = \frac{b}{2}Q$, and this means (because of (I′)) that K has constant Q-width b.

Petty and Crotty [1970] have proved a theorem involving some important properties of diametrical chords. They show that if K has constant Q-width b and also has an *equichordal point*, that is, a point $p \in \text{int}K$ such that all chords of K through p have the same Q-length, then K is homothetic to Q. Their proof proceeds by first noting that some diametrical chord of K passes through p (see Hammer [1963] for a proof of this, different from that of Petty and Crotty), from which it follows that all chords through p have Q-length b. This implies that all chords through p are diametrical, from which it can be deduced that K is centrally symmetric in p. It follows then from (I′) that K is homothetic to Q. Fujiwara [1916] has given a proof of the corresponding theorem for ordinary sets of constant width in two and three dimensions. We give a further discussion of equichordal points in Section 11.

To discuss the Minkowskian analogues of (V) and (VI) we call, similarly as in the Euclidean case, a chord [p,q] of $K \in \mathbf{C}^n$ a *Q-normal* of K at p if K has a supporting plane H at p such that [p,q] is Q-orthogonal to H. The following extension of (V) was proved by Eggleston [1965].

(V′) Let Q be a smooth and strictly convex gauge body. Then, a set $K \in \mathbf{C}^n$ is of constant Q-width if and only if any two parallel Q-normals of K coincide.

Without any restrictions on Q it is still true that a convex body with this 'coincidence normal property' must be of constant Q-width, but the converse may fail to be true.

The following extension of (VI) is easy to prove.

(VI′) Let Q be a smooth and strictly convex gauge body. Then, a set $K \in \mathbf{C}^n$ is of constant Q-width if and only if every chord [p,q] of K that is a Q-normal of K at p is also a Q-normal of K at q.

We proceed now to several other properties of ordinary convex bodies of constant width. Many of these properties are meaningful only in the case $n = 2$, and not all of them are characteristic.

Bückner [1937b] proved that a $K \in \mathbf{C}^2$ has constant width if and only if each pair of nonparallel normals of K intersect inside K. By projecting orthogonally into planes, he derived from this the result that $K \in \mathbf{C}^3$ has constant width if and only if the line perpendicular to each pair of nonparallel normals of K intersects K. Note that one may use here the easily established characterization: K has constant width if and only if each of its orthogonal projections into hyperplanes is a set of constant width of one less dimension. It is interesting to compare this to a more difficult theorem due to Süss (see Bonnesen-Fenchel [1934], p. 141): $K \in \mathbf{C}^3$ is a ball if and only if for some $p \in K$ all planes through p intersect K in plane sets of constant width.

As a rather nonintuitive property of normals of convex bodies, Kuiper [1964] has shown that the set of lengths of double normals of a convex body in E^n is of measure zero if $n \leqq 3$, while for each $n \geqq 4$ there exists a $K \in \mathbf{C}^n$ not of constant width such that the range of its width function coincides with the set of lengths of its double normals.

Another not easily visualized property of normals has been obtained by Heil [1979, 1980]. He has proved that every $K \in \mathbf{W}^3$ contains either a point through which infinitely many normals pass, or an open set of points through each of which pass at least 10 normals.

Hammer and Sobczyk [1953a,b] have made a thorough study of the diametrical chords of members of $\mathbf{W}^2$. An excellent summary of this work and its generalization to two-dimensional Minkowski spaces is found in Hammer [1963]. Other pertinent references in this regard are Hammer [1951, 1955] and Smith [1961]. Applying a very general parametric representation for the family of diametrical chords of a plane convex body, Hammer and Smith [1964] prove the following theorem:

(a) If $K \in \mathbf{W}^2$ is such that each of its diametrical chords bisects the circumference, then K is a circular disk.

With smoothness restrictions on K, this was proved by M.S. Klamkin (see the remarks in Hammer and Smith [1964]) and much earlier by Hirakawa [1933]. The theorem was also established by Besicovitch [1961] in the course of his proof of Mizel's conjecture (see Section 8). A proof of (a), together with an extensive collection of characteristic properties of the circle, may also be found in Miller [1969].

The following counterpart of (a), also proved with smoothness restrictions by Hirakawa [1933] and in full generality by Hammer and Smith [1964], is of interest.

(b) If $K \in \mathbf{W}^2$ is such that each of its diametrical chords bisects the area, then K is a circular disk.

Miller [1969] also includes a proof of (b).

Many characterizations of the ball among members of $\mathbf{W}^n$ are established by first proving that a convex body K with a given property must be centrally symmetric. Then, under the further assumption that K has constant width, (I) implies that K is a ball. Theorems (a) and (b) are examples where this approach is feasible.

Let K be a strictly convex member of $\mathbf{C}^2$. For $0 \leqq \theta \leqq 2\pi$ and $u = (\cos\theta, \sin\theta)$, the width of K in direction u may be conveniently expressed as a function of θ, $w(\theta) = w(K,u)$. The supporting line of K with outward normal u intersects bdrK in a single point $x = x(\theta)$, and for each θ the chord $[x(\theta), x(\theta + \pi)]$ is a diametrical chord of K. Let $l(\theta)$ be the length of the arc of bdrK from $x(\theta)$ to $x(\theta + \pi)$, traversed in the counterclockwise direction. Let $f(\theta)$ be the area of the part of K bounded by the above boundary arc from $x(\theta)$ to $x(\theta + \pi)$ and the diametrical chord $[x(\theta), x(\theta + \pi)]$. Extending a result of Bückner [1937b], Beretta and Maxia [1940a] prove that K has constant width if and only if

(2.1) $2f(\theta) - w(\theta)l(\theta) \equiv \text{constant}.$

In fact, the work of Beretta and Maxia enables one to conclude that for any strictly convex member of $\mathbf{C}^2$

(2.2) $2f'(\theta) - w(\theta)l'(\theta) \equiv 0.$

The fact that (2.1) characterizes members of $\mathbf{W}^2$ follows readily from (2.2). Furthermore, one sees immediately from (2.2) that all the diametrical chords of K are area bisectors if and only if they are all perimeter bisectors, for $f'(\theta) \equiv 0$ if and only if $l'(\theta) \equiv 0$. Thus theorems (a) and (b) are easy consequences of each other.

The work of Auerbach [1938] on Zindler curves bears a close relationship to the preceding. Zindler [1921] has shown that noncircular plane curves exist with the property that all chords bisecting the area have equal length and also bisect the perimeter. In the course of his investigation of a related class of curves, Auerbach proved the remarkable theorem that if a square is displaced in such a fashion that one diagonal describes the area bisectors of a Zindler curve C, then the other diagonal will describe the diametrical chords of a set of constant width having the same area as that enclosed by C. One may reverse this construction; starting with a plane convex body K of constant width one obtains the area bisectors of a Zindler curve by rotating each diametrical chord of K through a right angle about its midpoint. Hoschek [1974] has used this idea to construct nonplanar curves in E^3 analogous to Zindler curves by appropriately rotating the double normals of transnormal space curves (see Section 11 for a discussion of space curves of constant width and transnormal curves). Wegner [1976] found a general construction, of which Hoschek's is a special case, enabling him to extend the concept of a Zindler curve to curves in E^n, $n \geqq 3$.

Khassa [1977] has proved that if $K \in \mathbf{W}^2$ has the property that for each pair of parallel supporting lines of K the parallel chord midway between is a diametrical chord, then K is a circle. Using properties of outwardly simple line families, Falconer [1977a] proves the following more general result. If $K \in \mathbf{C}^2$ has the property that for each pair of its parallel supporting lines the parallel chord midway between is a diametrical chord, then K is centrally symmetric. Falconer also applies his theorems on line families to prove a result about semicircles inscribed in plane convex bodies; see Section 8 for this.

Heppes [1959] found the following characterization of members of $\mathbf{W}^2$ in terms of their chords: A plane convex body has constant width if and only if each of its chords is the longest chord of one of the two boundary arcs it determines.

In the course of their study of packings of pairs of similar convex sets into another similar set (see Section 9), Beck and Bleicher [1971] obtained a somewhat different characterization of plane bodies of constant width. For $p \in \text{bdr}K$, where $K \in \mathbf{C}^2$, define the *minimum width at p* to be the smallest distance between parallel supporting lines of K when one of them passes through p. Define K to have *constant minimum width* if the minimum width at p is the same for all $p \in \text{bdr}K$. Then Beck and Bleicher have proved that $K \in \mathbf{C}^2$ has constant width if and only if K is strictly convex and has constant minimum width.

While on the subject of characterizations of bodies of constant width, we should mention a beautiful one due to Minkowski, namely, $K \in \mathbf{C}^3$ has constant width if and only if all its orthogonal projections into planes have equal perimeter (see Bonnesen-Fenchel [1934], p. 136, and Firey [1965b], p. 39). In Section 8 we mention a variant of this, discovered by Firey [1959], and we take up some related matters, such as properties of sets of constant brightness, in Section 11.

3. Special convex bodies of constant width and approximations

In this section we discuss convex bodies of constant width, or classes of such bodies, that are for various reasons of particular interest. We also consider the problem of approximating (with respect to the Hausdorff metric) arbitrary n-dimensional convex bodies of constant width by the members of certain subclasses of $\mathbf{W}^n$. For a general survey on approximations of convex bodies see Gruber [1983].

Besides the solid sphere, the most notorious convex body of constant width is undoubtedly the Reuleaux triangle of width b, i.e. a set from $\mathbf{W}^2$ whose boundary consists of three congruent circular arcs of radius b. The Reuleaux triangle can also be characterized as a $K \in \mathbf{W}^2$ whose area is minimal if b(K) is given (see Section 7). It has been often noted (cf. Bonnesen-Fenchel [1934]) that Reuleaux triangles can be generalized to regular or arbitrary Reuleaux polygons. The latter are defined as plane convex bodies of constant width b whose boundary consists of a finite (necessarily odd) number of circular arcs of radius b. Any $K \in \mathbf{W}^2$ can be approximated arbitrarily closely by Reuleaux polygons (see Bonnesen-Fenchel [1934] and Bückner [1936]). Kallay [1974] has a quantitative version of this theorem, relating the number of sides of the approximating Reuleaux polygon to the Hausdorff distance of the two domains. The dependence of the area of m-sided Reuleaux polygons on m has been studied by Firey [1960]. He found that the areas of regular Reuleaux polygons of given width increase with m. Moreover, for a given m, the maximum of the areas of all m-sided Reuleaux polygons of fixed width is attained by the regular one, and the infimum equals (but is for $m > 3$ not attained) the area of the Reuleaux triangle. Simpler proofs of the same results have been given by Sallee [1970a]. A short didactic discussion of representations in polar coordinates of regular Reuleaux polygons and related curves has been given by Kearsley [1952]. Peterson [1973] used Reuleaux-type constructions to produce plane convex bodies of constant width with a given number of singular points. He also has shown that the singular points can be everywhere dense in the boundary. From the work of Silverman [1973] or Kallay [1974] it follows that Reuleaux polygons are indecomposable relative to $\mathbf{W}^2$ (see Section 6). On the other hand, the example of Kallay [1974] described in Section 6 is a member of $\mathbf{W}^2$ which is smooth and indecomposable relative to $\mathbf{W}^2$. The Thüring [1952] studied the Reuleaux triangle in its relationship to the theorem of Holditch.

Sallee [1969] has defined a kind of Reuleaux polygon with respect to a Minkowski metric; for the case of Reuleaux triangles see also Ohmann [1952] and Chakerian [1966]. Another generalization of Reuleaux polygons has been described by Hammer [1963]. In the Euclidean case it leads to plane convex bodies of constant width bounded by a finite number of circular arcs of possibly different radii (see also Rademacher-Toeplitz [1957], p. 167, and Gardner [1963]).

Further special plane convex bodies of constant width that have received some attention (already by Euler) are those whose boundary can be represented as involutes of hypocycloids or other curves; see Bonnesen-Fenchel [1934], pp. 131–132, and the more recent papers of Gere-Zupnik [1943], Burke [1966] and Kurbanov [1975]. The article of Varela Gill [1932] also deals with problems of this kind. Petty [1955] gives an analogue for Minkowski planes.

Three-dimensional convex bodies of constant width can be obtained by the rotation of axially symmetric two-dimensional ones. But already the simple construction of a Reuleaux triangle cannot be immediately copied in E^3 since the body resulting by the natural adjunction of spherical pieces to a tetrahedron is not of constant width. To force it to be of constant width some 'rounding off' of some of the curvilinear edges is required. This well-known process (cf. Bonnesen-Fenchel [1934], pp. 135–136) has recently been studied by Gray [1972] not only for tetrahedra but also for more general pyramids.

Bückner [1936] has described a kind of a three-dimensional analogue of Reuleaux polygons which he calls 'elementary bodies'. These are bodies $M \in \mathbf{W}^3$ that can be represented as the intersection of balls of radius b(M) with centers at a finite number of points or at circular arcs of radius b(M). He then shows that for any given $K \in \mathbf{W}^3$ one can find elementary bodies that approximate K to any desired degree of precision. Another construction of three-dimensional analogues of Reuleaux polygons has been given by Sallee [1970b]. He calls them 'Reuleaux polytopes' and shows also that they have the above-stated approximation property. In the latter paper Sallee also describes a modification procedure that enables one to construct from a given member of $\mathbf{W}^n$ others.

We now consider subclasses of $\mathbf{W}^n$ that have particular symmetries or boundaries of a high degree of regularity. Fillmore [1969] has shown that there are sets in $\mathbf{W}^n$ with analytic boundary and no non-trivial symmetry group. He also proves the existence of sets in $\mathbf{W}^n$ with analytic boundary and the same symmetry group as the regular n-simplex. (A gap in the proof concerning the existence of analogues of Reuleaux polygons in higher dimensions can apparently be closed by starting from a simplex and referring to the constructions of Rogers [1971] or Schulte [1981].) In the same paper Fillmore determines all those closed subgroups of SO(3) that may appear as symmetry groups of members of $\mathbf{W}^3$ with analytic boundary.

Wegner [1977] proved that any member of $\mathbf{W}^2$ can be approximated arbitrarily closely by particular sets from $\mathbf{W}^2$ with analytic boundary. A weaker result, with analyticity replaced by infinite differentiability, was previously treated by Tanno [1976], who showed in addition that if the given set from $\mathbf{W}^2$ has an axis of symmetry then the approximating bodies can be constructed so that they have the same property. The proofs of these theorems depend on a formula of Hammer [1955] that provides explicit polar coordinate representations of sets from $\mathbf{W}^2$. A more special representation of this kind that yields a subclass of $\mathbf{W}^2$ is due to Tennison [1976]. It can be used to construct explicitly members of $\mathbf{W}^2$ with boundaries of any desired degree of differentiability.

Instead of looking for very regular sets in $\mathbf{W}^n$ one can also try to find extremely singular ones. We have already mentioned that there exists a body in $\mathbf{W}^2$ whose set of singular points is everywhere dense in its boundary. The question whether this is also true for all dimensions greater than two has been raised by Melzak [1965] (Problem 35). Falconer [1981] gave an affirmative answer by showing not only that such bodies exist but also that for any $K \in \mathbf{W}^n$ ($n \geqslant 3$) there exist bodies of this kind that approximate K aribitrarily closely.

Various convex bodies of constant width have been constructed for the purpose of providing counterexamples to some more or less plausible statements. Besides the eight-dimensional body constructed by Rogers [1971] (see Section 9) we mention an example of Danzer [1957]. It is a $K \in \mathbf{W}^3$ of width b such that the minimal width of each of its two-dimensional sections is less than b.

4. Completions of convex bodies

In the theory of convex bodies a set $C \in \mathbf{C}^n$ is said to be *complete* if there is no $C' \in \mathbf{C}^n$ such that $C \subset C'$, $C \neq C'$ and $d(C') = d(C)$. (This definition is sometimes modified by permitting arbitrary bounded subsets of E^n instead of convex bodies. It is however easily seen that this generalization leads to the same class of complete sets.) If $K \in \mathbf{C}^n$ then any complete set C with $K \subset C$ and $d(K) = d(C)$ is called a *completion* of K. There are two classical theorems in this area, namely:

(A) A convex body is complete if and only if it is of constant width.

(B) Every convex body has at least one completion.

(A) is often referred to as the Theorem of Meissner, (B) as the Theorem of Pál. For proofs and the early history of these theorems see Bonnesen-Fenchel [1934]. A proof of (B) that leads to a completion after a finite number of geometric constructions has been described by Bückner [1936] (for $n = 2,3$). In the case $n = 2$ such a procedure can already be found in the paper of Reinhardt [1922]; see also Bückner [1937a] and Beretta-Maxia [1940a]. Sallee [1970b] gave another proof of (B) that represents the completion of K as an intersection of balls of radius d(K) with centers selected from K and an arbitrarily prescribed countable everywhere dense set in E^n.

Rademacher-Toeplitz [1957], p. 174, Mayer [1934c], and Bückner [1936] have established necessary and sufficient conditions in order that a given arc in the plane is part of the boundary of some two-dimensional convex body of constant width. Bückner [1936] also deals with the corresponding problem for $n = 3$. The crucial condition that appears in these characterizations is that the arc or surface must be a subset of the boundary of an intersection of (respectively 2 or 3-dimensional) balls of fixed radius R. Any $K \in \mathbf{W}^n$ with $b(K) = R$ has this property (cf. Section 5). Bückner [1936] points out that his completion theorems show that it is impossible to characterize convex bodies of constant width by local properties of their boundaries.

The usual proofs of (B) proceed by constructing a sequence $K \subset K_1 \subset K_2 \cdots$ so that $\cup K_i$ is complete and has the same diameter as K. It was noted by Scott [1981] that in addition the sets K_i can be selected so that they are inside the circumsphere of K. This leads to the result that any $K \in \mathbf{C}^n$ has a completion with the same circumradius as K. As an application Scott derives some new inequalities (cf. Section 7). Independently, Vrećica [1981] has also shown that there is a completion within the circumsphere, and has given a generalization for Minkowski spaces.

Groemer [1982] has investigated the general idea of completing (or reducing) a convex set with respect to a given function such as diameter, minimal width, circumradius, etc. It is proved that for most such functions there always exist extremal bodies (analogues of the complete sets defined above). Concerning the minimal width the following theorem is shown: An n-dimensional convex body K with smooth boundary is of constant width if and only if it contains no proper subset of equal minimal width (cf. also Heil [1978]). The question whether a corresponding theorem holds if the condition of smoothness is replaced by strict convexity has been raised by Heil (see Gruber-Schneider [1979], p. 161).

There are some results on the preservation of symmetry and regularity when forming a completion. Rogers [1971] has proved that any bounded subset of E^n of diameter d with symmetry group G is contained in a convex body of constant width d that is also invariant under G. Concerning the preservation of regularity, Falconer [1981] has shown that any compact set $X \subset E^n$ has a completion C such that a point $p \in \mathrm{bdr} C$ is singular if and only if $p \in X$ and there are at least two points in X at distance d(X) from p. Moreover, the dimensions of the convex cones with apex p generated by all these chords of length d(X) are the same with respect to X and C. For $X \in \mathbf{C}^n$, Schulte [1981] has proved essentially the same result with the additional feature that if X has symmetry group G then C can be chosen so that it is also invariant under G.

Boltjanskii [1970] has found a rather surprising characterization of plane convex sets whose completion is unique; see Section 9.

In Minkowski spaces statement (B) is still true. This follows, for example, immediately from the general theorem of Groemer [1982]. The 'if' part of statement (A) is also true (and rather obvious). The question whether in a Minkowski space completeness implies constant width has already been considered for $n = 2,3$ by Meissner [1911] and has been investigated thoroughly (and for all n) by Eggleston [1965]. If $n = 2$ the answer is affirmative. But already for $n = 3$ there are complete sets that are not of constant width, even if the gauge body is smooth and strictly convex. This fact points out an error in the articles of Meissner [1911] and Kelly [1949], and raises the problem of characterizing those Minkowski spaces for which the concepts of completeness and constant width coincide.

In the case when the gauge body is a parallelepiped P, Eggleston [1965] has shown that every complete set is homothetic to P (and therefore of constant width). Soltan [1977] has proved a converse of this statement, namely: If in a Minkowski space with gauge body Q every complete set is homothetic to Q, then Q is a parallelepiped. For the two-dimensional case see also Hammer [1963].

5. Intersection properties

First we mention some results on the representation of convex bodies of constant width as an intersection of balls. Following Eggleston [1965] we say that a set $C \in \mathbf{C}^n$ of diameter d has the *spherical intersection property* if C is the intersection of all balls B(d,p) of radius d with center $p \in C$. If some $K \in \mathbf{W}^n$ has width b we obviously have $K \subset \bigcap_{p \in K} B(b,p)$, and from the fact that K contains for each pair of parallel supporting planes a chord of length b orthogonal to these planes, it follows immediately that $\bigcap_{p \in K} B(b,p) \subset K$. Thus, any $K \in \mathbf{W}^n$ has the spherical intersection property. Conversely, if a $C \in \mathbf{C}^n$ is such that $C = \bigcap_{p \in C} B(b,p)$ one sees easily that C is complete (in the sense of Section 4) and therefore of constant width. Hence, any $C \in \mathbf{C}^n$ has the spherical intersection property if and only if it is of constant width. This theorem gives rise to the investigation of the relationship between convex bodies of constant width and so-called over-, super-, or hyper-convex sets, that is, intersections of balls of fixed radius with centers not necessarily in the set (see Mayer [1934c], Bückner [1936], and Blanc [1944]). Eggleston [1965] has found that in Minkowski spaces the (analogously defined) concepts of spherical intersection property and completeness are again equivalent. But, as already pointed out in Section 4, completeness does in general not imply that C has constant width. If K is a convex body of constant width b in a Minkowski space with gauge body Q, let t(K) denote the smallest number (possibly ∞) of homothetic copies $p + bQ$ of Q whose intersection is K. Some results on the possible values of t(K) have been obtained by Soltan [1979]. In particular, he finds $t(K) \leqq t(Q)$ and obtains necessary and sufficient conditions in order that t(K) is finite. In the two-dimensional case he determines also the gauge bodies that allow the possibility that t(K) is a given even number.

We describe now some results on intersection properties of the boundaries of plane convex bodies of constant width. If $K_1,K_2 \in \mathbf{W}^2$, let $\alpha(K_1,K_2)$ denote the number (possibly infinite) of connected components of $(\mathrm{bdr}K_1) \cap (\mathrm{bdr}K_2)$. Peterson [1973] has shown: If $(\mathrm{int}K_1) \cap (\mathrm{int}K_2) \neq \emptyset$ and if neither K_1 nor K_2 is contained in the other, then $\alpha(K_1,K_2)$ is either even or infinite. In particular, if $K \in \mathbf{W}^2$ and if D is a circular disc of diameter b(K), then the conditions $D \neq K$ and $(\mathrm{int}K) \cap (\mathrm{int}D) \neq \emptyset$, imply that $\alpha(D,K)$ is even or infinite. Peterson [1972] conjectured that this property completely characterizes plane convex bodies of constant width b. After an initial result of Goodey [1974] for plane convex bodies with smooth boundary, this conjecture was proved in full generality by Goodey and Woodcock [1978]. Recently Goodey [1982] announced the following more general result: If $K_1,K_2 \in \mathbf{C}^2$, and if for every translate K_2' of K_2 with $K_2' \neq K_1$ and $(\mathrm{int}K_1) \cap (\mathrm{int}K_2') \neq \emptyset$ the number $\alpha(K_1,K_2')$ is even or infinite, then $w(K_1,u) = w(K_2,u)$ for all directions u. In particular, if K_2 is of constant width b and $\alpha(K_1,K_2')$ is even or infinite (except when $K_1 = K_2'$ or $(\mathrm{int}K_1) \cap (\mathrm{int}K'_2) = \emptyset$), then K_1 is also of constant width b. This settles another conjecture of Peterson [19'2].

Finally we mention a Helly-type theorem of Chakerian and Sallee [1969]. If a collection $\mathbf{V}$ of translates of a plane body of constant width has the property that any two of its members intersect, then there is a set consisting of three points that meets every member of $\mathbf{V}$. The proof depends on a theorem concerning coverings of plane sets by three sets of constant width (see Section 9). Some analogous results for sets of constant relative width have recently been obtained by Loomis [1980]. For a comprehensive survey of related results of this type (up to 1963), see Danzer-Grünbaum-Klee [1963].

6. Curvature and mixed volumes

Suppose $K \in \mathbf{W}^n$ is such that bdrK is a twice continuously differentiable hypersurface, and let $R_1(u),\ldots,R_{n-1}(u)$ be the principal radii of curvature corresponding to the point of bdrK with outward normal u. Let $F_1(u) = R_1(u) + \cdots + R_{n-1}(u)$. If K has constant width b, then

$$(6.1) \quad F_1(u) + F_1(-u) = (n-1)b,$$

for all directions u (see Bonnesen-Fenchel [1934], p. 128). One can generalize this to all members of $\mathbf{W}^n$, without smoothness restrictions, by introducing the *first area function*, or *first curvature measure of the second kind*, $S_1(K,\omega)$, where ω ranges over Borel subsets of S^{n-1} (for the definition and fundamental properties of curvature measures, see the survey of Schneider [1979]). When $K \in \mathbf{W}^n$ has width b, then $K + (-K) = bB$, and it follows directly from the linearity and homogeneity of $S_1(K,\omega)$, as a function of K, that

$$(6.2) \quad S_1(K,\omega) + S_1(-K,\omega) = b\mu(\omega),$$

where $\mu(\omega)$ is the usual spherical Lebesgue measure of the Borel set $\omega \subset S^{n-1}$. The integral representation of $S_1(K,\omega)$ in terms of F_1 (Schneider [1979], p. 29) shows that (6.2) is equivalent to (6.1) when bdrK is sufficiently smooth. If $K \in \mathbf{C}^n$ satisfies (6.2) for all Borel sets $\omega \subset S^{n-1}$, then $S_1(K - K,\omega) = S_1(bB,\omega)$ for all such ω and the Aleksandrov-Fenchel-Jessen uniqueness theorem (see Schneider [1979], p. 43) implies that $2K^* = K - K = bB$. Thus (6.2) is a characteristic property of members of $\mathbf{W}^n$ having width b. See Chakerian [1967] for a discussion of the analogues of (6.1) and (6.2) in Minkowski spaces and references to the earlier literature.

When $K \in \mathbf{C}^2$ has a sufficiently smooth boundary, we shall denote the radius of curvature of bdrK at the point with outward normal $u = (\cos\theta, \sin\theta)$ by $R(\theta)$. If in addition K has constant width b (6.1) becomes

$$(6.3) \quad R(\theta) + R(\theta + \pi) = b.$$

Equation (6.3) is of course only a necessary condition on a given function $R(\theta)$, $0 \leqq \theta \leqq 2\pi$, in order that it be the radius of curvature of a member of $\mathbf{W}^2$. Kallay [1974] assigns a radius of curvature function to every member of $\mathbf{W}^2$ and gives necessary and sufficient conditions for a non-negative bounded and measurable function $R(\theta)$ to be the radius of curvature function of some member of $\mathbf{W}^2$. He uses this to obtain a characterization of the 'extreme bodies' among plane bodies of constant width 1, that is, those members of $\mathbf{W}^2$ of width 1 that are indecomposable with respect to Minkowski addition, relative to $\mathbf{W}^2$. Kallay shows that K is an extreme body in this sense if and only if its radius of curvature function takes only the values 0 or 1 for almost all θ. For example, Reuleaux polygons (see Section 3) of width 1 are among the extreme bodies. On the other hand, Kallay constructs a smooth $K \in \mathbf{W}^2$ whose radius of curvature takes both values 0 and 1 on sets of positive measure in every interval. He also uses his characterization theorem to prove a strengthening of the fact that members of $\mathbf{W}^2$ can be approximated by Reuleaux polygons (see Section 3). Kallay [1974, 1975] also generalizes these characterization theorems to families of plane convex bodies having a given width function.

Sz.-Nagy [1949] has given an elementary treatment of the difference body $K - K$ of a convex body K, including a proof of what amounts to (6.3) in the case of Minkowski planes. A counterpart of (6.3) in Minkowski planes is also given by Petty [1955]. Vincensini [1938] gave an extensive treatment of related matters, including the analogue of (6.1) in Minkowski spaces.

It is appropriate at this point to consider some special curvature properties of plane curves of constant width. In the following we shall assume that $C = \text{bdr}K$, where $K \in \mathbf{C}^2$, and that C is twice continuously differentiable. Without further restrictions on C, we know from the *Four Vertex Theorem* that C has at least four vertices, that is, four points where the curvature has a stationary value. In case C has constant width, that is $C = \text{bdr}K$ where $K \in \mathbf{W}^2$, condition (6.3) implies that the number of vertices, if finite, is of the form $4k + 2$, $k \geqq 1$. Bonnesen-Fenchel [1934], p. 132, has an early reference for this result. A proof with a minimum of smoothness assumptions is given in Beretta-Maxia [1940b]. See Vincensini [1938] for a discussion of the theorem that C must have at least six vertices if it has constant width.

The center of mass of a uniform mass distribution on C is the *perimeter centroid* and the center of mass of a distribution on C whose density at each point is equal to the curvature is the *curvature centroid* (also called the *Steiner point*). It is known that the locus of perimeter centroids of the outer parallel curves of C lie on a straight line, and that if C has constant width then the perimeter centroid and curvature centroid coincide (see Bose and Roy [1935] for this, and also Nádeník [1967] for an extension of this property to space curves of constant width). Su [1928] has associated with a convex curve C a *curvature axis* defined in terms of the mass distribution whose density is the curvature. Ganapathi [1934a] considers those curves whose outer parallel curves all have the same curvature axis and proves that each such curve has at least six vertices. Since the outer parallels of a curve of constant width satisfy the condition studied by Ganapathi, this shows again that any curve of constant width has at least six vertices. See Heil [1970] for related results for Minkowski planes.

Turning again to convex bodies in E^n, we note that in fact (6.1) is a consequence of the following more general result. If $K \in \mathbf{W}^n$ of width b has a sufficiently smooth boundary, the principal directions at the opposite ends of a diametrical chord are parallel and, with the convention $R_1(u) \leqq R_2(u) \leqq \cdots \leqq R_{n-1}(u)$, the corresponding principal radii of curvature satisfy

$$(6.4) \qquad R_k(u) + R_{n-k}(-u) = b, \quad k = 1,2,\ldots,n-1.$$

For a proof of this see Bonnesen-Fenchel [1934], p. 128, and for a discussion of the analogue in Minkowski spaces see Chakerian [1967]. Hsiao [1959] has proved the following generalization of (6.4) in E^3. Let $K \in \mathbf{W}^3$ have width b and a twice continuously differentiable boundary. Let $[p,p']$ be a diametrical chord of K, and let P be a plane containing $[p,p']$ and cutting bdrK in the curve $C = P \cap \text{bdr}K$. Let $\varkappa$ and $\varkappa'$ be the curvatures of C at p and p′ respectively, and let g and g′ be the Gaussian curvatures of bdrK at p and p′ respectively. Then

$$(6.5) \qquad \frac{\varkappa}{g} + \frac{\varkappa'}{g'} = b.$$

If moreover P contains a principal direction at p (and therefore also at p′) then (6.5) reduces to (6.4) when $n = 3$.

The relationships (6.4) for the principal radii of curvature lead to a system of linear equations connecting the mean projection measures of a body of constant width. The k-th *mean projection measure* (Quermassintegral) of $K \in \mathbf{C}^n$ may be defined in terms of mixed volumes by

$$W_k(K) = V(K,\ldots,K,B,\ldots,B),$$

where K appears $n - k$ times and B appears k times in the mixed volume (see Bonnesen-Fenchel [1934], p. 49). If bdrK is sufficiently smooth, there is an integral representation of $W_k(K)$ in terms of the $(n - k)$-th elementary symmetric function of the principal radii of curvature (Bonnesen-Fenchel [1934], p. 63. See Bonnesen-Fenchel [1934], p. 65, and Chakerian [1967] for the analogue in

Minkowski spaces). Using such integral representations in conjunction with (6.4), Dinghas [1940] obtained the following theorem: Let $K \in \mathbf{W}^n$ have constant width b and sufficiently smooth boundary. Then

$$(6.6)\quad W_{n-k}(K) = \sum_{i=0}^{k} (-1)^i \binom{k}{i} W_{n-i}(K) b^{k-i}, \quad k = 0,1,\ldots,n.$$

The equations (6.6) are not independent. Actually they are equivalent to the $[(n+1)/2]$ equations obtained for odd values of k. It follows from (6.6) that one can uniquely determine all the mean projection measures of a convex body K of constant width b in E^n given only $W_0(K), W_2(K), W_4(K), \ldots, W_{n-2}(K)$ if n is even, and $W_1(K), W_3(K), W_5(K), \ldots, W_{n-2}(K)$ if n is odd.

In the case of the plane, (6.6) reduces to the single independent relation $2W_1(K) = \pi b$. Since $p(K) = 2W_1(K)$ for plane convex bodies, this gives Barbier's theorem that for all plane convex bodies K of constant width b,

$$p(K) = \pi b.$$

In case $n = 3$ we obtain from (6.6) the two independent relations $2W_2 = bW_3$ and $2W_0 = b^3 W_3 - 3b^2 W_2 + 3bW_1$, where $W_k = W_k(K)$. Since for $K \in \mathbf{C}^3$ we have $W_0(K) = v(K)$, $3W_1(K) = s(K)$, $3W_2(K)/2\pi = \bar{w}(K) =$ the mean width of K, and $W_3(K) = 4\pi/3$, these two relations reduce to

$$\bar{w}(K) = b, \quad \text{and} \quad 2v(K) = bs(K) - \frac{2\pi}{3} b^3.$$

The first of these is obvious, while the second has been discovered by Blaschke [1915], whose work was the motivation for the later generalizations. Blaschke also derived an analogous formula for sets of constant width in three-dimensional spherical geometry. He obtained his results by substituting $-b$ for b in Steiner's formula for $v(K + bB)$ but, as remarked in Section 11, without giving a complete justification for this procedure. Santaló [1946] used Blaschke's idea to derive the generalization (6.6) and, applying (6.4), gave a justification for the technique. Santaló [1950] later derived the analogues of (6.6) for spherical and hyperbolic spaces.

Dinghas [1940] stated without proof that the validity of (6.6) for any body of constant width b can be derived from its validity for sufficiently smooth bodies using standard methods. Debrunner [1955] has given a short proof of (6.6), valid for all $K \in \mathbf{W}^n$ having width b, using only the most elementary properties of convex bodies, not requiring any differential-geometric apparatus. Groemer [1964] has also obtained a short elementary proof similar in character to Debrunner's. Chakerian [1982] has generalized (6.6) in the following way. Suppose K, L, M are members of $\mathbf{C}^n$ such that $K + L = M$. Let $V(K,k,;*)$ denote the mixed volume with K appearing k times and $*$ representing $n - k$ fixed entries, and let $V(L,i;M,k-i;*)$ be the mixed volume with L appearing i times, M appearing $k - i$ times, and $*$ representing the same $n - k$ fixed entries as before. Then,

$$(6.7)\quad V(K,k;*) = \sum_{i=0}^{k} (-1)^i \binom{k}{i} V(L,i;M,k-i;*)$$

Note that (6.6) results with $L = -K$, $M = bB$, and the remaining fixed entries set equal to B. If we let $L = -K$, $M = bQ$, and set the remaining entries equal to Q, where Q is a fixed gauge body, we obtain an analogue of (6.6) for sets of relative constant width.

Chakerian's proof of (6.7) is by induction, using the multilinearity of mixed volumes. However, it turns out that Groemer's proof of (6.6) generalizes virtually without change to give (6.7). It seems worthwhile to give this proof of (6.7) here since Groemer [1964] is unpublished. Assuming $M = K + L$, let $f(t) = V((1-t)L + tK, k; *)$ for $0 \leqq t \leqq 1$, where $*$ represents any fixed $n - k$ arguments in the mixed volume. For $0 \leqq t \leqq 1/2$ we have $(1-t)L + tK = (1-2t)L + tM$, so on $[0,1/2]$ $f(t)$ is equal to

$$(6.8) \quad V((1-2t)L + tM, k; *) = \sum_{i=0}^{k} \binom{k}{i} (1-2t)^i t^{k-i} V(L,i;M,k-i;*).$$

Thus $f(t)$ is a polynomial in t defined on $[0,1]$ that coincides on $[0,1/2]$ with the polynomial given by the sum in (6.8). Hence these two polynomials are identical on $[0,1]$ and, in particular, $f(1)$, which is the lefthand side of (6.7), is equal to what we get by setting $t = 1$ in the sum in (6.8). But this gives (6.7), as we wanted.

We might also note that for convex bodies satisfying $K_i + L_i = M_i$, $i = 1,\ldots,r$, the general expansion of $V(K_1,\ldots,K_r,*)$ indicated in Chakerian [1982], p. 59, can be established by the same kind of argument as above.

7. Inequalities

One of the best known extremum problems for sets of constant width is that of determining bounds on the volume when the width is given. Since, by a well-known theorem of Bieberbach (Bonnesen-Fenchel [1934], p. 76), a ball has maximum volume among all convex bodies of given diameter, we have, a fortiori, that the ball maximizes volume among all elements of $\mathbf{W}^n$ having given width. The problem of *minimizing* the volume with a given constant width is con-iderably more difficult and is unsolved in case $n \geqq 3$.

The solution of the minimum problem in case $n = 2$ is given by the Blaschke-Lebesgue theorem, which states that the Reuleaux triangle is the unique member of $\mathbf{W}^2$ having minimum area for a given width. Bonnesen-Fenchel [1934] reproduces Lebesgue's proof of the theorem and gives references to proofs by Blaschke and Fujiwara. Expositions of Lebesgue's proof may also be found in Beretta-Maxia [1940a] and Blatter [1981]. Mayer [1934a] later obtained the theorem as a by-product of his work on the minimal annulus associated with a plane set of constant width. More recently short proofs have been given by Eggleston [1952b, 1958b], Besicovitch [1963a], and Chakerian [1966]. A variant of Lebesgue's proof may be found in Yaglom-Boltyanskii [1961]. Ohmann [1952] has established the analogue of the Blaschke-Lebesgue theorem for Minkowski planes, the minimizing set being a certain analogue of the Reuleaux triangle based on a triangle that is equilateral relative to the given gauge body. Another proof of the Minkowski plane analogue has been given by Chakerian [1966].

If $K \in \mathbf{W}^2$ has area $a(K)$ and width b, then the Blaschke-Lebesgue theorem is equivalent to the inequality

$$(7.1)\quad a(K) \geqq \frac{\pi - \sqrt{3}}{2} b^2,$$

with equality holding if and only if K is a Reuleaux triangle. In dealing with an array of inequalities for plane convex bodies, Kubota and Hemmi [1953] independently discovered the above result of Ohmann and obtained the following inequality. Let the diameter d and minimal width Δ of $K \in \mathbf{C}^2$ satisfy $\Delta \geqq d\sqrt{3}/2$. Then

$$(7.2)\quad a(K) \geqq 3\Delta\left\{\sqrt{d^2 - \Delta^2} + \Delta\left(\sin^{-1}\frac{\Delta}{d} - \frac{\pi}{3}\right)\right\} - \frac{\sqrt{3}}{2} d^2,$$

with equality holding if and only if K is a generalized Reuleaux triangle as described by Ohmann. Note that if $K \in \mathbf{W}^2$ has width b, then (7.2) is valid with $\Delta = d = b$, yielding the inequality (7.1). Sholander [1952] has given a detailed treatment of inequalities involving a(K), d(K), Δ(K), and the perimeter p(K), similar to those examined by Kubota and Hemmi, where one holds two of these quantities fixed and extremizes a third. Santaló [1959] also gives an extensive treatment of inequalities of this type.

Meissner's three-dimensional analogue of the Reuleaux triangle (cf. Bonnesen-Fenchel [1934], pp. 135–136) has long been conjectured to have least volume among all members of $\mathbf{W}^3$ of equal constant width. Were this conjecture indeed correct, it would follow that the greatest lower bound for the volumes of members of $\mathbf{W}^3$ having width 1 would be

$$\frac{2\pi}{3} - \frac{\pi\sqrt{3}}{4} \cos^{-1}\left(\frac{1}{3}\right) \approx .42.$$

The best lower bound presently known appears to be $\pi(3\sqrt{6} - 7)/3 \approx .365$, given by Chakerian [1966]. By Blaschke's theorem relating the volume and surface area of a member of $\mathbf{W}^3$ (cf. Section 6), the problem of minimizing volume among three-dimensional sets of a given constant width is equivalent to minimizing the surface area.

Lower bounds for the volume of sets of given constant width in E^n, $n \geqq 3$, have been derived by Firey [1965a] and Chakerian [1966]. Danzer has conjectured that the volume is minimized by an element of $\mathbf{W}^n$ with the symmetry group of a regular simplex. Danzer's conjecture and other pertinent problems are discussed by Heil in Gruber-Schneider [1979], pp. 260–261; see also Heil [1978].

In general the insphere and circumsphere of a convex body are not concentric. If however K is a convex body of constant width b, then the insphere and circumsphere are concentric and their radii, r and R respectively, satisfy

$$(7.3)\quad r + R = b \quad \text{and} \quad b\left(1 - \sqrt{\frac{n}{2n+2}}\right) \leqq r \leqq R \leqq b\sqrt{\frac{n}{2n+2}};$$

see Eggleston [1958b] and Melzak [1960]. Minoda [1939b] derives the first relation in (7.3) when n = 2 and applies it to some inequalities for the area. A generalization to Minkowski spaces of the first relation in (7.3) has been given by Chakerian [1967].

For $K \in \mathbf{W}^n$ the set of points between the insphere and the circumsphere is called the *minimal shell* (in case n = 2, the *minimal annulus* associated with K. Given a shell S with inner radius r and outer radius R satisfying the relations (7.3) with a fixed $b > 0$, it is natural to try to determine those $K \in \mathbf{W}^n$ having S as their associated minimal shell and having either maximum or minimum volume. This problem has been investigated only in dimension n = 2. The plane sets of constant width with maximum area given a prescribed minimal annulus were determined by Bonnesen; see Bonnesen-Fenchel [1934], pp. 134–135, for a description of the solution and for references. Mayer [1934a] has given upper and lower bounds for the areas of plane sets of constant width with prescribed minimal annulus. While these bounds are not sharp, Mayer's lower bound has inequality (7.1) as a consequence and so yields another proof of the Blaschke-Lebesgue theorem. Mayer [1934b] gives a sketch of a proof that the minimum area, for a prescribed annulus, is attained by a certain Reuleaux-type polygon, as conjectured by Bonnesen, however a detailed proof does not appear to have been published. See Bonnesen-Fenchel [1934], p. 135, for a description of Bonnesen's conjectured solution of the minimum problem.

Let $K \in \mathbf{C}^n$ have diameter d and circumradius R. It has been proved by Scott [1981] that there exists a $\bar{K}$ in $\mathbf{W}^n$, with $K \subset \bar{K}$, having the same diameter and circumradius as K (see Section 4). He uses this to establish several inequalities, involving the factor $2R - d$, by first proving them for sets of constant width. By way of example, suppose K is a plane convex body with perimeter p, circumradius R, and diameter d. Let $\bar{K} \in \mathbf{W}^2$, with $K \subset \bar{K}$, be the set provided by Scott's theorem. Then $\bar{p} = \pi d$ and $R + \bar{r} = d$, where $\bar{p}$ is the perimeter and $\bar{r}$ the inradius of $\bar{K}$. It follows then that

$$(2R - d)\bar{p} = (R - \bar{r})(\pi d) = \pi(R^2 - \bar{r}^2).$$

As a consequence of (7.3), with n = 2, we have $(\sqrt{3} - 1)R \leqq \bar{r}$, from which it follows that $R^2 - \bar{r}^2 \leqq (2\sqrt{3} - 3)R^2$. The preceding equation therefore yields $(2R - d)\bar{p} \leqq (2\sqrt{3} - 3)\pi R^2$. Since $p \leqq \bar{p}$ we finally have one of Scott's inequalities:

$$(7.4) \quad (2R - d)p \leqq (2\sqrt{3} - 3)\pi R^2.$$

Equality holds only for Reuleaux triangles, since $\bar{r}/R$ takes its minimum value of $\sqrt{3} - 1$ for $\bar{K}$ only in this case. Using the fact that $p(K) \geqq \pi\Delta(K)$ (see Yaglom-Boltyanskii [1961], p. 76), we see from (7.4) that

$$(7.5) \quad (2R - d)\Delta \leqq (2\sqrt{3} - 3)R^2,$$

for all $K \in \mathbf{C}^2$, with equality only for Reuleaux triangles. The last inequality was derived by Scott [1978] together with some inequalities for members of $\mathbf{C}^2$ involving the factor $(\Delta - 2r)$. Scott [1981] gives a generalization of (7.5) to E^n in the form

$$(7.6) \quad (2R - d)\Delta \leqq \frac{2\sqrt{n+1}}{n}(\sqrt{2n} - \sqrt{n+1})R^2.$$

In the latter paper he also finds an upper bound on $(2R - d)a/R^3$ for plane convex bodies of area a, diameter d, and circumradius R, and proposes as a reasonable conjecture that the least upper bound is attained by Reuleaux triangles

The fact that $p(K) \geqq \pi\Delta(K)$ for $K \in \mathbf{C}^2$, used in the preceding paragraph, is related to an interesting question about closed curves. For a compact set S in E^n, let $\rho(S,x)$ denote the maximum of the distances from x to points of S. Reidemeister [1921] showed that a convex body K in E^n such that $\rho(K,x)$ is constant for $x \in \mathrm{bdr}K$ is a body of constant width. Alexander [1975] conjectured that if C is a rectifiable closed curve in E^n such that $\rho(C,x) \geqq \lambda$ for all $x \in C$, then the length of C is at least $\pi\lambda$, and C can have length $\pi\lambda$ only if it is a plane convex curve of constant width. This was proved by Sallee [1974] and Falconer [1977b]. We give here Falconer's elegant integral geometric proof of Alexander's conjecture.

By appropriately developing C into a plane, one can show that it is enough to prove the conjecture in case $C = \mathrm{bdr}K$, where $K \in \mathbf{C}^2$. Assuming $\rho(K,x) \geqq \lambda$ for all $x \in \mathrm{bdr}K$, let $n(x)$ be the number of intersections of bdrK with the circle of radius λ centered at x. From Santaló [1976], p. 112, we have

$$(7.7) \quad \int n(x)\,dx = 4\lambda p(K),$$

where the integration is with respect to Lebesgue measure over E^2. For x inside the outer parallel set of K at distance λ, but outside K, the hypothesis on K implies that $n(x) \geqq 2$ except on a set of measure zero. The set of such x has area $\lambda p(K) + \pi\lambda^2$, so the integral in (7.7) is at least $2(\lambda p(K) + \pi\lambda^2)$. Thus from (7.7) we have

$$(7.8) \quad 4\lambda p(K) \geqq 2(\lambda p(K) + \pi\lambda^2),$$

giving the desired result, $p(K) \geqq \pi\lambda$. In case $p(K) = \pi\lambda$ we must have equality in (7.8), implying that $n(x) = 0$ for x interior to K. Then it follows easily that $w(K,\theta) \leqq \lambda$ for all θ. But if $w(K,\theta) < \lambda$ for some θ, we would have $p(K) < \pi\lambda$, contrary to hypothesis. Hence $w(K,\theta) = \lambda$ for all θ, so $K \in \mathbf{W}^2$.

Note that the argument yields as a corollary Barbier's theorem that $p(K) = \pi b$ for all $K \in \mathbf{W}^2$ having width b.

In response to a conjecture of Herda, Chakerian [1974] proved that if C is a plane rectifiable simple closed curve, and $x \mapsto f(x)$, $x \in C$, is any involution of C without fixed points such that the distance from x to f(x) is at least λ for all $x \in C$, then C has length at least $\pi\lambda$, and C has length equal to $\pi\lambda$ if and only if it has constant width and the chords joining x to f(x) are the diametrical chords of C. Herda's conjecture is the special case where f is such that the length of the arc along C from x to f(x) is half the perimeter for all $x \in C$. See Herda [1974] for a survey of the numerous proofs and offshoots of his conjecture. Note that Alexander's conjecture provides a generalization of these results, since for each $x \in C$ we have f(x) at distance at least λ from x, so $\rho(x,C) \geqq \lambda$ for all $x \in C$.

Lutwak [1980] has considered extensions of the Herda conjecture to integral power means of the perimeter bisectors and the area bisectors of convex curves.

The fact that $p(K) \geqq \pi\Delta(K)$ for plane convex bodies may be generalized in the following way. Let $K \in \mathbf{C}^2$, and let $\alpha_1,\ldots,\alpha_r$ be fixed angles. Consider the product of the widths of K,

$$P(K,\theta) = w(K,\theta + \alpha_1) \cdots w(K,\theta + \alpha_r), \quad 0 \leqq \theta \leqq 2\pi.$$

From Hölder's inequality we have

$$\int_0^{2\pi} P(K,\theta)^{1/r} d\theta = \int_0^{2\pi} w(K,\theta + \alpha_1)^{1/r} \cdots w(K,\theta + \alpha_r)^{1/r} d\theta$$

$$\leqq \prod_{i=1}^{r} \left(\int_0^{2\pi} w(K,\theta + \alpha_i) d\theta \right)^{1/r}.$$

Since $\int_0^{2\pi} w(K,\theta + \alpha) d\theta = 2p(K)$, this leads to

$$\text{(7.9)} \quad \min_\theta P(K,\theta) \leqq \left(\frac{p(K)}{\pi} \right)^r,$$

with equality only when K has constant width. If $r = 2$, $\alpha_1 = 0$, $\alpha_2 = \pi/2$, then $P(K,\theta)$ is the area of the rectangle circumscribed about K with a side parallel to direction θ. Then the preceding inequality asserts that K admits a circumscribed rectangle of area at most $(p(K)/\pi)^2$. This was a conjecture of Leo Moser, first proved by J. Schaer (see Schaer-Wetzel [1972], and Chakerian-Klamkin [1973] for the proof and references). Later proofs were given by Hansen [1978] and Lutwak [1979]. The derivation above is based on that of Lutwak. A slightly stronger statement of this kind, involving the perimeter of the circumscribing rectangle, and generalizations to E^n have been found by Firey and Groemer [1964].

Inequality (7.9) shows that, for fixed perimeter, $\min_\theta P(K,\theta)$ is maximized by sets of constant width. In case $r = 1$, this becomes the trivial assertion that, with p(K) fixed, Δ(K) assumes its maximum value when K has constant width. Correspondingly, from the fact that $p(K) \leqq \pi d(K)$ we have that d(K) is minimized, with p(K) fixed, when K has constant width. This result may also be viewed as a special case of theorems of Sachs [1958] concerning the integral power means of the chord lengths of closed curves.

The following inequality is related to the above. Let K be a convex body in E^3, and let $\delta(K)$ be the *inner diameter* of bdrK, that is, the least upper bound of the lengths of geodesics joining pairs of points on bdrK. Using the fact that $p \leqq \pi d$ for plane convex curves, with equality only for curves of constant width, Makuha [1966] proves that $\delta(K) \leqq \pi d(K)/2$, with equality holding if and only if K is a convex body of revolution having constant width.

If K and L are convex bodies in E^n, and either K or L is centrally symmetric, then

$$\int h(L,u)\, dS(K,u) = \int h(-L,u)\, dS(K,u),$$

where the integration is over the sphere S^{n-1} with respect to the surface area measure of K (for background on surface area measures, see Schneider [1979]). Thus we have for the mixed volume

$$\text{(7.10)} \quad V(L,K,\ldots,K) = \frac{1}{2n} \int (h(L,u) + h(-L,u))\, dS(K,u)$$

$$= \frac{1}{2n} \int w(L,u)\, dS(K,u).$$

From this follows

$$(7.11)\quad \frac{1}{2n}\Delta(L)s(K) \leqq V(L,K,\ldots,K) \leqq \frac{1}{2n}d(L)s(K).$$

Setting $L = K - K$ in (7.11), where K is any convex body, we obtain

$$(7.12)\quad \frac{1}{n}\Delta(K)s(K) \leqq V(K-K,K,\ldots,K) \leqq \frac{1}{n}d(K)s(K).$$

In case $n = 2$, the righthand inequality in (7.12) was given by Kubota [1948], who showed that equality holds if and only if K has constant width. From Minkowski's inequalities for mixed volumes we have $V(-K,K,\ldots,K) \geqq v(K)$, so $V(K-K,K,\ldots,K) = v(K) + V(-K,K,\ldots,K) \geqq 2v(K)$. Thus from (7.12) we obtain for any convex body K,

$$(7.13)\quad v(K) \leqq \frac{1}{2n}d(K)s(K).$$

Comparison of (7.13) with the lefthand inequality in (7.11) yields an inequality of Petermann [1967] for convex bodies K and L, where K is centrally symmetric, namely

$$(7.14)\quad \Delta(L)v(K) \leqq d(K)V(L,K,\ldots,K).$$

With a closer analysis Petermann showed that equality holds in (7.14) if and only if L has constant width and K is a ball.

Inequality (7.13), valid for any convex body K, was derived by Firey [1965a] who also proved that equality holds only if K is a ball. In case K has constant width b the inequality becomes $v(K)/s(K) \leqq b/2n$. If as usual $K^* = \frac{1}{2}(K-K)$, then K^* is a ball of radius $b/2$, for which we have $v(K^*)/s(K^*) = b/2n$. Thus we see that a set K of constant width satisfies

$$(7.15)\quad v(K)/s(K) \leqq v(K^*)/s(K^*).$$

This was noted by Heil (see Gruber-Schneider [1979], p. 260) who went on to ask whether (7.15) holds for all convex bodies.

With regard to mixed volumes of sets of constant width, Godbersen [1938] has shown that if $K \in \mathbf{W}^n$, then

$$V(K,\ldots,K,-K,\ldots,-K) \leqq \binom{n}{k} v(K),$$

for $k = 0,1,\ldots,n$, where K appears k times and $-K$ appears $n-k$ times in the mixed volume. He also verified this inequality for certain other classes of convex bodies and conjectured that it holds for all $K \in \mathbf{C}^n$. Were this true, the result would imply directly the Rogers-Shephard inequality for the difference body (see Eggleston [1958b]), namely,

$$v(K-K) \leqq \binom{2n}{n} v(K).$$

For $K \in \mathbf{C}^2$, let again $K^* = \frac{1}{2}(K - K)$ and let p^*,a^*,R^*,r^* be the usual geometric quantities associated with K^*. We have $p^* = p = p(K)$, $R^* = d/2 = d(K)/2$, and $r^* = \Delta/2 = \Delta(K)/2$. The minimal annulus associated with K^* therefore has inner radius $\Delta/2$ and outer radius $d/2$. By Bonnesen's sharpening of the isoperimetric inequality (see Bonnesen-Fenchel [1934], p. 83) we then have

$$p^2 - 4\pi a^* \geqq \pi(d - \Delta)^2,$$

with equality only if K^* is a circular disk, that is, only if K has constant width. Rearrangement of this inequality gives

$$\text{(7.16)} \quad p^2 - 4\pi a \geqq \pi(d - \Delta)^2 + 4\pi(a^* - a),$$

with equality only for sets of constant width. This inequality was given by Ganapathi [1934b], with an interpretation of $(a^* - a)/2$ as the area associated with the locus of midpoints of the diametrical chords of K. From the Blaschke-Lebesgue theorem it obviously follows that $(a^* - a)/2$ is maximized by the Reuleaux triangle among all sets of equal constant width. For a further discussion of (7.16), see Vincensini [1938]. Ganapathi [1935] has given a formulation analogous to (7.16) of Minkowski's inequality in E^3, with perimeter replaced by the total mean curvature of K and area replaced by the surface area. The righthand side of the inequality becomes a term proportional to the signed surface area of the midpoint locus of the diametrical chords of K. Again, equality holds only for bodies of constant width.

Firey [1960] has proved some isoperimetric theorems for Reuleaux polygons, analogous to the classical theorems of Zenodorus for ordinary polygons. Sallee [1970a] later gave simplified proofs of the theorems. See Section 3 for a description of these results.

Taylor [1953] solves the isoperimetric problem of *minimizing* the area of those plane convex bodies of given perimeter constrained to lie in a fixed Reuleaux triangle.

In a loosely related vein, Minoda [1939a] treated the problem of minimizing the perimeters of plane closed curves inside which a given $K \in \mathbf{C}^2$ can be rotated, showing that the minimum is attained by the circumcircle of K and also in some cases by certain non-circular curves of constant width.

Motivated by considerations arising out of measure theory, Freilich [1951] obtained the following inequality which eludes a satisfactory characterization in terms of the categories we are using in this survey. Let $K \in \mathbf{W}^2$ have width $b \geqq 1$ and let $[p_1,q_1]$ and $[p_2,q_2]$ be parallel chords of K whose distance from each other is at most 1. Then the lengths of $[p_1,q_1]$ and $[p_2,q_2]$ differ by at most $2\sqrt{2b}$. Freilich also proves a generalization involving the measures of parallel cross-sections of convex bodies of constant width in E^n, $n \geqq 3$.

8. Inscribed and circumscribed bodies

A polytope P is *circumscribed* about $K \in \mathbf{C}^n$ if $K \subset P$ and each facet of P is contained in a supporting plane of K. A polytope is *inscribed* in K if each of its vertices belongs to bdrK. For convex bodies K and L generally, some authors say that L is circumscribed about K if $K \subset L$ and that L is inscribed in K if $L \subset K$; we shall deal with theorems involving these situations also (but without using this terminology). We begin with some observations about regular simplices circumscribed about bodies of constant width and use these to derive information about other types of circumscribed polytopes.

In what follows let T_n be a fixed regular simplex circumscribed about the unit ball B in E^n. Let K be a convex body of constant width b in E^n. For each rotation $g \in SO(n)$, let $T(g)$ be the regular simplex circumscribed about K and positively homothetic to gT_n. Let $T^*(g)$ be the regular simplex positively homothetic to $-gT_n$ circumscribed about K. By continuity, there exists some $g \in SO(n)$ such that $T = T(g)$ and $T^* = T^*(g)$ have the same edge length. Additionally, corresponding pairs of facets of T and T^* are parallel and distance b apart, so it is easy to see that the centroids of T and T^* coincide, and T^* is obtained by central reflection of T through its centroid. It follows that T is congruent to $(b/2)T_n$. Thus the argument shows that any set of constant width b admits a circumscribed regular simplex whose inscribed sphere has diameter b. This implies the theorem of Gale [1953] that each compact subset of E^n with diameter d can be covered by a simplex congruent to $(d/2)T_n$. The argument given above is based on one by Viet and Süss; see Danzer-Grünbaum-Klee [1963] for references on this and for further applications.

It is easy to estimate the variation in the sizes of regular simplices circumscribed about $K \in \mathbf{W}^n$. Since the inradius of K is at least $b(1 - \sqrt{n/(2n + 2)})$, the smallest regular simplex circumscribed about K must have inradius at least that large. Similarly, the largest regular simplex circumscribed about K has inradius at most $b\sqrt{n/(2n + 2)}$, the maximum possible circumradius of K. It is not difficult to see that any set of constant width b having an inscribed regular simplex of edge length b admits circumscribed regular simplices with the above extreme sizes.

Specializing to the case $n = 2$, we see from the preceding that an equilateral triangle of minimal width Δ contains no convex body of constant width greater than $\Delta/(3 - \sqrt{3})$. This was observed by Eggleston [1955a], or [1957] pp. 140–149, who further showed that every plane convex body of minimal width Δ contains a set of constant width equal to $\Delta/(3 - \sqrt{3})$.

With $K \in \mathbf{W}^n$ of constant width b, and T, T^* the circumscribed simplices described above, let $P = T \cap T^*$ and $P_n = T_n \cap (-T_n)$. Then P is a polytope congruent to $(b/2)P_n$ circumscribed about K. Since P_2 is a regular hexagon, we obtain as a consequence Pál's theorem (see Bonnesen-Fenchel [1934], p. 131) that each plane convex body of constant width admits a circumscribed regular hexagon. Since P_3 is a regular octahedron, we also have the observation of Gale [1953] that every body of constant width in E^3 admits a circumscribed regular octahedron.

It might be noted that Pál's theorem is a special case of the fact that every plane convex body admits a circumscribed affine image of a regular hexagon (cf. Grünbaum [1965]). It is easy to see that such a hexagon circumscribed about a body of constant width must in fact be regular.

In Section 7 we considered rectangles circumscribed about plane convex bodies. There are few interesting questions concerning the circumscribed boxes of a body of constant width since a convex body K in E^n has constant width if and only if all its circumscribed boxes are mutually congruent cubes. In fact, if $n \geqq 3$ we can deduce that K has constant width under the sole hypothesis that all its circumscribed boxes are cubes. For, let u_1 and u_2 be any two directions and let u_3 be orthogonal to both u_1 and u_2. By considering a circumscribed box with a pair of edges parallel to u_1 and u_3, we see that $w(K,u_1) = w(K,u_3)$. Similarly $w(K,u_2) = w(K,u_3)$. Thus $w(K,u_1) = w(K,u_2)$, so K has constant width. On the other hand, there exist plane convex bodies, not of constant width, all of whose circumscribed rectangles are squares. Indeed, a square is an example of such a set! These sets are discussed in Yaglom-Boltyanskii [1961], p. 97. Kárteszi [1959a, 1959b] has investigated polygons having this property.

While all circumscribed boxes of a body of constant width have the same volume, this property does not characterize the class $\mathbf{W}^n$. Convex bodies with this property have been characterized by Petty and McKinney [1974]. Yaglom and Boltyanskii [1961], p. 93, give a detailed discussion of plane convex bodies whose circumscribed rectangles all have the same perimeter.

Toranzos [1964] has given a characterization of sets of constant width in terms of circumscribed rhombuses.

Given a convex body K in E^3, for each direction u let $p(K,u)$ be the perimeter of the orthogonal projection of K into a plane orthogonal to u. If $p(K,u)$ is constant the body K is said to have *constant girth.* Then the theorem of Minkowski already mentioned in Section 2 is the statement that K has constant width if and only if it has constant girth. The product $p(K,u)w(K,u)$ is the lateral surface area of a cylinder circumscribed about K with generators parallel to u. Firey [1959] proved that K has constant width if and only if all circumscribed cylinders of K have equal lateral surface area.

We next turn to inscribed figures in sets of constant width. While every plane convex body admits infinitely many inscribed equilateral triangles and at least one inscribed square, Eggleston [1958a] has given an example of a plane convex body of constant width (a modified Reuleaux triangle) that admits no inscribed regular polygon with more than four sides. Eggleston [1959] also investigated the largest square that can fit inside a plane set of constant width b. He proved that if the largest square contained in a Reuleaux triangle of width b has side s, then every plane set of constant width b contains a square of side s, with only the Reuleaux triangle containing no larger square.

In connection with these matters it is worth mentioning Mizel's conjecture that if K is a plane convex body such that every rectangle with three vertices on bdrK also has its fourth vertex on bdrK, then K must be a circle. This was proved by Besicovitch [1961], with more elementary proofs provided later by Watson [1962] and Danzer [1963]. All these proofs proceed by first showing that K must have constant width.

Given any plane convex body K of constant width b, let s_1 and s_2 be, respectively, the sides of the smallest and the largest equilateral triangles inscribed in K. Let σ be the side of the largest equilateral triangle inscribed in a regular Reuleaux pentagon of width b. Then Eggleston and Taylor [1952] prove that

$$(\sqrt{3} - 1)b \leqq s_1 \leqq (\sqrt{3}/2)b \quad \text{and} \quad \sigma \leqq s_2 \leqq b.$$

One can have $s_1 = (\sqrt{3} - 1)b$ or $s_2 = b$ only for a Reuleaux triangle, $s_1 = (\sqrt{3}/2)b$ only for a circle, and $s_2 = \sigma$ only for the regular Reuleaux pentagon. The proof that $s_2 \geqq \sigma$ is by far the most difficult among these inqualities.

Klee conjectured that every plane convex body K of constant width b contains a semicircle of diameter b. Besicovitch [1963b], with restrictive assumptions on the regularity of bdrK, proved there exist three such semicircles in K. In the course of his investigations on minimal universal covers, Eggleston [1963] proved, without any regularity restrictions, that K contains a semicircle of diameter b. Cooke [1964] and Knuth [1969] proved, again without any regularity restrictions, that K contains three semicircles of diameter b. Falconer [1977a] has established the general result that any plane convex body with minimal width Δ contains three semicircles of diameter Δ.

The following result concerns polygons contained in plane sets of constant width. A regular hexagon H of sidelength $b/\sqrt{3}$ is partitioned into three congruent (irregular) pentagons by dropping perpendiculars from the center to alternate sides of H. If P is any such pentagon, Chakerian and Sallee [1969] proved that each plane body of constant width b contains a translate of P. They obtained a variant of Helly's theorem for sets of constant width (cf. Section 5) by using this result to prove that each plane set of diameter d can be covered by three translates of any plane convex body of constant width d (cf. Section 9). Another discussion of this matter is given in Chakerian [1969].

Let S be a centrally symmetric subset of maximum volume contained in the convex body K. Then the ratio $\mu_1(K) = v(S)/v(K)$ is a *measure of symmetry* as defined by Grünbaum [1963a], and is referred to as the Kovner-Besicovitch measure of symmetry. It is well-known that $\mu_1(K)$ reaches its minimum value of 2/3 for a triangle as K ranges over all plane convex bodies. Besicovitch [1951] proved that $\mu_1(K)$ attains its minimum value for the Reuleaux triangles as K ranges over all plane convex bodies of constant width. Eggleston [1952a], or [1957] pp. 130–140, generalized this result by proving the following conjecture of Besicovitch. Let $b/2 \leqq c \leqq b$, and let A be the outer parallel set at distance $b - c$ of a Reuleaux triangle of width $2c - b$. If K is any plane convex body of constant width b whose radius of curvature function takes only values between $b - c$ and c, then $\mu_1(K) \geqq \mu_1(A)$.

If M is a centrally symmetric convex body of minimum volume containing a given $K \in \mathbf{C}^n$, then the ratio $\mu_2(K) = v(K)/v(M)$ is also a measure of symmetry. In case $n = 2$, it is known that $\mu_2(K)$ achieves its minimum value of 1/2 for a triangle as K ranges over all plane convex bodies. It follows readily from the Blaschke-Lebesgue theorem (see Section 7) that $\mu_2(K)$ is minimized by the Reuleaux triangles as K ranges over $\mathbf{W}^2$. The proof is given in Chakerian [1966]. Crude lower bounds on $\mu_2(K)$ as K ranges over $\mathbf{W}^n$, $n > 2$, are also given there, however, as with μ_1, the best lower bounds are not known for $n \geqq 3$.

We mention one further result concerning measures of symmetry of sets of constant width. For any plane convex body K, let m_1 be the measure of the set of interior points of K that each bisect exactly one chord of K. Let $\mu_3(K) = m_1/a(K)$. It was shown by Chakerian and Stein [1966] that μ_3 is a measure of symmetry taking its minimum value for Reuleaux triangles as K ranges over $\mathbf{W}^2$.

9. Packings, coverings, lattice points

Let $K, K_1, K_2, \ldots$ be convex bodies in E^n. The collection $\{K_i\}$ is said to be a *packing* in K if $K_1 \subset K$ and $(\mathrm{int}K_i) \cap (\mathrm{int}K_j) = \emptyset$ (for all i,j with $i \neq j$), and a covering of K if $K \subset \cup K_i$. If $\{K_i\}$ consists only of one convex body the ensuing problems have been discussed in Section 8. Thus, we assume now that $\{K_i\}$ contains at least two bodies. A packing or covering $\{K_i\}$ is said to be *translative* if there are a $K_0 \in \mathbf{C}^n$ and $p_i \in E^n$ so that $K_i = K_0 + p_i$.

If G is an open convex subset of E^n and $p \in E^n$ then the condition $G \cap (G + p) \neq \emptyset$ is obviously equivalent to the existence of two points $x,y \in G$ so that $x = y + p$. This can also be stated as $\frac{1}{2}p \in G^*$ where G^* is again defined by $\frac{1}{2}(G + (-G))$. Correspondingly, the condition $G^* \cap (G^* + p) \neq \emptyset$ is equivalent to $\frac{1}{2}p \in G^{**} = G^*$. Hence,

$$(9.1) \quad G \cap (G + p) \neq \emptyset \Leftrightarrow G^* \cap (G^* + p) \neq \emptyset.$$

This relation enables one to reduce many translative packing problems for bodies of constant width to corresponding problems for balls (take $G = \mathrm{int}K_0$). For example, the problem of finding densest translative packings in E^n of convex bodies of constant width is equivalent to the problem of densest packings of balls. Similarly, if one is interested in the largest number of non-overlapping translates of a given $K_0 \in \mathbf{W}^n$ that can touch K_0 (i.e. meet K_0, but not $\mathrm{int}K_0$), only the case of balls has to be considered.

Relation (9.1) is already contained in the work of Minkowski [1904] and has been rediscovered repeatedly. There is no corresponding reduction for covering problems (cf. Groemer [1962]). For the latest developments on packings of balls see the survey of G. Fejes Tóth [1983]. From (9.1) and the inequalities discussed at the beginning of Section 7 one can infer that the density of a densest translative packing of plane convex bodies of constant width is always between $(\pi/\sqrt{3}) - 1 = 0.81379\ldots$ and $\pi/\sqrt{12} = 0.90689\ldots$.

We discuss now some results regarding a translative packing problem that cannot be reduced to a corresponding packing problem for balls by a trivial application of (9.1). It concerns the 'permeability' of a layer of two-dimensional convex bodies. For a given $K_0 \in \mathbf{W}^2$ let $\{K_i\} = \{K_0 + p_i\}$ be a packing in a strip of width t (i.e. in a closed set bounded by two parallel lines of mutual distance t). Assume that t is minimal for the given collection $\{K_i\}$, and let λ denote the infimum of the lengths of all continuous rectifiable curves connecting the two boundary lines without meeting interior points of any K_i. The permeability of $\{K_i\}$ is defined by

$$\alpha(\{K_i\}) = \frac{t}{\lambda}.$$

It has been shown by Hortobágyi [1976b] that for any such $\{K_i\}$

$$\alpha(\{K_i)\} \geq \frac{\sqrt{27}}{2\pi(t - b) + b\sqrt{27}}$$

where b is the width of K_0. Equality is possible if $(2/\sqrt{3})((t-b)/b)$ is an integer. This estimate has been improved by Florian [1978] to

$$\alpha(\{K_i\}) \geqq (t/b)\left(\frac{m\pi}{3} + 1 + \sin^{-1}\left(\frac{t}{b} - 1 - \frac{\sqrt{3}}{2}m\right)\right)^{-1}$$

where $m = [(2/\sqrt{3})((t-b)/b)]$ and equality is possible (for certain packings that are explicitly described) for all feasible values of t and b. More recently Florian [1982] has studied the possibility of relating permeability estimates for layers $\{K_0 + p_i\}$ to those for $\{K_0^* + p_i\}$.

(9.1) is, of course, irrelevant when dealing with packings that involve not only translates of a given body. A typical problem of this kind concerns the 'Newton number' of a given $K \in \mathbf{W}^2$, i.e. the largest number of congruent non-overlapping copies of K that touch K. It has been shown by L. Fejes Tóth [1967] that $N(T) \leqq 7$ for Reauleaux triangles T, and it has been proved by Schopp [1970] that actually $N(K) \leqq 7$ for all $K \in \mathbf{W}^2$. On the other hand, Hortobaflgyi [1976a] has shown that any 8 sets $K, K_1, \ldots, K_7$ from $\mathbf{W}^2$ of equal width can always be arranged so that each K_i touches K and any two K_i do not overlap. Thus, $N(K) = 7$ for all $K \in \mathbf{W}^2$.

The following packing problem is of a completely different nature. For a given $K \in \mathbf{C}^2$ let K_1, K_2 be two non-overlapping domains similar to and contained in K. It is easy to see that for $K \in \mathbf{W}^2$ the perimeters satisfy the inequality $p(K_1) + p(K_2) \leqq p(K)$. Beck and Bleicher [1971] have shown that there is a kind of converse of this statement, namely: If a set $K \in \mathbf{C}^2$ has the property that for all non-overlapping K_1, K_2 similar to and contained in K the inequality $p(K_1) + p(K_2) \leqq p(K)$ holds, then K is either of constant width or a regular polygon. The proof depends on an interesting characterization of the members of $\mathbf{W}^2$ (see Section 2). Another problem of this kind has been investigated by Lutwak [1982]. It concerns the packing of two closed curves of equal lengths in a circle; pairs of curves of constant width require a larger circle than other curves.

Let now Λ^n be an n-dimensional lattice in E^n, and Λ_0^n the lattice of all points of E^n with integral coordinates. If $K \in \mathbf{W}^n$ (or if K is an affine image of some set from $\mathbf{W}^n$) then K^* is a ball (or a solid ellipsoid). Hence, if one is interested in densest lattice packings of K, i.e. densest packings of the form $\{K + p : p \in \Lambda^n\}$ it follows from (9.1) that one may select the direction of one of the basis vectors of Λ^n arbitrarily. L. Fejes Tóth [1971] noticed this property of sets of constant width (in the case $n = 2$) and proposed to characterize all convex bodies that permit such 'direction invariant' densest packings (see also the comments of Bambah in his review of Fejes Tóth's paper in the Mathematical Reviews).

We turn now to several other results that associate convex bodies of constant width with lattice points. The first indication that this area might be of some interest appeared as a problem by Hammer [1966]. Subsequently Sallee [1969] proved that there exists an essentially unique set in $\mathbf{W}^2$ of maximal width that contains no points of Λ_0^2 in its interior. This set is a Reuleaux triangle whose width b is a root of a certain polynomial of fourth degree. Numerically, b is between 1.545 and 1.546. Moreover, it is noted that for the Minkowski planes with strictly convex gauge bodies the corresponding extremal sets are appropriately defined 'relative Reuleaux triangles'. Also discussed are sets from $\mathbf{W}^2$ of maximal width whose interiors avoid of a locally finite collection of convex sets in E^2. The resulting maximal sets, if they exist, are certain Reuleaux polygons (or "relative Reuleaux polygons" in the case of the Minkowski plane). The proofs of these results use an interesting procedure that enables one to construct new Reuleaux polygons from a given one by a replacement procedure for the boundary arcs. The following related result has been proved (using Sallee's theorem) by Elkington and Hammer [1976]: If g(c) denotes the minimum of the number of points from Λ_0^2 in a set $K \in \mathbf{W}^2$ of width greater than c, then

$$\left[\frac{c}{1.546}\right]^2 \leqq g(c) \leqq \frac{c^2}{2}(\pi - \sqrt{3}).$$

The book of J. Hammer [1977] is also of interest in this connection.

We discuss now several covering theorems. A famous (and for $n > 3$ unsolved) problem in this area is Borsuk's conjecture. Formulated as a covering theorem for convex sets it states that every n-dimensional convex body of diameter 1 can be covered by $n + 1$ convex bodies of diameter less than 1. Because of theorems (A) and (B) of Section 4 one may assume that all these sets are of constant width. Therefore, the assumption that a convex body have constant width appears frequently in attacks on Borsuk's conjecture. For example, the first published solution for $n = 3$ proceeds under this assumption (Eggleston [1955b], or [1957] pp. 72–92). An excellent survey of the literature on Borsuk's conjecture (up to 1963) is the paper by Grünbaum [1963b]. We review here those results that are primarily concerned with bodies of constant width (and not merely with constant width as a simplifying assumption for the proof of something else).

Melzak [1963] has shown that any set $K \in \mathbf{W}^2$ of width 1 can be covered by three subsets of diameter at most $\min\{e(K), \sqrt{3} - e(K)\}$, where e(K) is the edge-length of the largest equilateral triangle with vertices in bdrK. Schopp [1977] proved that a set $K \in \mathbf{W}^2$ can be covered by three circular discs of diameter p/6, where p is the perimeter of the smallest equilateral triangle covering K. Chakerian and Sallee [1969] have shown that every plane set of diameter 1 can be covered by three translates of any plane convex body of constant width not less than 0.9101.

There are some results that go in the opposite direction by showing that the covering sets cannot be too few or too small. In the case $n = 2$ Lenz [1956] has proved that at least one of three sets that cover a $K \in \mathbf{W}^2$ of width 1 must have diameter at least $\sqrt{3} - 1$, and greater than $\sqrt{3} - 1$ if K is not a Reauleaux triangle. For arbitrary n Lenz [1955] has shown that a convex body of constant width 1 cannot be covered by fewer than $n + 1$ sets of diameter less than 1. For $n = 2$ a striking result of the same type is due to Boltjanskii [1970]. It states that any $K \in \mathbf{C}^2$ with $d(K) = 1$ cannot be covered by two sets of diameter less than 1 if and only if the completion of K is unique (cf. Section 4). An interesting exposition of this and related covering problems is given by Boltjanskii and Gohberg [1980].

Further contributions in this area are problems no. 7, 8, 9 of Melzak [1968] and an example of Rogers [1971]. Rogers' example is a set $K \in \mathbf{W}^8$ that is invariant under the symmetry group determined by a regular simplex in E^8 but cannot be covered by nine convex bodies of diameter less than b(K) each having this same symmetry group. This contrasts with the fact, also proved by Rogers, that any $K \in \mathbf{W}^n$ with the symmetry group of the regular simplex can be covered by n + 1 sets of diameter less than b(K) if there are no symmetry requirements on the covering sets.

We conclude this section with the following Helly-type theorem of Buchman and Valentine [1982]. If a collection $\mathbf{F} \subset \mathbf{C}^n$ has the property that any subcollection of n + 1 (or fewer) members is contained in some convex body of constant width b, then there exists a convex body of constant width b that contains all members of **F**.

10. Rotors in polytopes

Let $P \in \mathbf{C}^n$ be a polytope and $K \in \mathbf{C}^n$. We say that K is a *rotor* of P if for each rotation ρ of E^n there exists a translation τ so that P is circumscribed about $\tau\rho K$ (as defined in Section 8). If P is an n-dimensional cube then any rotor of P is obviously a member of $\mathbf{W}^n$. However, there are convex polytopes that have rotors which are not of constant width. Although only convex bodies of constant width are the principal subject of this survey, we discuss here some pertinent references. For the early literature see Bonnesen-Fenchel [1934], pp. 139–140. A survey of many results in this area has been given by Goldberg [1960]. Also of interest regarding these matters is the book of Yaglom-Boltyanskii [1961] where many basic properties of rotors of equilateral triangles are discussed.

First we deal with the case n = 2. The m-sided convex polygons that permit non-circular rotors have already been described by Fujiwara [1915]. The same problem has been solved by Kameneckii [1947] (for m = 3) and Schaal [1962]. Rotors consisting of circular arcs have been investigated by Goldberg [1948, 1957]. The problem of finding for a given regular polygon with m sides rotors of least area has been treated for m = 3 by Weissbach [1972, 1977] (see also Yaglom-Boltyanskii [1961]) and in the general case by Focke [1969a] and Klötzler [1975]. Two other extremal problems for rotors of regular polygons have been solved by Focke [1969b] and Focke-Gensel [1971]. Rotors of spherical polygons have been described by Goldberg [1952, 1956]. For a discussion of several other concepts loosely related to two-dimensional rotors, see the papers of Fischer [1936], Minoda [1939a, 1941], Goldberg [1954, 1958, 1959] and Kárteszi [1959]. Some technical applications of rotors have been described by Schaal [1963]. It may be worth mentioning that the often discussed rotor of the Wankel engine does not have to be a rotor as defined in this section, and certainly not a Reuleaux triangle (see the study of Müller [1971] for the geometric intricacies of rotary-type engines).

For n = 3 Meissner [1918] has already determined all regular polytopes that have rotors different from a ball (these are the cube, tetrahedron and octahedron). He has also found representations of the corresponding rotors in terms of their support functions. Schneider [1971] succeeded in finding a complete characterization of all (not necessarily bounded) convex polytopes in E^n that have non-spherical rotors. His investigation depends to a large extent on the solution of a certain functional equation by spherical harmonics.

11. Concepts related to constant width

In this section we deal with some variants of bodies of constant width whose discussion does not fit well into the categories treated in previous sections. We also briefly mention extensions to non-Euclidean spaces.

Convex bodies of constant outer k-measure provide a natural generalization of bodies of constant width. If k is an integer satisfying $1 \leqq k \leqq n-1$, then $K \in \mathbf{C}^n$ is said to have *constant outer k-measure* if all its orthogonal projections into k-dimensional subspaces of E^n have equal k-dimensional volume. Thus bodies of constant outer 1-measure are precisely the members of $\mathbf{W}^n$. The bodies of constant outer (n − 1)-measure are better known as *convex bodies of constant brightness*. It is a well-known fact that there exist non-spherical convex bodies of constant brightness. For references see Bonnesen-Fenchel [1934]. Klee [1969b] discussed a phenomenon from physics related to bodies of constant width and bodies of constant brightness, and he posed the problem of constructing non-spherical bodies of constant outer k-measure for any k satisfying $1 < k < n-1$. In response to this, Firey [1970] has given a method for constructing for each such k a non-spherical convex body of revolution having constant outer k-measure. In case $n = 3$, Firey's construction gives for $k = 1$ the body obtained by revolving a Reuleaux triangle about a symmetry axis and for $k = 2$ an example of a body of constant brightness constructed by Blaschke [1956]. Generalizing equation (6.1), Firey [1970] also proves that if K has constant outer k-measure and a sufficiently smooth boundary then

$$(11.1) \quad F_k(u) + F_k(-u) = \text{constant},$$

where $F_k(u)$ is the k-th elementary symmetric function of the principal radii of curvature at the boundary point with outward normal u. He derives a condition on the principal radii of curvature that ensures constant outer k-measure.

Define the *k-girth of K in direction u* by $nV(K,\ldots,K,B,\ldots,B,[u])$ where, in this mixed volume, K appears k times, the unit ball B appears $n-k-1$ times, and [u] is a line segment of unit length parallel to u. One can prove that a sufficiently smooth body satisfying (11.1) must have constant k-girth. By applying Kubota's integral formulas (cf. Bonnesen-Fenchel [1934], p. 49), Firey [1970] has shown that any body of constant outer k-measure must have constant k-girth. Whether the converse is true is an open question for $1 < k < n-1$, although Firey has shown that the converse holds for bodies of revolution.

Regarding convex bodies of constant brightness in E^3, Berwald [1937] proved the following result. Let $K \in \mathbf{C}^3$, and for each $u \in S^2$ let $\lambda(K,u)$ be the mean perimeter of the cross-sections of K by planes orthogonal to u. Then K has constant brightness if and only if $\lambda(K,u)$ is independent of u.

The reports of Firey [1956b, 1967] and the paper of Firey and Grünbaum [1964] are valuable sources of information about sets of constant brightness and related matters. In particular, Firey [1956b, 1967] gives a technique for obtaining non-spherical convex bodies of constant brightness satisfying preassigned smoothness requirements. A method developed by Vincensini [1935, 1936, 1937, 1938] plays an important role here. Vincensini has shown how to construct sets having a preassigned difference body by a process of extending linear (in the sense of Minkowski addition) families of convex bodies. In case the difference body is a ball, the method yields ordinary bodies of constant width (see Vincensini [1956, 1957] for later applications of the technique, and Valette [1971] for a related study of Minkowski addition). Firey adapts Vincensini's theory of extensions to the construction of sets of constant brightness, replacing Minkowski addition by what he calls Blaschke addition of convex bodies. A Blaschke sum of convex bodies K_1,K_2 is a convex body K whose surface area function is the sum of the surface area functions of K_1 and K_2. If K_1 and K_2 both have constant brightness, then their Blaschke sum K has constant brightness (see Firey [1967]).

An outstanding unsolved problem in this area is whether balls are the only convex bodies of constant width that have also constant brightness. The question was answered affirmatively by Matsumura under the assumption that the boundary of the body is twice continuously differentiable (see Bonnesen-Fenchel [1934], pp. 140–141). But we are not aware of a proof of the result without smoothness restrictions on the boundary. Chakerian [1967] proves an analogue of Matsumura's theorem for three-dimensional Minkowski spaces, assuming smoothness restrictions on both the gauge body and the convex body under consideration.

See Chakerian [1967] for a definition of sets of constant k-girth in Minkowski spaces. Petty [1972, 1974] discusses the connection between Minkowski geometry and affine differential geometry and gives a treatment of sets of relative constant brightness from a more general viewpoint. We might mention here that the references Hirakawa [1935a] and Petty [1955] contain several results about sets of constant width in Minkowski planes.

Convex bodies of constant affine width, as defined in affine differential geometry and studied, e.g., by Süss [1927], appear to have received little attention in the past fifty years. An excellent summary of the basic ideas in case $n = 2$ can be found in Beretta-Maxia [1940a], including a discussion of some work of Hirakawa [1935b] relating the concept of affine width to that of relative width as defined in Section 1. See Heil [1967] for more recent work on the affine geometry of convex curves. Our discussion will be restricted to another affinely invariant counterpart of constant width, namely the concept of affine triangular width, or simply triangular width, as introduced by Sancho de San Román [1956]. For a given $K \in \mathbf{C}^2$ and each $p \in \text{bdr}K$ let $t(p)$ be the maximum area of triangles inscribed in K with one vertex at p. Then $t(p)$ is called the *triangular width of K at p*, and K is said to have *constant triangular width* if $t(p)$ is the same for all $p \in \text{bdr}K$. It is clear from the definition that the property of constant triangular width is an affine invariant. Sancho de San Román [1956, 1957] has proved the existence of plane bodies of constant triangular width other than ellipses and also has considered the natural generalization of the concept to higher dimensions. He later has shown that if $k = 3m \pm 1$ then regular k-gons have constant triangular width while they do not if $k = 3m$ (see Sancho de San Román [1965]). Garay de Pablo and Clavera Emperador [1964, 1967] prove that 4-gons and 5-gons of constant triangular width must be the affine images of regular polygons, but give examples of 6-gons and 7-gons of constant triangular width that are not affine images of regular polygons.

We consider next another affinely invariant concept, that of the affine image of a body of constant width. In order to discuss some characteristic properties of affine images of bodies of constant width, we need to introduce the definition of conjugate diametrical chords. Two diametrical chords (see Section 2), $[p_1,q_1]$ and $[p_2,q_2]$, of $K \in \mathbf{C}^2$ are said to be *conjugate* if there exist supporting lines of K through p_1 and q_1 parallel to $[p_2,q_2]$, and supporting lines through p_2 and q_2 parallel to $[p_1,q_1]$. The boundary of $K \in \mathbf{C}^2$ is a *P-curve* if each diametrical chord of K belongs to a conjugate pair. The centrally symmetric P-curves are known as *Radon curves* and are essentially the boundaries of those planar gauge bodies relative to which the relation of 'perpendicularity' is symmetric. That is, a centrally symmetric $Q \in \mathbf{C}^2$ with center at o is bounded by a Radon curve if and only if whenever a line L is Q-orthogonal to a line L′, then L′ is Q-orthogonal to L. Any member of $\mathbf{W}^2$ is a P-curve and, because of the affine invariance of the defining property, so is any affine image of a member of $\mathbf{W}^2$. Krautwald [1980] has proved that a $K \in \mathbf{C}^n$ is the affine image of a member of $\mathbf{W}^n$, $n \geqq 3$, if and only if each orthogonal projection of K into a two-flat is bounded by a P-curve. From this follows a characterization given by Chakerian [1965]: A set $K \in \mathbf{C}^n$, where $n \geqq 3$, is the affine image of a member of $\mathbf{W}^n$ if and only if all orthogonal projections into hyperplanes are affine images of bodies of constant width. Chakerian also proves that for $n \geqq 3$, a $K \in \mathbf{C}^n$ is the affine image of a body of constant width if and only if all rectangular boxes circumscribed about K have the same diameter. Krautwald [1980] gives a further characterization of affine images of bodies of constant width in terms of the smallest circumscribed parallelepiped and the largest inscribed cross-polytope. See Porcu [1963] for a number of results pertaining to P-curves and the affine and projective images of plane bodies of constant width.

We turn for a moment to an interesting question involving the polars of bodies of constant width. If $K \in \mathbf{W}^n$ has the origin o as an interior point then its polar body K° (see Eggleston [1958b]) has o as an *equireciprocal point*, that is, the sum of the reciprocals of the distances from o to p and from o to q is the same for all chords [p,q] of K° passing through o. If now S is the surface obtained from bdrK° by inversion with respect to S^{n-1}, then o is an equichordal point of S, that is, all chords of S passing through o have the same length. These considerations are of interest with regard to the famous unsolved problem of whether there exist plane convex bodies with more than one equichordal point. A discussion of this problem and a related problem about equireciprocal points is given by Klee [1969a, 1979]. Petty and Crotty [1970] have established the existence of Minkowski spaces of arbitrary dimension in which there are convex bodies with exactly two equichordal points (see Section 2 for another result on equichordal points obtained by Petty and Crotty). The surface S described above is for $n = 2$ the *pedal curve* of K. See Kelly [1957] for some properties of pedal curves of plane convex bodies of constant width.

Fujiwara [1914] extended the concept of constant width to space curves in the following way. Let C be a continuously differentiable closed curve in E^3, and suppose that the normal plane at each point $x \in C$ intersects C again in a single point y such that the length of [x,y] is a number b independent of x. Then C is a *space curve of constant width* b. Blaschke has shown, in answer to a question of Fujiwara, that such a curve always lies on the boundary of a convex body of constant width b (see Bonnesen-Fenchel [1934], p. 139, for early references). A short proof of this is given by Bückner [1936]. Bückner [1937b] also proves that any space curve of constant width b has length at least πb. Sancho de San Román [1949] and Müller [1976] have both studied general classes of space curves that include space curves of constant width as special cases.

Robertson [1964, 1967] introduced the concept of a transnormal manifold as a generalization of curves of constant width. Let V be a k-dimensional connected differentiable manifold imbedded in E^n. For each $x \in V$ let P(x) be the $(n - k)$-plane orthogonal to V at x. Then V is *r-transnormal* if for each $x \in V$ the normal plane P(x) intersects V in exactly r points and is orthogonal to V at each of those intersection points. The case $k = 1, n = 3, r = 2$ corresponds to a smooth curve C imbedded in E^3 in such a way that for each $x \in C$ the normal plane P(x) intersects C again in a single point y, with $P(y) = P(x)$. Any such curve is a space curve of constant width in the sense of Fujiwara, since the definition implies that the length of [x,y] is independent of $x \in C$. On the other hand, there do exist space curves of constant width in the sense of Fujiwara that are not transnormal. For a discussion of this and other properties of transnormal curves, see Wegner [1972, 1975]. These articles also give references to further work of Wegner on transnormal manifolds. Irwin [1967] has shown that there exist smooth simple closed curves in E^n, $n > 3$, that are r-transnormal with $r > 2$, while this is impossible when $n = 3$. Nádeník has taken another geometrically rich approach to the study of curves of constant width in even dimensional Euclidean spaces. Nádeník [1967] gives an excellent survey and extensive bibliography. For a further investigation along these lines and some additional references, see Šmakal [1973].

In a related vein, Varela Gil [1944] and Vivanti [1946] have investigated plane closed curves, possible self-intersecting, possessing a kind of double normal property. Sancho de San Román [1955] deals with a notion of constant affine width for closed curves in E^3.

Finally we consider work that has been done on generalizing the concept of constant width to non-Euclidean spaces. A subset K of the unit sphere S^2 is spherically convex if the shorter arc of any great circle joining two points of K is contained in K. Such a set has constant width β if it can be rotated on the sphere through 360° inside a fixed lune of angle β while always touching the two boundary arcs of the lune at their midpoints. Blaschke [1915] has shown that if $K \subset S^2$ is spherically convex with constant width β, then

$$(11.2) \quad p = k \tan\left(\frac{\beta}{2}\right),$$

where p is the length of bdrK and k its total geodesic curvature. Barbier's theorem follows from a similar formula for sets on a sphere of radius R by letting $R \to \infty$ in an appropriate fashion. Blaschke also derives an analogue of the Blaschke-Lebesgue theorem (see Section 7) for spherical sets of constant width. Bodies of constant width in a three-dimensional spherical space are defined analogously. Blaschke obtains for such sets relationships connecting the total mean curvature, surface area, and volume, analogous to identities in Euclidean space (see Section 6). Several of Blaschke's arguments involve an imaginative, albeit not completely justified, application of the formulas for geometric invariants associated with outer parallel bodies. Santaló [1950] proves the analogous relationships for sets of constant width in n-dimensional spherical and hyperbolic spaces (see Section 6 for the corresponding results in Euclidean spaces).

Santaló [1944b] derives an integral formula from which one can calculate the perimeter of a spherically convex set using an appropriately defined width function. This yields formula (11.2) in case the width is constant. A limiting argument leads Santaló to the following interesting question about plane convex bodies. For each $x \in K$, where $K \in \mathbf{C}^2$, let $n(x)$ be the number of normals to bdr$K$ passing through x. The average number of normals passing through a point of K is then

$$g(K) = \frac{1}{a(K)} \int_K n(x)\,dx,$$

where the integral is with respect to plane Lebesgue measure. Santaló notes that if $K \in \mathbf{W}^2$ then

$$g(K) \leqq 2\pi/(\pi - \sqrt{3}) < 4.46,$$

and asks for the least upper bound of $g(K)$ for $K \in \mathbf{C}^2$. See Santaló [1944a] for further results on spherical sets of constant width.

Santaló [1945] treats sets of constant width in hyperbolic geometry. He obtains an integral formula for the perimeter of any convex subset of the hyperbolic plane of Gaussian curvature -1 analogous to his formula in the case of S^2. From this he derives an analogue of (11.2) for sets of constant width β, namely

$$(11.3) \quad p = k \tanh\left(\frac{\beta}{2}\right),$$

where, as before, p is the perimeter and k the total geodesic curvature. Fillmore [1970] has established an analogue of Barbier's theorem for sets of constant width that are *horocycle convex*, rather than geodesically convex, in a hyperbolic plane of Gaussian curvature $-c^2$. Barbier's theorem in E^2 follows from this in the limit $c \to 0$.

References

The abbreviations Jbuch, MR, Zbl, refer to reviews of the papers in Jahrbuch über die Fortschritte der Mathematik, Mathematical Reviews, Zentralblatt der Mathematik.

Alexander, R.

1975 Problem 15. The geometry of metric and linear spaces (Proc. Conf., Michigan State Univ., East Lansing, Mich., 1974) p. 240. Lecture Notes in Math. Vol. 490, Springer, Berlin 1975. MR 53, 6419.

Auerbach, H.

1938 Sur un problème de M. Ulam concernant l'équilibre des corps flottants. Studia Math. **7** (1938), 121–142. Zbl. 18, p. 175.

Beck, A., M.N. Bleicher

1971 Packing convex sets into similar sets. Acta Math. Acad. Sci. Hung. **22** (3–4) (1971/72), 283–303. MR 45, 7613.

Benson, R.V.
1966 Euclidean geometry and convexity. McGraw-Hill Co., New York-Toronto-London 1966. MR 35, 844.
Beretta, L., A. Maxia
1940a Insiemi convessi e orbiformi. Univ. Roma e Ist. Naz. Alta Mat. Rend. Mat. (5) **1** (1940), 1–64. MR 1, p. 264.
1940b Sui vertici di un'orbiforme e sulle cuspidi della sua curva media. Boll. Un. Mat. Ital. (2) **2** (1940) 216–220. MR 2, p. 261.
Berwald, L.
1937 Integralgeometrie. XXV. Über die Körper konstanter Helligkeit. (Bemerkungen zu der vorstehenden Abhandlung.). Math. Z. **42** (1937), 737–738. Zbl. 16, p. 374.
Besicovitch, A.S.
1951 Measures of asymmetry of convex curves (II): Curves of constant width. J. London Math. Soc. **26** (1951), 81–93. MR 12, p. 850.
1961 A problem on a circle. J. London Math. Soc. **36** (1961), 241–244. MR 23, A1285.
1963a Minimum area of a set of constant width. Proc. Sympos. Pure Math., Vol. VII, pp. 13–14. Amer. Math. Soc., Providence, R.I., 1963. MR 27, 1878.
1963b On semicircles inscribed into sets of constant width. Proc. Sympos. Pure Math., Vol. VII, pp. 15–18. Amer. Math. Soc., Providence, R.I., 1963. MR 27, 2912.
Blanc, E.
1944 Sur une généralisation des domaines d'épaisseur constante. C.R. Acad. Sci. Paris **219** (1944), 662–663. MR 7, p. 475.
Blaschke, W.
1915 Einige Bemerkungen über Kurven and Flächen von konstanter Breite. Ber. d. Verh. d. Sächs. Akad. Leipzig **67** (1915), 290–297. Jbuch. 45, p. 731.
1956 Kreis und Kugel. Zweite Aufl., Walter de Gruyter Co., Berlin 1956. MR 14, p. 1123.
Blatter, C.
1981 Über Kurven konstanter Breite. Elem. Math. **36** (1981), 105–115.
Boltjanskii, V.G.
1970 The partitioning of plane figures into parts of smaller diameter. (Russian). Colloq. Math. **21** (1970), 253–263. MR 42, 2366.
Boltyanskii, V.G., I.T. Gohberg
1980 The Decomposition of Figures into Smaller Parts. (Translated and adapted from the Russian by H. Christoffers and T.P. Branson). Univ. of Chicago Press, Chicago, Ill.-London, 1980. MR 81a:52001.
Bonnesen, T., W. Fenchel
1934 Theorie der konvexen Körper. Ergebn. d. Math. u. ihrer Grenzgeb. Bd. 3, J. Springer Verl., Berlin 1934. Zbl. 8, p. 77.
Bontrager, M.M.
1969 Characterizations and properties of sets of constant width. Ed. D. Dissertation, Oklahoma State University, Stillwater, 1969.
Bose, R.C., S.N. Roy
1935 Some properties of the convex oval with reference to its perimeter centroid. Bull. Calcutta Math. Soc. **27** (1935), 79–86. Zbl. 14, p. 127.
Buchman, E.O., F.A. Valentine
1982 Any new Helly numbers? Amer. Math. Montly **89** (1982), 370–375.
Bückner, H.
1936 Über Flächen von fester Breite. Jber. Deutsch. Math.-Verein. **46** (1936), 96–139. Zbl. 14, p. 361.
1937a Ergänzungen zu meiner Arbeit: "Über Flächen von fester Breite". (Dieser Jahresber. d. D.M.V. **46**, Heft 5–8). Jber. Deutsch. Math.-Verein. **47** (1937), 55. Zbl. 16, p. 228.
1937b Zur Differenzialgeometrie der Kurven und Flächen fester Breite. Schr. d. Königsberg. gel. Ges. Jg. 14, H.1; Halle a. d. S., Max Niemeyer 1937, 22 pp. Zbl. 17, p. 188.
Burke, J.F.
1966 A curve of constant diameter. Math. Mag. **39** (1966), 84–85. MR 33, 1794.
Chakerian, G.D.
1965 The affine image of a convex body of constant breadth. Israel J. Math. **3** (1965), 19–22. MR 32, 382.
1966 Sets of constant width. Pacific J. Math. **19** (1966), 13–21. MR 34, 4986.

1967 Sets of constant relative width and constant relative brightness. Trans. Amer. Math. Soc. **129** (1967), 26–37. MR 35, 3545.
1969 Intersection and covering properties of convex sets. Amer. Math. Monthly **76** (1969), 753–766. MR 40, 3433.
1974 A characterization of curves of constant width. Amer. Math. Monthly **81** (1974), 153–155. MR 49, 1321.
1982 Mixed volumes and geometric inequalities. Convexity and Related Combinatorial Geometry. Proceedings of the Second University of Oklahoma Conference, pp. 57–62. Marcel Dekker, Inc., New York and Basel 1982.

Chakerian, G.D., M.S. Klamkin
1973 Minimal covers for closed curves. Math. Mag. **46** (1973), 55–61. MR 47, 2496.

Chakerian, G.D., G.T. Sallee
1969 An intersection theorem for sets of constant width. Duke Math. J. **36** (1969), 165–170. MR 39, 3392.

Chakerian, G.D., S.K. Stein
1966 Bisected chords of a convex body. Arch. Math. (Basel) **17** (1966), 561–565. MR 34, 6635.

Cooke, G.
1964 A problem in convexity. J. London Math. Soc. **39** (1964), 370–378. MR 29, 521.

Danzer, L.
1957 Über die maximale Dicke der ebenen Schnitte eines konvexen Körpers. Arch. Math. (Basel) **8** (1957), 314–316. MR 19, p. 1074.
1963 A characterization of the circle. Proc. Sympos. Pure Math., Vol. VII, pp. 99–100. Amer. Math. Soc., Providence, R.I. 1963. MR 27, 4141b.

Danzer, L., B. Grünbaum, V.Klee
1963 Helly's theorem and its relatives. Proc. Sympos. Pure Math., Vol. VII, pp. 101–180. Amer. Math. Soc., Providence, R.I. 1963. MR 28, 524.

Debrunner, H.
1955 Zu einem massgeometrischen Satz über Körper konstanter Breite. Math. Nachr. **13** (1955), 165–167. MR 17, p. 294.

Dinghas, A.
1940 Verallgemeinerung eines Blaschkeschen Satzes über konvexe Körper konstanter Breite. Rev. Math. Union Interbalkan. **3** (1940), 17–20. MR 2, p. 261.

Dubois, J.
1977 Sur la convexité et ses applications. Ann. Sci. Math. Québec **1** (1977), no. 1, 7–31. MR 58, 24002.

Eggleston, H.G.
1952a Measures of asymmetry of convex curves of constant width and restricted radii of curvature. Quart. J. Math., Oxford Ser. (2) **3** (1952), 63–72. MR 13, p. 768.
1952b A proof of Blaschke's theorem on the Reuleaux triangle. Quart. J. Math., Oxford Ser. (2) **3** (1952), 296–297. MR 14, p. 496.
1955a Sets of constant width contained in a set of given minimal width. Mathematika **2** (1955), 48–55. MR 17, p. 185.
1955b Covering a three-dimensional set with sets of smaller diameter. J. London Math. Soc. **30** (1955), 11–24. MR 16, p. 743.
1957 Problems in Euclidean Space: Applications of convexity. International Series of Monographs on Pure and Applied Mathematics, Vol. V. Pergamon Press, New York-London-Paris-Los Angeles, 1957. MR 23, A3228.
1958a Figures inscribed in convex sets. Amer. Math. Montly **65** (1958), 76–80. MR 20, 4235.
1958b Convexity. Cambridge Tracts in Mathematics and Mathematical Physics, No. 47. Cambridge Univ. Press, New York, 1959. MR 23, A2123.
1959 Squares contained in plane convex sets. Quart. J. Math., Oxford Ser. (2) **10** (1959), 261–269. MR 22, 8436.
1963 Minimal universal covers in E^n. Israel J. Math. **1** (1963), 149–155. MR 28, 4432.
1965 Sets of constant width in finite dimensional Banach spaces. Israel J. Math. **3** (1965), 163–172. MR 34, 583.

Eggleston, H.G., S.J. Taylor
1952 On the size of equilateral triangles which may be inscribed in curves of constant width. J. London Math. Soc. **27** (1952), 438–448. MR 14, 401.

Elkington, G.B., J. Hammer
1976 Lattice points in a convex set of given width. J. Austral. Math. Soc. Ser. A **21** (1976), no. 4, 504–507. MR 54, 3594.
Ewald, G.
1966 Von Klassen konvexer Körper erzeugte Hilberträume. Math. Ann. **162** (1965/66), 140–146. MR 32, 8255.
Ewald, G., G.C. Shephard
1966 Normed vector spaces consisting of classes of convex sets. Math. Z. **91** (1966), 1–19. MR 32, 4597.
Falconer, K.J.
1977a Applications of outwardly simple line families to plane convex sets. Math. Proc. Cambridge Philos. Soc. **82** (1977), no. 3, 353–356. MR 56, 3744.
1977b A characterization of plane curves of constant width. J. London Math. Soc. (2) **16** (1977), no. 3, 536–538. MR 57, 1272.
1981 Singularities of sets of constant width. Geometriae Dedicata **11** (1981), 187–193.
Fejes Tóth, G.
1982 New results in the theory of packing and covering. In convexity and its applications, ed. by Gruber and Wills, pp. 318–359 (Birkhäuser, Basel 1983).
Fejes Tóth, L.
1967 On the number of equal discs that can touch another of the same kind. Studia Sci. Math. Hungar. **2** (1967), 363–367. MR 36, 4440.
1971 Über Scheiben mit richtungsinvarianter Packungsdichte. Elem. Math. **26** (1971), 58–60. MR 45, 1045.
Fillmore, J.P.
1969 Symmetries of surfaces of constant width. J. Differential Geometry **3** (1969), 103–110. MR 40, 858.
1970 Barbier's theorem in the Lobachevski plane. Proc. Amer. Math. Soc. **24** (1970), 705–709. MR 40, 6365.
Firey, W.J.
1959 A note on a theorem of Knothe. Michigan Math. J. **6** (1959), 53–54. MR 21, 319.
1960 Isoperimetric ratios of Reuleaux polygons. Pacific J. Math. **10** (1960), 823–829. MR 22, 4014.
1965a Lower bounds for volumes of convex bodies. Arch. Math. (Basel) **16** (1965), 69–74. MR 31, 5152.
1965b The brightness of convex bodies. Tech. Report No. 19, Oregon State Univ., Corvallis, Oregon 1965.
1967 Blaschke sums of convex bodies and mixed bodies. Proc. Colloquium on Convexity (Copenhagen, 1965), pp. 94–101. Københavns Univ. Math. Inst., Copenhagen, 1967. MR 36, 3231.
1970 Convex bodies with constant outer p-measure. Mathematika **17** (1970), 21–27. MR 42, 2367.
Firey, W.J., H. Groemer
1964 Convex polyhedra which cover a convex set. J. London Math. Soc. **39** (1964), 261–266, MR 29, 2707.
Firey, W.J., B. Grünbaum
1964 Addition and decomposition of convex polytopes, Israel J. Math. **2** (1964), 91–100. MR 30, 5218.
Fischer, H.J.
1936 Kurven, in denen ein Drei-oder Vieleck so herumbewegt werden kann, dass seine Ecken die Kurve durchlaufen. Deutsche Math. **1** (1936), Zbl. 14, p. 409.
Florian, A.
1978 On the permeability of layers of discs. Studia Sci. Math. Hungar. **13** (1978), 125–132.
1980 Über die Durchlässigkeit einer Schicht konvexer Scheiben. Studia Sci. Math. Hungar. **15** (1980), 201–213.
Focke, J.
1969a Symmetrische n-Orbiformen kleinsten Inhalts. Acta Math. Acad. Sci. Hungar. **20** (1969), 39–68. MR 38, 6463.
1969b Die beste Ausbohrung eines regulären n-Ecks (English and Russian summaries). Z. Angew. Math. Mech. **49** (1969), 235–248. MR 39, 6167.
Focke, J., B. Gensel
1971 n-Orbiformen maximaler Randknickung. Beiträge Anal. Heft 2 (1971), 7–16. MR 46, 9862.

Freilich, G.
1951 On sets of constant width. Proc. Amer. Math. Soc. **2** (1951), 92–96. MR 12, p. 850.
Fujiwara, M.
1914 On space curves of constant breadth. Tôhoku Math. J. **5** (1914), 180–184. Jbuch. 45, p. 731.
1915 Über die einem Vielecke eingeschriebenen und umdrehbaren konvexen geschlossenen Kurven. Sci. Rep. Tôhoku Univ. **4** (1915), 43–55. Jbuch. 45, p. 1348.
1916 Über die Mittelkurve zweier geschlossener konvexer Kurven in bezug auf einen Punkt. Tôhoku Math. J. **10** (1916), 99–103. Jbuch. 46, p. 1117.
Gale, D.
1953 On inscribing n-dimensional sets in a regular n-simplex. Proc. Amer. Math. Soc. **4** (1953), 222–225. MR 14, p. 787.
Ganapathi, P.
1934a On a certain class of ovals. Math. Z. **38** (1934), 687–690. Zbl. 9, p. 124.
1934b The isoperimetric deficit of a convex closed curve. Math. Z. **38** (1934), 691–694. Zbl. 9, p. 124.
1935 A property of ovaloids of constant breadth. Tôhoku Math. J. **40** (1935), 421–424. Zbl. 11, p. 318.
Garay de Pablo, J., F. Clavera Emperador
1964 On polygons of constant affine width (Spanish). Rev. Acad. Ci. Zaragoza (2) **19** (1964), 75–78. MR 31, 2661.
1967 Consideraciones acerca de los poligonos de anchura afin constante. Actas 5 reun. anual mat. esp., Valencia 1964, pp. 94–98. Publ. Inst. "Jorge Juan" Mat., Madrid, 1967. MR 39, 7503.
Gardner, M.
1963 Mathematical games. Curves of constant width, one of which makes it possible to drill square holes. Scientific American, Feb. 1963, 148–156.
Gere, B.H., D. Zupnik
1943 On the construction of curves of constant width. J. Math. Phys. Mass. Inst. Tech. **22** (1943), 31–36. MR 5, p. 10.
Godbersen, C.
1938 Der Satz vom Vektorbereich in Räumen beliebiger Dimension. Diss. Göttingen 1938, 49 pp. Zbl. 20, p. 77.
Goldberg, M.
1948 Circular-arc rotors in regular polygons. Amer. Math. Monthly **55** (1948), 393–402. MR 10, 205.
1952 Rotors in spherical polygons. J. Math. Physics **30** (1952), 235–244. MR 13, p. 577.
1954 Rotors within rotors. Amer. Math. Monthly **61** (1954), 166–171. MR 15, p. 740.
1956 Basic rotors in spherical polygons. J. Math. Phys. **34** (1956), 322–327. MR 17, p. 655.
1957 Trammel rotors in regular polygons. Amer. Math. Monthly **64** (1957), 71–78. MR 18, p. 668.
1958 Rotors tangent to n fixed circles. J. Math. Phys. **37** (1958), 69–74. MR 20, 1438.
1959 Intermittent rotors. J. Math. Phys. **38** (1959/60), 135–140. MR 21, 5936.
1960 Rotors in polygons and polyhedra. Math. Comput. **14** (1960), 229–239. MR 22, 5934.
Goodey, P.R.
1974 Intersection of circles and curves of constant width. Math. Ann. **208** (1974), 49–58. MR 49, 6028.
1982 Intersections of convex bodies and surfaces. III. Geometry Symposium, Siegen, 14–17 July 1982. Mimeographed abstracts.
Goodey, P.R., M.M. Woodcock
1978 A characterization of sets of constant width. Math. Ann. **238** (1978), no. 1, 15–21. MR 80b:52007.
Gray, C.G.
1972 Solids of constant breadth. Math. Gaz. **56** (1972), no. 398, 289–292. MR 58, 7392.
Groemer, H.
1962 Eine Bemerkung über die Überdeckung des Raumes mit unsymmetrischen konvexen Körpern. Monatsh. Math. **66** (1962), 410–416. MR 30, 4192.
1964 Über die Quermassintegrale von konvexen Körpern konstanter Breite. (Unpublished).
1982 Extremal convex sets. To appear.
Gruber, P.M.
1983 Approximation of convex bodies. In Convexity and its applications, ed. by Gruber and Wills, pp. 131–162 (Birkhäuser, Basel 1983).

Gruber, P.M., R. Schneider
1979 Problems in geometric convexity. Contributions to Geometry (Proc. Geom. Sympos., Siegen 1978), pp. 255–278, Birkhäuser, Basel 1979. MR 82d:52001.
Grünbaum, B.
1963a Measures of symmetry for convex sets. Proc. Sympos. Pure Math., Vol. VII, pp. 233–270. Amer. Math. Soc., Providence, R.I. 1963. MR 27, 6187.
1963b Borsuk's problem and related questions. Proc. Sympos. Pure Math., Vol. VII, pp. 271–284. Amer. Math. Soc., Providence, R.I. 1963. MR 27, 4134.
1965 Self-circumference of convex sets. Colloq. Math. **13** (1965), 55–57. MR 30, 2396.
Hammer, J.
1966 Problem 5368, Amer. Math. Monthly **73** (1966), 206.
1977 Unsolved problems concerning lattice points. Research Notes in Mathematics, vol. 15. Pitman, London-San Francisco-Melbourne, 1977. MR 56, 16515.
Hammer, P.C.
1951 Convex bodies associated with a convex body, Proc. Amer. Math. Soc. **2** (1951), 781–793. MR 14, p. 678.
1954 Diameters of convex bodies. Proc. Amer. Math. Soc. **5** (1954), 304–306. MR 15, p. 819.
1955 Constant breadth curves in the plane. Proc. Amer. Math. Soc. **6** (1955), 333–334. MR 16, p. 950.
1963 Convex curves of constant Minkowski breadth. Proc. Sympos. Pure Math., Vol. VII, pp. 291–304. Amer. Math. Soc., Providence, R.I., 1963. MR 27, 4136.
Hammer, P.C., T.J. Smith
1964 Conditions equivalent to central symmetry of convex curves. Proc. Cambridge Philos. Soc. **60** (1964), 779–785. MR 30, 506.
Hammer, P.C., A. Sobczyk
1953a Planar line families I. Proc. Amer. Math. Soc. **4** (1953), 226–233. MR 14, p. 787.
1953b Planar line families II. Proc. Amer. Math. Soc. **4** (1953), 341–349. MR 15, p. 149.
Hansen, H.C.
1978 Et isoperimetrisk problem. Nordisk Math. Tidskr. **25/26** (1978), no. 3–4, 181–184, 210. MR 81a:52011.
Heil, E.
1967 Abschätzungen für einige Affininvarianten konvexer Kurven. Monatsh. Math. **71** (1967), 405–423. MR 37, 5796.
1970 Verschärfungen des Vierscheitelsatzes und ihre relativ-geometrischen Verallgemeinerungen. Math. Nachr. **45** (1970), 227–241. MR 42, 5158.
1978 Kleinste konvexe Körper gegebener Dicke. Preprint Nr. 453, Technische Hochschule Darmstadt, 1978.
1979 Existenz eines 6-Normalenpunktes in einem konvexen Körper. Arch. Math. (Basel) **32** (1979), no. 4, 412–416. MR 80j:52002a.
1980 Korrektur zu: "Existenz eines 6-Normalenpunktes in einem konvexen Körper". Arch. Math. (Basel) **33** (1979/80), no. 5, 496. MR 80j:52002b.
Heppes, A.
1959 On characterization of curves of constant width. Mat. Lapok **10** (1959), 133–135 (Hungarian, Russian and English summaries). MR 22, 1846.
Herda, H.
1974 A characterization of circles and other closed curves. Amer. Math. Monthly **81** (1974), 146–149. MR 48, 12290.
Hirakawa, J.
1933 On a characteristic property of the circle. Tôhoku Math. J. **37** (1933), 175–178. Zbl. 7, p. 318.
1935a On the relative breadth. Proc. Phys.-Math. Soc. Jap. III, **17** (1935), 137–144. Zbl. 11, p. 223.
1935b The affine breadth and its relation to the relative breadth. Japan. J. Math. **12** (1935), 43–50. Zbl. 12, p. 272.
Hortobágyi, I.
1976a Über die Anzahl der Scheiben konstanter Breite, die eine gegebene Scheibe Konstanter Breite berühren. Ann. Univ. Sci. Budapest. Eötvös Sect. Math. **19** (1976), 45–48 (1977). MR 58, 24004.
1976b Über die Durchlässigkeit einer aus Scheiben konstanter Breite bestehenden Schicht. Studia Sci. Math. Hungar. **11** (1976), no. 3–4, 383–387 (1978). MR 80c:52014.

Hoschek, J.
1974 Räumliche Zindlerkurven. Math. Balkanica **4** (1974), 253–260. MR 51, 9000.
Hsiao, E.K.
1959 Notes on differential geometry. Amer. Math. Monthly **66** (1959), 898–899. MR 21, 7540.
Irwin, M.C.
1967 Transnormal circles. J. London Math. Soc. **42** (1967), 545–552. MR 35, 4822.
Kallay, M.
1974 Reconstruction of a plane convex body from the curvature of its boundary. Israel J. Math. **17** (1974), 149–161. MR 50, 3110.
1975 The extremal bodies in the set of plane bodies with a given width function. Israel J. Math. **22** (1975), 203–207. MR 53, 9033.
Kameneckii, I.M.
1947 Solution of a geometric problem of L. Ljusternik. (Russian). Uspehi Mat. Nauk **2**, no. 2 (18), (1947), 199–202. MR 10, p. 60.
Kárteszi, F.
1959a Sur les figures convexes enveloppées par des carrées. Köz. Mat. Lapok **18** (1959), 1–6, 33–37 (Hungarian). MR 21, 5935.
1959b Über ein elementargeometrisches Problem. Ann. Univ. Sci. Budapest. Eötvös, Sect. Math. **2** (1959), 49–60. MR 22, 5933.
Kearsley, M.J.
1952 Curves of constant diameter. Math. Gaz. **36** (1952), 176–179. MR 14, p. 198.
Kelly, P.J.
1949 On Minkowski bodies of constant width. Bull. Amer. Math. Soc. **55** (1949), 1147–1150. MR 11, p. 386.
1957 Curves with a kind of constant width. Amer. Math. Monthly **64** (1957), 333–336. MR 19, p. 1073.
Khassa, D.S.
1977 Relation between maximal chords and symmetry for convex sets. J. London Math. Soc. **15** (1977), 541–546. MR 56, 3745.
Klee, V.
1969a Can a plane convex body have two equichordal points? Amer. Math. Monthly **76** (1969), 54–55.
1969b Is a body spherical if its HA-measurements are constant? Amer. Math. Monthly **76** (1969), 539–542.
1971 What is a convex set? Amer. Math. Monthly **78** (1971), 616–631. MR 44, 3202.
1979 Some unsolved problems in plane geometry. Math. Mag. **52** (1979) no. 3, 131–145. MR 80m:52006.
Klötzler, R.
1975 Beweis einer Vermutung über n-Orbiformen kleinsten Inhalts (English and Russian summaries). Z. Angew. Math. Mech. **55** (1975), no. 10, 557–570. MR 53, 3884.
Knuth, E.
1969 On a problem of Klee, Acta Math. Acad. Sci. Hungar. **20** (1969), 169–177. MR 39, 861.
Krautwald, W.
1980 Kennzeichnungen der affinen Bilder von Körpern konstanter Breite. J. Geometry **15**/2 (1980), 140–148. Zbl. 457, 52001.
Kubota, T.
1948 Einige Ungleichungen für die Eilinien und Eiflächen. Proc. Japan Acad. **24**, no. 7–8 (1948), 1–3. MR 14, p. 678.
Kubota, T., D. Hemmi
1953 Some problems of minima concerning the oval. J. Math. Soc. Japan **5** (1953), 372–389. MR 15, p. 981.
Kuiper, N.H.
1964 Double normals of convex bodies. Israel J. Math. **2** (1964), 71–80. MR 30, 4191.
Kurbanov, N.
1975 Curves of constant width. (Russian. Tajiki summary). Izv. Akad. Nauk Turkmen. SSR Ser. Fiz.-Tehn. Him. Geol. Nauk 1975, no. 6, 3–8. MR 54, 3580.
Lay, S.R.
1982 Convex sets and their applications. Pure and applied mathematics, "A Wiley-Interscience publication". J. Wiley and Sons, New York-Chichester-Brisbane-Toronto-Singapore 1982.

Lenz, H.
1955 Zur Zerlegung von Punktmengen in solche kleineren Durchmessers. Arch. Math. (Basel) **6** (1955), 413–416. MR 17, 887.
1956 Über die Bedeckung ebener Punktmengen durch solche kleineren Durchmessers. Arch. Math. (Basel) **7** (1956), 34–40. MR 17, p. 888.
Lewis, J.E.
1975 A Banach space whose elements are classes of sets of constant width. Canad. Math. Bull. **18** (1975), no. 5, 679–689. MR 58, 12289.
Loomis, P.
1980 Covering among constant relative width bodies in the plane. Ph.D. Dissertation, University of California, Davis, 1980.
Lutwak, E.
1979 On isoperimetric inequalities related to a problem of Moser. Amer. Math. Monthly **86** (1979), no. 6, 476–477. MR 80c:52010.
1980 Isoperimetric inequalities involving bisectors, Bull. London Math. Soc. **12** (1980), 289–295. MR 81k:52024.
1982 On packing curves into circles. Convexity and Related Combinatorial Geometry. Proceedings of the second University of Oklahoma Conference, pp. 107–111. Marcel Dekker, Inc., New York and Basel 1982.
Makuha, N.P.
1966 A connection between the inner and outer diameters of a general closed convex surface. (Russian). Ukrain. Geometr. Sb. Vyp. 2 (1966), 49–51. MR 35, 3550.
Mayer, A.E.
1934a Der Inhalt der Gleichdicke. Abschätzungen für ebene Gleichdicke. Math. Ann. **110** (1934), 97–127. Zbl. 9, p. 321.
1934b Über Gleichdicke kleinsten Flächeninhalts. Anz. Akad. Wiss. Wien Nr. 7 (Sonderdruck), 4 pp. (1934). Zbl. 8, p. 404.
1934c Eine Überkonvexität. Math. Z. **39** (1934), 512–531. Zbl. 10, p. 270.
Meissner, E.
1911 Über Punktmengen konstanter Breite. Vierteljahresschr. naturforsch. Ges Zürich **56** (1911), 42–50. Jbuch. 42, p. 91.
1918 Über die durch reguläre Polyeder nicht stützbaren Körper. Vierteljahresschr. naturforsch. Ges. Zürich **63** (1918), 544–551. Jbuch. 46, p. 1117.
Melzak, Z.A.
1960 A note on sets of constant width. Proc. Amer. Math. Soc. **11** (1960), 493–497. MR 23, A2124.
1963 A property of plane sets of constant width. Canad. Math. Bull. **6** (1963), 409–415. MR 28, 3368.
1965 Problems connected with convexity. Canad. Math. Bull. **8** (1965), 565–573. MR 33, 4781.
1968 More problems connected with convexity. Canad. Math. Bull. **11** (1968), 489–494. MR 38, 3767.
Miller, K.G.
1969 Characterizations of the circle and the circular disk. Senior thesis, Macalester College, 1969.
Minkowski, H.
1904 Dichteste gitterförmige Lagerung kongruenter Körper. Nachrichten d. K. Ges. d. Wiss. zu Göttingen, Math.-phys. Klasse 1904, 311–355 = Ges. Abh. Bd. 2, 1–42, Leipzig 1911.
Minoda, T.
1939a On the minimum of the perimeter of the circumrevolvable curve about an oval. Tôhoku Math. J. **45** (1939), 369–371. Zbl. 21, p. 67.
1939b On the circle circumscribed about and the circle inscribed in an oval, and the sphere circumscribed about and the sphere inscribed in an ovaloid. Tôhoku Math. J. **45** (1939), 372–376. Zbl. 21, p. 67.
1941 On certain ovals. Tôhoku Math. J. **48** (1941), 312–320. MR 10, p. 320.
Müller, H.R.
1971 Trochoidenhüllbahnen und Rotationskolbenmaschinen. Selecta Mathematica, III, pp. 119–138. Heidelberger Taschenbucher, 86. Springer, Berlin 1971. MR 58, 30783.
1976 Kurven konstanter Breite und Schiebkurven. Abh. Braunschweig. Wiss. Gesellsch. **26** (1976), 119–122. MR 55, 8965.

Nádeník, Z.
1967 Les courbes gauches de largeur constante. Czech. Math. J. **17** (92) (1967), 540–549. MR 36, 3242.
Ohmann, D.
1952 Extremalprobleme für konvexe Bereiche der euklidischen Ebene. Math. Z. **55** (1952), 346–352. MR 14, p. 76.
Petermann, E.
1967 Some relations between breadth and mixed volumes of convex bodies in R^n. Proc. Colloquium on Convexity (Copenhagen, 1965), pp. 229–233. Københavns Univ. Inst. Copenhagen, 1967. MR 35, 6033.
Peterson, B.B.
1972 Do self-intersections characterize curves of constant width? Amer. Math. Monthly **79** (1972), 505–506. Zbl. 257, 52004.
1973 Intersection properties of curves of constant width. Illinois J. Math. **17** (1973), 411–420. MR 47, 9418.
Petty, C.M.
1955 On the geometry of the Minkowski plane. Riv. Math. Univ. Parma. **6** (1955), 269–292. MR 18, p. 760.
1972 Isoperimetric problems. Proceedings of the Conference on Convexity and Combinatorial Geometry (Univ. of Oklahoma, Norman, Okla., 1971), (1972), 26–41. MR 50, 14499.
1974 Geominimal surface area. Geom. Dedicata **3** (1974), 77–97. MR 48, 12313.
Petty, C.M., J.M. Crotty
1970 Characterization of spherical neighborhoods. Canad. J. Math. **22** (1970), 431–435. MR 41, 2538.
Petty, C.M., J.R. McKinney
1974 Convex bodies with circumscribing boxes of constant volume. To appear.
Porcu, L.
1963 Proprietá metriche e affini di notevoli classi di ovali. Ist. Lombardo Accad. Sci. Lett. Rend. A **97** (1963), 899–929. MR 30, 2400.
Rademacher, H., O. Toeplitz
1957 The Enjoyment of Mathematics. Princeton Univ. Press, Princeton, N.J. 1957 (Translation of "Von Zahlen und Figuren", Springer, Berlin 1933). MR 18, p. 454.
Reidemeister, K.
1921 Über Körper konstanten Durchmessers. Math. Z. **10** (1921), 214–216. Jbuch. 48, p. 835.
Reinhardt, K.
1922 Extremale Polygone gegebenen Durchmessers. Jahresber. Deutsch. Math.-Verein. **31** (1922), 251–270.
Robertson, S.A.
1964 Generalized constant width for manifolds. Michigan Math. J. **11** (1964), 97–105. MR 29, 2815.
1967 On transnormal manifolds. Topology **6** (1967), 117–123. MR 34, 6793.
Rogers, C.A.
1971 Symmetrical sets of constant width and their partitions. Mathematika **18** (1971), 105–111. MR 44, 7432.
Sachs, H.
1958 Über eine Klasse isoperimetrischer Probleme. I, II, III. Wiss. Z. Martin-Luther-Univ. Halle-Wittenberg. Math.-Nat. Reihe **8** (1958/59), 121–126, 127–134, 345–350. MR 21, 7482/7483.
Sallee, G.T.
1969 The maximal set of constant width in a lattice. Pacific J. Math. **28** (1969), 669–674. MR 39, 2069.
1970a Maximal areas of Reuleaux polygons. Canad. Math. Bull. **13** (1970), 175–179. MR 42, 2368.
1970b Reuleaux polytopes. Mathematika **17** (1970), 315–328. MR 45, 5872.
1974 An inequality for curves. Unpublished manuscript.
Sancho de San Román, J.
1949 On closed skew curves, especially of constant width. (Spanish). Memorias de Matematica del Instituto "Jorge Juan" no. 10 (1949), 67 pp. MR 11, p. 201.
1955 Curvas alabeadas de anchura afin constante. Collect. Math. **8** (1955–1956), 85–98. MR 18, p. 760.

1956 Sobre un nuevo concepto de anchura de óvalos, invariante afin. Mat. Hisp.-Amer. (4) **16** (1956), 151–171. MR 18, p. 505.
1957 Un nuevo concepto de anchura afin de cuerpos ovales. Acad. Ci. Madrid **51** (1957), 229–244. MR 20, 2666.
1965 Poligonos de anchura afin constante. Fourth Annual Meeting Spanish Mathematicians, pp. 107–108. Univ. Salamanca, Salamanca 1965. MR 33, 1795.

Santaló, L.A.
1944a Propriedades de las figuras convexas sobre la esfera. Math. Notae **4** (1944), 11–40. MR 6, p. 15.
1944b Note on convex spherical curves. Bull. Amer. Math. Soc. **50** (1944), 528–534. MR 6, p. 100.
1945 Note on convex curves on the hyperbolic plane. Bull. Amer. Math. Soc. **51** (1945), 405–412. MR 7, p. 26.
1946 Sobre los cuerpos convexos de anchura constante en E^n. Portugaliae Math. **5** (1946), 195–201. MR 9, p. 526.
1950 On parallel hypersurfaces in the elliptic and hyperbolic n-dimensional space. Proc. Amer. Math. Soc. **1** (1950), 325–330. MR 12, p. 124.
1959 Sobre systemas completos de desigualidades entre tres elementos de una figura convexa plana. Math. Notae **17** (1959/61), 82–104. MR 26, 6857.
1976 Integral geometry and geometric probability. Encyclopedia of Mathematics and its Applications, Vol. 1. Addison-Wesley Publ. Co., Reading, Mass.-London-Amsterdam, 1976. MR 55, 6340.

Schaal, H.
1962 Prüfung einer Kreisform mit Hohlwinkel und Taster. Elem. Math. **17** (1962), 33–38. MR 25, 1489.
1963 Gleitkurven in regulären Polygonen. Zeitschr. Angew. Math. Mech. **43** (1963), 459–476. (English and Russian summaries). MR 28, 1800.

Schaer, J., J.E. Wetzel
1972 Boxes for curves of constant length. Israel J. Math. **12** (1972), 257–265. MR 47, 5726.

Schneider, R.
1971 Gleitkörper in konvexen Polytopen. J. Reine Angew. Math. **248** (1971), 193–220. MR 43, 5411.
1979 Boundary structure and curvature of convex bodies. Contributions to Geometry (Proc. Geom. Sympos., Siegen 1978), pp. 13–59, Birkhäuser, Basel 1979. MR 81i:52001.

Schopp, J.
1970 Über die Newtonsche Zahl einer Scheibe konstanter Breite. Studia Sci. Math. Hungar. **5** (1970), 475–478. MR 44, 3200.
1977 Überdeckungsprobleme ebener Bereiche von konstantem Durchmesser. Period. Polytechn. Mech. Engrg. **21** (1977), no. 2, 103–109. MR 58, 7408.

Schulte, E.
1981 Konstruktion regulärer Hüllen konstanter Breite. Monatsh. Math. **92** (1981), 313–322.

Scott, P.R.
1978 Two inequalities for convex sets in the plane. Bull. Austral. Math. Soc. **19** (1978), no. 1, 131–133. MR 80d:52005.
1981 Sets of constant width and inequalities. Quart. J. Math. Oxford (2) **32** (1981), 345–348.

Sholander, M.
1952 On certain minimum problems in the theory of convex curves. Trans. Amer. Math. Soc. **73** (1952), 139–173. MR 14, p. 787.

Silverman, R.
1973 Decomposition of plane convex sets II. Sets associated with a width function. Pacific J. Math. **48** (1973), 497–502. MR 54, 3581.

Sine, R.
1976 A characterization of the ball in R^3. Amer. Math. Monthly **83** (1976), 260–261. MR 53, 1404.

Sine R., V. Kreĭnovič
1979 Remarks on billiards. Amer. Math. Monthly **86** (1979), 204–206. MR 80k:28021.

Šmakal, S.
1973 Curves of constant width. Czech. Math. J. **23** (98) (1973), 86–94. MR 47, 4154.

Smith, T.J.
1961 Planar line families. Ph.D. Thesis, University of Wisconsin, 1961.

Soltan, V.P.
1977 A theorem on full sets (Russian). Dokl. Akad. Nauk. SSSR **234** (1977), no. 2, 320–322. MR 58, 2607.
1979 Bodies of constant width in finite-dimensional normed spaces (Russian). Mat. Zametki **25** (1979), no. 2, 283–291, 319. MR 80d:52004.
Su, B.
1928 On the curvature-axis of the convex closed curve. Sci. Rep. Tôhoku Univ. **17** (1928), 35–42. Jbuch. 54, p. 799.
Süss, W.
1927 Über affine Geometrie XL: Eiflächen konstanter Affinbreite. Math. Ann. **96** (1927), 251–260. Jbuch. 53, p. 713.
Sz.-Nagy, G.
1949 Zentralsymmetrisierung konvexer Körper. Publ. Math. Debrecen **1** (1949), 29–32. MR 11, p. 386.
Tanno, S.
1976 C^∞-approximation of continuous ovals of constant width. J. Math. Soc. Japan **28** (1976), no. 2, 384–395. MR 53, 1489.
Taylor, S.J.
1953 Some simple geometrical extremal problems. Math. Gaz. **37** (1953), 188–198. MR 15, p. 247.
Tennison, R.L.
1976 Smooth curves of constant width. Math. Gaz. **60** (1976), 270–272. Zbl. 341, 53002.
Thüring, R.
1952 Studien über den Holditchschen Satz. Verh. Naturforsch. Ges. Basel **63** (1952), 221–252. MR 14, p. 679.
Togliatti, E.
1968 Sulle curve piane di larghezza costante. Period. Mat. (4) **46** (1968), 359–371. MR 38, 6457.
Toranzos, F.A.
1964 Sobre un problema de B. Grünbaum. Rev. Un. Mat. Argentina **22** (1964), 103–106. MR 30, 507.
Toranzos, F.A., J.H. Nanclares
1978 Convexidad. Cursos, Seminarios y Tesis del PEAM, No. 4, PEAM, Universidad del Zulia, Maracaibo, 1978, 155 pp. MR 58, 24001.
Valette, G.
1971 Couples de corps convexes dont la difference des largeurs est constante. Collection of articles honoring P. Burniat, Th. Lepage and P. Libois. Bull. Soc. Math. Belg. **23** (1971), 493–499. MR 48, 1037.
Varela Gil, J.
1932 Sobre generación de curvas orbiformes. Atti del Congresso Internazionale dei Matematici, Bologna 1928. Vol. V, pp. 483–488, Bologna 1932.
1944 Curvas orbiformes multiples. Univ. Nat. Tucumán. Revista A. **4** (1944). MR 7, p. 168.
Vincensini, P.
1935 Sur les corps convexes admettant un domaine vectoriel donné. C.R. Acad. Sci. Paris **201** (1935), 761–763. Zbl. 12, p. 272.
1936 Sur le prolongement des séries linéaires de corp convexes. Applications. Rend. Circ. mat. Palermo **60** (1936), 361–372. Zbl. 17, p. 230.
1937 Sur les corps de largeur constante de l'espace à trois dimensions. C.R. Acad. Sci. Paris **204** (1937), 84–86. Zbl. 15, p. 368.
1938 Corps convexes. Series linéaires. Domaines vectoriels. Mem. des Sciences Math., fasc. XCIV, Paris 1938. Zbl. 21, p. 356.
1956 Sur l'application d'une méthode géométrique à l'étude de certains ensembles de corps convexes. Colloque sur les questions de réalité en géométrie, Liège 1955, pp. 77–79, Paris 1956. MR 17, p. 1123.
1957 Sur un invariant de partition de l'ensemble des corps convexes de l'espace euclidien a n-dimension. C.R. Acad. Sci. Paris **245** (1957), 132–133. MR 19, p. 573.
Vivanti, G.
1946 Sulle curve piane a normali doppie. Ist. Lombardo Sci. Lett. Rend. Cl. Sci. Mat. (3) **10** (**79**) (1946), 256–260. MR 10, p. 325.

Vrecica, S.
1981 A note on sets of constant width. Publications de L'institut Mathematique, Nouvelle série **29** (1981), 289–291.
Wajnstejn, D.
1938 Courbes à largeur constante. (Polish). Wiadom. mat. **45** (1938), 39–63. Zbl. 19, p. 140.
Watson, A.G.D.
1962 On Mizel's problem. J. London Math. Soc. **37** (1962). 307–308. MR 27, 4141a.
Wegner, B.
1972 Globale Sätze über Raumkurven konstanter Breite. Math. Nachr. **53** (1972), 337–344. MR 47, 5735.
1975 Globale Sätze über Raumkurven konstanter Breite II. Math. Nachr. **67** (1975), 213–223. MR 51, 11380.
1976 Verallgemeinerte Zindlerkurven in euklidischen Räumen. Math. Balkanica **6** (1976), 298–306 (1978). MR 80m:53002.
1977 Analytic approximation of continuous ovals of constant width. J. Math. Soc. Japan **29** (1977), no. 3, 537–540. MR 57, 4013.
Weissbach, B.
1972 Rotoren im regulären Dreieck. Publ. Math. Debrecen **19** (1972), 21–27 (1972). MR 49, 3685.
1977 Zur Inhaltsschätzung von Eibereichen. Wiss. Beitr. Martin-Luther-Univ. Halle-Wittenberg M 8 Beitr. Algebra Geom. **6** (1977), 27–35. MR 58, 7410.
Yaglom, I.M., V.G. Boltyanskii
1961 Convex figures. Holt, Rinehart and Winston, New York, 1961. MR 23, A1283.
Zindler, K.
1921 Über konvexe Gebilde, II. Teil. Monatsh. Math. Phys. **31** (1921), 25–56. Jbuch. 48, p. 833.

G.D. Chakerian
Department of Mathematics
University of California Davis
Davis CA 95616 (USA)

H. Groemer
Department of Mathematics
University of Arizona
Tucson AZ 85721 (USA)

J.H.H. Chalk

Algebraic Lattices

1. Introduction

In his recent article [5], P. Gruber has surveyed the development of the Geometry of Numbers since the publication of the books of Cassels [3] and Lekkerkerker [6]. As none of these sources deals specifically with lattices other than those of Minkowski-type (i.e. a $\mathbb{Z}$-module with N generators in $\mathbb{R}^N$) it seems worthwhile to trace the main developments there for lattices which have more algebraic structure. Even though these are often endowed with arithmetic properties as for example when $\mathbb{R}$ and $\mathbb{Z}$ are replaced by some field k and a ring O of integral elements in k, we shall use the term 'algebraic lattice'. A special case is the Leech lattice, which arose as a $\mathbb{Z}$-module of rank 24 in $\mathbb{R}^{24}$ and can now be interpreted in this way as an O-module of rank 12 in k^{12}, where k is the Eisenstein field $Q(\rho)$, $\rho = \exp(2\pi i/3)$. Such features of an algebraic lattice have in recent years been harnessed to deal with problems in other areas, e.g. Finite groups, Sphere packings and Codes. As we intend to keep mainly to the ideas and spirit of the Geometry of Numbers, the section devoted to these special algebraic lattices is brief and supplied only with a selection of general references. With this view, it is however appropriate to include details of the case when k is a field endowed with a non-archimedean valuation and, in particular, to survey the work of Armitage [1], [2] on the Riemann-Roch theorem. Although a number of other generalizations appear in the literature (e.g. Dubois [4]), we shall confine our attention to one sufficiently general to include most of these and which permits the study of non-commutative lattices. As a Minkowski lattice may be interpreted as a discrete subgroup Γ of the additive group G of $\mathbb{R}^N$ with the property that the factor group G/Γ has compact closure, it is natural to review the impact of the ideas of the Geometry of Numbers on discrete subgroups of topological groups and of Lie groups. But, overall, the recurrent themes are the fundamental theorems of Minkowski ([7], [5]) (§3.1) for convex bodies and the compactness theorem of Mahler ([5], Section 2).

2. Algebraic Lattices Associated to a Number-Field (or Division Algebra)

Let k be an algebraic number field of degree $[k:Q] = n$ and let O denote a ring* of integers of k. Let V be a vector space over k with $\dim_k V = m$.

Definition 1. (W. Feit [15]). *An algebraic lattice is a finitely generated O-module L with $L \subset V$ and with*

$$L \otimes_O k \cong V.$$

Remark. For $n = 1$, we may clearly take $V = Q^m$ and, for $n > 1$, it can be shown that $V \cong Q^N$, where $N = mn$. Thus, for $N \geq 1$, L may be interpreted as a special lattice Λ of Minkowski-type with $\Lambda \subset Q^N$.

* i.e. an order, not necessarily maximal, in the sense of Dedekind.

Examples of algebraic lattices with $m = n = 2$ were introduced into the Geometry of Numbers by Minkowski [20], but the first systematic treatment was due to Weyl [27] some forty years later. Broadly, the theme is 'arithmetic reduction theory' in the Geometry of Nunbers for suitably defined convex bodies and algebraic lattices. In the course of this, analogues of Minkowski's first and second fundamental theorems ([5], Section 3]) were given (cf. [27], p. 212). For comparison purposes it is useful to have a summary of his introduction to algebraic lattices (for earlier work, see the references in [27], p. 215). Let k be a field or, more generally, a not necessarily commutative division algebra of finite degree $[k:Q] = n$ over Q. Let V_k be a vector space with $\dim_k V_k = m$ which is to be viewed also as a vector space V_Q over Q with $\dim_Q V_Q = mn$. As $V_k \cong Q^{mn}$, let $\psi : \zeta \to X$ $(\zeta \in V_k, X \in Q^{mn})$ be a Q-linear mapping identifying V_k with Q^{mn}. Let Λ denote any (Minkowski) lattice (with non-zero determinant), which is contained in Q^{mn}. Then $L = \psi^{-1}\Lambda$ is clearly an O-module with rank n.

Definition 2. (i) *An element* $\delta \in k$ *is said to be a multiplier for* L if, *and only if,*

$$\zeta \to \delta\zeta \Rightarrow \delta\zeta \in L (\zeta \in L).$$

(ii) *Denote* $[k] = \{\delta \in k : \delta \textit{ is a multiplier for } L\}$.

He observes that [k] is a $\mathbb{Z}$-module in k/Q with rank n which satisfies the requirements of an order of k. Now, a change of k-basis in V_k induces a Q-linear mapping $\Lambda \to \Lambda'$ say in V_Q. Let $L' = \psi^{-1}\Lambda'$; then it is clear that L and L' have the same system of multipliers. Any two such lattices are said to be equivalent and to belong to the same class.

Definition 3. *A lattice L is said to belong to an order* [k] *of* k *if every element of* [k] *is a multiplier of L.*

A well-known theorem [25], asserts that the number of classes belonging to a given order is finite. In particular, for a given order [k], the set L_0 of m-tuples $(\xi_1, \ldots, \xi_m) \in [k]^m$ corresponds under the identification ψ, to a lattice Λ_0 in V_Q called the *unit lattice* for [k].

To introduce the Geometry, Weyl restricted k to a field and extended the vector space V_k to a commutative algebra A over the reals $\mathbb{R}$, defined a *gauge function* (convex distance function) and a *gauge body* $K : f(\xi) < 1 (\xi \in A)$ and proceeded to reduction theory for the function f. In particular, Minkowski's second theorem appeared in the shape

$$(1) \qquad (M_1 M_2 \cdots M_m)^n V(K) [L : L_0] \leq 2^{mn},$$

where the volume V(K) of K is suitably defined in $\mathbb{R}^{mn}$ and the M_i are the successive minima for the gauge function f with respect to L (cf. [27], Section 4, p. 212). Following a suggestion of K. Mahler, who set the foundations of the theory of star-bodies with respect to a Minkowski lattice in 1946 [19], K. Rogers and H.P.F. Swinnerton-Dyer [23] developed a similar theory for star-bodies with respect to algebraic lattices. This substantial contribution was realized by a modification to Weyl's definition of an algebraic lattice, it being desirable to reduce its arithmetic content and so avoid discussion of class-numbers. The modified definition enables one to work either with O-modules L or with corresponding Minkowski lattices of special type. My version of this is perhaps unnecessarily explicit, but it turns out to be useful in an application to quadratic forms [9].

Fix a $\mathbb{Z}$-basis $\rho_1 = 1, \rho_2,\ldots,\rho_n$ for O let *

(2) $\mathbf{x}^{(r)} = \rho_1^{(r)}\mathbf{z}_1 + \cdots + \rho_n^{(r)}\mathbf{z}_n \; (\mathbf{z}_k \in \mathbb{R}^m; 1 \leq k \leq n)$

for $r = 1,2,\ldots,n$, where the superscript '(r)' indicates the corresponding Q-conjugate in the field $k^{(r)}$ conjugate to k. (For $r = 1$, we take $k^{(1)} = k$ and usually omit the superscript.) Then

(3) $d = |\det[\rho_s^{(r)}]|^2$

is the absolute value of the discriminant of O. Let $V = V_N$ denote the N-dimensional real vector space of elements $\mathbf{X}$ of the form

(4) $\mathbf{X} = (\mathbf{x}^{(1)},\mathbf{x}^{(2)},\ldots,\mathbf{x}^{(n)}),$

where $N = mn$ and $\mathbf{x}^{(r)}$ is as defined in (2). Note that all elements of the $\mathfrak{O}$-module $M = \mathfrak{O}^m$ are given by (2) with $r = 1$, as $\mathbf{z}_1,\ldots,\mathbf{z}_n$ run through those of $\mathbb{Z}^m$; the corresponding elements $\mathbf{x}^{(r)}$ of the conjugate modules $M^{(r)}$ being determined by (2) in the same way.

Definition 4. (i) *The integer (or unit) lattice* L_0 *is the* $\mathfrak{O}$*-module* M *defined by* (2) *with* $r = 1$ *as* $\mathbf{z}_k$ runs through the elements of $\mathbb{Z}^m$, $(k = 1,2,\ldots,n)$. *Its discriminant* $d(L_0) = d^m$.

(ii) *The integer lattice* $\Lambda_0 = \Lambda_0(L)$ *is the lattice of all points* $\mathbf{X} \in V_N$ *of the form* (4), *where* $\mathbf{x}^{(r)}$ $(1 \leq r \leq n)$ *run through all elements arising from* (2) *with* $\mathbf{z}_k \in \mathbb{Z}^m$. *Its determinant* $d(\Lambda_0) = d^{m/2}$.

Although this defnition is less general* than Weyl's, we shall work with a generalization of Definition 4(ii) designed for the methods of the Geometry of Numbers.

Definition 5. *An algebraic lattice* Λ *is a linear transform of* $\Lambda = \Lambda(L)$ *of the type*

(5) $\mathbf{x}'^{(r)} = \mathbf{x}^{(r)}\mathbf{B}^{(r)}, \quad (r = 1,2,\ldots,n)$

where $B^{(r)}$ *is an* $m \times m$ *matrix, with entries* $b^{(r)}$ *say, of the form*

(6) $b^{(r)} = \rho_1^{(r)}\beta_1 + \cdots + \rho_n^{(r)}\beta_n, \quad (\beta_i \in \mathbb{R}, 1 \leq i \leq n)$

subject only to the condition

(7) $$\Delta = \left|\prod_{1 \leq r \leq n} \det \mathbf{B}^{(r)}\right| \neq 0.$$

Then

(8) $d(\Lambda) = \Delta d^{m/2}.$

* row-vectors are used, throughout.

* See [23], Theorem 1 and Section 2, for a discussion of this point.

Remarks. Weyl's commutative algebra A is viewed here as a *real* vector space V_N and, if k is a totally real field, Λ is a Minkowski lattice of special type in $\mathbb{R}^N$ with determinant $\Delta d^{m/2}$. If k is not totally real and has a real and t complex embeddings in $\mathbb{C}$ ($t \geq 1, n = s + 2t$), a linear transformation of the form

$$(9)\qquad \zeta \to \left(\frac{1}{2}(\zeta + \bar{\zeta}), \frac{1}{2i}(\zeta - \bar{\zeta})\right) \in \mathbb{R}^{2m},$$

applied to each of the t pairs

$$(10)\qquad \mathbf{x}^{(k)} \quad \text{and} \quad \mathbf{x}^{(l)} = \bar{\mathbf{x}}^{(k)}$$

of conjugate complex components of $\mathbf{X}$, will map Λ into a Minkowski lattice in $\mathbb{R}^{sm+(2t)m} = \mathbb{R}^{mn} = \mathbb{R}^N$ of determinant $2^{mt}d(\Lambda)$. As an illustration, we define a star body S first and then introduce the successive minima for S relative to an algebraic lattice of this type. In fact, a more general class of star bodies is obtained by working with V_N and Λ, rather than with *M* and *L* as in [27] and [23].

Definition 6. *For any real finite dimensional vector space* W *endowed with a metric, a distance function* W *is defined as a continuous mapping* $f: W \to \mathbb{R}$ *which satisfies*

$$(11)\qquad \text{(a) } f(x) \geq 0 \quad \textit{for all} \quad x \in W \quad \textit{and} \quad f(x) > 0 \quad \textit{for some} \quad x \in W,$$

$$(12)\qquad \text{(b) } f(tx) = |t| f(x) \quad \textit{for all} \quad x \in W \quad \textit{and for all} \quad t \in \mathbb{R}.$$

Then, for any $M > 0$,

$$S = \{x \in W : f(x) < M\}$$

is called a star body in W, (by (b) *with* $t = -1$, S *is symmetric in the orgin of* W.). *Further, a star body is a convex star body if, in addition to* (a) *and* (b), f *satisfies*

$$(13)\qquad \text{(c) } f(x + y) \leq f(x) + f(y), \quad \textit{for all} \quad x, y \textit{ in } W.$$

In the theory, S is required to be bounded in the sense that S meets every discrete subset of W in at most finitely many points. Note that, for all $\lambda > 0$,

$$\lambda S = \{x \in W, \lambda^{-1} x \in S\}$$

the dilatation of S in the ratio $\lambda : 1$, is a star body, or bounded star body, whenever S has these respective properties.

Now take $W = V_N$ and for simplicity, $\Lambda = \Lambda_0$. Let

$$(14)\qquad S : f(\mathbf{X}) < 1$$

be an arbitrary bounded star body in V_N. Suppose, for the moment, that k is totally real.

Definition 7. *For* $1 \leq k \leq m$, *let* $E_k = E_k(\delta_1, \ldots, \delta_k)$ *denote the vector space over* k *spanned by the elements* $\delta_1, \ldots, \delta_k$ *of the* $\mathfrak{O}$*-module* M, *where* $M = \mathfrak{O}^m$, *and put* $E_0 = \{\mathbf{O}\}$. *Then, given* E_{k-1}, *we may select an element* $\delta_k \in M$ *with* $\delta_k \notin E_{k-1}$ *for which* $f(\mathbf{X}_k)$ *is minimal, where*

$$(15)\qquad \mathbf{X}_k = (\delta_k, \delta_k^{(2)}, \ldots, \delta_k^{(n)}) \in \Lambda_0;$$

the bound being attained since $\lambda S (\lambda > 0)$ is a bounded set and Λ_0 is a discrete subset of V_N. Then

$$f(\mathbf{X}) \geq f(\mathbf{X}_k)$$

for all $\mathbf{X} = (\mathbf{x}, \mathbf{x}^{(2)}, \ldots, \mathbf{x}^{(n)}) \in \Lambda_0$ *such that* $\mathbf{x} \notin E_{k-1}$ $(\mathbf{x} \in M)$. *Define* $\mathbf{E}_k = (\delta_1, \ldots, \delta_{k-1}, \delta_k)$ *and set*

$$\mu_k = f(\mathbf{X}_k).$$

Thus, if $0 < \mu < \mu_k$, the star body μS cannot contain k points $\mathbf{Y}_1, \ldots, \mathbf{Y}_k$ of Λ_0 with $\mathbf{y}_1, \ldots, \mathbf{y}_k$ linearly independent over $\mathfrak{O}$, but if $\mu > \mu_k$ then it contains at least k such points. In particular, *the successive minima* μ_k (k = 1,2,...,m) *satisfy*

$$\mu_1 \leq \mu_2 \leq \cdots \leq \mu_m$$

and are uniquely determined.

This presentation has the advantage of maintaining some control over the magnitude of the conjugate $\mathbf{x}^{(r)}$ of $\mathbf{x} = \mathbf{x}^{(1)}$ in $\mathbf{X}$, in terms of the distance function $f(\mathbf{X})$, whereas in [27] and [25] the emphasis is centred on a distance function $f(\mathbf{x})$ defined on the vector space k^{*m}, where k^* is an extension of k (viewed as an algebra over Q) to an algebra over $\mathbb{R}$. Moreover, the ideas and theorems in [23], in so far as they apply to a general situation*, require only routine changes. Thus, for a given bounded star-body $S: f(X) < 1$, there exists a constant $C(S) > 0$ such that

(16) $\quad (\mu_1 \mu_2 \cdots \mu_m)^n C(S) \leq d(\Lambda_0).$

As there was no version of Minkowski's theorem on successive minima for a convex body for algebraic lattices of type Λ_0, I [10] modified Weyl's argument for (1) to give an inequality of the following form

(17) $\quad (\mu_1 \mu_2 \cdots \mu_m)^n V(K) \leq 2^N d(\Lambda_0),$

where K is a bounded convex star body in V_N and V(K) is the volume of K (defined in the natural way, using the isomorphism between V_N and $\mathbb{R}^N$).

As in the classical Geometry of Numbers, there is no difficulty in stating (and proving) inequalities such as (16) and (17) in an apparently more general form with Λ_0 replaced by Λ (cf. Definition 5). For, if Ω is an invertible $\mathbb{R}$-linear map of V_N onto itself, as represented by a block matrix $\operatorname{diag}(B^{(1)}, \ldots, B^{(n)})$ defined in (5) with $|\det \Omega| = \Delta$, then $\Lambda = \Omega \Lambda_0$ is an algebraic lattice and $S' = \Omega S$, where S as defined in (14), is also a bounded star-body in V_N. Moreover, the successive minima ν_k of

$$S' = \{\mathbf{X} \in V_N : g(\mathbf{X}) < 1\},$$

say and the lattice Λ, are given by $\nu_k = g(\Omega \mathbf{X}_k)$, where the $\mathbf{X}_k$ are defined in (15). Thus, in particular, (17) has the equivalent form:

(18) $\quad (\nu_1 \nu_2 \cdots \nu_m)^n V(K') \leq 2^N d(\Lambda).$

* e.g. the analogue of Mahler's Compactness theorem, (cf. [23], Theorem 4).

since convexity is also preserved. Of course the technical restriction that k be totally real is easily removed by mapping $V_N \to V'_N$, where the components of elements of V'_N are all real and $V_N \cong V'_N$ under linear transformations of the type in (9).

Relatively few specific problems, even familiar ones like the determination of lattice constants, have been settled for algebraic lattices. The computations are usually much more difficult. Rogers and Swinnerton-Dyer ([23], Sections 6, 7) applied their theory to a few such problems with success and, in particular, were able to simplify the proofs in the examples of Minkowski (which concern two complex linear forms in variables taking integral values in (i) Q(i), (ii) Q(ρ), $\rho = \exp(2\pi i/3)$). Another case, where $k = Q(i\sqrt{5})$ has class number 2 is interesting; they proved (with an immense amount of computing) that

$$(19) \qquad \max\{|\xi|,|\eta|\} \leq \left(\frac{2|\Delta|}{\sqrt{15}-3}\right)^{1/2}$$

is solvable in integers x, y ≠ 0, 0 in $Q(i\sqrt{5})$ for every pair of complex linear forms $\xi = ax + by$, $\eta = cx + dy$, with $\Delta = ad - bc \neq 0$ (and moreover, determined the cases of equality in (19)). There are other examples of this type (e.g. [22]), including one with m = 3, n = 2 ([23], Theorem 10), but it is not really surprising that interest in algebraic lattices has turned to other directions. Before discussing that, I should like to mention the absence of a counterpart for the Minkowski-Hlawka theorem ([5], Section 3.2) and report upon an application of (17) to a problem in a different area of Number theory. The latter concerns Cassel's lemma [8] on the height (or magnitude, suitably defined) of a zero $\mathbf{x} \in \mathbb{Z}^m - (\mathbf{o})$ of an idefinite quadratic form $f \in [\mathbb{Z}\mathbf{x}]$, which represents 0 non-trivially. Recently, Raghavan [21] established a corresponding estimate for the more general case where $\mathbb{Z}$ is replaced by an arbitrary order $\mathfrak{O}$ of an algebraic number-field k (for this he used Minkowski's first theorem for a parallopiped in V_N and a lattice of the type Λ_0). Davenport [13] obtained a refinement of Cassel's lemma by considering a pair of zeros of $f \in \mathbb{Z}[\mathbf{x}]$, linearly independent over Q. In [9], I extended Davenport's results to the case when $\mathbb{Z}$ is replaced by $\mathfrak{O}$. In fact, if $f(\mathbf{x}) = \sum_{i \leq k, l \leq m} f_{kl} x_k x_l$, $(\mathbf{x} = (x_1, x_2, \ldots, x_m))$, $m \geq 2$ and $f \in \mathfrak{O}[\mathbf{x}]$ is a non-degenerate form which represents 0 for some $\mathbf{x} \in \mathfrak{O}^m - (\mathbf{o})$, then there exist elements **a** and **b** in $\mathfrak{O}^m$, linearly independent over k, with $f(\mathbf{a}) = f(\mathbf{b}) = 0$ and satisfying

$$(20) \qquad 0 < \|\mathbf{a}\| \cdot \|\mathbf{b}\| \leq (3F)^{m-1} d^{m/n},$$

where the '*heights*' are defined by

$$(21) \qquad \|\mathbf{x}\| = \max_{1 \leq r \leq n} \max_{1 \leq k \leq m} |x_k^{(r)}|,$$

and

$$(22) \qquad F = \max_{1 \leq r \leq n} \sum_{1 \leq k,l \leq m} |f_{kl}^{(r)}|.$$

The key to the proof is an application of the inequality (17) for the successive minima of a suitably chosen parellopiped* relative to Λ_0. It is independent of

* This was not definable in terms of the distance functions in [27] and [23].

Davenport's proof and so provides an alternative verification for the special case $\mathfrak{O} = \mathbb{Z}$.

Another area where algebraic lattices have a major role is the theory of finite groups. If these lattices (cf. Definition 1) satisfy certain duality properties interest has focussed on their groups of automorphisms. In some cases, these groups are related to some of the well known finite simple groups (e.g. the Conway groups .0 and .1 and Suz, the Suzuki group; cf. [11]). The Leech lattice [16] is such an example, for though it was originally studied as a $\mathbb{Z}$-module in $\mathbb{R}^{24}$ (a special Minkowski lattice), it can be viewed as an $\mathfrak{O}$-module ($\mathfrak{O} \supset \mathbb{Z}, \mathfrak{O} \neq \mathbb{Z}$) with the required duality property. Now, for the general Minkowski lattice Λ one can always define its dual lattice Λ^* (e.g. as in [3], Chapter 1.5). But for an algebraic lattice (as in Definition 1), one wants the dual to be an algebraic lattice. For this, we suppose two things: (a) there is a Q-automorphism σ of k/Q with $\sigma^2 = 1$, and (b) there is a hermitian form, symmetric with respect to σ, on V. Then, the h-*dual* L_h^* of L is defined by

$$L_h^* = \{v \in V : h(u,v) \in \mathfrak{O} \text{ for all } u \in L\}.$$

Clearly, L_h^* is an $\mathfrak{O}$-module $\subset$ V. Thus, if h is non-degenerate L_h^* is an algebraic lattice. A lattice L is said to be *integral* with respect to h, if $h(u,v) \in \mathfrak{O}$ for all u,v in L.

Suppose now that L is integral with respect to a positive definite hermitian form h. Let $D(L)$, the h-*discriminant of* L, be defined as the determinant of the matrix of h. Then

$$[L_h^* L] = D(L)^2,$$

(as is easily verified by properties of invariants) and so L is *unimodular* if, and only if, $L_h^* = L$. Consider an example with $k = Q(\rho)$, $\rho = \exp(2\pi i/3)$ and $\mathfrak{O} = \mathbb{Z}[\rho]$. Since $\mathfrak{O}$ is a principal ideal domain, every finitely generated torsion-free $\mathfrak{O}$-module is free. Write $h(x,y) = (x,y)$, where now x and y belong to $V = k^m$ (say). Since h is positive definite, the group $G(L)$ of all automorphisms of L which preserve h, is a finite group. Now Feit [14] has, for example, listed all pair-wise non-isomorphic unimodular lattices of rank ≤ 12, which are *indecomposable* in the sense that $L = L_1 \oplus L_2$ and

$$L_1 \perp L_2 \Rightarrow L_1 = (0) \quad \text{or} \quad L_2 = (0),$$

with respect to h. In particular, the Leech lattice L_1 is a $\mathbb{Z}[\rho]$-module of rank 12 and $G(L_1)$ is Conway's group .0 ([18], [26]). J. Tits [26], in a largely expository but highly ingenious article explains the group-theoretical reasons for expecting a representation of L_1 in this and other similar ways. Indeed, he gives an alternative with L_1 as a free $\mathfrak{O}$-module of rank 3 where now $\mathfrak{O}$ is a maximal order in the quaternion algebra

$$\mathbb{H} = k_0(i,j,k), \quad k_0 = Q(\sqrt{5}), \quad i^2 = j^2 = -1, \quad k = ij = -ji,$$

and $L \subset \mathbb{H} \times \mathbb{H} \times \mathbb{H} = V$ (so that $\dim_Q V = 4 \cdot 2 \cdot 3 = 24$). On examining the centralizers of subgroups of Conway's group, it was noted that they are multiple coverings of some well-known groups, e.g. Aut $L_1 \cong$ double cover $\tilde{J}_2$ of the sporadic group of Hall-Janko if L_1 is viewed as a rank 3 module over $\mathbb{H}$; Aut $L_1 \cong$ double cover of $GL(2,\mathbb{F}_4)$, if L_1 is viewed as a rank 6 module over

$\mathbb{H}_0 = Q(i,j,k)$. Such lattices are of particular interest in the study of codes (c.f. [24], [17] where the term 'complex lattice' is favoured). Their relation to complex polytopes is studied in [12], cf. Chapter 12.

3. Algebraic Lattices in Ultrametric Spaces

Early applications of the Geometry of Numbers were mainly in the classical theory of algebraic number-fields (e.g. the finiteness of the class-number, the unit theorem and a bound for the discriminant) and usually featured a special case of Minkowski's first fundamental theorem. Later, it had a minor part in the theory of Adeles, where it was used in the wider setting, attributed to Blichfeld [5] (an arbitrary set $S \subset \mathbb{R}^N$, having a volume $V(S) > 1$, contains two distinct points vhose difference is a point of $\mathbb{Z}^N$). These applications provided impetus for the study of lattices in other types of vector space with a view to similar applications in other areas, such as algebraic function-fields. We shall consider *ultrametric spaces* where the usual archimedean valuation $|\cdot|$ or $\mathbb{R}$ or $\mathbb{C}$ is replaced by one with a stronger form of the triangle inequality.

Let k be a field with a non-archimedean valuation $|\cdot|$ (e.g. Q_p, the field of p-adic numbers with the p-adic valuation $|\cdot|_p$), a **valuation** being a mapping from k to $\mathbb{R}$ satisfying

(i) $|\alpha| \geq 0$ for all $\alpha \in k$ and $|\alpha| = 0 \Leftrightarrow \alpha = 0$.

(ii) $|\alpha\beta| = |\alpha||\beta|$ for all $\alpha, \beta \in k$.

(iii) $\exists c = c(k) > 0$ such that $|\alpha| \leq 1 \Rightarrow |1 + \alpha| \leq c$.

The valuation is said to be *non-archimedean* if (iii) holds with $c = 1$; otherwise it is said to be *archimedean*. If it is true with $c = 1$, it is readily shown that (iii) can be replaced by

(iii)′ $|\alpha + \beta| \leq \max(|\alpha|,|\beta|)$, for all α, β in k,

which is the strong form of the triangle inequality, referred to above. The valuation is said to be **discrete** if the set $\{|\alpha| : \alpha \in k\}$ of values is a discrete subset of $\mathbb{R}$. By putting $d(x,y) = |x - y|$, the valuation defines a metric on k and so induces a topology on k. Let $\tilde{k}$ denote the completion of k with respect to this metric and let

$$\mathfrak{O} = \{\alpha \in k : |\alpha| \leq 1\}.$$

By (ii) and (iii), it is apparent that $\mathfrak{O}$ is a ring; this is the ring of integers in k (e.g. if $k = Q$ and $|\cdot| = |\cdot|_p$, then $\mathfrak{O} = \{u/v : u,v \in \mathbb{Z}, p \nmid v\}$). K. Mahler [29] gave the first systematic account of the Geometry of Numbers in ultrametric space in 1941. Years earlier, (see, e.g. M.R.37, No. 192 for references) some analogue of Minkowski's theorem on linear forms had been recognized as a key to a proof of the Rieman-Roch theorem in algebraic function-fields. A summary of part of Mahler's long paper has already been published ([28], Section 2) and I append a few details from this source.

Let k_0 be an arbitrary field, $k = k_0(t)$ a transcendental extension of k_0 and let $\mathfrak{O}$ denote the ring $k_0[t]$ of polynomials in t. If $\alpha = u/v$ ($u \in \mathfrak{O}, v \in \mathfrak{O}, v \neq 0$), define

$$|\alpha| = q^{-\operatorname{ord} \alpha}$$

where ord $\alpha = \deg u - \deg v$ and $q > 1$ is some fixed constant, (usually taken to be card k_0 when k_0 is a finite field of q elements). Let $V = \tilde{k}^n$ be the n-dimensional vector space over the completion $\tilde{k}$ of k with respect to $|\cdot|$ so that in fact the elements of $\tilde{k}$ are formal Laurent series of the form

$$a_m t^m + \cdots + a_0 + a_{-1} t^{-1} + \cdots.$$

The valuation $|\cdot|$ on k extends, in the obvious way, to $\tilde{k}$ and inherits the properties (i), (ii) and (iii) (which are clearly satisfied by $|\cdot|$ on k).

Definition. (i) $\Lambda_0 = \mathfrak{O}^m$ *is the integer lattice in* $V = \tilde{k}^m$.
(ii) *A lattice* Λ *in* V *is the image of* Λ_0 *under an invertible* $\tilde{k}$-linear mapping λ of V *into itself. Its determinant* $d(\lambda)$ *is defined by*

$$d(\lambda) = |\det \lambda|.$$

(iii) $\|x - y\| = \max_{1 \leq i \leq n} |x_i - y_i|, \quad (x, y \in V; x_i, y_i \in \tilde{k}, 1 \leq i \leq m)$

defines a metric on V.

(iv) *A distance function* f *is as defined for a real vector space* (cf. *Definition* 6) *with* $\mathbb{R}$ *replaced by* $\tilde{k}$. *It is a convex distance function if* (c) *in Definition* 6 *is replaced by the strong triangle inequality*

$$f(x - y) \leq \max(f(x), f(y))$$

and then the set $K[\tau] = \{x \in V : f(x) < \tau\}$ *is said to be a convex body.*

We single out two theorems in Mahler's paper that indicate the force of the strong triangle inequality. First, he proved that every convex body K is a parallelopiped, i.e., it can be defined by $f(x) = |Ax|$ for some $A \in GL(m, \tilde{k})$. Incidentally, he deduced that

(23) $\quad V(K) = (|\det A|)^{-1}$,

by defining the 'volume' V(K) of K in terms of the rank of a k_0-module of points inside a suitable dilatation of K. Second, the *successive minima* $\sigma_1 \leq \sigma_2 \leq \cdots \leq \sigma_m$ of $K = K(1)$ *relative to* Λ being defined in the customary way so that $K(\tau) = \tau K$ contains at most $l - 1$ points of Λ linearly independent over k if $\tau < \sigma_l$ and contains at least l such points if $\tau > \sigma_l$, the counterpart of Milkowski's second theorem takes the shape of an equality:

(24) $\quad \sigma_1 \sigma_2 \cdots \sigma_m V(K) = d(\Lambda)$.

Armitage [28] introduced an arbitrary finite algebraic extension L of k with degree $[L:k] = n$ (say), (i.e. an algebraic function-field of one variable) and applied local divisor theory to L. Let $\mathfrak{O} \supset \mathfrak{o}$ denote the integral closure of $\mathfrak{o}$ in L and let $\mathscr{D}(L)$ denote the group of divisors of L. Then any divisor $\mathfrak{a} \in \mathscr{D}(L)$ can be expressed in the form

$$\mathfrak{a} = \mathfrak{a}_f \mathfrak{a}_i,$$

where $\mathfrak{a}_f$, $\mathfrak{a}_i$ denote products of powers of finite, infinite prime divisors of L, respectively. It turns out that $\mathfrak{a}_f$ corresponds to an algebraic lattice $\Lambda(\mathfrak{a})$ in $V = \tilde{k}^m$

and $\mathfrak{a}_k$ corresponds to a convex body K in V and a proof of the Riemann-Roch theorem is achieved with the aid of two theorems of Mahler (including (24)) on successive minima. It is timely to remark that, due to recent simplifications in elementary proofs of the Riemann hypothesis for algebraic function fields, defined over a finite field, k_0, little more than the Riemann-Roch theorem itself is now required.

4. Lattices and Algebraic Lattices (Generalized)

We shall indicate two different but intimately related generalizations of a Minkowski lattice, discuss their connections with the Geometry of Numbers and include some examples of generalized algebraic lattices. First, a Minkowski lattice can be viewed as a discrete subgroup $H = \Lambda$ of the additive group G of $\mathbb{R}^N$ with the property that the factor group G/H has compact closure. This leads to the theory of discrete subgroups of topological groups ([43], [40]) and of Lie groups ([50], [41]). Second, if we view $\Gamma = GL(n,\mathbb{Z})$ as a group of automorphisms of Λ, then Γ is also a discrete subgroup of $G = GL(n,\mathbb{R})$ and the objects of study are now the discrete subgroups of G. This leads to the theories of linear algebraic groups [46], algebraic groups [31] and arithmetic groups [30]. (For lack of a precise definition we shall regard arithmetic groups as in [50] as generalized algebraic lattices). One of the major subjects of study has been the 'arithmeticity' of discrete subgroups ([31], [36], [37], [38]); culminating in the classification by G.A. Margulis. We begin with an example of the simplest type for n = 2 which is not commutative. A *Fuchsian group* is a discrete subgroup H of $G = SL(2,\mathbb{R})$, or, on identifying $\pm g$ $(g \in G)$, the corresponding group H′ of Möbius transformations on the Poincaré upper-half-plane $\mathscr{H}$ $(\operatorname{Im} z > 0$, $z = x + iy)$. Here it is possible to construct a fundamental region $\mathscr{R}$ (analogous to a lattice cell in Λ), e.g. by Dirichlet's procedure [34] and, in fact, one can infer that G/H has compact closure if, and only if, there is such an $\mathscr{R}$ for which $\partial\mathscr{R}$ is a closed non-euclidean polygon. It is also known that arithmetic subgroups of G correspond to groups of units of a ring of integers in an indefinite rational quaternion algebra ([42] Chapter 9, [33], [47], [49], [49]). M. Tsuji [48] introduced the Geometry of Numbers into Fuchsian groups, by giving an analogue of Minkowski's first theorem (for more general discontinuous groups, Siegel [44] gave an analogue for ellipsoids). Siegel [45] in his studies on the symplectic group $Sp(2m,\mathbb{R})$ generalized Fuchsian groups by introducing an upper half-space $\mathscr{H}_m$, defined by

$$\{X + iY : X, Y \in M(m,\mathbb{R}) \text{ are symmetric, } Y \text{ positive definite}\};$$

in fact, subgroups of $G = Sp(2m,\mathbb{R})$ are mapped by general Möbius transformations of the form $\tau(Z) = (AZ + B)(CZ + D)^{-1}$, $(Z = X + iY)$ into subgroups of the corresponding group G′, acting on $\mathscr{H}_m$. For discrete subgroups H′ of G′, he invented explicitly defined subsets of $\mathscr{H}_m$ (the 'Siegel domains') which contained a fundamental region for G′/H′ and which could, for certain applications, serve in place of a fundamental region. This, however, has been supplemented by far-reaching developments in the theory of linear algebraic groups (where we return to H and G). For example, if $G = GL(n,\mathbb{R})$, $H = GL(n,\mathbb{Z})$ and $K = O(f)$, then G/H is a homogeneous space isomorphic to the space of Minkowski lattices in $\mathbb{R}^n$ and the homogeneous space of (right) cosets $K\backslash G$ is isomorphic to the space of positive definite quadratic forms on $\mathbb{R}^n$; here, K denotes the orthogonal group which leaves invariant the special form $f = \sum_{1 \le i \le} x_i^2$ on $\mathbb{R}^n$. It is now well-known (cf. [30]) that

$$GL(n,\mathbb{R}) = \mathscr{S}_{t,u} GL(n,\mathbb{Z}),$$

where $\mathscr{S}_{t,u}$ is a 'Siegel set', defined as follows. If A is the group of diagonal matrices with positive entries and N is the group of upper triangular matrices with eigenvalues = 1, then

$$\mathscr{S}_{t,u} = K.A_t.N_u$$

where $A_t = \{a \in A : a_{ii} \le t a_{i+1,i+1}, 1 \le i \le n-1\}$, $N_u = \{n \in N : |n_{ij}| \le u, 1 \le i \le j \le n\}$. In fact, by using classical estimates from the Geometry of Numbers, one can actually take $t = 2/\sqrt{3}$, $u = 1/2$; the process is reversible in the sense that much of the classical Geometry of Numbers can be re-formulated in the language of algebraic groups and one can for example deduce Mahler's compactness theorem in this setting. It was Chabauty [32] who first established a version of this in the context of topological groups. This was refined and augmented by A.M. Macbeath and S. Swierczkowski [35], who describe the situation as follows. Let $\{H_n\}$ be a sequence of (Minkowski) lattices in $\mathbb{R}^n$ with the properties

(i) $\exists$ neighbourhood U_e of the zero-vector e such that, for all n,

$$H_n \cap U_e - \{e\},$$

(ii) $d(H_n)$ is bounded above.

There then exists a subsequence $\{H_n'\}$ of $\{H_n\}$ which converges to some fixed lattice H.

More generally, let G be a locally compact topological group (and in this work, it is assumed that G is a countable union of compact sets) and let H be a discrete subgroup of G. Then G/H, the space of left cosets H_x of H has an induced Haar measure $\mu = \mu(G/H)$, inherited from G. If $\mu(G/H) < \infty$, the group is called a *lattice* and, if G/H is compact, then H is certainly a lattice and is called a *bounded lattice*. In the classical case, the constant $\mu(G/H)$ is just the determinant d(H) of H. Now a sequence $\{H_n\}$ of subgroups H_n of G is called *uniformly discrete* if

$$H_n \cap U_e = \{e\}$$

for some neighbourhood U_e of the identity e of G and all n. Now, a sequence $\{H_n\}$ is said to *converge* to a subgroup H of G if, for any compact set C and *any such* U_e, the inclusions

$$H \cap C \subset H_n U_e \quad \text{and} \quad H_n \cap C \subset HU_e$$

hold for all sufficiently large n. Chabauty (*loc. cit.*) showed that a uniformly discrete sequence $\{H_n\}$ of subgroups H_n of G has a subsequence converging to a discrete subgroup H of G and that

(25) $\mu(G/H) \le \liminf \mu(G/H_n)$.

In the classical case, we know that

(26) $\mu(G/H) = \lim \mu(G/H_n)$

and Chabauty showed that, under certain conditions, (26) was still true. The problem was settled by Macbeath and Swierczkowski in [35], where they proved that (26) holds if, and only if, G has a compact set of generators (or H is finitely generated). Siegel [43] had defined a fundamental region for G/H and investigated its properties; [35] contains some refinements and a nice property of connectivity is established in [39]. In fact, the foundations and general lines for future progress had already been set by Siegel in his classic memoirs ([43], [44]; cf. Coll. Works).

Acknowledgements

I wish to thank Professor Peter Gruber for the invitation to prepare a general article on Algebraic Lattices highlighting the Geometry of Numbers. I hope that this will serve as an Addendum to his recent survey [5]. This work has been carried out at the Universities of Nottingham and Toronto and I am grateful to S.E.R.C. (U.K.) and N.S.E.R.C (Canada) for their financial support.

References

Section 1

1. Armitage, J.V. 'Algebraic Function and an Analogue of the Geometry of Numbers: The Riemann-Roch Theorem', Arch. der Math., XVIII (1967), 383–393.
2. Armitage, J.V. 'The Product of N linear forms in a Field of Series and the Riemann Hypothesis for Curves' (Univ. de Bordeaux, 1969), 17–27; Paris Soc. Math. France, 1971.
3. Cassels, J.W.S. 'An Introduction to the Geometry of Numbers' (Grundl. Math. Wiss., **99**, Berlin, Springer 1959).
4. Dubois, E. 'Théorèmes de Transfert en Geometrie des Nombres sur un Anneau d'Adeles de Q', C.R. Acad. Sci. Paris, Ser. A–B **283** (1976), A803–A806.
5. Gruber, P.M. 'The Geometry of Numbers-, Proc. Geometry Symposium (Siegen, 1978), Birkhäuser, Basle, 1979 186–225.
6. Lekkerkerker, G.C. 'Geometry of Numbers', Bibl. Math., **8** (Groningen, Wolters-Noordhoff et Amsterdam: North Holland, 1969).
7. Minowski, H. Gesammelte Abhandlungen, I, II Teubner, Leipzig, Berlin, 1911.

Section 2

8. Cassels, J.W.S. 'Bounds for the Least Solution of Homogeneous Quadratic Equations', Proc. Cambridge Phil. Soc., **51** (1955), 262–264; **52** (1956), 604 (cf. H. Davenport, **ibid, 53** (1957), 539–540).
9. Chalk, J.H.H. 'Linearly Independent Zeros of a Quadratic Form over Number-fields', Monatsh. Math., **90** (1980), 13–25.
10. Chalk, J.H.H. 'Algebraic Lattices', C.R. Math. Rep. Acad. Sci. Canada, II (1980), No. 1, 5–10.
11. Conway, J.H. 'Three Lectures on Exceptional Groups', Proc. Conference Lond. Math. Soc., (Finite Simple Groups, Ch. VII, 215–224) (Acad. Press, 1971); (see also Inventiones Math., **7** (1969) 117–142).
12. Coxeter, H.S.M. 'Regular Complex Polytopes' (Cambridge Univ. Press, 1974).
13. Davenport, H. 'Homogeneous Quadratic Forms', Mathematika, **18** (1971), 1–4.
14. Feit, W. 'Some Lattices over $Q(\sqrt{-3})$', J. Algebra, **52** (1978), 248–263.
15. Feit, W. 'On Integral Representations of Finite Groups', Proc. London Math. Soc. (3), **29** (1974), 633–683.

16. Leech, J. 'Some Sphere Packing in higher space', Canad. J. of Math., **16** (1964), 657–682 (see [1] for further references).
17. Leech J. and N.J.A. Sloane, 'Sphere Packing and Error-correcting Codes', Canad. J. of Math., **23** (1971), 718–745.
18. Linsey, J.H. 'A Correlation between $PSU_4(3)$, the Suzuki group and the Conway group', Trans. Amer. Math. Soc., **157** (1971), 189–204.
19. Mahler, K. 'On Lattice points in n-dimensional Star-bodies I, Existence Theorems', Proc. Roy. Soc. London A, **187** (1946), 151–187; Proc. Kon. Ned. Akad. Wet. (1946), 331–343, 444–454, 524–532, 622–631.
20. Minkowski, H. 'Diophantische Approximationen', (Leipzig, 1907), Ch. 6.
21. Raghavan, S. 'Bounds for Minimal Solutions of Diophantive Equations', Nachr. Akad. Wiss., Gottingen, Math. Phys. Kl, **9** (1975), 109–114.
22. Rogers, K. 'Complex Homogeneous Linear Forms', Proc. Camb. Phil. Soc., **52** (1955), 35–38.
23. Rogers, K., and H.P.F. Swinnerton-Dyer, 'The Geometry of Numbers over Algebraic Number-fields', Trans. Amer. Math. Soc., **88** (1958), 227–242.
24. Sloane, N.J.A. 'Codes over GF(4) and Complex Lattices', J. Algebra, **52** (1978), 168–181.
25. Steinitz, E. 'Rechteckige Systeme und Moduli in Algebraischen Zahlenkorper', Math. Annalen, **71** (1912), 328–354.
26. Tits, J. 'Quaternions over $Q(\sqrt{-3})$, Leech's lattice and the Sporadic group of Hall-Janko', J. Algebra, **63** (1980), 56–75.
27. Weyl, H. 'Theory of Reduction for Arithmetic Equivalence', Trans. Amer. Math. Soc., **48** (1940), 126–164; **51** (1942), 203–231.

Section 3

28. Armitage, J.V. 'Algebraic Functions and an Analogue of the Geometry of Numbers: The Riemann-Roch theorem', Arch. Math., XVIII (1967), 383–393.
29. Mahler, K. 'An Analogue to Minkowski's Geometry of Numbers in a Field of Series', Ann. of Math., II, **42** (1941), 481–522.

Section 4

30. Borel, A. 'Introduction aux Groupes Arithmétiques', (Hermann, Paris, 1969).
31. Borel A., and Harish-Chandra, 'Arithmetic Subgroups of Algebraic Groups', Annals of Math., **75** (1962), 485–535.
32. Chabauty, C. 'Limite d'ensembles et Géomètrie des Nombres', Bull. Soc. Math. France, **78** (1950), 143–151.
33. Gel'fand, I.M., M.I. Graev, I.I. Pjateckiĭ-Šapiro, 'Representation Theory and Automorphic Functions', (Saunders, 1969, Ch. 2).
34. Macbeath, A.M. 'Fuchsian Groups' (Lecture Notes, Queen's College, Dundee, 1961).
35. Macbeath, A.M., and S. Swierczkowski, 'Limits of Lattices in a Compactly Generated Group', Canad. J. of Math., **12** (1960), 427–437.
36. Margulis, G.A. 'On the Arithmeticity of Discrete Subgroups', Soviet Math. Dokl., **10** (1969), 900–902.
37. Margulis, G.A. 'Arithmeticity of Non-uniform Lattices', Funkcional Anal. i Priložen., **7(3),** (1973), 88–89.
38. Mostow, G.D. 'Arithmetic Subgroups of Groups with Radical', Annals of Math., **93** (1971), 409–438.
39. Oler. N. 'The Connectness of Fundamental Sets', J. London Math. Soc., **43,** (1968), 111–114.
40. Pontrigin, L.S. 'Topological Groups', (Gordon and Breach N.Y., 1966), Ch. 3, Sections 22–24.
41. Raghunathan, M.S. 'Discrete Subgroups of Lie Groups', (Springer, Berlin, 1972).
42. Shimura, G. 'Introduction to the Arithmetic Theory of Automorphic Functions', (Princeton Univ. Press, 1971).
43. Siegel, C.L. 'Discontinuous Groups', Annals of Math. **44** (1943), 674–689.
44. Siegel, C.L. 'Some Remarks on Discontinuous Groups', Annals of Math., **46** (1945), 708–718, (cf. Theorem 2).
45. Siegel, C.L. 'Symplectic Geometry', (Acad. Press, N.Y., 1964).

46. Springer, T.A. 'Linear Algebraic Groups', (Birkhäuser, Basle, 1981).
47. Takeuchi, K. 'A characterisation of Arithmetic Fuchsian groups', J. Math. Soc. Japan, No. 4, **27** (1975), 600–612.
48. Tsuji, M. 'Theorems in the Geometry of Numbers for Fuschian Groups', J. Math. Soc. Japan, **4** (1952), 189–193.
49. Vigneras, M.-F. 'Quaternions', No. 800, Lecture Notes in Mathematics, (Springer, Berlin 1980).
50. Weil, A. 'On Discrete Subgroups of Lie Groups', Annals of Math., **72** (1960), 364–384; II, **75** (1962), 578–602.

J.H.H. Chalk
Department of Mathematics
University of Toronto
Toronto M5S 1A1 (Canada)

H.S.M. Coxeter (Toronto)

The Twenty-Seven Lines on the Cubic Surface

1. Introduction

Instead of attempting to prove that the general cubic surface contains 27 lines, I refer the reader to an excellent account in the treatise of Miller, Blichfeldt and Dickson [1916, pp. 343–344]. The early history of this famous arrangement of lines is described in Section 2. Here I am indebted to L. Kollros, who edited the collected works of Schläfli [1858, p. 216]. I make consistent use of Schläfli's "epoch-making" notation, even though it has the disadvantage of specializing one of the 36 double sixes. A completely symmetrical notation, in which the number 27 arises as 3^3 instead of $12 + 15$, was devised by Philip Hall [see Coxeter 1930, p. 396], improved by Frame [1938, p. 660] and perfected by Beniamino Segre [1942, p. 3; see also Coxeter 1974, p. 119].

For a complete bibliography up to 1911, see the Cambridge Tract by Archibald Henderson [1911, pp. 96–100], who also wrote a biography of Bernard Shaw.

Section 3 sketches some more recent developments, showing how the cubic surface arises as the case $n = 3$ of a family of n-ic surfaces in n dimensions. Section 4 briefly mentions Segre's discovery that, when the 27 lines are all real, the surface forms a single non-orientable closed manifold of characteristic -5. Section 5 introduces the representation of the 27 lines by the vertices of a uniform polytope 2_{21} in Euclidean 6-space. Section 6 extends an idea of Burnside which leads, in Section 7, to new coordinates for the vertices of 2_{21}. Finally, Section 8 provides simple instructions for drawing a highly symmetrical projection of this polytope.

2. Cayley, Salmon, Steiner and Schläfli

Since a cubic surface has twenty terms in its equation

$$A_0x_0^3 + \cdots + A_3x_3^3 + B_{01}x_0^2x_1 + \cdots + B_{32}x_3^2x_2 \\ + C_0x_1x_2x_3 + C_1x_2x_3x_0 + C_2x_3x_0x_1 + C_3x_0x_1x_2 = 0,$$

just one such surface can be drawn through nineteen points of sufficiently general position in real or complex projective (or affine) space. The above equation shows that a line usually intersects the surface in three points. If the surface contains four points on a line, it must contain the whole line. Thus four linear conditions are imposed on a line if it is to lie on a given cubic surface. In 1849, such considerations persuaded Arthur Cayley that, since there are ∞^4 lines in space, the cubic surface should contain a finite number of them. He communicated this idea to George Salmon, who replied that this finite number is 27.

In 1854 there was an equally fruitful correspondence between Jacob Steiner and Ludwig Schläfli in Switzerland. Steiner showed that the nineteen points that suffice to determine a cubic surface can be replaced by one point and nine lines: the intersections $\alpha_\mu \cdot \beta_\nu$ of two sets of three planes, $\alpha_1\alpha_2\alpha_3$ and $\beta_1\beta_2\beta_3$. Since each of the 27 lines lies on 5 such planes, there are altogether 45 planes, each containing a triangle. Since such a triangle is the complete intersection of its plane with the surface, these 45 planes are tritangent planes of the surface. Moreover, the number of such pairs of "conjugate trihedra" is 120. In his reply, Schläfli described his famous *double six*

$$N = \begin{bmatrix} a_1 & a_2 & a_3 & a_4 & a_5 & a_6 \\ b_1 & b_2 & b_3 & b_4 & b_5 & b_6 \end{bmatrix}$$

which is a special arrangement of twelve lines (on the surface) such that, when arranged as a 2-by-6 matrix, two of them intersect if and only if they occur in different rows and different columns. Any two columns determine a pair of planes, $a_\mu b_\nu$ and $a_\nu b_\mu$, whose intersection $c_{\mu\nu}$ $(=c_{\nu\mu})$ intersects all the four lines and therefore must lie entirely on the surface. In this way Schläfli obtained his notation

$$a_1,\ldots,a_6;\quad b_1,\ldots,b_6;\quad c_{12},\ldots,c_{56}$$

for the 6 + 6 + 15 lines, which "has remained unimproved upon up to the present time" [Henderson 1911, p. 21].

The lines c_{12} and c_{13} are skew (non-intersecting) because both intersect the two skew lines a_1 and b_1; similarly c_{12} and a_3 (or b_3) are skew, both intersecting b_1 and b_2 (or a_1 and a_2). To see that c_{12} intersects c_{34}, we may look for its point of intersection with the plane a_3b_4; since this point is on neither a_3 nor b_4, it must be on c_{34}(the remaining intersection of the plane with the surface). It follows that there are altogether 36 double sixes: the one denoted by N above, and 15 + 20 such as

$$N_{12} = \begin{bmatrix} a_1 & b_1 & c_{23} & c_{24} & c_{25} & c_{26} \\ a_2 & b_2 & c_{13} & c_{14} & c_{15} & c_{16} \end{bmatrix}$$

and

$$N_{123} = \begin{bmatrix} a_1 & a_2 & a_3 & c_{56} & c_{46} & c_{45} \\ c_{23} & c_{13} & c_{12} & b_4 & b_5 & b_6 \end{bmatrix}.$$

The 45 triangles (or tritangent planes) consist of 30 such as $a_1b_2c_{12}$ and 15 such as $c_{12}c_{34}c_{56}$.

In that same letter to Steiner, Schläfli classified the non-singular cubic surfaces in real space according to the number of (real) lines and planes:

(a) 27 lines and 45 planes,
(b) 15 lines and 15 planes,
(c) 7 lines and 5 planes,
(d) 3 lines and 13 planes,
(e) 3 lines and 7 planes.

Case (b) is especially interesting because the fifteen lines $c_{\mu\nu}$ and fifteen planes such as $c_{12}c_{34}c_{56}$ are related like the *duads* and *synthemes* of Sylvester [1844, p. 92; 1861, p. 268]; in fact, their section by an arbitrary plane is the self-dual configuration 15_3 [Coxeter 1968, pp. 135–138, 151–153].

In 1856, Schläfli submitted a manuscript to the Quarterly Journal of Mathematics. The editor handed it to Cayley, who thanked Schläfli for it and added "I hope I have your permission to make a translation of the paper for the Journal" [Schläfli 1858]. Along with the above results, that paper includes the fact that the 120 Steiner trihedral pairs fall into 40 triads such as

$$\begin{bmatrix} a_1 & b_3 & c_{13} \\ c_{12} & a_2 & b_1 \\ b_2 & c_{23} & a_3 \end{bmatrix}, \quad \begin{bmatrix} a_4 & b_6 & c_{46} \\ c_{45} & a_5 & b_4 \\ b_5 & c_{56} & a_6 \end{bmatrix}, \quad \begin{bmatrix} c_{14} & c_{26} & c_{35} \\ c_{36} & c_{15} & c_{24} \\ c_{25} & c_{34} & c_{16} \end{bmatrix},$$

each triad comprising all the 27 lines [Henderson 1911, pp. 39–42]. Schläfli also posed a challenge, asking for a direct proof (independent of the cubic surface) for the Double Six Theorem:

> *Given five lines* b_1, b_2, b_3, b_4, b_5 *having just one transversal* a_6 (that is, one line instersecting all the five), *let* a_ν *denote the second transversal to all the* b's *except* b_ν; *then the five lines* a_1, a_2, a_3, a_4, a_5 *have a transversal* (b_6).

Since that time, many proofs have appeared, notably two by Baker [1911].

3. Geiser, del Pezzo and Du Val

It was observed by Henderson [1911, p. 25] that, among the 27 lines, there are 216 pairs of skew lines, 720 triples, 1080 quadruples, 432 + 216 quintuples, and 72 sextuples (halves of the 36 double sixes).

It was observed by Du Val [1933, p. 69] that, for $n = 4, 5, 6, 7, 8$, the lines skew to $n - 3$ skew lines are related like the lines on the n-ic *del Pezzo surface* in n dimensions. This generalization of the cubic surface has the property that its hyperplane sections are represented by the cubic curves passing through $9 - n$ points of general position in plane. By projecting such an n-ic surface (with $3 < n < 9$) from a point on itself, one obtains the $(n - 1)$-ic surface. Stretching this notion to the case when $n = 3$, one projects the cubic surface from a point on itself to obtain a double plane branching along a general quartic curve whose 28 bitangents arise from the 27 lines along with the tangent plane at the centre of projection [Geiser 1869, p. 129; Coxeter 1928]. The two types of octavic del Pezzo surfaces arise from the two types of quintuples on the cubic surface: one type (such as $b_1b_2b_3b_4b_5$) having only one transversal (a_6), and the other (such as $c_{16}c_{26}c_{36}c_{46}c_{56}$) having two transversals (a_6 and b_6).

4. Segre's generalized polyhedra

Beniamino Segre [1942, pp. 51, 52, 92, 98–111] showed that, in Schläfli's cases (a), (b), (c), (e), the cubic surface is a single closed non-orientable manifold, while in case (d) it consists of two pieces, one of which, being oval, contains no line. The real lines decompose the surface (or, in the last case, its non-oval piece) into a "map" of polygonal regions: say p_3 triangles, p_4 quadrangles, p_5 pentagons, and p_6 hexagons. Suppose this map has V vertices, E edges and F faces, so that

$$\sum p_\nu = F, \qquad \sum \nu p_\nu = 2E;$$

then the Euler–Poincaré characteristic is

$$\chi = V - E + F,$$

as in the following table:

Case	Lines	V	E	F	χ	p_3	p_4	p_5	p_6
(a)	27	135	270	130	-5	10	30 + 60	30	0
(b)	15	45	90	42	-3	6	18	18	0
(c)	7	9	18	8	-1	4	0	0	4
(d) or (e)	3	3	6	4	1	4	0	0	0

5. Jordan, Gosset, Schoute and Todd

Schäfli's notation shows at once that the group of automorphisms of his arrangement of 27 lines includes the symmetric group $\mathfrak{S}_6$ on the subscripts 1,2,...,6 and, commuting with that, the transposition of the letters a and b (leaving all the $c_{\mu\nu}$ invariant). In fact, the transpositions (a b) and (μ ν) interchange the rows of the double sixes N and $N_{\mu\nu}$. Since the 36 double sixes represent the cosets of this subgroup $\mathfrak{S}_6 \times \mathfrak{C}_2$, the order of the whole group, first described by Jordan [1869], is

$$6! \times 2 \times 36 = 51840.$$

The simplest way to generate this group is by six involutions: the five obvious transpositions

$$(1\ \ 2), \quad (2\ \ 3), \quad (3\ \ 4), \quad (4\ \ 5), \quad (5\ \ 6)$$

and the interchange of one more double six,

$$N_{123}$$

[Coxeter 1932, p. 164]. These six involutions can be associated with the vertices of the graph

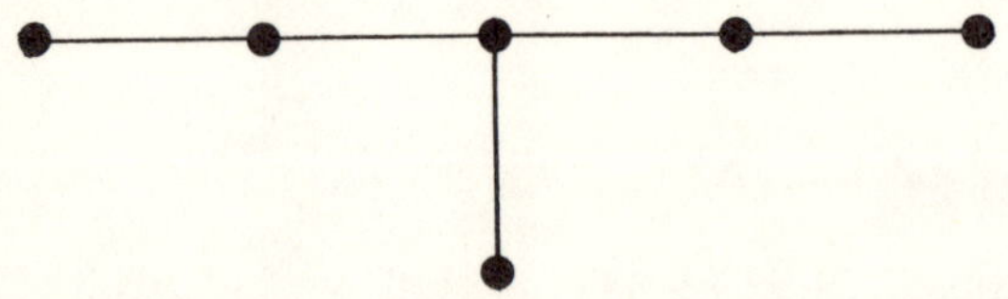

in such a way that two of them, say A and B, satisfy ABA = BAB or AB = BA according as the corresponding vertices are, or are not, joined by an edge [Coxeter 1974, pp. 2, 117, 148].

Since two interesting mirrors generate a dihedral group $\mathfrak{D}_n$ if the angle between them is π/n, it is natural to represent those six involutions in Euclidean 6-space by an appropriate kaleidoscope, namely, by reflections in six hyperplanes intersecting at angles $\pi/3$ and $\pi/2$ in the manner indicated by the graph. The five edges, representing five subgroups $\mathfrak{D}_3$, suggest denoting the group by the symbol $[3^{2,2,1}]$ [Coxeter 1940, p. 472].

Such kaleidoscopes were used by Cartan [1927, p. 230] in his investigation of the simple Lie groups. This particular one is relevant to the Lie group E_6. The images of a point lying on all the mirrors except the first consist of 27 points whose convex hull is the uniform polytope 2_{21}, first described by Gosset [1900, p. 47]. Its connection with the 27 lines was noticed by Schoute [1910] and explained by Todd [1932]. The 27 vertices of this six-dimensional polytope represent the 27 lines in such a way that two vertices are adjacent (joined by an edge) if and only if the corresponding lines are *skew*. In fact, these 27 points (like the 5 vertices of a regular pentagon) are a two-distance set: the $\binom{27}{2}$ pairs of them form 216 edges (representing the 216 pairs of skew lines) and 135 diagonals (representing the 135 pairs of intersecting lines, or the vertices of the 45 triangles, which are the points of contact of the 45 tritantent planes). Along with its 27 vertices and 216 edges, 2_{21} has 720 triangular faces, 1080 tetrahedra, 432 + 216 α_4's (four-dimensional simplexes), 72 α_5's and 27 β_5's (five-dimensional cross polytopes) [Coxeter 1973, pp. 203, 211], representing the sets of 3, 4, 5, 6 skew lines and, for each of the 27 lines, the ten that intersect it: five intersecting pairs, represented by the five pairs of opposite vertices of the octahedron-analogue β_5.

6. An affine interpretation for Burnside's algebra

Burnside [1911, pp. 485–488] used symbols for the 27 lines as elements of an algebra. This amounts to representing the lines by 27 points in an affine space of n dimensions ($2 \le n \le 6$), such that the 45 triangles representing the tritangent planes all have the same centroid. Using this centroid as the origin, he applies Schläfli's symbols for the 27 lines to the position vertors of the 27 points, so that

$$\mathrm{a}_1 + \mathrm{b}_2 + \mathrm{c}_{12} = \cdots = \mathrm{c}_{12} + \mathrm{c}_{34} + \mathrm{c}_{56} = \cdots = 0.$$

Consequently the 135 "double twos" such as

$$\begin{bmatrix} \mathrm{a}_1 & \mathrm{a}_2 \\ \mathrm{b}_1 & \mathrm{b}_2 \end{bmatrix}, \quad \begin{bmatrix} \mathrm{a}_1 & \mathrm{c}_{23} \\ \mathrm{a}_2 & \mathrm{c}_{13} \end{bmatrix}, \quad \begin{bmatrix} \mathrm{b}_1 & \mathrm{c}_{23} \\ \mathrm{b}_2 & \mathrm{c}_{13} \end{bmatrix}, \quad \begin{bmatrix} \mathrm{c}_{12} & \mathrm{c}_{14} \\ \mathrm{c}_{23} & \mathrm{c}_{34} \end{bmatrix}$$

all appear as parallelograms.

Choosing as a basis an arbitrary set of six points (or vectors) $\mathrm{a}_1,\ldots,\mathrm{a}_6$, let us define

$$\mathrm{s} = (\mathrm{a}_1 + \cdots + \mathrm{a}_6)/3$$

(Burnside's $s/3$). Then the remaining 21 vectors are

$$\mathrm{b}_\nu = \mathrm{a}_\nu - \mathrm{s}, \quad \mathrm{c}_{\mu\nu} = \mathrm{s} - \mathrm{a}_\mu - \mathrm{a}_\nu.$$

7. Cartesian coordinates for the 6-dimensional polytope 2_{21}

The polytope 2_{21} arises as the special case when $\mathrm{a}_1\mathrm{a}_2\mathrm{a}_3\mathrm{a}_4\mathrm{a}_5\mathrm{a}_6$ is a regular simplex α_5 in a hyperplane of Euclidean 6-space, while the origin is placed so that all the 135 parallelograms are squares. Considering the parallelogram $\mathrm{a}_1\mathrm{a}_2\mathrm{b}_2\mathrm{b}_1$, we see that the sides a_1a_2 and a_1b_1 should have the same length:

$$(a_1 - a_2)^2 = (a_1 - b_1)^2 = s^2.$$

Avoiding fractions, we may take $a_1,\ldots,a_6$ to be

$$(x + 6, x, x, x, x, x),\ldots,(x, x, x, x, x, x + 6),$$

so that $s = 2(x + 1)(1, 1, 1, 1, 1, 1)$, $a_1 - a_2 = 6(1, -1, 0, 0, 0, 0)$ and the condition $s^2 = (a_1 - a_2)^2$ becomes $24(x + 1)^2 = 72$,

$$x = -1 \pm \sqrt{3}.$$

Choosing the positive value, so that $s = 2\sqrt{3}(1, 1, 1, 1, 1, 1)$, we find

$$a_1 = (\sqrt{3}+5, \sqrt{3}-1,\ldots, \sqrt{3}-1),\ldots, a_6 = (\sqrt{3}-1,\ldots, \sqrt{3}-1, \sqrt{3}+5),$$
$$b_1 = (-\sqrt{3}+5, -\sqrt{3}-1,\ldots, -\sqrt{3}-1),\ldots, b_6 = (-\sqrt{3}-1,\ldots, -\sqrt{3}-1, -\sqrt{3}+5),$$
$$c_{12} = (-4, -4, 2, 2, 2, 2),\ldots, \qquad c_{56} = (2, 2, 2, 2, -4, -4).$$

These 27 points represent the 27 lines in such a way that two lines are skew or intersecting according as the distance between the corresponding two points is $6\sqrt{2}$ or 12. In other words, these coordinates describe a 2_{21} of edge $6\sqrt{2}$ and circumradius $4\sqrt{3}$. For coordinates referred to other Cartesian frames, see Elte [1912, pp. 109–111] and Coxeter [1928, pp. 3, 6; 1940, p. 469].

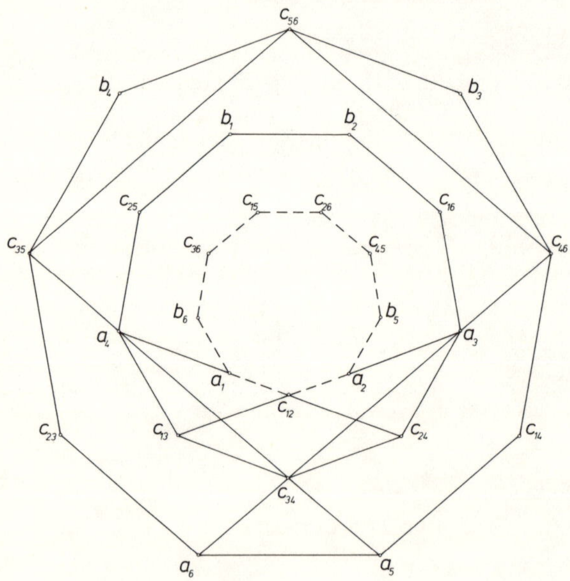

Fig. 1: The three concentric enneagons

8. The enneagonal projection of 2_{21}

After referring to Salmon's remark that the 27 lines are the complete intersection of the cubic surface with a surface of order nine, Baker [1933, p. 15] mentioned

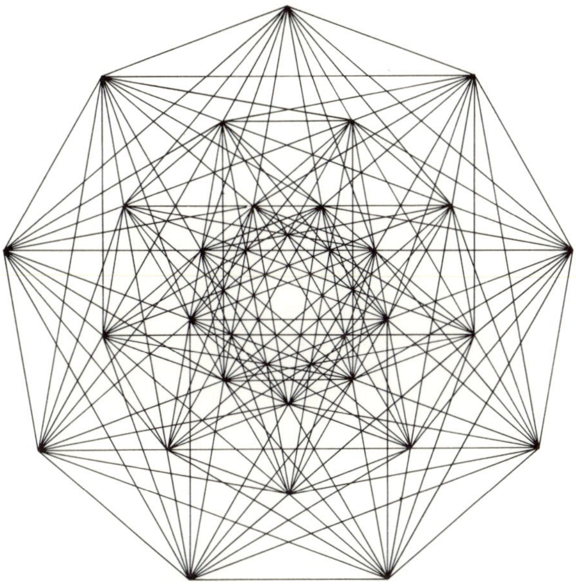

Fig. 2: An orthogonal projection of 2_{21}

my observation that the latter may be taken to consist of nine tritangent planes* given by the columns of the matrix

$$\begin{bmatrix} a_5 & c_{14} & c_{46} & b_3 & c_{56} & b_4 & c_{35} & c_{23} & a_6 \\ b_1 & c_{25} & a_4 & c_{13} & c_{34} & c_{24} & a_3 & c_{16} & b_2 \\ c_{15} & c_{36} & b_6 & a_1 & c_{12} & a_2 & b_5 & c_{45} & c_{26} \end{bmatrix}.$$

The first row of this matrix consists of a cycle of nine lines such that each is skew to its four neighbours (two before and two after) but intersects each of the remaining four [Coxeter 1930, p. 419]. By drawing a regular enneagon, associating its vertices with these nine symbols, and completing parallelograms to construct two smaller enneagons (as in Figure 1), we obtain an eminently symmetrical example of the arrangement of 27 points described in Section 6. When all the parallelograms have been completed, as in Figure 2, the result is an orthogonal projection of the vertices and edges of 2_{21} onto a suitable plane [Coxeter 1940, pp. 460–462]. It is, perhaps, unfortunate that many of the parallelograms, such as $c_{15}c_{16}c_{26}c_{25}$, have been foreshortened into sets of four collinear points.

Figure 3 shows how two opposite facets α_5 form a simplicial prism $\alpha_5 \times \alpha_1$, representing a double six. Finally, Figure 4 shows an inscribed "double prism" $\alpha_2 \times \alpha_2$ (the Cartesian product of two equilateral triangles) representing the nine lines of a Steiner trihedral pair [Coxeter 1940, p. 466]. By rotating this drawing through $\pm 2\pi/9$, we could obtain two other such prisms, completing one of the forty triads of trihedral pairs.

* Other such sets of nine planes had already been found by Schläfli, as we saw near the end of Section 2.

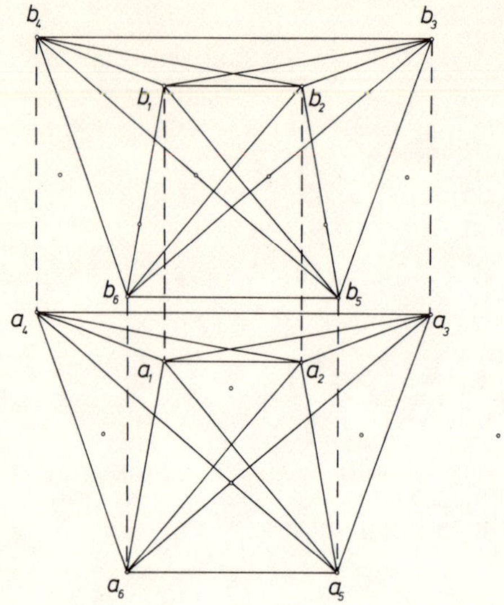

Fig. 3: The simplicial prism $\alpha_5 \times \alpha_1$ representing a double six

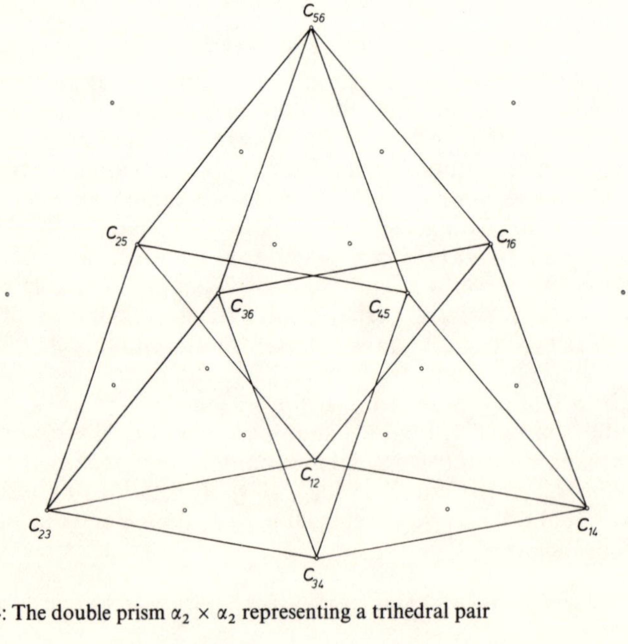

Fig. 4: The double prism $\alpha_2 \times \alpha_2$ representing a trihedral pair

References

Baker, H.F.
1911 *A geometrical proof of the theorem of a double six of straight lines*, Proc. Royal Soc., A **84,** 597–602.
1933 *Principles of Geometry*, vol. VI. Cambridge University Press.
William Burnside
1911 *Theory of Groups of Finite Order*. Cambridge University Press.
Élie Cartan
1927 *La géométrie des groupes simples*, Annali di Mat. (4), **4,** 209–256.
Coxeter, H.S.M.
1928 *The pure Archimedean polytopes in six and seven dimensions*, Proc. Camb. Phil. Soc. **24,** 1–9.
1930 *The polytopes with regular-prismatic vertex figures*, Part 1, Phil. Trans. Royal Soc., A **229,** 329–425.
1932 *Ibid.*, Part 2, Proc. London Math. Soc. (2), **34,** 126–189.
1940 *The polytope 2_{21}, whose twenty-seven vertices correspond to the lines on the general cubic surface*, Amer. J. Math., **62,** 457–486.
1968 *Twelve Geometric Essays*. Southern Illinois University Press, Carbondale, Il.
1973 *Regular Polytopes*. Dover, New York.
1974 *Regular Complex Polytopes*. Cambridge University Press.
Patrick Du Val
1933 *On the directrices of a set of points in a plane*, Proc. London Math. Soc. (2), **35,** 23–74.
Elte, E.L.
1912 *The Semiregular Polytopes of the Hyperspaces*. Gebroeders Hoitsema, Groningen.
Frame, J.S.
1938 *A symmetric representation of the twenty-seven lines on a cubic surface by lines in a finite geometry*, Bull. Amer. Math. Soc., **44,** 658–661.
Geiser, C.F.
1869 *Ueber die Doppeltangenten einer ebenen Curve vierten Grades*, Math. Ann. **1,** 129–138.
Thorold Gosset
1900 *On the regular and semi-regular figures in space of n dimensions*, Messenger of Math., **29,** 43–48.
Archibald Henderson
1911 *The Twenty-seven Lines upon the Cubic Surface*. Cambridge University Press.
Camille Jordan
1869 *Sur l'équation aux vingt-sept droites des surfaces du troisiéme degré*. J. de Math. (2), **14,** 147–166.
Miller, G.A., H.F. Blichfeldt and L.E. Dickson
1916 *Theory and Applications of Finite Groups*. Wiley, New York.
Ludwig Schläfli
1858 *An attempt to determine the twenty-seven lines upon a surface of the third order, and to divide such surfaces into species in reference to the reality of the lines upon the surface*, Quarterly J. Math. **2,** 55–56; *Gesammelte Mathematische Abhandlungen*, **2,** 198–218 (Birkhäuser, Basel, 1953).
Schoute, P.H.
1910 *On the relation between the vertices of a definite six-dimensional polytope and the lines of a cubic surface*, K. Akad. Wetenschappen te Amsterdam, Proceedings of the Section of Sciences, **13,** 375–383.
Beniamino Segre
1942 *The Non-singular Cubic Surfaces*, Clarendon Press, Oxford.
Jacob Steiner
1857 *Über die Flächen dritten Grades*, J. für r. und a. Math. **53,** 133–141.
Sylvester, J.J.
1844 Collected Mathematical Papers, vol. I (Cambridge University Press, 1904).
1861 *Ibid.*, vol. II (Cambridge University Press, 1907).
Todd, J.A.
1932 *Polytopes associated with the general cubic surfaces*, J. London Math. Soc., **7,** 200–205.

H.S.M. Coxeter
Department of Mathematics
University of Toronto
Toronto M5S 1A1 (Canada)

W. Fenchel

Convexity Through the Ages*

Cependent, les théories ont leurs commencements: des allusions vagues, des essais inachevés, des problèmes particuliers; et même lorsque ces commencements importent peu dans l'état actuel de la Science, on aurait tort de les passer sous silence.

F. Riesz, 1913.

1. In his treatise "On the sphere and cylinder" Archimedes defines a convex arc as a plane curve which lies on one side of the line joining its endpoints and all chords of which lie on the same side of it. Analogously he defines a convex surface bounded by a plane curve. His determination of arc lengths is based on certain postulates. One of these is: If one of two convex arcs with common endpoints lies between the other and the line joining the endpoints, the length of the first arc is smaller than that of the second. The determination of surface areas is founded on an analogous postulate.

In his treatise "On the equilibrium of plane figures" Archimedes sets up postulates on the centroid of a plane figure. One of these states that a convex figure contains its centroid. In the other relevant writings which have been preserved he seems however not to use this postulate but the related one: If a plane (not necessarily convex) figure lies on one side of a line, its centroid lies on the same side of the line.

When in the seventeenth century Archimedes' methods were taken up again, convexity and the postulates quoted played still a role, for instance in the work of Fermat. But the use of intuitive limiting processes and the development of the Calculus let them recede to the background and even be forgotten. That this in any case holds of the postulates on arc length and surface area, which now had to be considered as theorems, may be inferred from the following: In 1841 A. Cauchy discovered: The perimeter of a closed convex curve equals π times the mean value of the lengths of the orthogonal projections of the curve onto the lines through a point. The surface area of a closed convex surface equals 4 times the mean value of the areas of the orthogonal projections of the surface onto the planes through a point. In a more elaborate paper of 1850 Cauchy mentions as corollary only that if a closed convex curve is contained in a circle its perimeter is smaller than that of the circle, and the 3-dimensional analogue. The general statement of Archimedes on surface area is proved in a paper by H. Minkowski of 1901 on a generalization of the notion of surface area, and in a later paper it is also obtained as an immediate consequence of the rediscovered Cauchy formula.

* The following is essentially a translation of a talk given at the celebration of the centenary of the Danish Mathematical Society in 1973. Readers are kindly requested to take into consideration that the audience did not consist of experts in convexity. (The original version appeared in "Dansk Matematisk Forening 1923–1973", Copenhagen 1973, p. 103–116.)

2. Simple convexity properties of special sets, curves and functions certainly played a role in numerous investigations in Analysis and Geometry. To give a detailed account of these is infeasable. In the sequel attention is called to some classical ideas and results which have led to intensive research, and the development of these fields is briefly described. A strictly chronological account is not aimed at since it would be rather confusing. Completeness in any sense is not achieved; several important aspects such as the theory of convex functions, various generalizations, and the most recent contributions are not mentioned, and many more contributors could have been named.

3. A well-known early example of the occurance of convexity, though originally in a more implicite role, is Euler's theorem on polyhedra: The numbers v of vertices, e of edges, and f of faces of a closed polyhedron satisfy $v - e + f = 2$. In many proofs, in the older ones often tacitely, it is assumed that the polyhedron is convex. This holds of Euler's proof of 1753 and the more than 100 years older one of Descartes (only known from an annotation by Leibniz discovered in 1860). That the statement does not hold for arbitrary polyhedra and that the condition for its validity is of a topological nature was first cleared up in the last century.

In the investigations of which the theorem is a part Euler aims at a classification of the polyhedra. The "type" of a polyhedron may be given by a table listing for each face its vertices in cyclical order. Two polyhedra are of the same type, are isomorphic, if their tables, possibly after a renumbering of the vertices and faces of one of them, are identical. What Euler did may in these terms be described as a study of the properties of such tables. To mention one of his results: The number of faces with an odd number of vertices is even.

There appeared many investigations about this topic, in particular in the second half of the last century. In most of them only convex polyhedra are considered, presumably often because the problems otherwise seemed insuperable. The convexity assumption added however also aspects of independent interest. A concluding result, obtained by E. Steinitz in 1920, states that certain rather obvious necessary conditions for a table to come from a convex polyhedron are also sufficient.

Interest arose also in the analogues of the convex polyhedra in higher-dimensional spaces, the convex polytopes. The first contributions are to be found in a comprehensive treatise by L. Schläfli of 1852, published not until 1901, long after his death. In the meantime and later, in particular in the last thirty years, many papers and several books have been devoted to the subject. Much of the work is centred around the following problem, solved for $n = 3$ by Steinitz in 1906: to find the conditions an n-tuple $(h_0,\dots,h_{n-1})$ of positive integers has to satisfy in order that there exists a convex polytope in $\mathbb{R}^n$ with h_i faces of dimensions i. In its generality the problem seems inaccessible at present, but remarkable partial results have recently been obtained.

4. In a paper on statics of 1798 J. Fourier points out that the methods then available are not applicable to equilibrium problems for mechanical systems where the bodies are subjected to one-sided constraints because inequalities occur as conditions. Twenty-five years later he returns to the subject in a couple of communications to the Académie des Sciences. He observes that in general the inequalities in question are linear and that a theory of systems of such inequalities should be developed. In the case of three variables he interpretes the set of solutions as the intersection of half-spaces, thus as a convex polyhedron, and discusses some examples. In 1826 he describes a method to solve systems of linear inequalities.

Presumably statics problems of the kind considered by Fourier were taken up rather soon. The following one is solved in L. Poinsot's "Eléments de statique", in any case in the 8-th edition of 1842 (the first inaccessible to me appeared in 1804). A rigid body constrained to one side of a plane and subjected to forces is in equilibrium if and only if the body touches the plane, the resultant of the forces is directed orthogonally toward the plane, and the line of action of the resultant intersects the plane at a point belonging to the convex hull of the set of points of contact. This is obtained by proving that every half-plane containing the points of contact must also contain that point of intersection.

Fourier's challenge concerning systems of linear inequalities seems to have been taken up much later. In 1873 P. Gordan had reason, in another context, to prove a theorem on such systems. The first more systematic treatment, however not using the geometrical interpretation, is to be found in several notes of 1894 and a paper of 1902 by a physicist, J. Farkas.

Independently H. Minkowski dealt with systems of linear inequalities in his book "Geometrie der Zahlen" the main part of which appeared in 1896. He considers systems of the form $Ax \geq 0$ where A is a given real $m \times n$-matrix, x a column with n elements, and where the inequality is required for every element of the resulting column. If the system has non-zero solutions, they make up an unbounded convex pyramid in $\mathbb{R}^n$. Minkowski proves that there are finitely many "extreme" solutions, points on the edges of the pyramid, such that every solution is a linear combination of these with non-negative coefficients. Inhomogeneous systems $Ax \geq b$ with a given column b he reduces to the homogeneous case by adjoining a new variable. It is strange that Minkowski here does not use a geometrical terminology. There is however hardly any doubt that these investigations have inspired him to introduce later on the notion "extreme point" of a convex set. This is a point of the set which is not interior to any segment belonging to the set; equivalently, the set remains convex if the point is removed. He shows that a closed and bounded convex set in $\mathbb{R}^3$ is the convex hull of the set of its extreme points; more precisely, every point of the set belongs to a (possibly degenerate) tetrahedron the vertices of which are extreme points. In 1940 M. Krein and D. Milman succeeded in generalizing this to a large class of infinite-dimensional topological vector spaces in the slightly weaker form: A compact convex set is the closure of the convex hull of its extreme points. This is a well-known important tool in Functional Analysis.

In the first third of this century several authors have developed the theory of systems of linear inequalities further, often using convexity arguments. To give an example: The theorem of Gordan mentioned above may be formulated as follows. In addition to the system of inequalities $Ax > 0$ one considers the system of equations $A'y = 0$ with the conditions $y \geq 0$, $y \neq 0$, where A' denotes the transpose of the matrix A. The statement is that precisely one of the two systems has solutions. To see this one has only to interprete the rows of A as points $a^1,\ldots,a^m$ in $\mathbb{R}^n$. That the first system has a solution x is equivalent to the existence of a hyperplane H with normal vector x through the origin such that all of the points lie in one of the open half-spaces bounded by H, and this means that the origin is exterior to the convex hull of the m points. That the second system has a solution $y = (y_1,\ldots,y_m)$ means that the origin is the centroid of the points a^i with masses y_i, hence that the origin belongs to the convex hull.

A comprehensive treatment of linear systems in which the signs $=$, $\geq$, $>$, enter arbitrarily is to be found in Th. Motzkin's thesis of 1936. In the late forties the interest in linear inequalities rose strongly due to the emergence of the linear programming problems. Such a problem consists in the determination of the maximum or minimum of a linear function with variables subject to linear inequality conditions. The theoretical study as well as the development of numerical methods led to many investigations on systems of linear inequalities in which the interpretation of the solution sets as convex polytopes plays a prominent role. This has certainly contributed to the recent rise of the theory of convex polytopes mentioned earlier.

5. In 1812 Cauchy proved: If corresponding faces of two isomorphic convex polyhedra are congruent, the polyhedra are (properly or improperly) congruent. His elegant proof runs essentially as follows. If the angles at every pair of corresponding edges of the two polyhedra P and P′ are equal, the statement is easily seen to be true. Suppose this is not the case. Mark every edge of P at which the angle is different from that of the corresponding edge of P′ by + or − according as the second angle is greater or smaller than the first one. The main lemma states that if one goes once around a vertex at which marked edges end, one would have at least four changes of sign. A marking of the edges with this property contradicts however Euler's theorem on polyhedra.

Cauchy's result shows in particular that a convex polyhedron can not be deformed continuously such that each face remains congruent with itself. In a report of 1845 to the Académie des Sciences, on a paper on bending of surfaces submitted by O. Bonnet, Cauchy claims that an immediate consequence of his theorem on polyhedra is that a closed convex surface is rigid in the sense that it does not admit isometric deformations, and he says that Lagrange long ago had called his attention hereto. (By the way, this was already claimed earlier, namely by Euler about 1770, even wrongly for arbitrary closed surfaces, and by F. Minding in 1838.)

J.H. Jellett was the first who attempted to prove the statement, even somewhat more. In a paper of 1854 he introduces the notion of "infinitesimal bending" of a surface, meaning a deformation of it under which the lengths of the curves on it are stationary in the beginning. Jellett finds conditions which infinitesimal bendings of pieces of convex surfaces have to satisfy, but they are not sufficient to prove the rigidity in his sense of closed convex surfaces. The first proof is due to H. Liebmann; it appeared in 1900.

Liebmann's theorem is however not the analogue of Cauchy's. It was conceivable that there exist two isometric, but non-congruent closed convex surfaces which can not be deformed into each other by an isometric deformation.

This problem was taken up by H. Weyl in 1916. He added a new aspect: the question of the existence of a closed convex surface with a given intrinsic metric. More precisely he states: On the unit sphere let there be given a positive definite quadratic differential form (satisfying certain differentiability conditions) with positive Gauss curvature. Then there exists one and, up to orthogonal transformations, only one closed convex surface which admits a bijection onto the unit sphere such that its metric fundamental form at each point equals the prescribed form at the corresponding point of the unit sphere. Weyl proves, however, only the existence and under the assumption that the given differential form deviates little from the metric form of the sphere. He indicates briefly how he intends to prove the general statement. On the occasion of his seventieth birthday in 1955 a collection of Weyl's papers, among these the one under discussion, was edited. In a remark added by him in 1955 he says that he had published the paper in this unfinished form because he was drafted by the German army and "... ob es mir damals gelungen wäre, wirklich zum Ziel zu kommen, (ich glaubte es)—wer will das heute sagen?" Weyl did not return to the subject, apart from a simplified proof of the theorem of Liebmann mentioned above.

The first essential contribution thereafter is due to S. Cohn-Vossen who in 1927 for smooth, piecewise analytic and some years later for three times differentiable convex surfaces proved the uniqueness theorem analogous to Cauchy's for polyhedra: Two isometric closed convex surfaces are congruent. The proof does not follow the way sketched by Weyl. It may be considered as analogous to Cauchy's, but can not be obtained from it by a limiting process. A simpler proof under weaker assumptions (non-negative instead of positive Gauss curvature) based on integral relations was found by G. Herglotz in 1943.

The existence problem with analytic data was solved by H. Lewy in 1938 on the basis of his former investigations on elliptic partial differential equations of Monge-Ampère type. In 1953 L. Nirenberg succeeded in proving Weyl's existence statement completely and as Weyl says "... auf dem Wege ..., der mir vorgeschwebt hatte".

In the meantime the subject was, however, taken up from a more general, purely geometrical point of view. An arbitrary closed convex surface may be provided with a metric by defining the distance of two of its points to be the greatest lower bound of the lengths of the curves on the surface joining the points. About 1940 A.D. Aleksandrov and several collaborators investigated this intrinsic metric thoroughly. The most difficult part was to find the properties equivalent to the local convexity of the surface. In 1942 Aleksandrov was able to prove: Given a metric on the unit sphere satisfying these local conditions, there exists a closed convex surface for which it is the intrinsic metric. He proves this first by an interesting topological argument for convex polyhedra the intrinsic metric of which can easily be characterized. He uses that a convex polyhedron is determined by its intrinsic metric, a sharpening of Cauchy's theorem provable by Cauchy's method. The general case is dealt with by a rather involved passage to the limit. The uniqueness problem still required a considerable effort. After having solved it under restrictive assumptions, A.V. Pogorelov was able to prove in 1952 that every closed convex surface is uniquely determined, up to orthogonal transformations, by its intrinsic metric. Hence, what started with Cauchy's paper has been rounded off most satisfactorily.

6. Several other early contributions to convexity theory have to be mentioned because they played a role in the later development.

This holds of some of J. Steiner's papers. In 1838 he introduced the notion of curvature centroid of a closed convex curve, that is, its centroid for a mass distribution the density of which is the curvature. Its generalization to arbitrary dimension, now called Steiner point, has turned out to be a useful tool in some recent investigations. A generalization of his formula of 1840 for the volume bounded by two parallel convex surfaces is basic for Minkowski's theory of convex bodies to be described later on. Some of Steiner's ideas concerning isoperimetric problems have proved to be fruitful.

In his book "Geometrie der Lage" of 1847 G.K.C. von Staudt considers subsets of the projective plane and the projective space which are now called convex. Such a set K may be defined as follows: There exists a line (plane) P not meeting K, and for any two points of K that segment joining them which does not meet P is contained in K. The set L of lines (planes) not meeting K has as von Staudt observes precisely the dualistic properties: There exists a point p, namely any one belonging to K, which is not contained in any of the lines (planes) of L, and the pencil determined by two lines (planes) of L is divided by these lines (planes) into two parts of which the one not containing the line (plane) through p belongs entirely to L. Consequently a correlation, in particular a polarity, maps L onto a convex point set. This leads to a one-to-one mapping of the set of closed convex sets onto itself, involutory if a polarity is applied.

This section of von Staudt's book has apparently remained unnoticed. His result was, however, rediscovered. In the third part of his comprehensive paper "Bedingt konvergente Reihen and konvexe Systeme", 1913, 1914, 1916, Steinitz gives an exact treatment of the general duality property of convex sets in a projective space of arbitrary finite dimension. A more special version, which has turned out to be of great importance and which will be described below, had already appeared in Minkowski's posthumous paper "Theorie der konvexen Körper, insbesondere Begründung ihres Oberflächenbegriffs".

7. The first thorough investigations of sets, curves, and surfaces characterized exclusively by convexity properties are due to H. Brunn. Most important in this respect are his first papers, "Über Ovale and Eiflächen", 1887, and "Über Curven ohne Wendepunkte", 1889. Intuitively obvious properties are listed without proofs, and modern requirements of exactness are not always satisfied. But many interesting results are obtained. To mention one example: If every intersection of a convex body with a plane has a centre, the body is an ellipsoid. This and similar results of Brunn have inspired numerous characterizations of circles, spheres, ellipses, ellipsoids. One due to W. Blaschke, 1916, has found several applications: If every circumscribed cylinder of a convex body touches the body in a plane curve, the body is an ellipsoid.

The most important of Brunn's theorems may be formulated as follows: Consider a convex body K and a line L. For every plane orthogonal to L and intersecting K replace the intersection by a circle having the same area and its centre on L. Then the union of these circles is a convex body. In other words, the square root of the area of the intersection is a concave function of a coordinate on L. This function is linear if and only if K is a truncated cone (especially a cone or a cyclinder) with an arbitrary convex base (orthogonal to L).

The first complete proof of the last statement and the carrying out of a proof by induction, sketched by Brunn, of the theorem for arbitrary finite dimensions are due to Minkowski. He saw also the wide applicability of the theorem.

8. Minkowski's interest in convexity originated, strange to say, from the Theory of Numbers. In his theory of the reduction of positive definite quadratic forms he needed a theorem of Ch. Hermite of 1850. In Minkowski's geometrical formulation it states that an ellipsoid in $\mathbb{R}^n$ with centre at the origin and volume greater than a number depending only on n contains at least one lattice point (point with integral coordinates) different from the origin. In 1891 Minkowski realized that the relevant properties of the ellipsoid are its convexity and its central symmetry, and he proved: Every closed bounded convex set in $\mathbb{R}^n$ with the origin as centre and volume greater than 2^n contains at least one lattice point different from the origin.

A simplified proof due to G.D. Birkhoff, communicated in a paper by H.F. Blichfeldt of 1914 about a generalization of the theorem, runs as follows. It is sufficient to show that the convex set K contains distinct points x and y such that the coordinates of $y - x$ are even integers. Indeed, because of the central symmetry K will contain $-x$ and consequently $\frac{1}{2}(y - x)$ because of the convexity. The hyperplanes which are parallel to the coordinate hyperplanes and pass through the points with even coordinates divide the space into hypercubes with edges of length 2. Choose one of them, H, and carry the finitely many closed hypercubes having a non-empty intersection with K by translations onto H. The translated intersections are closed subsets of H the sum of the volumes of which is greater than or equal to the volume 2^n of H. Hence, they cannot be mutually disjoint. A common point of two of them originates from two points x and y of K with the property required.

By specializing and improving the theorem Minkowski found many important results in the Theory of Numbers and founded therewith the branch of it which he gave the name "Geometrie der Zahlen" and expounded in the book with this title and several papers. Minkowski's theorem, sharpenings, variants, and generalizations play still a prominent part in this field.

9. Besides in their applications to number theory Minkowski became more and more interested in convex sets for their own sake. In the years 1897–1904 he proved several important theorems to be mentioned later on. A systematic exposition of the basic properties of closed bounded convex sets in $\mathbb{R}^3$, in particular those with interior points, the convex bodies, he gave in the posthumous paper quoted in Section **6**. It was published in 1911 in Vol. II of the "Gesammelte Abhandlungen". A minor part of its content had been dealt with earlier in papers by J. Hjelmslev, C. Carathéodory and, especially, Brunn. Among the topics dealt with by Minkowski are the existence of a supporting plane through every boundary point, the separation by a plane of convex bodies with disjoint interiors, extreme points, linear combinations of convex sets, introduction of the distance function and the support function of a convex body.

Further contributions are due to Carathéodory, 1911, and Steinitz. Both were induced to consider convex sets by problems from Analysis. In the large paper quoted in Section **6** Steinitz studies arbitrary (not necessarily closed or bounded) convex sets.

10. Let K be a convex body in $\mathbb{R}^n$ with the origin o in its interior. Minkowski defines the distance function F of K as follows. Let x denote an arbitrary point different from o and ξ the point at which the half-line from o through x intersects the boundary of K. Then $F(x) > 0$ is the number for which $x = F(x)\xi$. By definition $F(o) = 0$. The function F is convex and positively homogeneous (that is, $F(tx) = tF(x)$ for $t \geq 0$). Every function with these properties is the distance function of a convex body. Clearly, $F(x) \leq 1$ if and only if $x \in K$.

The support function H of a convex body K may be defined by

$$H(u) = \sup_{x \in K} u \cdot x, \quad u \in \mathbb{R}^n,$$

where $u \cdot x$ denotes the inner product. If the equation of a hyperplane not passing through o with normal vector u is written $u \cdot x = 1$, then $H(u) \leq 1$ if and only if the hyperplane does not intersect the interior of K. Now, if o is an interior point of K, we have $H(u) > 0$ for $u \neq o$ and H has the same properties as F. Consequently it is the distance function of a convex body K^0, the polar body of K. Its support function turns out to be the distance function of K. One has therefore an involutory map of the set of convex bodies with o in the interior onto itself. This is von Staudt's duality brought about by the polarity with respect to the unit sphere.

Minkowski himself did not make any applications of it. The first one seems to be that of A. Haar in a paper of 1918 on Tchebychev approximation. Profound applications in the Geometry of Numbers were initiated by K. Mahler in 1939.

In his "Geonetrie der Zahlen" Minkowski uses the distance function F of a convex body in $\mathbb{R}^n$ to define a metric by letting the distance from a point x to a point y be $F(y - x)$. If $F(-x) = F(x)$ is assumed, F is what is now called a norm in the vector space $\mathbb{R}^n$. The support function H of the body may then be considered as a norm in the dual space, that is, the space of linear forms $u: x \mapsto u \cdot x, x \in \mathbb{R}^n$. In this interpretation the polarity plays now an important role in the theory of infinite-dimensional vector spaces.

Inspired by work of F. Riesz on l^p-spaces, E. Helly in 1921 carried some of Minkowski's ideas and results over to spaces of number sequences. A year later S. Banach introduced the notion of what is now called a Banach space, a normed vector space which, considered as a metric space, is complete. The following basic theorem, known as the Hahn-Banach theorem, was proved by Helly in his more special setup, in full generality by H. Hahn in 1927 and independently by Banach in 1929: Let F be a positively homogeneous convex function on a real vector space V, and let f be a linear form defined on a linear subspace U of V such that $f(x) \leq F(x)$ for $x \in U$. Then f can be extended to a linear form on V satisfying the same inequality for $x \in V$. In geometrical terminology: The graph of F in $V \times \mathbb{R}$ is a convex cone with the vertex at the origin, and the graph of f is a subspace lying under or on the surface of the cone. The theorem then states that there exists a hyperplane supporting the cone and passing through the graph of f. Application to a normed vector space with the norm as the function F yields the existence of non-trivial continuous linear forms.

Neither in the paper mentioned nor in his book "Théorie des opérations linéaires" of 1932, which played such a decisive role in the development of the field, Banach discusses the close connection with convexity theory. But others, not the least his collaborators, in particular S. Mazur, explained about the same time the connection between the Hahn-Banach theorem and questions about existence of supporting hyperplanes through the boundary points of convex sets and of hyperplanes separating disjoint convex sets.

It turned out that the notion of normed vector space, in particular of Banach space, is not general enough to deal with important problems of Functional Analysis. A more comprehensive class of topological vector spaces, which has found many applications, was introduced by J. von Neumann in 1935. Here convexity enters essentially. It is required that the origin and consequently every point has convex neighborhoods which form a basis for the neighborhood system. In these "locally convex" topological vector spaces the theorems mentioned on convex sets and the Krein-Milman theorem quoted in Section **4** are valid. The Hahn-Banach theorem implies the existence of continuous linear forms which makes a duality theory possible. These investigations and their further development got a very useful setting in the theory of pairs of vector spaces in duality founded in the early forties by J. Dieudonné and G.W. Mackey.

Another specialization of von Staudt's duality has also found applications. The convex sets to be considered are the epigraphs of convex functions defined on convex subsets of $\mathbb{R}^n$. If f is such a function, its epigraph is the set of points in $\mathbb{R}^n \times \mathbb{R}$ which lie on or above the graph of f. The polarity with respect to a paraboloid of revolution in $\mathbb{R}^n \times \mathbb{R}$ gives rise to an involutory bijection of the set of epigraphs of lower semicontinuous convex functions onto itself. Also this has been carried over to infinite-dimensional vector spaces. For a pair of vector spaces in duality one obtains a one-to-one correspondence of the lower semicontinuous convex functions in the two spaces.

11. Since work of Helly was mentioned above, it seems appropriate to quote the famous theorem named after him: If any $n + 1$ of a family of compact convex sets in $\mathbb{R}^n$ have a non-empty intersection, then this holds for the entire family. Helly discovered the theorem in 1917 but did not publish his proof until 1923. The first proof to appear was J. Radon's of 1921. Numerous later papers deal with proofs, applications, and generalizations of the theorem and with related subjects.

12. Minkowski's more profound results on convex bodies, which, as mentioned, mainly date back to the years 1987–1904, are to a great extent based on the theorem of Brunn quoted in Section **7**. For a suitable formulation of the theorem one has to use the notion of a linear family of convex bodies. Let $K_o,\ldots,K_m$ be convex bodies in $\mathbb{R}^n$. The linear family determined by them consists of the convex bodies

$$K(\lambda_o,\ldots,\lambda_m) = \lambda_o K_o + \cdots + \lambda_o K_m$$

where $\lambda_o,\ldots,\lambda_m$ run through the nonnegative numbers. The volume of $K(\lambda_o,\ldots,\lambda_m)$ turns out to be a homogeneous polynomial of degree n of the λ's (generalization of a formula of Steiner mentioned in Section **6**). Its coefficients are essentially Minkowski's "mixed volumes" of the bodies $K_o,\ldots,K_m$. The general Brunn-Minkowski theorem states that the n-th root of the volume of $K(\lambda_o,\ldots,\lambda_m)$ is a concave function of the λ's. It is an immediate consequence of the following: The n-th root of the volume of

$$K = (1 - \nu)K_0 + \nu K_1, \quad 0 \leq \nu \leq 1,$$

is a concave function f of ν and linear if and only if K_0 and K_1 are homothetic.

As consequences of the theorem for dimensions 2 and 3 Minkowski obtained several inequalities between the mixed volumes of two or three bodies. For arbitrary dimension and two bodies K_0 and K_1 this can be done by using that the right derivative of the concave function f at $\nu = 0$ is greater than or equal to the slope of the chord joining the endpoints of the graph of f. It follows also that equality implies that the bodies are homothetic. If, in particular, K_1 is the unit ball, the surface area of K_0 is n times one of the mixed volumes. Thus, one obtains an inequality between the volume and the surface area of K_0 with equality if and only if K_0 is also a ball. This solves the classical "isoperimetric problem" for convex bodies: Among all convex bodies with given volume the ball and only it has minimal surface area. Other inequalities between mixed volumes are obtained by using that the second derivative of the concave function f is nonpositive. Specializations yield also here solutions of extremum problems of geometrical interest.

Furthermore, Minkowski studied the following problem: Given unit vectors $\xi_1,\ldots,\xi_N$ in $\mathbb{R}^3$ and positive numbers $v_1,\ldots,v_N$, does there exist a convex polyhedron the faces of which are normal to the ξ_i and have areas v_i? He shows that the easily obtainable necessary condition

$$v_1\xi_1 + \cdots + v_N\xi_N = 0$$

is also sufficient for the existence. The uniqueness up to translations is a simple consequence of one of the Minkowski inequalities and the condition for equality. The analogous problem for convex bodies with a smooth boundary was also solved by Minkowski. Let there be given a positive continuous function F on the unit sphere such that the centroid of the mass distribution with density F on the sphere is the origin. Then there exists one and, up to translations, only one closed convex surface such that for every unit vector ξ its reciprocal Gauss curvature (suitably interpreted) at the point with outer normal vector ξ equals $F(\xi)$.

In 1910 D. Hilbert brought Minkowski's investigations in connection with certain elliptic partial differential equations on the unit sphere and proved Minkowski's inequalities by means of his theory of integral equations.

In the following decades many papers dealt with more elementary proofs and sharpenings of the inequalities, in particular those connected with the isoperimetric and related extremum problems. Of the older publications may be mentioned Blaschke's book "Kreis und Kugel" of 1916, further papers and a book from the twenties by T. Bonnesen containing among other things quite elementary proofs of sharpened inequalities.

As mentioned above, Minkowski considered in this context only dimensions 2 and 3. The main part of the theory, including the two uniqueness and existence theorems, could easily be generalized to arbitrary dimension. It was, however, clear that there must be valid inequalities between the mixed volumes which are not consequences of the Brunn-Minkowski theorem. Another challenge was to subordinate the theorems on the determination of convex polyhedra and smooth convex bodies to one theorem valid for arbitrary convex bodies. Satisfactory results in both directions were achieved in the second half of the thirties, mainly through the effort of A.D. Aleksandrov. In the theorem last mentioned a nonnegative additive set function on the unit sphere, satisfying certain conditions, has to be given instead of the product of the principal radii of curvature in the smooth case. Here a natural question arises: Are analogous theorems valid for set functions replacing the other elementary symmetric functions of the principal radii of curvature? So far only the case of the sum of the radii has been dealt with successfully. The rather involved necessary and sufficient conditions the corresponding set function has to satisfy were found independently by W.J. Firey and C. Berg in the late sixties.

Finally let me mention that the Brunn-Minkowski theorem in a weakened form remains valid for non-convex sets. This was discovered by L.A. Ljusternik in 1935. His result and proof were improved later on and applied to the isoperimetric problem under very general assumptions. The most far-reaching results (also for non-euclidean spaces) were achieved by E. Schmidt in the late forties.

W. Fenchel
Sønderrengen 110
DK-2860 Søborg (Denmark)

Peter M. Gruber

Approximation of convex bodies

1. Introduction

Approximation of convex bodies by either smooth convex bodies or polytopes has been considered frequently in convexity. There are two incentives for it. On the one hand approximation is used as a tool for investigations ranging from classical results on mixed volumes to questions concerning the ε-entropy of spaces of convex bodies, on the other hand there is an intrinsic geometric interest in the approximation problem itself.

Approximation of convex bodies is a book with many chapters, some of which will be considered in the following, for example bounds and asymptotic estimates for distances from best approximating polytopes, approximation by random polytopes and approximation by special convex bodies such as simplices and balls.

A good reference for the literature up to 1953 is the book of Fejes Tóth [53]. While we have tried to put more emphasis on modern developments it has frequently been necessary to mention older results too. We are not claiming that this survey is complete, but it is hoped that the more important topics are presented.

For concepts and notation not explained below see any of the standard treatises on convexity, e.g. Bonnesen and Fenchel [34], Hadwiger [57], Eggleston [57], Valentine [64], Grünbaum [67] or Leichtweiß [80].

2. Dense sets of convex bodies

Let $\mathscr{C}$ denote the space of all *convex bodies* of the d-dimensional Euclidean space E^d. These are compact convex subsets of E^d with non-empty interiors. For $d = 2$ we prefer to speak of *convex disks* instead. The 'natural' topology on $\mathscr{C}$ is most easily introduced by specifying a metric on $\mathscr{C}$ which induces this topology. One possible choice is the *Hausdorff-metric* or *Blaschke's Nachbarschaftsmaß* δ^H, defined by

$$\delta^H(C,D) = \max\left\{\sup_{x\in C}\inf_{y\in D}|x-y|, \sup_{y\in D}\inf_{x\in C}|x-y|\right\} \quad \text{for} \quad C,D\in\mathscr{C},$$

where $|\cdot|$ denotes the Euclidean norm on E^d. Note that the metric δ^H can be extended to the space of all closed subsets of E^d, admitting the value $+\infty$. We assume that $\mathscr{C}$ is endowed with the topology induced by δ^H. Let $\langle\cdot,\cdot\rangle$ be the inner product on E^d. Given a convex body C define its *support function* h_C as follows:

$$h_C(u) = \sup_{x\in C}\langle u,x\rangle \quad \text{for} \quad u\in S^{d-1}.$$

Here S^{d-1} stands for the $(d-1)$-dimensional unit sphere of E^d. Each convex body is determined uniquely by its support function. For $d = 2$ there is a one-to-one correspondence between the points $u \in S^1$ and the angles $\varphi \in [0,2\pi[$. In this case we prefer to write $h_C(\varphi)$ instead of $h_C(u)$.

For the definitions of polytopes, vertices, edges, faces, facets and regular polytopes see any of the texts mentioned in the introduction.

2.1. Polytopes of special types. It is well known and easy to prove that the set of all convex polytopes is dense in $\mathscr{C}$. Ewald raised the question whether there exist nonnegative integers p,q such that the set of all polytopes for which each vertex belongs to at most p edges and each facet contains at most q vertices, is dense in $\mathscr{C}$ (see Larman and Rogers [76]). The answer is affirmative for $d = 3$ if (p,q) is any of (3,6), (4,4), (6,3), as shown by Bokowski [77]. For $d = 4$ and $d \geq 5$ the problem was settled in the affirmative by Bokowski and Schulz [78] and Bokowski and Mani [82], respectively. Obviously, Ewald's original problem can be generalized in various directions.

A problem due to Shephard [66] is to find out whether the set of polytopes all facets of which are projective images of some fixed regular polytope (of dimension $d - 1$) is dense in $\mathscr{C}$. It is well known that the set of simplicial polytopes (i.e. all facets of which are simplices) is dense in $\mathscr{C}$. Shephard proves that the set of polytopes whose facets are projective images of cubes (of dimension $d - 1$) is dense in $\mathscr{C}$, but leaves open the corresponding problem for cross-polytopes. Most other cases, which are necessarily of dimension $d \leq 5$, are settled by Shephard.

Given a polytope P let f(P) denote the vector $(f_0, f_1, \ldots, f_{d-1})$ where f_i is the number of faces of P of dimension i and call it *f-vector* of P. Perles and Shephard [70] investigate the dimensions of the affine hulls of the sets of f-vectors of dense sets of polytopes. For the set of simplicial polytopes this dimension equals $[d/2]$. For each $k = 1, \ldots, [d/2]$ Perles and Shephard specify a countable dense set of polytopes for which the dimension of the affine hull of the f-vectors is k.

An approximation result for polytopes of a different kind is due to Kömhoff and Shephard [74].

2.2 Zonotopes and zonoids. A *zonotope* Z is a convex body which can be represented in the form

$$Z = S_1 + \cdots + S_n (= \{x_1 + \cdots + x_n : x_k \in S_k\}),$$

where n is a nonnegative integer and the S_k's are line segments. Zonotopes and *zonoids*, which are limits in the sense of the Hausdorff-metric of sequences of zonotopes, play an important role in various areas, e.g. in the theory of vector measures and in tiling problems. Thus it is of interest to find out whether any (necessarily) centrally symmetric convex body can be approximated by zonoids.

The set of zonoids with the center at the origin O is closed in the space of convex bodies symmetric in O and nowhere dense. For related but deeper results consult Lindquist [75] and Weil [76].

The support function h_Z of a zonoid Z with center at the origin can be represented in the form

$$(1) \qquad h_Z(u) = \int_{S^{d-1}} |\langle u,x\rangle| \, d\rho(x),$$

where ρ is a suitable nonnegative symmetric Borel measure on S^{d-1}. Here ρ is called symmetric if $\rho(A) = \rho(-A)$ for any Borel set $A \subset S^{d-1}$. The convex bodies symmetric in O the support functions of which can be represented in the form (1) with ρ a signed symmetric Borel measure on S^{d-1} are called *generalized zonoids.* A result of Blaschke [16] yields that all convex bodies with center at O which are of sufficiently high differentiability class, are generalized zonoids. This implies in particular that the generalized zonoids are dense in the space of all convex bodies symmetric in the origin. For related results on zonoids see Weil [76].

Quantitative approximation results for zonotopes are contained in a paper by Betke and McMullen [82].

2.3 Smooth convex bodies. A convex body is smooth if its boundary is a surface of class $\mathscr{C}^1$. One readily proves that the smooth convex bodies are dense in $\mathscr{C}$. We state some refinements of this.

Call a convex body *analytic* or *polynomial* if it can be represented in the form

$$\{x : \varphi(x) \leq 1\},$$

where φ is a real analytic or polynomial function on E^d. It was proved by Minkowski [03] that the analytic convex bodies are dense in $\mathscr{C}$. Other proofs of this are due to Radon [36] and Bonnesen and Fenchel [34]. Hammer [63] showed that the set of polynomial convex bodies (employing convex polynomials) is dense in $\mathscr{C}$. He also indicated how to find polynomial convex bodies approximating any given convex body.

A slightly different view is taken by Firey [74] who showed that the set of convex bodies with algebraic support functions is dense in $\mathscr{C}$.

Weil [73] showed that any convex body C can be approximated by a sequence $C_1, C_2, \ldots$ of convex bodies with supporting functions of class $\mathscr{C}^\infty$ and positive principal curvatures, such that the principal curvatures of $C_1, C_2, \ldots$ tend to the principal curvatures of C almost everywhere in the sense of the surface area measure.

2.4 Convex bodies of constant width. A convex body C is of *constant width* if

$$h_C(u) + h_C(-u) = \text{const for } u \in S^{d-1}.$$

Convex bodies of constant width have attracted considerable interest (see Chakerian and Groemer [83] for a recent survey).

A result of Tanno [76] yields that the set of convex disks of constant width (the boundaries of) which are of class $\mathscr{C}^\infty$, are dense in the space of all convex disks of constant width. Given any convex disk of constant width which is symmetric in a line, the approximating convex disks of constant width and of class $\mathscr{C}^\infty$ can be chosen such as to be symmetric in this line too. These results continue to hold if the class $\mathscr{C}^\infty$ is replaced by the class of real analytic functions, as shown by Wegner [77]. This was extended to general dimensions by Schneider [82].

A convex disk is called a *Reuleaux polygon* if it is of constant width v and the boundary consists of finitely many circular arcs of radius v. The Reuleaux polygons are dense in the space of all convex disks of constant width (see Bonnesen and Fenchel [34]). A quantitative result of this type is due to Kallay [74] (cf. Chakerian and Groemer [83]).

2.5 None of the above, but in this section. Klee [60] considered the problem to approximate an unbounded closed convex set C by unbounded polyhedra P and proved that either $\delta^H(C,P) = +\infty$ for each polyhedron P or C can be approximated arbitrarily closely by polyhedra.

A convex body C is called *indecomposable* if it cannot be represented in the form $C = D + E$ with convex bodies D,E not homothetic to C (see Shephard [64] and Berg [69]). Since the simplicial polytopes are indecomposable the set of indecomposable convex bodies is dense in $\mathscr{C}$.

The approximation of compact starshaped sets by either smooth or polyhedral sets was investigated by Toranzos [79]. His results were partly anticipated by Guggenheimer [77].

A result on the approximation on certain locally convex compact sets on Riemannian manifolds by locally convex compact sets of class $\mathscr{C}^\infty$ is due to Bangert [78].

Erdös and Vincze [82] consider the approximation of convex disks by 'generalized ellipses with many foci'.

3. General properties of best approximating polytopes

For $n = d + 1, d + 2, \ldots$ let $\mathscr{P}_n$ be the set of all convex polytopes having at most n vertices. Given a convex body C define $\mathscr{P}^i_n(C)$ to be the set of polytopes of $\mathscr{P}_n$ contained in C and let similarly $\mathscr{P}^c_n(C)$ denote the set of polytopes of $\mathscr{P}_n$ containing C. We will frequently write $\mathscr{P}^i_n, \mathscr{P}^c_n$ for $\mathscr{P}^i_n(C), \mathscr{P}^c_n(C)$. Let δ be any metric or other measure of distance on $\mathscr{C}$ and let C be a convex body. A polytope $P \in \mathscr{P}_n$ (or $\mathscr{P}^i_n, \mathscr{P}^c_n$) is called a *best approximating polytope* of C in $\mathscr{P}_n$ (or $\mathscr{P}^i_n, \mathscr{P}^c_n$) in the sense of δ if

$$\delta(C,P) = \inf_{Q \in \mathscr{P}_n} \delta(C,Q) = \delta(C,\mathscr{P}_n)$$

(and analogously for $\mathscr{P}^i_n, \mathscr{P}^c_n$). For the metrics considered in this paper, best approximating polytopes always exist.

Before stating some properties of best approximating polygons and polytopes we give a short list of

3.1 Metrics on $\mathscr{C}$. In the setting of approximation of convex bodies besides δ^H several further metrics and other measures of deviation are used.

Recall the definition of δ^H:

$$\delta^H(C,D) = \max\left\{\sup_{x \in C} \inf_{y \in D} |x - y|, \sup_{y \in D} \inf_{x \in D} |x - y|\right\} \quad \text{for} \quad C,D \in \mathscr{C}$$

(see Blaschke [16], Fejes Tóth [53], Sendov [79]). Related to δ^H is the metric δ^E:

$$\delta^E(C,D) = \sup_{x \in C} \inf_{y \in D} |x - y| + \sup_{y \in D} \inf_{x \in D} |x - y| \quad \text{for} \quad C,D \in \mathscr{C}$$

(see Eggleston [57], [61]). Note that δ^H and δ^E can be defined for any other norm on E^d than $|\ |$.

Let μ and σ denote Lebesgue measure and ordinary surface area measure in E^d

respectively. For a set C in E^d the surface area (or perimeter in case d = 2) of the boundary bd C of C is denoted by $\sigma(C)$.

The *symmetric difference metric* δ^S and the *surface deviation* δ^P (also called *perimeter deviation* in the planar case; not a metric) are defined by

$$\delta^S(C,D) = \mu(C\Delta D) \quad \text{for} \quad C,D \in \mathscr{C},$$
$$\delta^P(C,D) = \sigma(C \cup D) - \sigma(C \cap D) \quad \text{for} \quad C,D \in \mathscr{C}$$

(see Fejes Tóth [53], Eggleston [57] and Shephard and Webster [65]).

The map $C \to h_C$ transforms the Hausdorff metric into the maximum norm:

$$\delta^H(C,D) = \max_{u \in S^{d-1}} |h_C(u) - h_D(u)| = \|h_C - h_D\|_\infty \quad \text{for} \quad C,D \in \mathscr{C}.$$

This suggests to define for $p \geq 1$ the p-*metric* δ^p on $\mathscr{C}$ by

$$\delta^p(C,D) = \left(\int_{S^{d-1}} |h_C(u) - h_D(u)|^p d\sigma(u) \right)^{1/p} = \|h_C - h_D\|_p \quad \text{for} \quad C,D \in \mathscr{C}$$

(see Davis [82] and McClure and Vitale [75]). Note that for d = 2 and $C \subset D$ or $D \subset C$ we have $\delta^1(C,D) = \delta^P(C,D)$.

For other measures of deviation we refer to sections 10.1 to 10.3 and the article of Shephard and Webster [65].

3.2 Dowker-type theorems. Let C be a convex disk. A classical result of Dowker [44] shows that

$$2\delta^S(C,\mathscr{P}^i_n) \leq \delta^S(C,\mathscr{P}^i_{n-1}) + \delta^S(C,\mathscr{P}^i_{n+1}) \quad \text{for} \quad n = 4,5,\ldots,$$
$$2\delta^S(C,\mathscr{P}^c_n) \leq \delta^S(C,\mathscr{P}^c_{n-1}) + \delta^S(C,\mathscr{P}^c_{n+1}) \quad \text{for} \quad n = 4,5,\ldots$$

The problem to prove analogous results for other cases too was posed by Fejes Tóth [53]. It was shown by Eggleston [57], [61] that

$$2\delta^S(C,\mathscr{P}_n) \leq \delta^S(C,\mathscr{P}_{n-1}) + \delta^S(C,\mathscr{P}_{n+1}) \quad \text{for} \quad n = 4,5,\ldots$$

Eggleston's result was considerably refined by G. Fejes Tóth [77] by using a metric on $\mathscr{C}$ more general then δ^S. Molnar [55] and Eggleston [57], [61] derived similar results for δ^P:

$$2\delta^P(C,\mathscr{P}^i_n) \leq \delta^P(C,\mathscr{P}^i_{n-1}) + \delta^P(C,\mathscr{P}^i_{n+1}) \quad \text{for} \quad n = 4,5,\ldots,$$
$$2\delta^P(C,\mathscr{P}^c_n) \leq \delta^P(C,\mathscr{P}^c_{n-1}) + \delta^P(C,\mathscr{P}^c_{n+1}) \quad \text{for} \quad n = 4,5,\ldots,$$
$$2\delta^P(C,\mathscr{P}_n) \leq \delta^P(C,\mathscr{P}_{n-1}) + \delta^P(C,\mathscr{P}_{n+1}) \quad \text{for} \quad n = 4,5,\ldots$$

Corresponding inequalities for δ^E do not hold, as shown by Eggleston [57], [61]. The case of δ^H seems to be still open.

Dowker-type theorems for convex sets on a sphere are given by Fejes Tóth [58].

3.3 Symmetry of best approximating polygons. Let C be a convex disk with k-fold rotatory symmetry. G. Fejes Tóth and L. Fejes Tóth [73] proved that there exist best approximating polygons in $\mathscr{P}^i_{kn}$ and $\mathscr{P}^c_{kn}$ with respect to both δ^S and δ^P also having k-fold rotatory symmetry. A more general result of G. Fejes Tóth [77] implies the corresponding result for $\mathscr{P}_{kn}$ and δ^S.

3.4 Equioscillation properties of best approximating polygons. A familiar theorem from the theory of Chebychev-approximation can be stated as follows: If, among all polynomials of degree $\leq n$ a polynomial p best approximates a continuous function f on [0,1] in the sense of the maximum norm, then there are points $0 \leq \tau_0 < \tau_1 < \cdots < \tau_n \leq 1$ such that

$$\|f - p\|_\infty = \varepsilon(-1)^i(f(\tau_i) - p(\tau_i)) \quad \text{for} \quad i = 0,\ldots,n$$

where $\varepsilon = \pm 1$ is fixed. Analogous 'equioscillation' results hold for best approximating polygons of convex disks.

Let C be a convex disks. By a result of Kenderov [80], [81] for any best approximating polygon $P \in \mathscr{P}_n$ in the sense of the metric δ^H there are angles $0 \leq \varphi_1 < \cdots < \varphi_n \leq 2\pi$, corresponding to the edges of P, and angles $0 \leq \psi_1 < \cdots < \psi_n \leq 2\pi$, corresponding to the vertices of P such that

$$\delta^H(C,P) = \sup_{\varphi \in [0,2\pi]} |h_C(\varphi) - h_P(\varphi)| = \begin{Bmatrix} h_C(\varphi_i) - h_P(\varphi_i) \\ h_P(\psi_i) - h_C(\psi_i) \end{Bmatrix} \quad \text{for} \quad i = 1,\ldots,n.$$

This means that each vertex of P has distance $\delta^H(C,P)$ from C and each edge of P 'penetrates' the same distance into C. Kenderov's result was generalized by Zhivkov [82] to Hausdorff Metrics definded by means of norms other than the Euclidean norm.

If C is a strictly convex disk and $P \in \mathscr{P}_n^c$ a best approximating polygon of C with respect to δ^S (i.e. P has minimal area among all polygons of $\mathscr{P}_n^c$) then a result of Trost [49] shows that each edge of P touches C at its midpoints. More difficult to prove is the following result of Eggleston [61]: Let $P \in \mathscr{P}_n$ be a best approximating polygon of a strictly convex disk C in the sense of the symmetric difference metric. Then each edge v_i,v_{i+1} of P meets the boundary of C in two points w_i,w_{i+1} such that $|v_i - w_i| = |v_{i+1} - w_{i+1}| = |v_i - v_{i+1}|/4$. Related results are due to Stammler and Weissbach [79].

These 'equioscillation theorems' do not determine the best approximating polygons uniquely, but in special cases they are of some use.

3.5. Higher dimensional results. It is easy to show that for the metrics δ^S, δ^P the best approximating polytopes $P \in \mathscr{P}_n^i$ of a convex body C (i.e. the polytopes $P \in \mathscr{P}_n^i(C)$ having maximal volume and surface area respectively) have all their vertices on the boundary of C.

The following are some results for the case of the symmetric difference metric. Given a convex disk C and one of its best approximating polygons $P \in \mathscr{P}_n^i$, an old result of Steiner shows that for any three consecutive vertices of P, say v_{i-1},v_i,v_{i+1}, the line segment $v_{i-1}v_{i+1}$ is parallel to a supporting line of C at v_i (see Fejes Tóth [39]). Fejes Tóth [39] generalized the theorem of Steiner to dimensions $d \geq 2$: Let $P \in \mathscr{P}_n^i$ be a best approximating polytope of a convex body C, let v be a vertex of P and denote the facets of P containing v by $F_1,\ldots,F_m$. Choose exterior normal vectors n_k of F_k for all k such that $|n_k|$ equals the surface area of F_k. Then $n_1 + \cdots + n_m$ is an exterior normal vector of C at v. Fejes Tóth [53] indicated that for smooth convex bodies the best approximating polytopes $P \in \mathscr{P}_n^i$ are simplicial.

A result similar to the above but with δ^P in place of δ^S is contained in the paper [40] of Fejes Tóth.

4. Bounds and asymptotic results for convex disks

Fejes Tóth's book [53] contains a wealth of results on approximation of convex disks and 3-dimensional convex bodies. Thus an overlap with his book could not be avoided in this and the following section. Moreover, many results which Fejes Tóth indicated without proofs have been completely proved in the time elapsed since his book was first published. Some of these will be mentioned below.

4.1 *The Hausdorff metric* is considered first. The author has been unable to find out who originally proved that for any convex disk C

$$\delta^H(C,\mathscr{P}^i_n),\ \delta^H(C,\mathscr{P}^c_n),\ \delta^H(C,\mathscr{P}_n) \leq \frac{\text{const}}{n^2} \quad \text{for} \quad n = 3,4,\ldots$$

(but see section 5 for the general case). More precise upper bounds are due to Popov [68], [70]:

$$\delta^H(C,\mathscr{P}^i_n) \leq \sigma(C)\frac{\sin(\pi/n)}{2n} \quad \text{for} \quad n = 3,4,\ldots,$$

$$\delta^H(C,\mathscr{P}^c_n) \leq \sigma(C)\frac{\tan(\pi/n)}{2n} \quad \text{for} \quad n = 3,4,\ldots,$$

$$\delta^H(C,\mathscr{P}_n) \leq \sigma(C)\frac{\sin(\pi/n)}{2(1+\cos(\pi/n))n} \quad \text{for} \quad n = 3,4,\ldots.$$

A conjecture of Sendov and Popov asserts that among all convex disks of given perimeter the regular (n + 1) -gon is worst approximated by polygons with n vertices. For results related to this conjecture see Ivanov [34], Popov [68] and Kenderov [73].

Let C be a convex disk with support function $h = h(\varphi)$ of class $\mathscr{C}^2$. Its radius of curvature ρ can be expressed in the form $\rho = h + h''$. The asymptotic formulas

$$\delta^H(C,\mathscr{P}^i_n),\ \delta^H(C,\mathscr{P}^c_n) \sim \frac{1}{8}\left(\int_0^{2\pi} \rho^{1/2}d\varphi\right)^2 \frac{1}{n^2} \quad \text{as} \quad n \to \infty,$$

$$\delta^H(C,\mathscr{P}_n) \sim \frac{1}{16}\left(\int_0^{2\pi} \rho^{1/2}d\varphi\right)^2 \frac{1}{n^2} \quad \text{as} \quad n \to \infty$$

were indicated by Fejes Tóth [73] in slightly different forms since he prefers to use the curvature κ instead of ρ. Note that

$$\int_0^{2\pi} \rho^\alpha d\varphi = \int_{\text{bd}C} \kappa^{1-\alpha}d\sigma$$

where σ denotes arc-length. A proof of the first two formulas was given by McClure and Vitale [75] and the last follows from the first one since one can show generally that for smooth convex disks C

$$\delta^H(C,\mathscr{P}_n) \sim \frac{1}{2}\delta^H(C,\mathscr{P}^i_n) \quad \text{as} \quad n \to \infty.$$

The Cauchy-Schwarz inequality yields

$$\left(\int_0^{2\pi} \rho^{1/2} d\varphi\right)^2 \leq \int_0^{2\pi} \rho d\varphi \int_0^{2\pi} d\varphi = 2\pi\sigma(C),$$

where equality holds if and only if ρ is constant. Thus among all convex disks of equal perimeter and with support functions of class $\mathscr{C}^2$, the circles are asymptotically worst approximated.

It seems to be difficult to find best approximating polygons of a convex disk let alone very special cases. A problem which is more easily accessible is to construct 'asymptotically best approximating' sequences of polygons. McClure and Vitale [75] specify one such construction by means of the *method of empirical distribution*: Let C be a convex disk with support function of class $\mathscr{C}^2$ and positive radius of curvature ρ. Put

$$\gamma^H(\tau) = \int_0^{\tau} \rho^{1/2} d\varphi \Big/ \int_0^{2\pi} \rho^{1/2} d\varphi \quad \text{for} \quad 0 \leq \tau \leq 2\pi.$$

γ^H is continuous and strictly increasing from 0 to 1 as τ ranges from 0 to 2π. For each $n = 3,4,\ldots$ the angles $0 < \tau_1 < \cdots < \tau_n < 2\pi$ where

$$\tau_k = \gamma^{H^{-1}}\left(\frac{k}{n+1}\right) \quad \text{for} \quad k = 1,\ldots,n$$

are defined uniquely. Let P_n^i denote the polygon inscribed to C whose vertices are at the boundary points of C at which the exterior normals of C have directions $\tau_1,\ldots,\tau_n$. Similarly, define P_n^c to be the polygon circumscribed to C such that the edges of P_n^c touch C and have exterior normals in directions $\tau_1,\ldots,\tau_n$. Then

$$\frac{\delta^H(C,P_n^i)}{\delta^H(C,\mathscr{P}_n^i)} \to 1, \frac{\delta^H(C,P_n^c)}{\delta^H(C,\mathscr{P}_n^c)} \to 1 \quad \text{as} \quad n \to \infty.$$

Actually this is only a special case of the result of McClure and Vitale [75].

4.2 The symmetric difference metric. Generalizing a result of Blaschke [23] for inscribed triangles it was proved by Sas [39] that for any convex disk C

$$\delta^S(C,\mathscr{P}_n^i) \leq \mu(C)\left(1 - \frac{n}{2\pi}\sin\frac{2\pi}{n}\right) \sim \frac{2\pi^2\mu(C)}{3}\frac{1}{n^2} \quad \text{as} \quad n \to \infty,$$

where the equality sign holds if and only if C is an ellipse. For a nice proof of this see Fejes Tóth [53]. It follows from the result of Sas that

$$\delta^S(C,\mathscr{P}_n) \leq \frac{\text{const}}{n^2} \quad \text{for} \quad n = 3,4,\ldots.$$

Upper bounds for $\delta^S(C,\mathscr{P}_n^c)$ and $\delta^S(C,\mathscr{P}_n)$ comparable to those of Popov in the Hausdorff metric case are still missing to the best of our knowledge.

G. Fejes Tóth [82] proved that for all convex disks C of area 1 and perimeter

$$\sigma(C) \leq \frac{4n}{\cos(\pi/n)},$$

$\delta^S(C,\mathscr{P}_n^i)$ is minimal if C has n-fold rotatory symmetry and is bounded by n congruent circular arcs.

Assume now that C is a convex disk with support function of class $\mathscr{C}^2$ and radius of curvature ρ. Then

$$\delta^S(C,\mathscr{P}_n^i) \sim 2\delta^S(C,\mathscr{P}_n^c) \sim \frac{1}{12}\left(\int_0^{2\pi} \rho^{2/3}d\varphi\right)^3 \frac{1}{n^2} \quad \text{as} \quad n \to \infty,$$

$$\delta^S(C,\mathscr{P}_n) \sim \frac{1}{32}\left(\int_0^{2\pi} \rho^{2/3}d\varphi\right)^3 \frac{1}{n^2} \quad \text{as} \quad n \to \infty.$$

This was stated without proof by Fejes Tóth [53] and proofs of the first two asymptotic results were provided by McClure and Vitale [75]. Let κ denote the curvature of C. The integral

$$\int_0^{2\pi} \rho^{2/3}d\varphi = \int_{\mathrm{bd}C} \kappa^{1/3}d\sigma$$

is called the *affine length* of C, a concept which goes back to Pick. It is invariant for area preserving affinities of C, see Blaschke [23]. The affine isoperimetric inequality of Blaschke [23] shows that

$$\left(\int_0^{2\pi} \rho^{2/3}d\varphi\right)^3 \leq 8\pi^2\mu(C),$$

where equality holds if and only if C is an ellipse. Thus among all convex disks of equal area with support functions of class $\mathscr{C}^2$ the ellipses are asymptotically worst approximated by polygons.

Since it does not seem to be easy to explicitely construct best approximating polygons, McClure and Vitale [75] specified two methods for constructing 'asymptotically best approximating' polygons. Let C be a convex disk with support function of class $\mathscr{C}^2$ and positive radius of curvature ρ. The method of empirical distribution was mentioned already in section 4.1. We indicate the modifications needed for the present case. In place of γ^H take

$$\gamma^S(\tau) = \int_0^{\tau} \rho^{2/3}d\varphi \Big/ \int_0^{2\pi} \rho^{2/3}d\varphi \quad \text{for} \quad 0 \leq \tau \leq 2\pi$$

and define τ_k, P_n^i and P_n^c as before with γ^H replaced by γ^S. Then

$$\frac{\delta^S(C,P_n^i)}{\delta^S(C,\mathscr{P}_n^i)} \to 1, \frac{\delta^S(C,P_n^c)}{\delta^S(C,\mathscr{P}_n^c)} \to 1 \quad \text{as} \quad n \to \infty.$$

The second method, called *method of balance of local errors*, is more geometric in spirit and may be described as follows. Let $Q_n^i \in \mathscr{P}_n^i$ be a polygon, all n vertices of which are on the boundary of C and such that the n 'lunae' cut off from C by the edges of Q_n^i have equal areas. Then

$$\frac{\delta^S(C,Q_n^i)}{\delta^S(C,\mathscr{P}_n^i)} \to 1 \quad \text{as} \quad n \to \infty.$$

Next choose an interior point p of C and let $Q_n^c \in \mathscr{P}_n^c$ be a polygon circumscribed to C such that all of its n edges touch C and consider the n triangles defined by p and the edges of Q_n^c. Let Q_n^c be chosen such that the areas of the parts of these triangles not contained in C are equal. Then

$$\frac{\delta^S(C,Q_n^c)}{\delta^S(C,\mathscr{P}_n^c)} \to 1 \quad \text{as} \quad n \to \infty.$$

4.3 The perimeter deviation. A result of Fejes Tóth [53] on inscribed and circumscribed polygons yields that for each convex disk C

$$\delta^P(C,\mathscr{P}_n^i),\ \delta^P(C,\mathscr{P}_n^c),\ \delta^P(C,\mathscr{P}_n) \le \frac{\text{const}}{n^2} \quad \text{for} \quad n = 3,4,\ldots.$$

This could also be deduced from the results of 4.1 by means of Cauchy's perimeter formula. Schneider [71′] derived the following more precise results:

$$\delta^P(C,\mathscr{P}_n^i) \le \sigma(C)\left(1 - \frac{n}{\pi}\sin\frac{\pi}{n}\right),$$

$$\delta^P(C,\mathscr{P}_n^c) \le \sigma(C)\left(\frac{n}{\pi}\tan\frac{\pi}{n} - 1\right),$$

where in the first inequality we have equality if and only if C is a circular disk.

Now suppose that C is a convex disk with support function of class $\mathscr{C}^2$ and radius of curvature ρ. Then

$$\delta^P(C,\mathscr{P}_n^i) \sim \frac{1}{2}\delta^P(C,\mathscr{P}_n^c) \sim \frac{1}{24}\left(\int_0^{2\pi} \rho^{1/3}d\varphi\right)^3 \frac{1}{n^2} \quad \text{as} \quad n \to \infty,$$

$$\delta^P(C,\mathscr{P}_n) \sim \frac{1}{24}\left(\int_0^{2\pi} \rho^{1/3}d\varphi\right)^3 \frac{1}{n^2} \quad \text{as} \quad n \to \infty.$$

All three asymptotic formulas were stated by Fejes Tóth [53]. Proofs of the first two are due to McClure and Vitale [75].

As in the cases of δ^H and δ^S McClure and Vitale [75] described two methods for constructing asymptotically best approximating polygons. Let C be a convex disk having support function of class $\mathscr{C}^2$ and with positive radius of curvature ρ. In the method of empirical distribution let

$$\gamma^P(\tau) = \int_0^\tau \rho^{1/3}d\varphi \Big/ \int_0^{2\pi} \rho^{1/3}d\varphi \quad \text{for} \quad 0 \le \tau \le 2\pi$$

and define τ_k, P_n^i and P_n^c as before in 4.1 and 4.2 where γ^H and γ^S are now replaced by γ^P. Then

$$\frac{\delta^P(C,P_n^i)}{\delta^P(C,\mathscr{P}_n^i)} \to 1,\ \frac{\delta^P(C,P_n^c)}{\delta^P(C,\mathscr{P}_n^c)} \to 1 \quad \text{as} \quad n \to \infty.$$

The method of balance of local errors can be described in the following way. Let $Q_n^i \in \mathscr{P}_n^i$ have all its n vertices on the boundary of C such that for the edges of Q_n^i the differences in length of the boundary arc of C defined by the endpoints of the edge, and the edge are all equal. Then

$$\frac{\delta^P(C,Q_n^i)}{\delta^P(C,\mathscr{P}_n^i)} \to 1 \quad \text{as} \quad n \to \infty.$$

In order the obtain asymptotically best circumscribed polygons one has to choose an interior point p of C and then define Q_n^c as follows. Let Q_n^c be a polygon all edges of which touch C. For each edge consider the difference of the lengths of the edge and of the boundary arc of C contained in the triangle with vertex p and base the edge. Supposing that all these differences are equal

$$\frac{\delta^P(C,Q_n^c)}{\delta^P(C,\mathscr{P}_n^c)} \to 1 \quad \text{as} \quad n \to \infty.$$

4.4 The p-metric. Here only a few results are known. Presumably this is due to the fact that $\delta^p(C,D)$ for $1 < p < \infty$ cannot be easily interpreted in geometric terms. Since for $C,D \in \mathscr{C}$

$$\delta^p(C,D) = \left(\int_0^{2\pi} |h_C(\varphi) - h_D(\varphi)|^p d\varphi\right)^{1/p} \leq (2\pi\delta^H(C,D)^p)^{1/p}$$
$$\leq (2\pi)^{1/p}\delta^H(C,D),$$

we infer from previous results that for any convex disk C

$$\delta^p(C,\mathscr{P}_n^i),\ \delta^p(C,\mathscr{P}_n^c),\ \delta^p(C,\mathscr{P}_n) \leq \frac{\text{const}}{n^2} \quad \text{for} \quad n = 3,4,\ldots$$

McClure and Vitale [75] show that for convex disks C with support function of class $\mathscr{C}^2$ and radius of curvature ρ

$$\delta^p(C,\mathscr{P}_n^i) \sim \frac{1}{8}(2p+1)\left(\int_0^{2\pi} \rho^{p/(2p+1)} d\varphi\right)^{(2p+1)/p} \frac{1}{n^2} \quad \text{as} \quad n \to \infty,$$

$$\delta^p(C,\mathscr{P}_n^c) \sim \frac{1}{2}\left(\int_0^1 \tau^p(1-\tau)^p d\tau\right)^{1/p} \left(\int_0^{2\pi} \rho^{p/(2p+1)} d\varphi\right)^{(2p+1)/p} \frac{1}{n^2}$$
$$\text{as} \quad n \to \infty.$$

They also given two constructions for asymptotically best approximating polygons (cf. 4.1–4.3). Let C be a convex disk and assume that in addition of having a support function of class $\mathscr{C}^2$, the radius of curvature ρ of C is positive. The method of empirical distribution is as described previously where the distribution function is now defined in the form

$$\gamma^p(\tau) = \int_0^\tau \rho^{p/(2p+1)} d\varphi \Big/ \int_0^{2\pi} \rho^{p/(2p+1)} d\varphi \quad \text{for} \quad 0 \leq \tau \leq 2\pi.$$

The method of balance of local errors is analogous to the cases for δ^S, δ^P as exhibited in 4.2, 4.3.

4.5 None of the above, but in this section. A simple algorithm for inscribing polygons into convex disks is described by Johnson and Vogt [80]. We also refer to Dobkin and Snyder [79].

Button and Wilker [78] consider the following method of approximating convex bodies C. Assume that C is not a polytope. For any polytope P_1 containing C, let $P_1, P_2, \ldots$ be a sequence of polytopes with intersection C such that $P_{n+1} = P_n \cap H_n$ for $n = 1,2,\ldots$ where H_n is a supporting half space of C. Denote the width of $P_n \backslash P_{n+1}$ in the direction orthogonal to be boundary plane of H_n by w_n. The number

$$e(C;P_1,P_2,\ldots) = \inf\left\{\alpha \geq 0 : \sum_1^n w_n^\alpha < \infty\right\}$$

is called the *cutting exponent*. If it is large the sequence $P_1, P_2, \ldots$ approximates C well.

A result of Button and Wilker [78] shows that

$$\sum_1^\infty w_n \leq \pi \sigma(P_1),$$

thus implying $e(C;P_1,P_2,\ldots) \leq 1$ for each convex disk C. See also Taylor [76]. Button and Wilker [78] also show that the cutting exponents of convex disks of class $\mathscr{C}^2$ and with positive curvature can take any value in [1/2,1].

There exist several papers on the approximation of convex functions on [0,1] by convex polygonal functions, convex splines or rational functions. References are the papers of Azarin and Barmin [76], [76a], Beatson [81], Bojanic and Roulier [75], Bulanov [78], Bulanov and Hatamov [78], Hatamov [77], Petrushev [77] and Robva [77]. See also the book of Sendov [79] on Hausdorff approximation.

5. Bounds and asymptotic results for general convex bodies

While here the emphasis is on results for general dimensions the reader is referred to Fejes Tóth [53] for a synopsis of 3-dimensional results.

5.1 The Hausdorff metric. Dudley [74] proved that for each convex body C

$$\delta^H(C,\mathscr{P}_n^i) \leq \frac{\text{const}}{n^{2/(d-1)}} \quad \text{for} \quad n = d+1, d+2, \ldots$$

Bronštein and Ivanov [75] and Betke and Wills [78] later and independently gave alternative proofs. It is from the proof of Theorem 3 of the paper of Schneider and Wieacker [81] that one can deduce that for any convex body of class $\mathscr{C}^2$ there exists a positive constant such that

$$\delta^H(C,\mathscr{P}_n^i) \geq \frac{\text{const}}{n^{2/(d-1)}} \quad \text{for} \quad n = d+1, d+2, \ldots$$

Note that for each convex body C

$$\delta^H(C,\mathscr{P}_n) \leq \delta^H(C,\mathscr{P}_n^i),\ \delta^H(C,\mathscr{P}_n^c) \leq \text{const}\ \delta^H(C,\mathscr{P}_n)$$
$$\text{for} \quad n = d+1, d+2, \ldots$$

and that for sufficiently often differentiable convex bodies C

$$\delta^H(C,\mathscr{P}_n^i) \sim \delta^H(C,\mathscr{P}_n^c) \sim 2\delta^H(C,\mathscr{P}_n) \quad \text{as} \quad n \to \infty.$$

Considering the upper and lower bounds and the asymptotic results in the planar case, it is plausible to conjecture that for convex bodies of higher differentiability classes there are asymptotic results in general dimensions too. Using differential geometric tools Schneider [81] derived for convex bodies C of class $\mathscr{C}^3$ and with positive Gaussian curvature κ the asymptotic formula

$$\delta^H(C,\mathscr{P}_n^i) \sim \left(\frac{\vartheta_{d-1}}{2\pi_{d-1}} \int_{\mathrm{bd}C} \kappa^{1/2} d\sigma\right)^{2/(d-1)} \frac{1}{n^{2/(d-1)}} \quad \text{as} \quad n \to \infty.$$

The constants π_k and ϑ_k are the k-dimensional volume of the unit ball B^k in E^k and the density of the least dense covering of E^k by unit balls. The only explicitely known value of ϑ_k is $\vartheta_2 = 2\pi/27^{1/2}$. This yields for $d = 3$ the formula

$$\delta^H(C,\mathscr{P}_n^i) \sim \frac{1}{27^{1/2}} \int_{\mathrm{bd}C} \kappa^{1/2} d\sigma \frac{1}{n} \quad \text{as} \quad n \to \infty,$$

indicated already by Fejes Tóth [53]. As in 4.1 the Cauchy Schwarz inequality implies that among all convex bodies of class $\mathscr{C}^3$ and with positive Gaussian curvature having equal surface area, the balls are asymptotically worst approximable.

5.2 *The symmetric difference metric.* Since for any convex body C

$$\delta^S(C,\mathscr{P}_n) \le \text{const}\ \sigma(C)\delta^H(C,\mathscr{P}_n) \quad \text{for} \quad n = d+1, d+2, \ldots$$

and correspondingly for $\delta^S(C,\mathscr{P}_n^i)$, $\delta^S(C,\mathscr{P}_n^c)$ the results of 5.1 imply that

$$\delta^S(C,\mathscr{P}_n^i),\ \delta^S(C,\mathscr{P}_n^c),\ \delta^S(C,\mathscr{P}_n) \le \frac{\text{const}}{n^{2/(d-1)}} \quad \text{for} \quad n = d+1, d+2, \ldots$$

for each convex body C. If C is a convex body of class $\mathscr{C}^2$ a theorem of Gruber and Kenderov [82] shows that

$$\delta^S(C,\mathscr{P}_n) \ge \frac{\text{const}}{n^{2/(d-1)}} > 0 \quad \text{for} \quad n = d+1, d+2, \ldots$$

Also in the case of the symmetric difference metric it is highly probable that there are asymptotic formulas for $\delta^S(C,\mathscr{P}_n^i)$, $\delta(C,\mathscr{P}_n^c)$, $\delta(C,\mathscr{P}_n)$ for C sufficiently smooth. For $d = 3$ the following result was made plausible by Fejes Tóth [53].

$$\delta^S(C,\mathscr{P}_n^i) \sim \frac{9}{5}\,\delta^S(C,\mathscr{P}_n^c) \sim \frac{1}{4\cdot 3^{1/2}}\left(\int_{\mathrm{bd}C} \kappa^{1/4} d\sigma\right)^2 \frac{1}{n} \quad \text{as} \quad n \to \infty.$$

Blaschke's affine isoperimetric inequality shows that

$$\left(\int_{\mathrm{bd}C} \kappa^{1/4} d\sigma\right)^2 \le 12\pi\mu(C)$$

where equality holds if and only if C is an ellipsoid (see Blaschke [23]). Hence only the ellipsoids are asymptotically worst approximable among all sufficiently smooth convex bodies of equal measure in E^3.

The last remark is not too surprising, since a result of Macbeath [51] (which generalizes a theorem of Blaschke [23] on inscribed simplices) says that among all convex bodies C of equal volume $\delta^S(C,\mathscr{P}^i_n)$ is maximal if and only if C is an ellipsoid.

5.3 None of the above, but in this section. For the definition of cutting exponents see 4.5. Button and Wilker [78] prove that for each convex body C the cutting exponent is at most d and that there is a convex body with cutting exponent greater than $(d-1)/2$.

Rogers [53] derived a lower bound for $\delta^S(B^d,Q_n)$ where B^d denotes the unit ball of E^d and Q_n a polytope contained in B^d having at most n facets.

Rivlin and Shapiro [61] consider the problem to find the best uniform approximation by affine linear functions of $\xi^2 + \eta^2$ where (ξ,η) ranges in a convex disk.

Numerical algorithms for the approximation of convex functions on $[0,1]^d$ by piecewise affine linear functions are considered by Sonnevend [80], [81].

6. Uniqueness of best approximating polytopes

It follows essentially from Blaschke's selection theorem that $\mathscr{C}$ is a locally compact space. Hence a version of Baire's category theorem shows that in $\mathscr{C}$ a subset of first category is 'small' compared with the whole space, where a set is of first category if it is a countable union of nowhere dense sets. We say that *most* convex bodies have some property if this property holds for all convex bodies except those from a set of first category in $\mathscr{C}$.

In approximation theory a question which turns up as soon as existence is guaranteed is uniqueness. Unfortunately uniqueness results in approximation of convex bodies are very scarce. From the example of the ball one may see that best approximating polytopes need not be unique.

6.1 Most convex disks have unique best approximating polygons. Kenderov [80], [81] and Gruber and Kenderov [82] proved that for most convex disks the best approximating polygons in $\mathscr{P}_n$ with respect to δ^H and δ^S are unique. The corresponding results for inscribed and circumscribed polygons are easily proved along the same lines. While most convex disks have unique best approximating polygons, it seems to be difficult to give explicit examples. Presumably the semicircle provides an example for both metrics. An extension of the above mentioned uniqueness result for the Hausdorff metric to Hausdorff metrics defined by means of norms different from the Euclidean norm was obtained by Zhivkov [82].

6.2 Uniqueness results in higher dimensions. If one considers inscribed polytopes only, then most convex bodies have unique best approximating polytopes in $\mathscr{P}^i_n$ with respect to δ^S. This is proved in the same way as the corresponding result for convex disks. It is strongly conjectured that analogous results hold in all other cases too.

7. Approximation of typical convex bodies and classification by approximation properties

7.1 *The approximation behaviour of most convex bodies is irregular.* Given real functions φ, ψ on N such that

$$0 < \varphi(n), \psi(n) = o(1/n^{2/(d-1)} \quad \text{as} \quad n \to \infty,$$

it was proved by Gruber and Kenderov [82] that for most convex bodies C

$$\delta^H(C,\mathscr{P}_n),\ \delta^S(C,\mathscr{P}_n) < \varphi(n) \text{ for infinitely many n,}$$
$$\delta^H(C,\mathscr{P}_n),\ \delta^S(C,\mathscr{P}_n) > \psi(n) \text{ for infinitely many n.}$$

Recalling the results of section 5, this shows that the approximation behaviour of most convex bodies is rather 'irregular'. A result of Schneider and Wieacker [81] constitutes a more general form of 'one fourth' of this theorem.

The idea underlying the proof of Gruber and Kenderov [82] can be extended to many different situations ranging from numerical analysis to Diophantine approximation and the theory of Hausdorff measures.

7.2 *Approximability types of convex bodies.* Let $\mathscr{H}$ denote the set of Hausdorff functions, i.e. nonnegative increasing real functions on $[0, +\infty[$ continuous on the right and equal to 0 only at the origin. We now follow Schneider and Wieacker [81]. For a convex body C define the *approximability number* $\alpha(C)$ by

$$\alpha(c) = \inf\left\{\alpha > 0\colon \liminf_{n\to\infty} n\delta^H(C,\mathscr{P}^i_n)^\alpha > 0\right\}$$

and the *approximability* $\mathscr{A}(C)$ by

$$\mathscr{A}(C) = \left\{h \in \mathscr{H} : \liminf_{n\to\infty} nh(\delta^H(C,\mathscr{P}^i_n)) > 0.\right\}$$

These notions are in analogy to Hausdorff dimension, and a generalized concept of dimension introduced by Larman for subsets of metric spaces (cf. Schneider and Wieacker [81] for references). A point f of a convex body C is called a *farthest point* if there exists a point $a \in E^d$ such that

$$|f - a| = \sup_{x\in C} |x - a|.$$

It follows from the results of 5.1 that $\alpha(C) \le (d-1)/2$ for each convex body C where equality holds if C is of class $\mathscr{C}^3$ and with positive Gaussian curvature. A result mentioned in 7.1 yields $\alpha(C) = 0$ for most convex bodies C.

Schneider and Wieacker [81] show that for a Hausdorff function $h \in \mathscr{H}$ there exists a convex disk C with

$$0 < \liminf_{n\to\infty} nh(\delta^H(C,\mathscr{P}^i_n)) < +\infty$$

if and only if

$$\liminf_{\tau\to 0} \frac{h(\tau)}{\tau^{1/2}} > 0.$$

From this it follows that the approximability numbers of convex disks can take any value of [0,1/2].

The set of farthest points is related to the approximability of a convex body. In particular a result of Schneider and Wieacker [81] shows that $\alpha(C)$ is at least half the Hausdorff dimension of the set of farthest points of C. This yields that for convex bodies C of class $\mathscr{C}^2$ one has $\alpha(C) = (d-1)/2$, thus refining a result mentioned above. Concluding this section we point out that according to a communication of Burton problem (b) of Schneider and Wieacker [81] has a negative answer.

8. Approximation of convex bodies by random polytopes

The theory of random polytopes is well established and offers many interesting facets (see Santalò [76] and Baddeley [77] for surveys). One of these facets will be described in detail below.

8.1 The measure space. Let C be a convex body and ν a probability measure on C, absolutely continuous with respect to Lebesgue measure μ. In many cases μ will be equal to μ/μ (C). For $n = d+1, d+2, \ldots$ consider the map

$$(x_1,\ldots,x_n) \to \operatorname{conv}\{x_1,\ldots,x_n\} \quad \text{for} \quad (x_1,\ldots,x_n) \in C^n,$$

where conv denotes the convex hull. Neglecting a set of n-tuples of measure 0 this is a map of C^n onto $\mathscr{P}^i_n(C)$. To each set of polytopes of $\mathscr{P}^i_n(C)$ for which the inverse image in C^n is ν^n-measurable assign the ν^n-measure of its inverse image. This makes $\mathscr{P}^i_n(C)$ into a probability space. The expected volume and surface area of a random polytope in C having at most n vertices are thus given by

$$\int_C \cdots \int_C \mu(\operatorname{conv}\{x_1,\ldots,x_n\}) d\nu(x_1)\ldots d\nu(x_n)$$

and

$$\int_C \cdots \int_C \sigma(\operatorname{conv}\{x_1,\ldots,x_n\}) d\nu(x_1)\ldots d\nu(x_n)$$

respectively. For the case of the uniform distribution one obtains the expressions

$$\frac{1}{\mu(C)^n} \int_C \cdots \int_C \mu(\operatorname{conv}\{x_1,\ldots,x_n\}) d\mu(x_1)\ldots d\mu(x_n)$$

and

$$\frac{1}{\mu(C)^n} \int_C \cdots \int_C \sigma(\operatorname{conv}\{x_1,\ldots,x_n\}) d\mu(x_1)\ldots d\mu(x_n).$$

8.2 Approximation of convex disks by inscribed random polygons. Let C be a convex disk with continuous curvature κ and endowed with the uniform distribution. The expected area and perimeter of random polygons of $\mathscr{P}^i_n$ are given by

$$\mu(C) - \frac{\Gamma(8/3)(12\mu(C))^{2/3}}{10} \int_{\mathrm{bd}C} \kappa^{1/3} d\sigma \frac{1}{n^{2/3}} + 0\left(\frac{1}{n}\right) \quad \text{as} \quad n \to \infty,$$

and

$$\sigma(C) - \frac{\Gamma(2/3)(12\mu(C))^{2/3}}{12} \int_{bdC} \kappa^{4/3} d\sigma \frac{1}{n^{2/3}} + 0\left(\frac{1}{n}\right) \quad \text{as} \quad n \to \infty,$$

as shown by Rényi and Sulanke [63], [64] (but see Buchta [84]). A generalization of the result for the perimeter, which provides a simpler proof of the result of Rényi and Sulanke was given by Schneider and Wieacker [80]. From the result of Rényi and Sulanke one obtains for the expected distances in the sense of the metrics δ^S and δ^P of C to a random polygon $P \in \mathscr{P}^i_n(C)$ the expressions

$$\delta^S\colon \frac{\Gamma(8/3)(12\mu(C))^{2/3}}{10} \int_{bdC} \kappa^{1/3} d\sigma \frac{1}{n^{2/3}} + 0\left(\frac{1}{n}\right) \quad \text{as} \quad n \to \infty,$$

and

$$\delta^P\colon \frac{\Gamma(2/3)(12\mu(C))^{2/3}}{10} \int_{bdC} \kappa^{4/3} d\sigma \frac{1}{n^{2/3}} + 0\left(\frac{1}{n}\right) \quad \text{as} \quad n \to \infty.$$

As in 4.2 the affine isoperimetric inequality of Blaschke [23] implies that among all convex disks C of equal area the ellipses alone are asymptotically worst approximated in the sense of δ^S by inscribed random polygons with at most n vertices as $n \to \infty$. A result of Groemer [74] shows more precisely, that for each n only the ellipses are worst approximable by inscribed random polygons with at most n vertices among all convex disks of equal area.

The case of C a square was also considered by Rényi and Sulanke [64] while the case of general polygons is due to Buchta [83]. Suppose that C is a convex polygon with angles $\varphi_1,\ldots,\varphi_m$. Then the expected distances with respect to δ^S,δ^P of C to a random polygon $P \in \mathscr{P}^i_n(C)$ are

$$\delta^S\colon \frac{2m\mu(C)}{3}\frac{\log n}{n} + 0\left(\frac{1}{n}\right) \quad \text{as} \quad n \to \infty,$$

$$\delta^P\colon \frac{(2\pi\mu(C))^{1/2}}{4} \sum_1^m \left(\frac{2}{(\sin\varphi_k)^{1/2}} - I(\varphi_k)\right)\frac{1}{n^{1/2}} + o\left(\frac{1}{n^{1-\varepsilon}}\right) \quad \text{as} \quad n \to \infty,$$

where $\varepsilon > 0$ is arbitrary and

$$I(\varphi) = \int_0^{1/\sin\varphi} (((1 + (-\tau + \cot\varphi)^2)^{1/2} - (1 + \cot^2\varphi)^{1/2})/\tau^{3/2})d\tau.$$

As before, also here uniform distribution was chosen.

8.3 *Approximation of convex disks by circumscribed random polygons.* For the moment we leave the path begun in 8.1 to describe work of Carlsson and Grenander [67]. Let C be a convex disks. Assume that the boundary of C is represented in the form

$$x = x(\varphi),\ 0 \le \varphi \le 2\pi,$$

where the angle φ denotes the direction of the normal of C at $x(\varphi)$ and such that x is of class $\mathscr{C}^2$ and the radius of curvature ρ is of class $\mathscr{C}^1$. Let an absolute continuous distribution on $[0,2\pi]$ be given such that its density δ is positive and of class $\mathscr{C}^1$. For an ordered sample of n angles, $0 \leq \varphi_1 \leq \cdots \leq \varphi_n \leq 2\pi$ say, denote by $P(\varphi_1,\ldots,\varphi_n)$ the (possibly unbounded) polygon containing C the edges of which have exterior normals in directions $\varphi_1,\ldots,\varphi_n$ and touch C. Then the distance of C to $P(\varphi_1,\ldots,\varphi_n)$ in the sense of the metric δ^S can be written in the form

$$\gamma_n(\varphi_1,\ldots,\varphi_n) + \delta_n(\varphi_1,\ldots,\varphi_n)$$

where for the expectation of γ_n we have the expression

$$\frac{1}{4}\int_0^{2\pi} \frac{\varphi^2}{\delta^2}\,d\varphi\,\frac{1}{n^2} + o\left(\frac{1}{n^2}\right) \quad \text{as} \quad n \to \infty,$$

further

$$n^2\gamma_n \to \frac{1}{4}\int_0^{2\pi} \frac{\rho^2}{\delta^2}\,d\varphi \quad \text{in probability as } n \to \infty,$$

and

$$n^2\delta_n \to 0 \quad \text{in probability as } n \to \infty,$$

and δ_n does not have finite expectation. Using Hölder's inequality one can show that

$$\int_0^{2\pi} \frac{\rho^2}{\delta^2}\,d\varphi \geq \left(\int_0^{2\pi} \rho^{2/3}\,d\varphi\right)^3.$$

If ρ is positive the minimum is attained for

$$\delta = \frac{\rho^{2/3}}{\int_0^{2\pi} \rho^{2/3}\,d\varphi}.$$

8.4 Approximation of convex bodies by inscribed random polytopes. Let C be a convex body in E^3 of class $\mathscr{C}^3$ with Gaussian curvature κ and endowed with the uniform distribution. The expectation of $\delta^S(C,P)$ where $P \in \mathscr{P}^i_n(C)$ is given by

$$\frac{35\mu(C)^{1/2}}{48}\int_{\mathrm{bd}C} \kappa^{1/4}\,d\sigma\,\frac{1}{n^{1/2}} + o\left(\frac{1}{n^{1/2}}\right) \quad \text{as} \quad n \to \infty,$$

as proved by Wieacker [78].

Similarly to 5.2 Blaschke's affine isoperimetric inequality implies that among all convex bodies of equal volume the ellipsoids alone are asymptotically worst approximable by random polytopes. A very precise adjunct to this asymptotic extremal property of ellipsoids is the following result of Groemer [74] for general d. For each n among all convex bodies of equal volume the ellipsoids are the worst approximable ones in the sense of the metric δ by inscribed random polytopes with at most n vertices, assuming uniform distribution.

8.5 Approximation of balls by inscribed random polytopes. The following is a result of Wieacker [78]. Let B^d denote the unit ball in E^d endowed with the uniform distribution. The expected distances with respect to the metrics δ^S and δ^P of B^d to a random polytope $P \in \mathscr{P}_n^i(B^d)$ are given by

$$(d^2 + d + 2)\alpha \frac{1}{n^{2/(d+1)}} + o\left(\frac{1}{n^{2/(d+1)}}\right) \quad \text{as} \quad n \to \infty,$$

and

$$d(d^2 - 1)\alpha \frac{1}{n^{2/(d+1)}} + o\left(\frac{1}{n^{2/(d+1)}}\right) \quad \text{as} \quad n \to \infty$$

where

$$\alpha = \frac{(d^2 + 1)d(d + 1)^{2/(d+1)}\Gamma((d^2 + 1)/(d + 1))\pi_d^{(d+3)/(d+1)}}{d(d + 3)(d + 1)!\pi_{d-1}^{2/(d+1)}}.$$

8.6 None of the above, but in this section. There exist many results related to the above. The following may serve as an example. Let C be a triangle endowed with the uniform distribution. Then the expected area of a random triangle in C is equal to $\mu(C)/12$. For other results of this type see Kendall and Moran [63], Santaló [76] and Efron [65].

The result of Groemer [76] referred to in 8.2 and 8.4 has been generalized by Schöpf [77].

9. Approximation by particular convex bodies

In this section some results on the approximation of convex bodies by particular convex bodies such as simplices and balls will be collected. Although some planar results are mentioned, the emphasis is on results for general dimensions. In this area it is difficult to draw a borderline between approximation results and those results which are not. Thus the following reflects my personal view much more than this was the case in previous sections.

9.1 Approximation by simplices. A theorem of Gale [53] shows that each convex body of diameter 1 can be covered by a regular simplex of edge-length at most

$$(d(d + 1))^{1/2}.$$

For a refinement of this result see Süss [55].

Results of Sas [39], Blaschke [23] and Macbeath [51] show that for each convex body C the simplex S of maximal volume contained in C satisfies

$$\mu(S) \geq \frac{(d + 1)^{(d+2)/2}}{d^{d/2}d!\pi_d} \mu(C),$$

where equality holds if and only if C is an ellipsoid. This result was complemented by the following result of McKinney [74] for centrally symmetric convex bodies C. Let S be a simplex contained in C of maximal volume and having one vertex at the center of C. Then

$$\mu(S) \geq \frac{1}{d!\pi_d}\,\mu(C),$$

where equality holds if and only if C is an ellipsoid.

Let C be a smooth strictly convex body. If S is a simplex, a result of Nandares and Toranzos [77] shows that there exists a homothetic copy T of S contained in C and such that all vertices of T are on the boundary of C. (T is a *homothetic* image of S if $T = \{\lambda x + c \mid x \in S\}$ for suitable $\lambda > 0$ and $c \in E^d$.) Further, for any boundary point p of C there exists a regular simplex in C, all vertices of which are boundary points of C, one of them coinciding with p.

Sangwine-Yager [82] considered the following problem. Let C be a convex body and S a fixed simplex circumscribed to B^d such that all facets of S touch B^d. For any rotation r of E^d let S(r) denote the homothetic image of r(S) circumscribed to C such that all facets of S(r) touch C. The mean values of $\mu(S(r))$ and $\sigma(S(r))$ as r ranges over the group of rotations of E^d endowed with its Haar measure, are at least

$$\mu(S)\mu(C)/\pi_d$$

and

$$\sigma(S)\sigma(C)/\pi_d$$

respectively, where equality holds in both cases if and only if C is a ball. A more general result for the planar case is also due to Sangwine-Yager [80].

Let C be a convex body of given volume endowed with the uniform distribution (cf. 8.1). It was proved by Groemer [73], [74] that the expected volume of a random simplex contained in C is minimal if and only if C is an ellipsoid. (See 8.2 and 8.4 for generalizations of this.) For the unit ball this expectation has been evaluated by Kingman [69]. It is

$$\frac{2\pi_{d+1}^{d}\pi_{d^2+2d}}{(d+1)!\pi_d^{d+1}\pi_{(d+1)^2}}\left(= \pi_d 2^{-d}\binom{d+1}{(d+1)/2}^{d+1}\Big/\binom{(d+1)^2}{(d+1)^2/2}\right).$$

Thus the values for $d = 1,2,3$ are $2/3$, $35/(48\pi)$, $12\pi/715$. The problem to find the expected volume of a random simplex contained in a given convex body is often called *Sylvester's problem*. For more information about it the reader is referred to Kendall and Moran [63], Klee [69], Reed [74] and the references cited there.

9.2 Approximation by boxes. A box is a parallelepiped generated by d orthogonal vectors. It was shown by Hadwiger [55] that for any convex body C there exist boxes A,B with parallel edges such that $A \subset C \subset B$ and

$$\frac{\mu(B)}{d!} \leq \mu(C) \leq d^d \mu(A)$$

(see also Macbeath [51′], Radiszewski [53], Kosiuski [57] and Freeman and Shapira [75]).

Let C be a convex body and denote the unit cube by P. For any rotation r of E^d

let B(r) be the box containing C such that all of its facets touch C and the edges of B(r) are parallel to those of r(P). The mean values of $\mu(B(r))$ and $\sigma(B(r))$ over the group of rotations of E^d are minimal if and only if C is a ball. These results are due to Chakerian [72] and Schneider [72]. We also refer to Radiszewski [59], Heil [72], Lutwak [77] and Sangwine-Yager [80].

9.3 Approximation by cubes. Kakutani [42] proved that for any convex body C there is a circumscribed cube all facets of which touch C. Alternative proofs and generalizations of this were given by Mira Fernandes [43], Yamabe and Yujobo [50] and Yang [55].

Aščepkov [80] indicated a construction for the maximal cube in a given domain.

9.4 Approximation by cross polytopes. A *cross polytope*, also called *generalized octahedron*, is a convex body of the form

$$\left\{c + \tau_1 u_1 + \cdots + \tau_d u_d : \sum_1^d \left|\frac{\tau_b}{\sigma_b}\right| \leq 1\right\},$$

where $c \in E^d$, $\sigma_1, \ldots, \sigma_d > 0$ and $u_1, \ldots, u_d \in E^d$ are linearly independent; $\{u_1, \ldots, u_d\}$ are said to correspond to the cross polytope.

A theorem of Hadwiger, Larman and Mani [78] says that for any smooth strictly convex body C and orthogonal vectors $u_1, \ldots, u_d$ there is a cross polytope corresponding to $u_1, \ldots, u_d$ and contained in C such that all of its vertices are boundary points of C. For $d \geq 3$ this does not continue to hold if the assumptions about C are omitted. It was proved by Kramer [80] that for C a smooth convex body and for linearly independent vectors $\{u_1, \ldots, u_d\}$ there exists a cross polytope corresponding to $\{u_1, \ldots, u_d\}$ contained in C whose vertices are on the boundary of C.

9.5 Approximation by circular disks and balls. The *Steiner point* of a convex body is the centroid of its curvature (see e.g. Schneider [71]). Denote by C a convex disk. A theorem of Davis [82] shows that among all circles in the plane the circle with center at the Steiner point of C and perimeter equal to that of C approximates C best in the sense of the metric δ^2.

By a classical theorem of Jung [01] each convex body of diameter 1 is contained in a ball of radius

$$\left(\frac{d}{2d+1}\right)^{1/2}$$

(cf. Hadwiger [57] and Leichtweiß [80]).

The *minimal width* of a convex body is the minimum of the widths of the parallel strips containing it. It was shown by Blaschke [14] and Steinhagen [21] that a convex body of minimal width 1 contains a ball of radius at least

$$\frac{1}{2d^{1/2}} \quad \text{for d odd}$$

and

$$\frac{(d+2)^{1/2}}{2d+2} \quad \text{for d even.}$$

A series of results on inscribed and circumscribed balls of 3-dimensional polytopes exhibiting extremal properties of regular polytopes due to Fejes Tóth and Florian are surveyed by Florian [79].

9.6 *Approximation by ellipsoids.* John [48] proved that for each convex body C there are concentric and homothetic ellipsoids E and F such that $E \subset C \subset F$ and

$$\mu(F) \leq d^d \mu(E).$$

If we assume further that C is centrally symmetric then E,F can be chosen such that

$$\mu(F) \leq d^{d/2} \mu(E)$$

holds. For an alternative proof of the first result see Leichtweiß [59].

A theorem on inscribed and circumscribed ellipsoids of 3-dimensional polyhedra is contained in Fejes Tóth [48].

An algorithm to determine the ellipse of minimal area that covers a given convex disk was found by Silverman and Titterington [80].

9.7 *None of the above, but in this section.* The best approximation of a convex disk C by the translates of a convex disk D in the sense of the metric δ^2 is the translate of D by s where s is the Steiner point of the convex disk $C - D = \{x - y : x \in C, y \in D\}$. For a proof and related results see Davis [82].

The problem to approximate a convex body by convex bodies which are symmetric in some sense has been considered frequently in the literature. For a review of older results of this type see the interesting survey of Grünbaum [63]. Some references on symmetry measures for centrally symmetric convex bodies in the setting of functional analysis are given by Kadets [75].

Let C be a convex disk and P a convex polygon containing C such that each edge of P touches C. For $0 \leq \varphi \leq 2\pi$ define $P(\varphi)$ to be the polygon circumscribed to C such that all edges of $P(\varphi)$ touch C and the set of the exterior normals to the edges of $P(\varphi)$ coincides with the set of the exterior normals of P rotated by the angle φ in positive direction. Then a result of Sangwine-Yager [80] says

$$\int_0^{2\pi} \mu(P(\varphi))\,d\varphi \geq 2\mu(C)\mu(P).$$

Croft [80] posed the problem to find for each regular polytope P among all regular polytopes of a different type the ones of maximal volume contained in P. He settled this problem in many cases.

The problem to approximate convex disks by circular rings was investigated by Vincze [47], [74].

10. Miscellanea

The following is a sample of theorems on convex bodies which can be considered as approximation results.

10.1 The Banach-Mazur metric. Let E,F be normed linear spaces of finite dimension d. (Essentially a normed space of dimension d can be identified with E^d if in E^d a convex body, centrally symmetric in the origin, is selected, namely the unit ball of the normed space.) The metric δ^{BM}, defined by

$$\delta^{BM}(E,F) = \log \inf\{\|1\|\cdot\|1^{-1}\| : 1{:}E \to F \quad \text{linear, onto}\}$$
$$\text{for E,F normed spaces of dimensions d}$$

is called the *Banach-Mazur metric* on the space of equivalence classes of isometric normed linear spaces of dimension d (see e.g. Kadets [75]). It follows from the result of John [48], referred to in 9.6, that for any pair of d-dimensional normed spaces E,F

$$\delta^{BM}(E,F) \leq \log d.$$

A deep recent result of Gluškin [81] shows the existence of a positive constant independent from d such that for any d

$$\delta^{BM}(E,F) \geq \log d - \text{const}$$

for suitable normed spaces E,F of dimension d.

10.2 Affine images of a convex body contained in a given convex body. Given two centrally symmetric convex bodies C,D a theorem of Thomas [79] says that there exists an affinity a such that $a(D) \subset C$ and

$$\mu(C) \leq d^d \mu(a(D)) \quad \text{or} \quad |\det a| \geq \frac{\mu(C)}{d^d \mu(D)},$$

where det a means the determinant of the linear transformation assigned to the affinity a. Thomas conjectures that this is true for all convex bodies and shows that in the case of arbitrary convex bodies C,D an affinity a can be chosen such that $a(D) \subset C$ and

$$\mu(C) \leq \frac{d^d(2d)!}{2^d(d!)^2}\, \mu(a(D)).$$

Macbeath [51′] introduced the distance δ^{MB} on the space of classes of affine-equivalent convex bodies by

$$\delta^{MB}(C,D) = \log\ \sup\{|\det ab| : a,b \ \text{affine},\ a(D) \subset C,\ b(C) \subset D\} \ \text{for}\ C,D \in \mathscr{C}$$

and investigated sup δ^{MB}. See also Bronštein [78].

10.3 The Hausdorff metric and Chebycheff centers. The *Chebycheff center* of a subset C of E^d is the point c_C such that

$$\sup_{x \in C} |x - c_C| = \inf_{y \in E} \sup_{x \in C} |x - y|.$$

(Note that c_C is unique if it exists at all, since the underlying norm is the Euclidean norm.) Define δ^{BK} on C by

$$\delta^{BK}(C,D) = \frac{|c_C - c_D|}{\delta^H(C,D)} \quad \text{for} \quad C,D \in \mathscr{C}.$$

Borwein and Keener [80] investigate sup $\delta^{BK}(C,D)$ when C,D range over various sets of pairs of convex bodies.

10.4 Steiner approximation of convex bodies. For hyperplanes $H_1,H_2,\ldots,H_n (n \geq d+1)$ let s_n denote the point of E^d for which the sum of the squares of the distances to $H_1,\ldots,H_n$ is minimal. s_n exists and in general is unique. It can be considered as a substitute for the solution of the corresponding linear equations which may not exist. We remark that this notion plays an important role in surveying. A modern tool for finding s_n is the Moore-Penrose inverse for matrices.

Now, following Davis [82], let $u_1,u_2,\ldots \in S^{d-1}$ be a sequence of unit vectors uniformly distributed on S^{d-1} in the sense of uniform distribution theory (cf. e.g. Hlawka [79]). Given a convex body C, let $H_1,H_2,\ldots$ denote the hyperplanes of support of C with exterior normal vectors $u_1,u_2,\ldots$. Define s_n as in the previous paragraph. Then s_n tends to the Steiner point of C as $n \to \infty$. Thus the Steiner point of a convex body C can be considered in some sense a best approximating point of C.

10.5 Lattice points in convex bodies. The problem to determine the number of points with integer coordinates in a convex body or on its boundary has frequently been considered in number theory; in particular geometry of numbers (see Cassels [59], Lekkerkerker [69] and Gruber [79]), lattice point problems in the sense of Hadwiger and Wills (see Betke and Wills [79]) and lattice point problems in the sense of Gauss and Landau (see Walfisz [57] and Fricker [82]) deserve to be mentioned. It is from the last area that we state some results.

A classical result of Jarník [25] shows that for a strictly convex disk C which contains the origin in its interior the number of points with integer coordinates on the boundary of λC does not exceed

$$\frac{3\sigma(C)}{(2\pi)^{1/3}} \lambda^{2/3} + 0(\lambda^{1/3}) \quad \text{as} \quad \lambda \to \infty,$$

where both the exponent and the coefficient of the leading term are best possible. For related results of this type and an extension to d = 3 see Swinnerton-Dyer [74], Diviš [76] and Chaix [77]. A corresponding result for general d was given by Andrews [63].

A result of Hlawka [50] shows that for a convex body C of sufficiently high differentiability class and with positive Gaussian curvature which contains the origin in its interior, the number of points with integer coordinates in λC is equal to

$$\mu(C)\lambda^d + 0(\lambda^{\frac{d(d-1)}{d+1}}) \quad \text{as} \quad \lambda \to +\infty.$$

For related results and refinements see Randol [66], Colin de Verdière [77] and Guillemin [79].

11. Final remarks and acknowledgement

This paper contains results on the approximation of convex bodies. The selection of results is necessarily subjective, in particular in sections 9 and 10. While applications of approximation abound in the literature, it seems to be difficult even to locate the more important ones since they are frequently hidden in proofs of results altogether unrelated to approximation problems. Therefore we did not insert a section on applications.

I am obliged to many colleagues who supported this work through valuable oral communications and by providing preprints and other unpublished work. Of these the surveys of Müller [80] and Wieacker [78] deserve particular mention.

For many suggestions I am obliged to Professors Schneider and Schnitzer, Dr. Buchta and Mr. Müller.

References

G.E. Andrews
[63] A lower bound for the volume of strictly convex bodies with many boundary lattice points, Trans. Amer. Math. Soc. **106** (1963) 270–279

L.T. Aščepkov
[80] On the construction of the maximum cube inscribed in a given domain, Ž. Vyčisl. Mat. Fiz. **20** (1980) 510–513

V.S. Azarin and V.I. Barmin
[76] Über eine Approximation vermittels stückweise linearer Funktionen, Theor. Funkts. Funkts. Anal. Prilzh. **25** (1976) 5–14
[76a] Die Approximation vermittels stückweise linearer Funktionen, Mat. Collect., Acad. Sci. Ukr. SSR, Inst. Mat. (1976) 25–26

A. Baddeley
[77] A fourth note on recent research in geometrical probability, Advances Appl. Probability **9** (1977) 824–860

V. Bangert
[78] Über die Approximation von lokal konvexen Mengen, Manuscripta Math. **25** (1978) 397–420

D. Barnette
[72] Dense sets of polytopes in P^3, Aequationes Math. *7* (1972) 28–35

R.K. Beatson
[81] Convex approximation by splines, SIAM J. Math. Anal. **12** (1981) 549–559

Chr. Berg
[72] Shephard's approximation theorem for convex bodies and the Milman theorem, Math. Scand. **25** (1969) 19–24

U. Betke and P. McMullen
[82] Estimating the sizes of convex bodies from projections, Manuscript 1982

U. Betke and J.M. Wills
[79] Diophantine approximation of convex bodies, Manuscript 1979
[79′] Stetige und diskrete Funktionale konvexer Körper, Proc. Geom. Sympos., Siegen (1978) 226–237, Birkhäuser, Basel 1979

W. Blaschke
[14] Über den größten Kreis in einer konvexen Punktmenge, J.-ber. Deutsch. Math. Verein. **23** (1914) 369–374

[16] Kreis und Kugel, Veit, Leipzig 1916, Chelsea, New York 1955, de Gruyter, Berlin 1956
[23] Differentialgeometrie II, Affine Differentialgeometrie, Springer, Berlin 1923
R. Bojanic and J. Roulier
[75] Approximation of convex functions by convex splines and convexity preserving continuous linear operators, Rev. Anal. numér. Theor. Approximation **3** (1974) 143–150 (1975)
J. Bokowski
[77] Konvexe Körper approximierende Polytopklassen, Elemente Math. **32** (1977) 88–90
J. Bokowski and P. Mani-Levitska
[82] Oral communication
J. Bokowski and Chr. Schulz
[78] Dichte Klassen konvexer Polytope, Math. Z. **160** (1978) 173–182
T. Bonnesen and W. Fenchel
[34] Theorie der konvexen Körper, Springer, Berlin 1934, 1974, Chelsea, New York 1948
J. Borwein and L. Keener
[80] The Hausdorff metric and Čebyšev centres, J. Approximation Theory **28** (1980) 366–376
E.M. Bronštein
[78] ε-entropy of affine-equivalent convex bodies and Minkowski's compactum, Optimizacija **22** (39) (1978) 5–11, 155
E.M. Bronštein and L.D. Ivanov
[75] The approximation of convex sets by polyhedra, Sibir. Mat. Ž. **16** (1975) 1110–1112 = Siber. Math. J. **16** (1975) 852–853
C. Buchta
[82] Stochastische Approximation konvexer Polygone, Submitted
[84] Das Volumen von Zufallspolyedern im Ellipsoid, Submitted
A.P. Bulanov
[78] Approximation by rational functions of convex functions with given modulus of continuity, Math. USSR Sbornik **34** (1978) 1–24
A.P. Bulanov and A. Hatamov
[78] On rational approximation of convex functions with a given modulus of continuity, Analysis Math. **4** (1978) 237–246
L. Button and J.B. Wilker
[78] Cutting exponents for polyhedral approximations to convex bodies, Geometriae dedicata **7** (1978) 415–426
S. Carlsson and U. Grenander
[67] Statistical approximation of plane convex sets, Skandinav. aktuarietidskr. **50** (1967) 113–127
J.W.S. Cassels
[59] An introduction to the geometry of numbers, Springer, Berlin 1959
H. Chaix
[77] Sur les points frontières des compacts strictement convexes de $\mathbb{R}^2$ appartenent à un réseau, C.R. Acad. Sci. Paris Sér A–B **285** (1977) A887–A889
G.D. Chakerian
[72] The mean value of boxes circumscribed about a convex body, Israel J. Math. **12** (1972) 249–256
G.D. Chakerian and H. Groemer
[83] Convex bodies of constant width, In: Convexity and its Applications, Birkhäuser, Basel 1983, To appear
G.D. Chakerian and J.R. Sangwine-Yager
[79] A generalization of Minkowski's inequality for plane convex sets, Geometriae dedicata **8** (1979) 437–444
Y. Colin de Verdière
[77] Nombre de points entiers dans une famille homothétique des domains de R, Ann. Sci. Ecole Norm. Sup. (4) **10** (1977) 559–575
H.T. Croft
[80] On maximal regular polyhedra inscribed in a regular polyhedron, Proc. London Math. Soc. III, **41** (1980) 279–296
P.J. Davis
[82] Lemoine approximation and Steiner approximation of convex sets, Manuscript 1982

P.J. Davis, R.A. Vitale and E. Ben-Sabar
[77] On the deterministic and stochastic approximation of regions, J. Approximation Theory **21** (1977) 60–88
B. Diviš
[76] On the lattice points on strictly convex surfaces, J. Number Theory **8** (1976) 298–307
D.P. Dobkin and L. Snyder
[79] On a general method for maximizing and minimizing among certain geometric problems. Proc. 20th Annual Sympos. on Foundations of Computer Sci., San Juan (1979) 9–17 IEEE, New York 1979
C.H. Dowker
[44] On minimum circumscribed polygons, Bull. Amer. Math. Soc. **50** (1944) 120–122
R. Dudley
[74] Metric entropy of some classes of sets with differentiable boundaries, J. Approximation Theory **10** (1974) 227–236, Corr., ibid, **26** (1979) 192–193
B. Efron
[65] The convex hull of a random set of points, Biometrika **52** (1965) 331–343
H.G. Eggleston
[57] Approximation to plane convex curves. I. Dowker-type theorems, Proc. London Math. Soc. (3) (1957) 351–377
[58] Convexity, Cambridge Univ. Press, Cambridge 1958, 1966
[61] Problems in Euclidean space: Applications of convexity, Pergamon, New York 1961
P. Erdö and I. Vincze
[82] On the approximation of convex, closed plane curves by multifocal ellipses,
G. Fejes Tóth
[77] On a Dowker-type theorem of Eggleston, Acta Math. Acad. Sci. Hungar. **29** (1977) 131–148
[82] Oral communication 1982
G. Fejes Tóth and L. Fejes Tóth
[73] Remark on a paper of C.H. Dowker, Period. Math. Hungar. **3** (1973) 271–274
L. Fejes Tóth (= L. Fejes)
[39] Über zwei Maximumaufgaben bei Polyedern, Tôhoku Math. J. **46** (1939) 79–83
[40] Über ein extremales Polyeder, Mat. Termeszettudományi Értesitö **59** (1940) 476–479
[43] Einige Extremaleigenschaften des Kreisbogens bezüglich der Annäherung durch Polygone, Acta Univ. Szeged. Sect. Sci. Math. **10** (1943) 164–173
[48] On ellipsoids circumscribed and inscribed to polyhedra, Acta Univ. Szeged. Sect. Sci. Math. **11** (1948) 225–228
[53] Lagerungen in der Ebene, auf der Kugel und im Raum, Springer, Berlin 1953, 1972
[58] Annäherung von Eibereichen durch Polygone, Math.-phys. Semesterber. **6** (1958/59) 253–261
[60] Über eine Volumenabschätzung für Polyeder, Monatsh. Math. **64** (1960) 374–377
[80] Approximation of convex domains by polygons, Stud. Sci. Math. Hungar. **15** (1980) 133–138
Wm. J. Firey
[74] Approximating convex bodies by algebraic ones, Arch. Math. **25** (1974) 424–425
A. Florian
[79] Neuere Entwicklungen über reguläre Polyeder, Proc. Geom. Sympos., Siegen (1978), 238–247, Birkhäuser, Basel, 1979
H. Freeman and R. Shapira
[75] Determining the minimum area enclosing rectangle for an arbitrary closed curve, Comm. ACM **18** (1975) 409–413
F. Fricker
[82] Einführung in die Gitterpunktlehre, Birkhäuser, Basel 1982
D. Gale
[53] On inscribing n-dimensional sets in a regular n-simplex, Proc. Amer. Math. Soc. **4** (1953) 222–225
O. Giering
[61] Ausfüllung von Eilinien durch einbeschriebene gleichseitige und gleichwinkelige n-Ecke, Elem. Math. **16** (1961) 78–81
E.D. Gluškin
[81] Diameter of the Minkowski compactum is approximately equal to n, Funkt. Analiz. Ego Prilozhen. **15** (1981) 72–73 = Funct. Anal. Appl. **15** (1981) 57–58

H. Groemer
[73] On some mean values associated with a randomly selected simplex in a convex set, Pacific J. Math. **45** (1973) 525–533
[74] On the mean value of the volume of a random polytope in a convex set, Arch. Math. **25** (1974) 86–90
P.M. Gruber
[79] Geometry of numbers, Proc. Geom. Sympos., Siegen (1978) 186–225, Birkhäuser, Basel 1979
[81] Approximation of convex bodies, C.R. Acad. Bulgare Sci. **34** (1981) 621–622
P.M. Gruber and P. Kenderov
[82] Approximation of convex bodies by polytopes, Rend. Circolo Mat. Palermo **31** (1982) 195–225
B. Grünbaum
[63] Measures of symmetry for convex sets, Convexity, Proc. Sympos. Pure Math **7**, Seattle (1961) 233–270, Amer. Math. Soc., Procidence R.I. 1963
[67] Convex polytopes, Wiley, New York 1967
H. Guggenheimer
[77] Applicable geometry. Global and local convexity, Krieger, Huntington N.Y. 1977
V. Guillemin
[79] Some classical theorems in spectral theory revisited, Seminars on singularities of solutions of linear partial differential equations 219–259, Princeton Univ. Press, Princeton 1979
H. Hadwiger
[41] Überdeckung ebener Bereiche durch Kreise und Quadrate, Comment. Math. Helv. **13** (1941) 195–200
[55] Volumschätzung für die einen Eikörper überdeckenden und unterdeckenden Parallelotope, Elem. Math. **10** (1955) 122–124
[57] Vorlesungen über Inhalt, Oberfläche und Isoperimetrie, Springer, Berlin 1957
H. Hadwiger, D.G. Larman and P. Mani
[78] Hyperrhombs inscribed to convex bodies, J. Combinat. Theory B **24** (1978) 290–293
P.C. Hammer
[63] Approximation of convex surfaces by algebraic surfaces, Mathematika **10** (1963) 64–71
A. Hatamov
[77] Über rationale Approximation konvexer Funktionen, Mat. Zametki **21** (1977) 355–370
E. Heil
[72] Eine Verschärfung der Bieberbachschen Ungleichung und einige andere Abschätzungen für ebene Bereiche, Elem. Math. **27** (1972) 4–8
E. Hlawka
[50] Integrale auf konvexen Körpern I, Monatsh. Math. **54** (1950) 1–36
[79] Theorie der Gleichverteilung, Bibl. Inst., Mannheim 1979
R. Ivanov
[74] Approximation of convex n-gons by inscribed (n-1) -gons, Proc. Conf. Bulgar. Math. Soc., Vidin (1973) 115–122, Izdat Bulgar. Akad. Nauk, Sofia 1974
V. Jarnik
[25] Über die Gitterpunkte auf konvexen Kurven, Math. Z. **24** (1925) 500–518
F. John
[48] Extremum problems with inequalities as subsidiary conditions, Courant Anniversary volume 187–204, New York 1948
H.H. Johnson and A. Vogt
[80] A geometric method for approximating convex arcs, SIAM J. Appl. Math. **38** (1980) 317–325
H. Jung
[01] Über die kleinste Kugel, die eine räumliche Figur einschließt, J. reine angew. Math. **123** (1901) 241–257
M.I. Kadets
[75] The geometry of normed spaces, Itogi Nauk Tekhn., Mat. Analiz. **13** (1975) 99–128 = J. Soviet Math. **7** (1977) 953–973
S. Kakutani
[42] A proof that there exists a circumscribing cube around any bounded closed convex set in $\mathbb{R}^3$, Ann. of Math. (2) **43** (1942) 739–741

M. Kallay
[74] Reconstruction of a plane convex body from the curvature of its boundary, Israel J. Math. **17** (1974) 149–161.
M.G. Kendall and P.A.P. Moran
[63] Geometrical probability, Griffin, London 1963
P. Kenderov
[73] On an optimal property of the square, C.R. Acad. Bulgare. Sci. **26** (1973) 1143–1146
[80] Approximation of plane convex compacta by polygons, C.R. Acad. Bulgare. Sci. **33** (1980) 889–891
[82] Polygonal approximation of plane convex compacta, Manuscript 1982
J.F.C. Kingman
[69] Random secants of a convex body, J. Appl. Prob. **6** (1969) 660–672
V. Klee
[59] Some characterizations of convex polyhedra, Acta Math. **102** (1959) 79–107
[69] What is the expected volume of a simplex whose vertices are chosen at random from a given convex body? Amer. Math. Monthly **76** (1969) 286–288
A. Kosiuski
[57] A proof of an Auerbach-Mazur-Ulam theorem on convex bodies, Colloquium Math. **4** (1957) 216–218
M. Kömhoff and G.C. Shephard
[74] Approximation problems for combinatorial isomorphism classes of convex polytopes, Geometriae dedicata **3** (1974) 139–153
H. Kramer
[80] Hyperparallelograms inscribed to convex bodies, Mathematica (Cluj) **22** (45) (1980) 67–70
D.G. Larman and C.A. Rogers
[76] Durham symposium on the relation between infinite dimensional and finite dimensional convexity, Bull. London Math. Soc. **8** (1976) 1–33
D. Lazar
[47] Sur l'approximation des courbes convexes par des polygones, Acta Univ. Szeged. Sect. Sci. Math. **11** (1947) 129–132
K. Leichtweiß
[59] Über die affine Exzentrizität konvexer Körper, Arch. Math. **10** (1959) 187–199
[80] Konvexe Mengen, Springer, Berlin 1980
C.G. Lekkerkerker
[69] Geometry of numbers, North-Holland, Amsterdam 1969
N.F. Lindquist
[75] Approximation of convex bodies by sums of line segments, Portugal. Math. **34** (1975) 233–240
E. Lutwak
[77] Mixed width-integrals of convex bodies, Israel J. Math. **28** (1974) 249–253
A.M. Macbeath
[51] An extremal property of the hypersphere, Proc. Cambridge Philos. Soc. **47** (1951) 245–247
[51′] A compactness theorem for affine equivalence-classes of convex regions, Canad. J. Math. **3** (1951) 54–61
D.E. McClure and R.A. Vitale
[75] Polygonal approximation of plane convex bodies, J. Math. Anal. Appl. **51** (1975) 326–358
J.R. McKinney
[74] On maximal simplices inscribed in a central convex set, Mathematika **21** (1974) 38–44
H. Minkowski
[03] Volumen und Oberfläche, Math. Ann. **57** (1903) 447–495 = Ges. Abh. II 230–276
A. de Mira Fernandes
[43] Funzioni continue sopra una superficie sferica, Portugal. Math. **4** (1943) 69–72
J. Molnár
[55] On inscribed and circumscribed polygones of convex regions, Mat. Lapok **6** (1955) 210–218
J. Müller
[80] Approximation konvexer Körper durch konvexe Polytope, Diplomarbeit, Techn. Univ., Wien 1980

I. Najfeld
[78] The method of least squared residues for the approximation of circular and spherical forms, Report, Brown & Sharpe, Providence R.I. 1978
J.H. Nandares and F.A. Toranzos
[77] Inscribing simplices in convex bodies, Rev. Un. Mat. Argentina **28** (1977/78) 215–219
M.A. Perles and G.C. Shephard
[70] Dense sequences of convex polytopes, Mathematika **17** (1970) 28–38
P.P. Petrušev
[76] On the rational approximation of functions with convex r-th derivative, Acta Math. Acad. Sci. Hungar. **28** (1976) 315–320
V.A. Popov
[68] Approximation of convex figures, C.R. Acad. Bulgare. Sci. **21** (1968) 993–995
[70] Approximation of convex sets, Bulgar. Akad. Nauk. Otdel. Mat. Fiz. Nauk. Izv. Mat. Inst. **11** (1970) 67–80
[75] Approximation of convex functions by algebraic polynomials in Hausdorff metric, Serdica **1** (1975) 385–398
V.A. Popov and P.P. Petrušev
[77] Die genaue Ordnung der besten gleichmäßigen Approximation konvexer Funktionen vermittels rationaler Funktionen, Mat. Sbornik n. Ser. **103** (145) (1977) 285–292
K. Radziszewski
[53] Sur un problème extrémal relatif aux figures inscrites et circonscrites aux figures convexes, Ann. Univ. Mariae Curie-Sklodowska A, **6** (1953) 5–18
[56] Sur un functionnelle définie sur les ovales, Ann. Univ. Mariae Curie-Sklodowska A, **10** (1956) 57–59
J. Radon
[36] Annäherung konvexer Körper durch analytisch begrenzte, Monatsh. Math. Phys. **43** (1936) 340–344
B. Randol
[66] A lattice point problem, Trans. Amer. Math. Soc. **121** (1966) 257–268
W.J. Reed
[74] Random points in a simplex, Pacific J. Math. **54** (1974) 183–198
A. Rényi and R. Sulanke
[63] Über die konvexe Hülle von n zufällig gewählten Punkten, Z. Wahrscheinlichkeitstheorie u. verw. Geb. **2** (1963) 75–84
[64] Über die konvexe Hülle von n zufällig gewählten Punkten II, Z. Wahrscheinlichkeitstheorie u. verw. Geb. **3** (1964) 138–147
T.J. Rivlin
[79] Approximation by circles, Computing **21** (1979) 93–104
T.J. Rivlin and H.S. Shapiro
[61] A unified approach to certain problems of approximation and minimization, SIAM J. **9** (1961) 670–699
C.A. Rogers
[53] The volume of a polyhedron inscribed in a sphere, J. London Math. Soc. **28** (1953) 410–416
E.A. Robva
[77] Die Approximation konvexer Funktionen vermittels rationaler Interpolationsfunktionen mit einer festen Anzahl von Polen, Dokl. Akad. Nauk BSSR **21** (1977) 781–783
J.R. Sangwine-Yager
[80] The mean value of the area of polygons circumscribed about a plane convex body, Israel J. Math. **37** (1980) 351–363
[82] The mean quermassintegrals of simplices circumscribed about a convex body, Manuscript 1982
[82′] The mixed volume of analogous polytopes and a ball, Manuscript 1982
L.A. Santaló
[76] Integral geometry and geometric probability, Addison-Wesley, Reading Mass. 1976
E. Sas
[39] Über eine Extremaleigenschaft der Ellipsen, Compositio Math. **6** (1939) 468–470
R. Schneider
[71] On the Steinerpoints of convex bodies, Israel J. Math. **9** (1971) 241–249

[71′] Zwei Extremalaufgaben für konvexe Bereiche, Acta Math. Acad. Sci. Hungar. **22** (1971) 379–383

[72] The mean surface area of the boxes circumscribed about a convex body, Ann. Polon. Math. **25** (1972) 325–328

[81] Zur optimalen Approximation konvexer Hyperflächen durch Polyeder, Math. Ann. **256** (1981) 289–301

[82] Oral communication

R. Schneider and J.A. Wieacker

[80] Random polytopes in a convex body, Z. Wahrscheinlichkeitstheorie u. verw. Geb. **52** (1980) 69–73

[81] Approximation of convex bodies by polytopes, Bull. London Math. Soc. **13** (1981) 149–156

P. Schöpf

[77] Gewichtete Volumsmittelwerte von Simplices, welche zufällig in einem konvexen Körper des R^n gewählt werden, Monatsh. Math. **83** (1977) 331–337

Bl. Sendov

[79] Hausdorff approximations, Bulgar. Akad. Nauk, Sofia 1979

G.C. Shephard

[64] Approximation problems for convex polyhedra, Mathematika **11** (1964) 9–18

[66] Approximations by polytopes with projectively regular facets, Mathematika **13** (1966) 189–195

G.C. Shephard and R.J. Webster

[65] Metrics for sets of convex bodies, Mathematika **12** (1965) 73–88

B.W. Silverman and D.M. Titterington

[80] Minimum covering ellipses, SIAM J. Sci. Stat. Comp. **1** (1980) 401–409

G. Sonnevend

[80] On the optimization of sequential n-step methods for the approximation of convex sets. Proc. Sem. Inst. Banach Centre (1980), To appear

[81] Optimal passive and sequential algorithms for the approximation of convex functions in $L_p([0,1]^s), p = 1, \infty$. Proc. Conf. Constructive Function Theory, Varna (1981), To appear

L. Stammler and B. Weissbach

[79] Halbierungssätze zur Gestaltabweichung ebener Figuren, Wiss. Beiträge Martin-Luther-Univ. Halle-Wittenberg M **12** (1979) 55–65

P. Steinhagen

[21] Über die größte Kugel in einer konvexen Punktmenge, Abh. Math. Sem. Univ. Hamburg **1** (1921) 15–26

W. Süss

[55] Über die kleinste Figur, die jede Figur gegebenen Durchmessers einschließt, Abh. Math. Sem. Univ. Hamburg **20** (1955) 115–116

H.P.F. Swinnerton-Dyer

[74] The number of lattice points on a convex curve, J. Number Theory **6** (1974) 128–135

S. Tanno

[76] C^∞-approximation of continuous ovals of constant width, J. Math. Soc. Japan **28** (1976) 384–395

H. Taylor

[76] Cutting corners is not so easy, Amer. Math. Monthly **83** (1976) 568–569

R.H.K. Thomas

[79] Pairs of convex bodies, J. London Math. Soc. (2) **19** (1979) 144–146

F.A. Toranzos

[79] Approximation of compact starshaped sets by particular families, Rev. Un. Mat. Argentina **29** (1979/80) 49–54

E. Trost

[49] Über eine Extremalaufgabe, Niew. Arch. Wiskunde (2) **23** (1949) 1–3, Corr., ibid. (2) **23** (1950) 131–132

F.A. Valentine

[64] Convex sets, McGraw-Hill, New York 1964, Krieger, Huntington N.Y. 1976

I. Vincze

[47] Über den Minimalkreisring einer Eilinie, Acta Univ. Szeged. Sect. Sci. Math. **11** (1947) 133–138

[74] Über Kreisringe, die eine Eilinie einschließen, Stud. Sci. Math. Hungar. **9** (1974) 155–159 (1975)

A. Walfisz

[57] Gitterpunkte in mehrdimensionalen Kugeln, Panstwowe Wyd. Nauk., Warszawa 1957

B. Wegner

[77] Analytic approximation of continuous ovals of constant width, J. Math. Soc. Japan **29** (1977) 537–540

W. Weil

[73] Ein Approximationssatz für konvexe Körper, Manuscripta Math. **8** (1973) 335–362

[75] Einschachtelung konvexer Körper, Arch. Math. **26** (1975) 666–669

[76] Kontinuierliche Linearkombination von Strecken, Math. Z. **148** (1976) 71–84

J.A. Wieacker

[78] Einige Probleme der polyedrischen Approximation, Diplomarbeit, Univ., Freiburg i. Br. 1978

H. Yamabe and Z. Yujobo

[50] On the continuous functions defined on a sphere, Osaka Math. J. **2** (1950) 19–22

C.T. Yang

[55] On theorems of Borsuk-Ulam, Kakutani-Yamabe-Yujobo and Dyson II, Ann of Math. **62** (1955) 271–283

N.V. Zhivkov

[82] Plane poloygonal approximation of bounded convex sets, C.R. Acad. Bulgare Sci., To appear

V.I. Zorkal'tsev

[78] An approximation of a convex compact set by a polyhedron, Appl. Math., Work Collect., Irkutsk **1978** (1978) 53–61

P.M. Gruber
Institut für Analysis, Technische Mathematik
und Versicherungsmathematik
Technische Universität Wien
Gusshausstrasse 25–27
A–1040 Wein (Austria)

Kurt Leichtweiß*, Stuttgart

Geometric convexity and differential geometry

The geometry of smooth convex hypersurfaces in euclidean n-space R_n on the one hand and the geometry of the boundary of arbitrary convex bodies in R_n on the other are closely related (see [1] §17). The former belongs to differential geometry, the latter to geometric convexity. Some theorems have a differential geometric version as well as a convexity version; these depend on each other.

It is the aim of this lecture to illuminate these relations by suitable examples. Therefore we selected the following topics:

(I) Volume-product of convex body and polar body

(II) Convexity of the quermassintegrals of certain arrays of convex bodies

(III) Uniqueness of a convex body (up to translation) if a curvature function is given.

(I): In 1917 W. Blaschke ([2] and [3]) proved for n = 2 and n = 3 that, for the product of the volume of a smooth convex body K with positive curvature and the volume of its polar body K* (with respect to its centroid):

$$V(K)V(K^*) \leq V(B)V(B^*), \tag{1}$$

where B denotes the n-unit ball with centre in the origin.

Equality holds if and only if bdK is an ellipse resp. an ellipsoid. Elementary computation shows that (1) belongs indeed to affine geometry, because V(K)V(K*) is invariant under arbitrary non-degenerate affine operations on K.

Later L.A. Santalo [4] extended (1) to the n-dimensional case. He was able to show the invariance of V(K)V(K*) by a suitable integral-geometric interpretation of V(K*) under arbitrary (non-degenerate) affine operations on K (compare [5], Th. 3.2). He dispensed with the assumption that K has a smooth boundary and positive curvature and obtained a purely convex-geometric result.

Unfortunately until now we only know a proof of (1) by approximation of K by convex bodies L (in the sense of Hausdorff-metric), which are bounded by convex hypersurfaces with positive curvature (Eihyperflächen) and of differentiability class C^3. From this the inequality (1) and the problem of equality in (1) would belong to differential-geometry, which is somewhat amazing. The proof (Blaschke) runs as follows:

If the center of gravity of L *is the origin of* R_n, *then*

$$\int_S \frac{\xi\, d\omega}{(h^*(\xi))^{n+1}} = \int_S (r(\xi))^{n+1}\xi\, d\omega = (n+1)\int_S \int_0^{r(\xi)} \xi\rho^n\, d\rho\, d\omega$$

$$= (n+1)\int_L x\, dV = 0 \tag{2}$$

* Lecture held at the convexity-conference in Vienna 1981

where $\xi \in S = \mathrm{bdB}$, $d\omega$: *element of surface of* S, h^*: *support function of* L^*, $r = 1/h^*$: *radius function of* L.

If we define K′ by $K' = (h^*)^{n+1}$, then a well-known result of Minkowski (see [1], p. 121, resp. [6], §10) shows that K′ is the Gauss-curvature of bdL′ where L′ is a suitable convex body of differentiability class C^4, by a theorem of A.V. Pogorelov ([7], p. 33, Th. 1).

Minkowski and Böhmer [8] called bdL* the curvature-image ("Krümmungsbild") of bdL′. On L′ we apply the following affine isoperimetric inequality, which was proved by Blaschke for n = 2 and n = 3 and which holds also for general n:

$$(3) \qquad (V(B))^2 \geq \frac{(0'_{aff})^{n+1}}{n^{n+1}(V(L'))^{n-1}}$$

For the affine-surface $0'_{aff}$ of L′ the following formulae hold [11]:

$$(4) \qquad 0'_{aff} = \int_{\mathrm{bdL}'} (K')^{\frac{1}{n+1}} d0' = \int_S \frac{d\omega}{(h^*(\xi))^n} = \int_S (r(\xi))^n d\omega = nV(L)$$

$$(5) \qquad 0'_{aff} = \int_{\mathrm{bdL}'} h^* d0' = nV(L^*,L',\ldots,L')$$

So by using Minkowski's inequalities for the mixed volume $V(L^*,L',\ldots,L')$:

$$(6) \qquad V(L^*,L',\ldots,L') \geq (V(L^*))^{\frac{1}{n}}(V(L'))^{\frac{n-1}{n}}$$

(compare also [12], p. 274) indeed

$$(7) \qquad V(B)V(B^*) = (V(B))^2 \geq \frac{V(L)(V(L^*,L'\ldots,L'))^n}{(V(L'))^{n-1}} \geq V(L)V(L^*)$$

Obviously equality holds in (7), if and only if equality holds in (3) and (6), i.e. if L′ is bounded by a hyperellipsoid and if L′ and L* are homothetic. Then L* as well as L are bounded by a hyperellipsoid with the origin as centre, and vice versa equality holds in (7) because of the affine- invariance of $V(L)V(L^*)$.

We remark that for a centrally symmetric convex body K the (best possible) inequality

$$(8) \qquad V(K)V(K^*) \geq V(P)V(P^*)$$

(P: parallelo-epiped) is conjectured.

(8) was proved for n = 2 by Mahler [13]. The proof for n = 3 by Guggenheimer [14] does not seem valid (compare also [15] and [16]).

(II): To prove extremal properties of certain convex hypersurfaces one often uses Steiner-symmetrization (in a hyperplane of R_n). The volume remains unchanged and the surface area does not increase. More generally one can investigate the consequences if Steiner-symmetrization is applied to the "curvature-integrals"

$$(9) \qquad C_\nu(L) = \int_{\mathrm{bdL}} H_\nu d0 \quad (\nu = 0,\ldots,n-1)$$

of a convex hypersurface bdL of differentiability-class C^3, where $H_0 = 1$ and H_{ν}, are the normed ν'th elementary-symmetric functions of the positive principal curvatures of bdL ($\nu' = 1,\ldots,n-1$). Also in this case it can be shown that the $C_\nu(L)$ do not increase, as we make clear using an idea of K. Voss ([17]), §10) as follows (compare also [18], §103):

If one connects bdL *and the convex hypersurface* $\mathrm{bd}\bar{L}$ (*with same* C_ν!), *reflected in the symmetry-hyperplane* A, *"linearly" by the so-called "symmetrization-array"* $\{\mathrm{bdL}(\tau)\}_{\tau\in[0,1]}$ with

$$(10)\qquad x(\tau) = x + \tau(\bar{x} - x) = x + \tau we$$

$x \in \mathrm{bdL}$, $\bar{x} \in \mathrm{bd}\bar{L}$, e: *constant unit vector orthogonal to* A, *then* $\mathrm{bdL}(\frac{1}{2})$ *is the convex hypersurface generated by symmetrization in* A. *Therefore it suffices to show the convexity of the functions* $C_\nu(L(\tau))$ *for* $\nu = 1,\ldots,n-1$.

For this we write the $C_\nu(L(\tau))$ as

$$(11)\qquad (-1)^\nu(n-1)!C_\nu(L(\tau)) = \int\limits_{\mathrm{bdL}(\tau)} (\xi(\tau),\underbrace{d\xi(\tau),\ldots,d\xi(\tau)}_{\nu},\underbrace{dx(\tau),\ldots,dx(\tau)}_{n-1-\nu})$$

where $\xi(\tau)$ is the inner normal (unit vector) of bdL(τ).
For the second derivative we obtain

$$(12)\qquad \frac{d^2C(L(\tau))}{d\tau^2} = \frac{(n-1-\nu)(\nu+1)}{n-1}\int\limits_{\mathrm{bdL}(\tau)} c^{ik}_{\nu+1}(\tau)\frac{\partial w}{\partial u^i}\frac{\partial w}{\partial u^k} < \xi(\tau), \quad e >^2 d0(\tau)$$

$$\geq 0$$

because $c^{ik}_{\nu+1}(\tau), i,k = 1,\ldots,n-1$ are the coefficients of a positive definite quadratic form. Here equality for all τ holds in the case $0 \leq \nu \leq n-2$, if and only if the $\partial w/\partial u^i, i = 1,\ldots,n-1$ vanish, i.e. if (by (10)) bdL and $\mathrm{bd}\bar{L}$ are equivalent under translations, or, what is the same, if L is symmetric with respect to a hyperplane parallel to A.

To conclude: Under Steiner-symmetrization in A all curvature-integrals of bdL, $0 \leq \nu \leq n-2$ strictly decrease if L does not have a symmetry-hyperplane parallel to A.

It is remarkable that the last mentioned theorem has a convex-geometric version, as can be seen in H. Hadwiger's book "Vorlesungen über Inhalt, Oberfläche and Isoperimetrie", Springer, 1957 (see p. 261, Zusatz IV, resp. [12], Th. 18.5): The curvature integrals $C_\nu(L)$ of a convex hypersurface are replaced by the n-fold quermassintegrals

$$(13)\qquad nW_{\nu+1}(K) = nV(\underbrace{K,\ldots,K}_{n-1-\nu},\underbrace{B,\ldots,B}_{\nu+1}) \quad 0 \leq \nu \leq n-1$$

of an arbitrary convex body K.

Instead by (10), the symmetrization-array $\{K(\tau)\}_{\tau\in[0,1]}$ can be defined by

$$(14)\qquad K(\tau) = \bigcup_{G\perp A} \{(1-\tau)(G\cap K) + \tau(G\cap \bar{K})\}$$

where G is an arbitrary line orthogonal to A, and $\bar{K}$ is the reflection of K with respect to A.

The array (14) is the special case of a family $\{K(\tau)\}_{\tau\in[0,1]}$ of compact convex bodies with the following convexity properties (where G denotes an arbitrary line orthogonal to A):

(15) $G \cap K(\lambda_1\tau_1 + \lambda_2\tau_2) \subseteq \lambda_1(G \cap K(\tau_1)) + \lambda_2(G \cap K(\tau_2))$

and a fortiori

(16) $K(\lambda_1\tau_1 + \lambda_2\tau_2) \subseteq \lambda_1 K(\tau_1) + \lambda_2 K(\tau_2)$

$\lambda_1 \geq 0, \quad \lambda_2 \geq 0, \quad \lambda_1 + \lambda_2 = 1, \quad \tau_1, \tau_2 \in [0,1]$

which is called "vollkonvex" by Hadwiger.

Hadwiger then showed that for a "vollkonvex" family the functions $W_{\nu+1}(K(\tau))$, $0 \leq \nu \leq n-1$ are also convex, where in the cases $0 \leq \nu \leq n-2$ we have linearity if and only if K(0) and K(1) only differ by an additive segment T through the origin and orthogonal to A. For the proof Hadwiger used induction arguments and Kubota's recursion formulae for the quermassintegrals (see [19], and [12], Th. 18.6 and 18.7):

(17) $$W_{\nu+1}(K) = \frac{1}{nV(B^{n-1})} \int_S W_\nu^{(n-1)}(p_\xi(K))\, d\omega, \quad 0 \leq \nu \leq n-1$$

where B^{n-1} denotes the $(n-1)$-dimensional unit-ball, and p_ξ is the orthogonal projection in direction of ξ.

In the special case of the symmetrization-family (14) T shrinks to a point, i.e. $K(0) = K$ and $K(1) = \bar{K}$ are translation-equivalent, and we have again:

> *All* $W_\nu(K)$, $1 \leq \nu \leq n-1$, *strictly decrease under Steiner-symmetrization in* A, *if* K *has no symmetry-hyperplane parallel to* A.

(III): In (I) we used an existence theorem of Minkowski (compare [1], p. 121, resp. [6], §10). The proof of this existence theorem is based on the following uniqueness theorem of Minkowski, due to the Fredholm-alternative for the solution of certain partial differential equations (analogous to the solution of Weyl's embedding problem):

> *If* bdL *denotes a convex hypersurface of differentiability-class* C^3, *then it is defined up to a translation by its Gauss-curvature as function of the inner normal* ξ *(unit-vector) of* bdL.

More generally S.S. Chern showed 1959 ([20], Th. 2):

> *If* bdL *and* bd$\bar{L}$ *are two convex hypersurfaces of differentiability-class* C^3 *in* R_n, *which have, at all points with the same normal unit-vector* ξ, *the same value of the normed* ν-th *elementary-symmetric functions* P_ν *of the principal radii of curvature*:

(18) $P_\nu(\xi) = \bar{P}_\nu(\xi) \quad \nu$ fixed; $\quad 1 \leq \nu \leq n-1$

> *then* bdL *and* bd$\bar{L}$ *are the same up to a translation.*

His proof is based (for $\nu > 1$) on the generalized Minkowskian integral-formula:

(19) $$\int_S \{h(\xi)(\bar{P}_\nu(\xi) - P_{\nu-1\,1}(\xi)) + \bar{h}(\xi)(P_\nu(\xi) - \bar{P}_{\nu-1\,1}(\xi))\}\, d\omega = 0.$$

Here h resp. $\bar{h}$ are the support functions of bdL resp. bd$\bar{L}$. Further the "mixed discriminants" $P_{\nu-1\,1}$ resp. $\bar{P}_{\nu-1\,1} = P_{1\,\nu-1}$ are the normed coefficients of $x^{\nu-1}\bar{x}$ resp. $x\bar{x}^{\nu-1}$ of the polynomial representation of the determinant

(20) $\quad \| \delta_i^k - l_{ij}e^{jk}x - \bar{l}_{ij}e^{jk}\bar{x} \|$

(l_{ij}) resp. $(\bar{l}_{ij})$ are the coefficient matrices of the 2nd fundamental form of bdL resp. bd$\bar{\text{L}}$: (e^{ij}) denotes the matrix of coefficients of the (common) third fundamental form of bdL resp. bd$\bar{\text{L}}$. Using an inequality of L. Garding ([21], Th. 5) we have

(21) $\quad P_{\nu-1\,1}(\xi) \geq P_\nu^{\frac{\nu-1}{\nu}}(\xi)\bar{P}_\nu^{\frac{1}{\nu}}(\xi),$

i.e. together with (18) we have

(22) $\quad P_{\nu-1\,1}(\xi) \geq \bar{P}_\nu(\xi)$ and $\bar{P}_{\nu-1\,1}(\xi) \geq P_\nu(\xi).$

Further, we can assume without loss of generality that the support functions of bdL and bd$\bar{\text{L}}$ satisfy:

(23) $\quad h(\xi) > 0$ and $\bar{h}(\xi) > 0.$

From (22), (23) and (19) it follows that equality always holds in (21). Using this Garding has shown that the second fundamental forms of bdL and bd$\bar{\text{L}}$ are proportional at corresponding points; and so, because of (18), they are equal. Because a convex hypersurface is defined by its second fundamental form (depending on ξ) up to a translation, we see that bdL and bd$\bar{\text{L}}$ are identical up to a translation. For $\nu = 1$ one needs a different proof. It is remarkable that Chern's differential-geometric theorem also has a convex-geometric version, which has been known since 1937 and which was found by A.D. Aleksandrov ([22], §7, Th. 1). To formulate this version, we have to generalize the $P_\nu(\xi)$ for the boundaries of an arbitrary compact convex set K, which is possible by W. Fenchel and B. Jessen [23] as follows:

If the convex hypersurface bdL *is of differentiability clsss* C^2, *then* $P_\nu(\xi)$ *is interpreted as the density of a Radon-measure on the unit-sphere* S, *and we have*:

(24) $\quad \int_S h'(\xi)P_\nu(\xi)\,d\omega = nVK',\underbrace{L,\ldots,L}_{\nu},\underbrace{B,\ldots,B}_{n-1-\nu})$

where K′ *denotes an arbitrary compact convex set in* R_n *with support function* h′ (*compare* [1], 37.(5) *and* 38.(7)).

Then, by a (uniquely possible) continuation of the positive-linear, monotone and continuous functional given on S by $h' \mapsto nV(K',\underbrace{K,\ldots,K}_{\nu},\underbrace{B,\ldots,B}_{n-1-\nu})$ to an analogous functional on S which is defined for all continuous functions on S, one can define a Radon-measure μ_ν with the property

(25) $\quad \int_S h'\,d\mu_\nu = nV(K',\underbrace{K,\ldots,K}_{\nu},\underbrace{B,\ldots,B}_{n-1-\nu}) = nW_{n-1-\nu}(K',\underbrace{K,\ldots,K}_{\nu})$

For K this μ_ν replaces $P_\nu(\xi)$, $1 \leq \nu \leq n-1$.

Aleksandrov calls μ_ν the ν-th curvature function of K and shows: If K and $\bar{\text{K}}$ are at least $(\nu+1)$-dimensional compact convex sets of R_n with the same ν-th curvature functions:

(26) $\mu_\nu = \bar{\mu}_\nu \quad \nu \text{ fixed}, \quad 1 \leq \nu \leq n-1,$

then K and $\bar{K}$ are identical up to a translation.

His proof, which is based on ideas of Minkowski for $\nu = n-1$, uses the fact that because of (26) in the "generalized Minkowskian inequalities of first type" for the "mixed quermassintegrals":

$$(27) \quad W_{n-1-\nu}(\underbrace{K,\dots,K}_{\nu},\bar{K}) \geq W_{n-1-\nu}^{\frac{\nu}{\nu+1}}(K)\,W_{n-1-\nu}^{\frac{1}{\nu+1}}(\bar{K}) \quad 1 \leq \nu \leq n-1$$

equality must hold (notice the formal analogy between (21) and (27)!). Because (27) is a consequence of the concavity (Aleksandrov) of the function

$$W_{n-1-\nu}^{\frac{1}{\nu+1}}((1-\tau)K + \tau\bar{K}), \quad 0 \leq \tau \leq 1$$

this function must be linear. From this it follows, by theorems of W. Fenchel 1936 and J. Favard 1933, that K and $\bar{K}$ are homothetic (=generalized theorem of Brunn-Minkowski, compare [21], Th. 24.1 and 24.2 or [24]).

From this and (26) it follows indeed that K and $\bar{K}$ are identical up to a translation, which is what we wanted to show.

References

[1] Bonnesen, T. und W. Fenchel: Theorie der konvexen Körper. Springer Berlin (1974).

[2] Blaschke, W.: Über affine Geometrie VII. Ber. Verh. Kgl. Sächs. Akad. Wiss. Leipzig **69**, S. 306–318 (1917).

[3] Heil, E.: Abschätzungen für einige Affininvarianten konvexer Kurven. Monatsh. Math. **71**, S. 405–423 (1967).

[4] Santalo, L.A.: Un invariante afin para los cuerpos convexos del espacio de n dimensiones. Portug. Math. **8**, S. 155–161 (1949).

[5] Santalo, L.A.: Integral geometry in projective and affine spaces. Ann. of Math. **51**, S. 739–755 (1950).

[6] Minkowski, H.: Volumen und Oberfläche. Math. Ann. **57**, S. 447–495 (1903).

[7] Pogorelov, A.V.: The Minkowski-multidimensional problem. V.H. Winston and Sons. Washington D.C. (1978).

[8] Böhmer, P.: Über elliptisch-konvexe Ovale. Math. Ann. **60**, S. 256–262 (1905).

[9] Blaschke, W.: Über affine Geometrie I. Ber. Verh. Kgl. Sächs. Akad. Wiss. Leipzig **68**, S. 217–239 (1916).

[10] Blaschke, W.: Über affine Geometrie VI. Ber. Verh. Kgl. Sächs. Akad. Wiss. Leipzig **69**, S. 207–225 (1917).

[11] Blaschke, W.: Vorlesungen über Differentialgeometrie II. Springer Berlin (1923).

[12] Leichtweiss, K.: Konvexe Mengen. Dt. Verl. Wiss. Berlin (1980).

[13] Mahler, K.: Ein Minimalproblem für konvexe Polygone. Math. (Zutphen) B **7**, S. 118–127 (1939).

[14] Guggenheimer, H.: Polar reciprocal convex bodies. Israel J. Math. **14**, S. 309–316 (1973) mit einer Korrektur im Israel J. Math. **29** (1978).

[15] Mahler, K.: Ein Übertragungsprinzip für konvexe Körper. Cas. Mat. Fys. **68**, S. 93–102 (1939).

[16] Bambah, R.P.: Polar reciprocal convex bodies. Proc. Cambr. Phil. Soc. **51**, S. 377–378 (1954).

[17] Voss, K.: Einige differentialgeometrische Kongruenzsätze füı geschlossene Flächen und Hyperflächen. Math. Ann. **131**, S. 180–218 (1956).

[18] Leichtweiss, K.: Elementare Differentialgeometrie. Springer Berlin (1973).

[19] Kubota, T.: Über konvex-geschlossene Mannigfaltigkeiten im n-dimensionalen Raum. Sci. Rep. Tohoku Univ. **14**, S. 85–99 (1925).

[20] Chern, S.S.: Integral formulas for hypersurfaces in euclidean space and their applications to uniqueness theorems. J. Math. Mech. **8**, S. 947–955 (1959).
[21] Garding, L.: An inequality for hyperbolic polynomials. J. Math. Mech. **8**, S. 957–965 (1959).
[22] Aleksandrov, A.D.: Neue Ungleichungen zwischen den gemischten Volumina und ihre Anwendungen. Mat. Sbornik N.S. **2**, S. 1205–1238 (1937).
[23] Fenchel, W. and B. Jessen: Mengenfunktionen und konvexe Körper. Kgl. Dansk. Vid. Sels. Math. -fys. Meddel. XVI **3**, S. 3–31 (1938).
[24] Leichtweiss, K.: Zum Beweis eines Eindeutigkeitssatzes von A.D. Aleksandrov. E.B. Christoffel. Birkhäuser Basel (1981).

K. Leichtweiss
Mathematisches Institut B
Universität Stuttgart
Pfaffenwaldring 57
C–7000 Stuttgart 80 (FRG)

Peter McMullen and Rolf Schneider

Valuations on convex bodies

The investigation of functions on convex bodies which are valuations, or additive in Hadwiger's sense, has always been of interest in particular parts of geometric convexity, and it has seen some progress in recent years. The occurrence of valuations in the theory of convex bodies can be traced back to the notion of volume in two essentially different ways. Firstly, the volume of convex bodies, being the restriction of a measure, is itself a valuation. This valuation property carries over to the functions which are deduced from volume in the Brunn-Minkowski theory, namely to mixed volumes, quermassintegrals, surface area functions, and others. Hadwiger's celebrated characterizations of the quermassintegrals by the valuation and other properties were the culmination of a series of papers on valuations and at the same time the starting point for various subsequent investigations of functionals with similar properties.

A different way from volume to more general valuations was opened by Hilbert's third problem and the solution given to it by Dehn. Motivated by the problem whether the notion of volume for three-dimensional polytopes can be introduced, in analogy to the plane case, by elementary dissection and congruence arguments, without the use of limit processes, Hilbert asked whether two three-dimensional polytopes of equal volume are necessarily equivalent by dissection. Dehn's negative answer was essentially achieved by constructing special valuations which must attain the same value on equidissectable polytopes, and by exhibiting pairs of convex polytopes with equal volume but different values of these functionals. Dehn's set of necessary conditions for equidissectability was proved to be also sufficient only many years later, and in the course of the investigation centring around this and related questions, much information on valuations was gained. Thus the dissection theory of polytopes (which still has to offer some deep open problems) is intimately tied up with valuation theory, and every dissection result has implications on valuations.

A third range for applications of valuations in convexity is seen in questions of combinatorial geometry, where the Euler characteristic on unions of convex bodies is a useful device. The Euler characteristic also plays a role in certain extension procedures for quermassintegrals and other functionals to non-convex sets.

Still another class of valuations arises from the counting of lattice points in convex bodies.

The following survey collects and describes the various examples of valuations on convex bodies that have been treated in the literature, and it presents the known results, mostly without proofs. The emphasis is, first, on the interrelations between simple valuations and dissections, which requires a fairly far-going description of the algebraic arguments on which the progress in equidissectability relies, and second, on characterization theorems for special valuations of geometric interest. Some open problems will also be mentioned at appropriate places.

The section headings are as follows.

§1. Preliminaries

By a *valuation*, or an *additive* functional, on a class $\mathscr{S}$ of sets we understand a function φ on $\mathscr{S}$ satisfying

(1.1) $\varphi(K \cup L) + \varphi(K \cap L) = \varphi(K) + \varphi(L)$

whenever $K,L,K \cup L$ and $K \cap L$ are elements of $\mathscr{S}$. Here we assume that φ takes its values in an abelian group, and we always suppose that $\varphi(\emptyset) = 0$. Often the class $\mathscr{S}$ will be *intersectional*, which means that $K,L \in \mathscr{S}$ implies $K \cap L \in \mathscr{S}$. If $\mathscr{S}$ is an intersectional class, we let $U(\mathscr{S})$ denote the lattice consisting of all finite unions of elements of $\mathscr{S}$.

For the classes $\mathscr{S}$ occurring most frequently in the following, we introduce special notation. Let $\mathbb{E}^d$ be d-dimensional euclidean vector space, with scalar product $\langle \cdot,\cdot \rangle$ and norm $\|\cdot\|$, and let $\Omega = \Omega^{d-1} := \{x \in \mathbb{E}^d : \|x\| = 1\}$ be its unit sphere. By $\mathscr{K}^d$ we denote the class of compact convex subsets of $\mathbb{E}^d$. The elements of $\mathscr{K}^d$ will be called *convex bodies*, which differs slightly from common usage (in particular, the empty set $\emptyset$ is a convex body, which is convenient when valuations are considered). On $\mathscr{K}^d \setminus \{\emptyset\}$ we have the vector or Minkowski addition $+$, defined by

$$K + L := \{x + y : x \in K, y \in L\},$$

and the usual Hausdorff metric ρ, defined by

$$\rho(K,L) := \min\{\rho \geq 0 : K \subseteq L + \rho B, \quad L \subseteq K + \rho B\},$$

with B the unit ball.

The important class $U(\mathcal{K}^d)$ is Hadwiger's *convex-ring*. By $\mathcal{P}^d \subset \mathcal{K}^d$ we denote the class of convex polytopes, and the elements of $U(\mathcal{P}^d)$ are the *polyhedra*. We write $\mathcal{K}^d_d \subset \mathcal{K}^d$ for the subset of d-dimensional convex bodies, and the elements of $\mathcal{P}^d_d := \mathcal{K}^d_d \cap \mathcal{P}^d$ will also be called d-polytopes.

Sometimes it will be convenient to consider also relatively open convex bodies. By a relatively open convex body we understand the relative interior, relint (i.e., the interior with respect to the affine hull) of a convex body. Let $\mathcal{K}^d_{ro}$ denote the set of all relatively open convex bodies in $\mathbb{E}^d$ and $\mathcal{P}^d_{ro}$ the subset of relatively open polytopes, and observe that $\mathcal{P}^d \subset U(\mathcal{P}^d_{ro})$.

For other types of sets to be considered we shall introduce special notation when it seems appropriate. A polytope is called *rational* if its vertices have rational coordinates (with respect to the standard basis of $\mathbb{E}^d$), and it is a *lattice polytope* if its vertices belong to the *integer lattice* $\mathbb{Z}^d$ consisting of all points in $\mathbb{E}^d$ with integer coordinates. A *polyhedral cone* with apex 0 is the intersection of finitely many closed halfspaces each having 0 in its boundary. The intersection of such a cone with the unit sphere Ω^{d-1} is called a *spherical polytope*. Some time we will also mention polytopes in hyperbolic spaces.

If φ is a valuation on $\mathcal{S}$ and $\mathcal{S}$ is a lattice, that is, closed under finite unions and finite intersections, then (1.1) and an easy induction argument yield

$$(1.2) \quad \varphi(K_1 \cup \cdots \cup K_m) = \sum_{r=1}^{m} (-1)^{r-1} \sum_{i_1 < \cdots < i_r} \varphi(K_{i_1} \cap \cdots \cap K_{i_r})$$

for $K_1,\ldots,K_m \in \mathcal{S}$. In general, the function φ defined on an arbitrary class $\mathcal{S}$ is said to satisfy the *inclusion-exclusion principle* if (1.2) holds whenever $K_1,\ldots,K_m, K_1 \cup \cdots \cup K_m, K_{i_1} \cap \cdots \cap K_{i_r} \in \mathcal{S}$. Clearly any valuation on $\mathcal{S}$ which can be extended, as a valuation, to the lattice generated by $\mathcal{S}$, satisfies the inclusion-exclusion principle. We shall consider such extensions in §5.

In the former literature, in particular in the work of Hadwiger (see [1957]), valuations are usually called *additive functionals*. This should not be confused with the notion of Minkowski additivity. A function φ on $\mathcal{K}^d$ or $\mathcal{P}^d$ (with values in an abelian group) is called *Minkowski additive* if $\varphi(\emptyset) = 0$ and

$$\varphi(K + L) = \varphi(K) + \varphi(L) \quad \text{for} \quad \emptyset \neq K, L \in \mathcal{K}^d \text{ resp. } \mathcal{P}^d.$$

Every Minkowski additive function is also a valuation, since

$$(1.3) \quad (K \cup L) + (K \cap L) = K + L$$

if K,L and $K \cup L$ are non-empty convex bodies. This fundamental relation, which appears surprisingly late in the literature (apparently not before Sallee [1966], p. 77; see also Hadwiger [1971]), can also be interpreted as saying that the identical mapping of $\mathcal{K}^d$ into itself is a valuation. (Here we admit a commutative semigroup with cancellation law, namely $\mathcal{K}^d$ with Minkowski addition, as the range of a valuation. This is not an essential difference, since any such semi-group can be embedded in an abelian group.) Since the mapping $\varphi: K \mapsto h(K,\cdot)$, where

$$h(K,u) := \max\{\langle x,u\rangle : x \in K\} \quad \text{for} \quad u \in \Omega^{d-1}$$

defines the *support function* of $K \neq \emptyset$ (restricted to Ω^{d-1}), is Minkowski additive, it is also a valuation, with values in the space of real continuous functions on Ω^{d-1}.

Let C be a fixed convex body. If $K,L \in \mathscr{K}^d$, then

$$(K \cup L) + C = (K + C) \cup (L + C),$$

and if $K \cup L$ is convex, then also

$$(K \cap L) + C = (K + C) \cap (L + C)$$

(see Hadwiger [1957], p. 144). Hence, if φ is a valuation on $\mathscr{K}^d$, then the function φ_C defined by

$$\varphi_C(K) := \varphi(K + C) \quad \text{for} \quad K \in \mathscr{K}^d$$

is also a valuation on $\mathscr{K}^d$. Thus the interplay between convexity and Minkowski addition yields new valuations from old ones, a remark which will be of importance in §§3 and 10.

The above passage from φ to φ_C is an example of the following obvious result.

(1.4) **Lemma.** *Let φ be a valuation on $\mathscr{K}^d$, and let $f: \mathscr{K}^d \to \mathscr{K}^d$ be a map which satisfies $f(K \cup L) = f(K) \cup f(L)$ and $f(K \cap L) = f(K) \cap f(L)$ if $K,L,K \cup L \in \mathscr{K}^d$. Then $\varphi \circ f$ is a valuation.*

In the above example, $f(K) = K + C$. Another example is given by $f(K) = K \cap C$, where C is a fixed closed convex set. A third one is given by $f(K) = \alpha(K)$, where $\alpha: \mathbb{E}^d \to \mathbb{E}^d$ is an affine map.

The following notion is useful in the investigation of valuations on polytopes. For a hyperplane $H \subset \mathbb{E}^d$ let H^+ and H^- be the two closed halfspaces bounded by H. A function φ on $\mathscr{P}^d$ or $\mathscr{K}^d$ is called a *weak valuation* if $\varphi(\emptyset) = 0$ and

$$\varphi(K) + \varphi(K \cap H) = \varphi(K \cap H^+) + \varphi(K \cap H^-)$$

for every hyperplane H and every K in the domain of φ. Sallee [1968] shows (among related and more general results) that every weak valuation on $\mathscr{P}^d$ is a valuation; see also Groemer [1978] for the case where φ takes its values in a real vector space. The following example (due to Groemer, private communication) shows that a weak valuation on $\mathscr{K}^d$ need not be a valuation. For $K \in \mathscr{K}^2$, define $\varphi(K) = 1$ if 0 (the origin of $\mathbb{E}^2$) lies in the boundary of K and is a one-sided, but not a two-sided, limit of singular points of K, let $\varphi(K) = 2$ if 0 is a two-sided limit of singular points of K, and $\varphi(K) = 0$ otherwise. Clearly φ is a weak valuation on $\mathscr{K}^2$. But it is not a valuation, since one easily finds $K_1,K_2 \in \mathscr{K}^2$ with $\varphi(K_1) = \varphi(K_2) = \varphi(K_1 \cap K_2) = 1$ and $\varphi(K_1 \cup K_2) = 0$. We do not know an example of a rigid motion invariant weak valuation which is not a valuation.

A different view on valuations is often useful. It is motivated by the well-known procedure of integration theory which transposes additivity (of set functions) into linearity (on vector spaces). Let $\mathscr{S}$ be a class of subsets of some set S. The characteristic function of an element $K \in \mathscr{S}$ will be denoted by K^*, that is,

$$K^*(x) = \begin{cases} 1 & \text{if} \quad x \in K \\ 0 & \text{if} \quad x \in S \setminus K. \end{cases}$$

By $V(\mathscr{S})$ we denote the real vector space which is generated by the functions

K^*, $K \in \mathscr{S}$. If $K, L, K \cup L$ and $K \cap L \in \mathscr{S}$, then

$$(K \cup L)^* + (K \cap L)^* = K^* + L^*,$$

thus the map $K \mapsto K^*$ is a valuation. In particular, if $\mathscr{S}$ is intersectional, then (1.2) shows that $K^* \in V(\mathscr{S})$ for all $K \in U(\mathscr{S})$.

Now suppose that $\bar{\varphi}$ is a map from $V(\mathscr{S})$ into an abelian group which satisfies

$$(1.5) \quad \bar{\varphi}(K^* + L^*) = \bar{\varphi}(K^*) + \bar{\varphi}(L^*).$$

Defining $\varphi(K) := \bar{\varphi}(K^*)$ for those $K \subset S$ for which $K^* \in V(\mathscr{S})$, we get

$$\begin{aligned}\varphi(K \cup L) + \varphi(K \cap L) &= \bar{\varphi}((K \cup L)^*) + \bar{\varphi}((K \cap L)^*)\\ &= \bar{\varphi}((K \cup L)^* + (K \cap L)^*) = \bar{\varphi}(K^* + L^*)\\ &= \bar{\varphi}(K^*) + \bar{\varphi}(L^*) = \varphi(K) + \varphi(L),\end{aligned}$$

provided that $\bar{\varphi}$ is defined in each case. Thus φ is a valuation. In particular, if $\mathscr{S}$ is intersectional, then this yields a valuation on $U(\mathscr{S})$.

Vice versa, if a valuation φ on $\mathscr{S}$ with values in some real vector space is given, one might try to define

$$(1.6) \quad \bar{\varphi}(f) := \sum \alpha_i \varphi(K_i) \quad \text{for} \quad f = \sum \alpha_i K_i^* \in V(\mathscr{S}) \quad (\alpha_i \in \mathbb{R})$$

as in integration theory (where usually φ is a measure defined on a ring of subsets). If this is possible, then $\bar{\varphi}$ thus defined clearly satisfies (1.5). But, in general, the right-hand side of (1.6) does not only depend on the function f, but on its special representation. We shall return to these questions in §§2 and 5.

We conclude these preliminaries with a most important definition. A valuation φ on a class of convex subsets of either $\mathbb{E}^d$ or Ω^d is called *simple* if $\varphi(K) = 0$ whenever $\dim K < d$.

I. Classical examples and general results

§2. The Euler characteristic

Before treating valuations from a general point of view, it seems appropriate to review the more familiar classical examples occurring in the theory of convex bodies. The simplest (non-zero) valuation on $\mathscr{K}^d$ is clearly the function χ defined by

$$(2.1) \quad \chi(K) := \begin{cases} 1 & \text{if} \quad K \neq \emptyset \\ 0 & \text{if} \quad K = \emptyset \end{cases} \qquad \text{for} \quad K \in \mathscr{K}^d.$$

While this function is of no interest when restricted to convex bodies, it is a non-trivial question whether χ can be extended, as a valuation, to the convex-ring $U(\mathscr{K}^d)$. The answer is in the affirmative, since the Euler characteristic as defined, for example, in singular homology theory, is a valuation satisfying (2.1). In a well-known and influential paper, Hadwiger [1955a] gave an entirely elementary existence proof which is independent of topology (the uniqueness is trivial by (1.2)). His construction proceeds by induction with respect to the dimension. The existence for $d = 0$ being trivial, suppose that the existence of χ has been proved in dimension d − 1. For a unit vector $u \in \Omega^{d-1}$ and for $\lambda \in \mathbb{R}$, let

(2.2) $$H_{u,\lambda} := \{x \in \mathbb{E}^d : \langle x,u \rangle = \lambda\}$$

be the hyperplane through λu orthogonal to u. Then put

(2.3) $$\chi(K) := \sum_{\lambda \in \mathbb{R}} \left[\chi(K \cap H_{u,\lambda}) - \lim_{\mu \downarrow \lambda} \chi(K \cap H_{u,\mu}) \right] \quad \text{for} \quad K \in U(\mathscr{K}^d),$$

where on the right-hand side χ denotes the (unique) Euler characteristic which by the inductive assumption exists in (d − 1)-dimensional affine spaces. The sum is finite, and χ thus defined turns out to be a valuation.

By an easy argument, Hadwiger [1955a] also deduced the existence of a valuation χ on the finite unions of closed spherically convex subsets of Ω^{d-1} with $\chi(A) = 1$ if A is spherically convex and contained in an open hemisphere.

The above existence proof for the Euler characteristic on $U(\mathscr{K}^d)$ is reproduced in Hadwiger [1957]. Different variants of the construction are found in Hadwiger [1959], [1968b], [1969c], Hadwiger-Mani [1972].

We also mention an article of Hadwiger [1974a] which contains an elementary treatment of the Euler characteristic for polygons in the plane.

Once the existence of the Euler characteristic on the convex-ring is known, it is, mainly through exploitation of the inclusion-exclusion principle (1.2), a useful tool in combinatorial geometry, see Hadwiger [1947], [1955a], [1968b], Klee [1963]. Klee's paper put the Euler characteristic in a lattice-theoretic setting. This general treatment of valuations and the Euler characteristic in combinatorial and algebraic terms has been further developed by Rota [1971], see also [1964]. Rota [1971], p. 231, makes a very interesting (though somewhat vague and perhaps too optimistic) remark on a conceivable connexion between valuations on $U(\mathscr{P}^d)$ and the problem of finding necessary and sufficient conditions for a lattice to be isomorphic to the face lattice of a convex polytope.

If an element K of the convex-ring $U(\mathscr{K}^d)$ is represented as the union of k convex sets, then (1.2) gives trivial lower and upper bounds for the value $\chi(K)$ of the Euler characteristic in terms of d and k alone. The problem of finding sharp bounds has been posed by Hadwiger-Mani [1974], and they have treated a related problem. A complete solution was given by Eckhoff [1980].

The recursive definition (2.3) works equally well for relatively open convex bodies, and this is often convenient, especially when polyhedra are considered. If we use (2.3) for $K \in U(\mathscr{P}^d_{ro})$, then this yields a valuation χ on the unions of relatively open convex polytopes which evidently satisfies

(2.4) $$\chi(P) = (-1)^{\dim P} \quad \text{for} \quad P \in \mathscr{P}^d_{ro}.$$

This extended Euler characteristic was considered by Lenz [1970] and Groemer [1972], and in special cases also by Hadwiger [1969c], [1973], who, however, preferred to use a different sign for odd-dimensional P, so that he did not get a valuation on $U(\mathscr{P}^d_{ro})$. It appears that the insistence on prescribing $\chi(P) = 1$ for every relatively open, non-empty polytope P, slightly complicates the investigation in Hadwiger [1973] and also in Hadwiger-Mani [1972].

Since every polytope is the disjoint union of the relative interiors of its faces, (2.4) and the additivity of χ immediately yield the well-known Euler relation.

Variants of the recursion formula (2.3) can also be used to extend the Euler characteristic to a linear functional on a vector space, as described in §1. Let us first consider the real vector space $V(\mathscr{P}^d)$ consisting of the finite linear combinations of indicator functions of convex polytopes. We are to show the existence of a real linear functional $\bar{\chi}$ on $V(\mathscr{P}^d)$ such that $\bar{\chi}(\mathscr{P}^*) = 1$ for $\emptyset \neq P \in \mathscr{P}^d$ (the uniqueness is clear). For $d = 1$, such a linear functional $\bar{\chi}_1$ is evidently given by

$$\text{(2.5)} \quad \bar{\chi}_1(f) := \sum_{\lambda \in \mathbb{R}} \left[f(\lambda) - \lim_{\mu \downarrow \lambda} f(\mu) \right], \quad f \in V(\mathscr{P}^1).$$

Let $d \geq 2$, suppose that the existence has already been proved in dimension $d - 1$, and call this functional $\bar{\chi}_{d-1}$. Consider $\mathbb{E}^{d-1}$ as a linear subspace of $\mathbb{E}^d$, and let $u \in \mathbb{E}^d$ be a unit vector not in $\mathbb{E}^{d-1}$. Let $f \in V(\mathscr{P}^d)$ be given and define

$$\tilde{f}(x,\lambda) := f(x + \lambda u) \quad \text{for} \quad x \in \mathbb{E}^{d-1}, \quad \lambda \in \mathbb{R}.$$

Two types of induction are possible:

(a) Define the projection $\pi_1 f$ of f on to $\mathbb{E}^{d-1}$ by

$$(\pi_1 f)(x) := \bar{\chi}_1(\tilde{f}(x,\cdot)) \quad \text{for} \quad x \in \mathbb{E}^{d-1},$$

and then put

$$\bar{\chi}_d(f) := \bar{\chi}_{d-1}(\pi_1 f).$$

(b) Define

$$(\pi_2 f)(\lambda) := \bar{\chi}_{d-1}(\tilde{f}(\cdot,\lambda)) \quad \text{for} \quad \lambda \in \mathbb{R}$$

and put

$$\bar{\chi}_d(f) := \bar{\chi}_1(\pi_2 f).$$

In each case it is easy to see that $\bar{\chi}_d$ thus defined has the desired properties.

A procedure equivalent to method (a) was employed by Hadwiger [1960] and also by Groemer [1972], who apparently did not know Hadwiger's paper. Method (b), again in a different but equivalent form, was used by Lenz [1970]. He generalized it as follows. By the basis theorem of linear algebra, the linear functional $\bar{\chi}_1$ on $V(P^1)$ has a linear extension, also called $\bar{\chi}_1$, to the vector space $\mathbb{R}^{\mathbb{R}}$ of all real functions on $\mathbb{R}$. Choose a basis $e_1,\ldots,e_d$ of $\mathbb{E}^d$ and then identify $\mathbb{E}^k$ with the subspace spanned by $e_1,\ldots,e_k$. If now method (b) is applied, one gets a linear functional $\bar{\chi}_d$ on the vector space of all real functions on $\mathbb{E}^d$ which satisfies $\bar{\chi}_d(K^*) = 1$ for $\emptyset \neq K \in \mathscr{K}^d$. Thus the definition $\chi(A) := \bar{\chi}_d(A^*)$ for $A \subset \mathbb{E}^d$ extends the Euler characteristic, as a valuation, from $\mathscr{K}^d$ to the system of all subsets of $\mathbb{E}^d$. This extension, of course, which depends on the extension of $\bar{\chi}_1$ and the choice of the basis, is highly arbitrary and therefore of little geometric interest, the more so since, as Lenz shows, it cannot be translation invariant.

The essential point of Groemer's [1972] paper is the introduction of a vector space A^d of real functions on $\mathbb{E}^d$ with a pseudonorm such that A^d contains $V(\mathscr{P}^d)$ as a proper dense subspace, and $\bar{\chi}$ has a unique continuous linear extension to A^d. The elements of A^d are called "approximable" functions. The system $\mathscr{S}_A$ of subsets of $\mathbb{E}^d$ whose characteristic functions are approximable contains the convex-ring $U(\mathscr{K}^d)$, and, for instance, the relative interiors of convex bodies and thus also the boundaries of convex bodies. Unfortunately, $\mathscr{S}_A$ is not intersectional. Some properties of the extended Euler characteristic on A^d are proved in Groemer [1972], and further invariance properties in Groemer [1973].

The fact that the Euler characteristic has a unique linear extension to the vector space generated by the characteristic functions of convex bodies and of their relative interiors, has been utilized by Groemer [1975] to show the existence and some properties of an Euler characteristic on certain systems of subsets of convex surfaces. This generalizes earlier work of Hadwiger-Mani [1972] which is concerned with spherical polyhedra.

More general results on Euler characteristics for subsets of convex surfaces could also be deduced, as Groemer remarks, from the following elegant result of Groemer [1974]. Let $\mathscr{S}$ be a system of subsets of a set S. The class $\mathscr{S}$ is called *separable* if to any two disjoint sets $A,B \in \mathscr{S}$ there exists a pair $X,Y \subset S$ such that: $X \cap C \in \mathscr{S}$ and $Y \cap C \in \mathscr{S}$ for every $C \in \mathscr{S}$, $A \subset X$, $A \cap Y = \emptyset$, $B \subset Y$, $B \cap X = \emptyset$, $X \cup Y = S$, and $Z \cap X \neq \emptyset, Z \cap Y \neq \emptyset$ for $Z \in \mathscr{S}$ only if $Z \cap X \cap Y \neq \emptyset$. Then Groemer shows:

(2.6) **Theorem.** *Let* S *be a set and let* $\mathscr{S}$ *be a separable intersectional class of subsets of* S. *There exists exactly one linear functional on the vector space* $V(\mathscr{S})$ *such that* $\chi(C^*) = 1$ *for every nonempty set* C *of* $\mathscr{S}$.

An example of a class $\mathscr{S}$ satisfying the assumptions of the theorem is the system of compact convex subsets of a locally convex topological vector space. But it should be noted that Groemer's theorem and its proof are purely combinatorial.

§3. Volume and valuations derived from it

Any measure on a ring of subsets of $\mathbb{E}^d$ containing $\mathscr{K}^d$ which is finite on $\mathscr{K}^d$ yields a real valued valuation on $\mathscr{K}^d$. In particular, restriction of the Lebesgue measure gives the volume V, the most familiar example of a simple valuation. In an axiomatic treatment of euclidean (or noneuclidean) geometry one might prefer, instead of taking Lebesgue measure for granted, to introduce the notion of volume for simple geometric figures, like polytopes, in an elementary geometric way. (For the plane case, compare Hilbert [1899], chap. IV.) As mentioned in the introduction, the attempts to do this have initiated a deeper study of simple valuations in general. For a description of these geometric approaches to volume and the difficulties involved, we refer the reader to the books by Hadwiger [1957], Boltianskii [1978], Böhm-Hertel [1980].

We first fix some more notation. By $B = B^d$ we denote the unit ball $\{x \in \mathbb{E}^d : \|x\| \leq 1\}$ of $\mathbb{E}^d$ and by $\varkappa(d)$ its volume. The ordinary spherical Lebesgue measure on the unit sphere $\Omega = \Omega^{d-1}$ is denoted by σ, thus $\sigma(\Omega^{d-1}) = d\varkappa(d)$.

In the theory of convex bodies, the valuation property of volume carries over to a series of other functions which are derived from volume in a natural way. The source of this is the fact, a special case of lemma (1.4), that for any fixed convex body C, the function $K \mapsto V(K + C)$ is also a valuation on $\mathscr{K}^d$. This leads at once to a valuation property of mixed volumes. As is well known, the volume of a linear combination $\lambda_1 K_1 + \cdots + \lambda_k K_k$ of convex bodies $K_1,\ldots,K_k \in \mathscr{K}^d$ with real coefficients $\lambda_1,\ldots,\lambda_k \geq 0$ can be expressed as a polynomial

$$(3.1) \quad V(\lambda_1 K_1 + \cdots + \lambda_k K_k) = \sum \lambda_{i_1} \cdots \lambda_{i_d} V(K_{i_1},\ldots,K_{i_d})$$

with $V(K_{i_1},\ldots,K_{i_d})$ symmetric in the indices and depending only on $K_{i_1},\ldots,K_{i_d}$. It is often convenient to write (3.1) in the form

$$(3.2) \quad V(\lambda_1 K_1 + \cdots + \lambda_k K_k) = \sum \binom{d}{r_1 \cdots r_k} \lambda_1^{r_1} \cdots \lambda_k^{r_k} V(K_1,r_1;\ldots;K_k,r_k),$$

where

$$\binom{d}{r_1 \cdots r_k} := \begin{cases} \dfrac{d!}{r_1! \cdots r_k!} & \text{if } \sum_{j=1}^{k} r_j = d, \quad r_j \geq 0, \\ 0, & \text{otherwise.} \end{cases}$$

Here, as in other cases, we use the abbreviation

$$(3.3) \quad f(K_1,r_1;\ldots;K_k,r_k) := f(\underbrace{K_1,\ldots,K_1}_{r_1 \text{ times}},\ldots,\underbrace{K_k,\ldots,K_k}_{r_k \text{ times}})$$

with $r_1 + \cdots + r_k = m$ for any function f of m variables, and we also write

$$(3.4) \quad f(K_1,\ldots,K_p,\mathscr{C}) := f(K_1,\ldots,K_p,L_{p+1},\ldots,L_m)$$

where $\mathscr{C}$ stands for the $(m-p)$-tuple $(L_{p+1},\ldots,L_m)$. With this notation, (3.2) implies

$$(3.5) \quad V(\lambda_1 K_1 + \cdots + \lambda_k K_k,p;\mathscr{C}) = \sum \binom{p}{r_1 \cdots r_k} \lambda_1^{r_1} \cdots \lambda_k^{r_k} V(K_1,r_1;\ldots;K_k,r_k;\mathscr{C}).$$

Now if $p \in \{1,\ldots,d\}$ and a $(d-p)$-tuple $\mathscr{C} = (K_{p+1},\ldots,K_d)$ of convex bodies is fixed, then the function φ defined by

$$(3.6) \quad \varphi(K) := V(K,p;\mathscr{C}) \quad \text{for} \quad K \in \mathscr{K}^d$$

is a valuation. This follows immediately from the fact that the function

$$K \mapsto V(\lambda K + \lambda_{p+1} K_{p+1} + \cdots + \lambda_d K_d)$$

is a valuation on $\mathscr{K}^d$, and that $(d!/p!)\varphi$ is the coefficient of $\lambda^p \lambda_{p+1} \cdots \lambda_d$ in the polynomial expansion of the latter expression.

In particular, this applies to the quermassintegrals $W_0,\ldots,W_d$, or the intrinsic volumes $V_0,\ldots,V_d$, respectively, defined by

$$(3.7) \quad W_r(K) := V(K,d-r;B,r) =: \frac{\varkappa(r)}{\binom{d}{r}} V_{d-r}(K).$$

Since $W_d(K) = \varkappa(d)$ if $K \neq \emptyset$, this gives $V_0(K) = 1(=\chi(K))$.

It appears that the valuation property of the quermassintegrals was first pointed out by Blaschke [1937], §43. Later it played an important role in the work of Hadwiger, to be reviewed later. Curiously, the valuation property of the general mixed volume (3.6) is not mentioned in the standard textbooks treating mixed volumes.

As a special consequence of the foregoing, we mention an identity for mixed volumes. Let $\mathscr{C}$ be a $(d-2)$-tuple of convex bodies, and let $K,L \in \mathscr{K}^d$ be such that $K \cup L$ is convex. Writing $V(K,K,\mathscr{C}) =: v(K)$ and $V(K,L,\mathscr{C}) =: v(K,L)$ for the moment and using the valuation property of v, the expansion (3.5), the identity (1.3) and again the expansion (3.5), we get

$$\begin{aligned} v(K) + v(L) + 2v(K \cup L, K \cap L) &= v(K \cup L) + v(K \cap L) \\ &\quad + 2v(K \cup L, K \cap L) \\ &= v((K \cup L) + (K \cap L)) = v(K + L) \\ &= v(K) + v(L) + 2v(K,L), \end{aligned}$$

thus

(3.8) $V(K,L,\mathscr{C}) = V(K \cup L, K \cap L, \mathscr{C})$.

Identity (3.8) was first observed by Groemer [1977a] p. 160, who proved it in a more indirect way.

Above, the quermassintegrals were defined as specialized mixed volumes or, equivalently, by means of the so-called *Steiner formula*

(3.9) $$V(K_\rho) = \sum_{j=0}^{d} \binom{d}{j} \rho^j W_j(K) = \sum_{i=0}^{d} \varkappa(d-i) \rho^{d-i} V_i(K),$$

where $K_\rho = K + \rho B$ is the outer parallel body of K at distance $\rho \geq 0$. A different approach comes from integral geometry. Let SO_d denote the rotation group of $\mathbb{E}^d$ and ν its invariant measure, suitably normalized. For given $r \in \{0,\ldots,d\}$, choose an r-dimensional linear subspace F_r of $\mathbb{E}^d$, and let δF_r denote its image under the rotation $\delta \in SO_d$. Further, let $\mathscr{E}_r^d$ be the space of r-flats in $\mathbb{E}^d$ with rigid motion invariant measure μ_r, also suitably normalized. Then for $K \in \mathscr{K}^d$ the formulae

(3.10) $$W_{d-r}(K) = a_{d,r} \int_{SO_d} V_r(K|\delta F_r) \, d\nu(\delta),$$

where $K|\delta F_r$ denotes the image of K under orthogonal projection on to δF_r, and

(3.11) $$W_r(K) = b_{d,r} \int_{\mathscr{E}_r^d} V_0(K \cap E_r) \, d\mu_r(E_r)$$

are valid. Here $a_{d,r}$ and $b_{d,r}$ are positive constants depending only on d and r. For proofs and generalizations of (3.10), (3.11) (and for explicit values of the constants) the reader may consult Hadwiger [1957], chap. 6, or Santaló [1976], §§13, 14. Note that V_r in (3.10) is just r-dimensional volume, while V_0 in (3.11) is the Euler characteristic. Thus in either formula the quermassintegrals are derived from a more elementary valuation. Hadwiger [1957] uses formulae similar to (3.10) to give a recursive definition for the quermassintegrals and later proves (3.9).

We remark that any measure λ on SO_d and any valuation φ on $\mathscr{K}^d$ for which each function $\delta \mapsto \varphi(K|\delta F_r)$ is λ-integrable ($K \in \mathscr{K}^d$), yields a new valuation ψ_r by means of the definition

(3.12) $$\psi_r(K) := \int_{SO_d} \varphi(K|\delta F_r) \, d\lambda(\delta), \quad K \in \mathscr{K}^d.$$

The valuation property carries over because of Lemma (1.4) (see the third example given there). Similarly, a measure λ_r on $\mathscr{E}_r^d$ and a valuation φ on $\mathscr{K}^d$ for which each function $E_r \mapsto \varphi(K \cap E_r)$ is λ_r-integrable ($K \in \mathcal{K}^d$), gives the new valuation φ_r defined by

$$(3.13)\quad \varphi_r(K) := \int_{\mathcal{E}_r^d} \varphi(K \cap E_r)\, d\lambda_r(E_r), \quad K \in \mathcal{K}^d.$$

For $\lambda_r = \mu_r$ and for continuous valuations φ, these associated valuations φ_r play an important role in Hadwiger's generalization of the principal kinematic formula of integral geometry, see Hadwiger [1956], [1957], p. 241.

Let us return to mixed volumes. Using their properties (Minkowski additivity and uniform continuity in each argument), one easily deduces from the Riesz representation theorem that, for given convex bodies $K_1,\ldots,K_{d-1} \in \mathcal{K}^d$, there exists a unique (positive) measure $S(K_1,\ldots,K_{d-1};\cdot)$ on the Borel sets of the unit sphere Ω of $\mathbb{E}^d$ such that

$$(3.14)\quad V(K,K_1,\ldots,K_{d-1}) = \frac{1}{d}\int_\Omega h(K,u)\, dS(K_1,\ldots,K_{d-1};u) \quad \text{for} \quad K \in \mathcal{K}^d,$$

where $h(K,\cdot)$ denotes the support function of K. This measure, which is called the *mixed area function* of $K_1,\ldots,K_{d-1}$, was introduced independently by Fenchel-Jessen [1938] and Aleksandrov [1937], see also Busemann [1958]. In particular, one writes (with the same notation as in (3.3))

$$S_p(K;\cdot) := S(K,p;B,d-1-p;\cdot)$$

for $p = 0,\ldots,d-1$ and calls this the p-*th order area function* of K. Clearly we have

$$(3.15)\quad S_p(K;\Omega) = dW_{d-p}(K).$$

$S_{d-1}(K;\cdot)$ has a simple geometric meaning: For a Borel set $\omega \subset \Omega$, $S_{d-1}(K;\omega)$ is the surface area ((d − 1)-dimensional Hausdorff measure) of the set of boundary points of K at which there exists an outer unit normal vector falling in ω. From this special measure, one gets back the general mixed area function by means of the polynomial expansion

$$(3.16)\quad S_{d-1}(\lambda_1 K_1 + \ldots + \lambda_{d-1}K_{d-1};\cdot) = \sum \lambda_{i_1} \cdots \lambda_{i_{d-1}}\, S(K_{i_1},\ldots,K_{i_{d-1}};\cdot).$$

Either from this representation (and the obvious valuation property of S_{d-1}), or from (3.14) and the valuation property of mixed volumes, it is clear that each function

$$K \mapsto S(K,p;K_{p+1},\ldots,K_{d-1};\cdot),$$

and in particular each S_p, is a valuation on $\mathcal{K}^d$ (with values in the vector space of signed Borel measures on Ω). Apparently the valuation property of the S_p was first pointed out and used by Schneider [1975a].

The area functions of order p can also be obtained from a local version of the Steiner formula (3.9). For $K \in \mathcal{K}^d$ let $p(K,\cdot)\colon \mathbb{E}^d \to K$ denote the metric projection, that is, $p(K,x)$ is the point in K nearest to x. Then for a Borel set $\omega \subset \Omega$ and for $\rho > 0$ consider the "local parallel set"

$$(3.17)\quad B_\rho(K,\omega) := \left\{ x \in \mathbb{E}^d : 0 < \|x - p(K,x)\| \le \rho \quad \text{and} \quad \frac{x - p(K,x)}{\|x - p(K,x)\|} \in \omega \right\}.$$

If V denotes Lebesgue measure, then

$$(3.18)\quad V(B_\rho(K,\omega)) = \frac{1}{d}\sum_{i=0}^{d-1}\binom{d}{i}\rho^{d-i}S_i(K;\omega).$$

In particular, the map $K \mapsto V(B_\rho(K,\cdot))$ is a valuation.

Similarly, if one defines, for a Borel set β in $\mathbb{E}^d$ and for $\rho > 0$,

$$(3.19)\quad A_\rho(K,\beta) := \{x \in \mathbb{E}^d : 0 < \|x - p(K,x)\| \le \rho \quad \text{and} \quad p(K,x) \in \beta\},$$

then $V(A_\rho(K,\cdot))$ is a measure and one has a polynomial expansion

$$(3.20)\quad V(A_\rho(K,\beta)) = \frac{1}{d}\sum_{i=0}^{d-1}\binom{d}{i}\rho^{d-i}C_i(K,\beta).$$

If $I_\rho(K,\beta,\cdot)$ denotes the characteristic function (on $\mathbb{E}^d$) of the set $A_\rho(K,\beta)$, then the map $K \mapsto I_\rho(K,\beta,\cdot)$ is a valuation (see Schneider [1978], p. 106). It follows that $K \mapsto V(A_\rho(K,\cdot))$ and hence each function $K \mapsto C_i(K,\cdot)$ is a valuation with values in the vector space of signed Borel measures (with compact support) on $\mathbb{E}^d$. The measures $C_0(K,\cdot),\ldots,C_{d-1}(K,\cdot)$ are Federer's curvature measures. They were introduced (for more general sets than convex bodies) by Federer [1959]. For a unified treatment of the measures S_i and C_i on $\mathscr{K}^d$ along the lines sketched above, see Schneider [1978]; further references are contained in the survey article Schneider [1979]. Far-reaching generalizations of the curvature measures and the area functions, in the form of measures on the Borel subsets of $\mathscr{K}^d$, have recently been proposed and investigated by Wieacker [1982].

In much the same way as the notion of volume, combined with Minkowski addition, leads to mixed volumes, the notion of centroid is the source of a series of vector valued functionals. For $K \in \mathscr{K}^d$, let

$$z(K) := \int_K x \, dV(x),$$

so that, for $\dim K = d$, the point $z(K)/V(K)$ is the centre of gravity of K. The vector z(K) will be called the *moment vector* of K. We have a polynomial expansion

$$(3.21)\quad z(\lambda_1K_1 + \cdots + \lambda_kK_k) = \sum \lambda_{i_1}\cdots\lambda_{i_{d+1}} z(K_{i_1},\ldots,K_{i_{d+1}}),$$

where the vector valued coefficients are assumed symmetric in their indices. These coefficients will be called *mixed moment vectors*. The expansion (3.21) was (for $d = 3$) already noticed by Minkowski [1911], §23. A more thorough study of mixed moment vectors was undertaken by Schneider [1972a,b]. In the same way as for mixed volumes, one shows that each function

$$K \mapsto z(K,p:\mathscr{C})$$

($p \in \{1,\ldots,d+1\}$ and the $(d+1-p)$-tuple $\mathscr{C}$ of convex bodies fixed) is a valuation on $\mathscr{K}^d$. Also the other properties of mixed moment vectors are analogous to those of mixed volumes, but observe that, while the mixed volume is invariant under translation of any of its arguments, one has

$$(3.22)\quad z(K_1 + t,K_2,\ldots,K_{d+1}) = z(K_1,K_2,\ldots,K_{d+1}) + \frac{1}{d+1}V(K_2,\ldots,K_{d+1})t.$$

Specialization of mixed moment vectors yields the so-called *quermassvectors* defined by

$$(3.23)\quad q_r(K) := \frac{d+1}{d+1-r}\, z(K,d+1-r;B,r)$$

for $r = 0,\ldots,d$. Note that (3.22) implies

$$(3.24)\quad q_r(K+t) = q_r(K) + W_r(K)t,$$

and that a Steiner formula,

$$(3.25)\quad z(K+\rho B) = \sum_{j=0}^{d} \binom{d}{j} \rho^j q_j(K),$$

is valid. Using Federer's curvature measures defined above, one has an integral representation

$$(3.26)\quad q_r(K) = \frac{1}{d} \int_{\partial K} x \, dC_{d-r}(K,x), \quad r = 1,\ldots,d$$

(compare Schneider [1972a], p. 123 and the remark on p. 129), which shows that $q_r(K)/W_r(K)$ is the centroid of the mass distribution on ∂K defined by the curvature measure $C_{d-1}(K;\cdot)$ In particular, q_1 is the area centroid, and

$$(3.27)\quad s(K) := \frac{q_d(K)}{\varkappa(d)}$$

is the so-called *Steiner point* of K. It can also be represented by

$$(3.28)\quad s(K) = \frac{1}{\varkappa(d)} \int_{\Omega} h(K,u)u \, d\sigma(u).$$

This shows that s is a Minkowski additive function on $\mathscr{K}^d$. References concerning this remarkable point can be found in Schneider [1972a] p. 128–129; others will be given in §13.

The quermassvectors satisfy integral geometric relations analogous to (3.11). Vice versa, this yields an alternative approach to these valuations: One may define s directly by (3.28) and then q_r, $0 \le r \le d$, by means of

$$(3.29)\quad q_r(K) = b_{d,r} \int_{\mathscr{E}_r^d} s(K \cap E_r) \, d\mu_r(E_r)$$

(with $b_{d,r}$ as in (3.11)). The valuation property of q_r is then obvious from (3.28), (3.29) and Lemma (1.4). In this way, the quermassvectors were introduced by Hadwiger-Schneider [1971].

In the above discussion of classical valuations we have stressed the existence of polynomial expansions, since this will be an essential point in the investigation of general valuations in later chapters. We have also mentioned a few more general constructions for valuations with a view to the problem of representing general, and characterizing special, valuations which will be the topic of chapter IV.

We conclude this section with a look at spherical space. Valuations on spherical polytopes, besides being interesting in themselves, also enter the investigation of valuations on euclidean polytopes. The spherical volume on Ω^{d-1}, which we denote by σ, yields a simple valuation on the spherically convex polytopes, say. It gives rise to the angle functions, which will play an important role in later discussions. For a (convex) polyhedral set P in $\mathbb{E}^d$ and non-empty faces $F \subset G$ of P, we denote by $\beta(F,G)$ and $\gamma(F,G)$ the internal and external angles, respectively, of G at its face F, measured in aff G and normalized so that the total angle is 1 (see, e.g., Grünbaum [1967], p. 297 and p. 308, for definitions). We also define $\beta(F,F) = 1 = \gamma(F,F)$ and $\beta(F,G) = 0 = \gamma(F,G)$ if $F \not\subseteq G$.

The Steiner formula (3.9), applied to a polytope $P \in \mathscr{P}^d$, yields an explicit representation of the quermassintegrals involving external angles, namely

$$(3.30)\qquad \frac{\binom{d}{r}}{\kappa(d-r)} W_{d-r}(P) = V_r(P) = \sum_{F^r} \gamma(F^r, P) V_r(F^r)$$

for $r = 0,1,\ldots,d$, where the sum extends over the r-dimensional faces F^r of P; note that $V_r(F^r)$ is just the r-dimensional volume of F^r. Similar formulae exist for the r-th order area functions and for the curvature measures (see Schneider [1978], (4.9) and (3.7)), as well as for the quermassvectors (Schneider [1972a], p. 125).

Now let us consider a spherically convex polytope $P \subset \Omega = \Omega^{d-1}$. It is the intersection of Ω with a convex polyhedral cone C with apex 0. We define

$$(3.31)\qquad \varphi_r(P) = \varphi_r(C) := \sum_{F^r} \beta(A,F^r)\gamma(F^r,C)$$

for $r = 0,\ldots,d$, where the sum extends over all r-faces F^r of C and A denotes the face of apices of C. By definition of the angles, $\varphi_d(P)$ is the normalized spherical volume of P, while $\varphi_0(P)$ is the normalized spherical volume of the polar set of P (the intersection of Ω with the polar cone of C). If $A = \{0\}$ (so that P lies in an open hemisphere) and $r \geq 1$, then $\beta(A,F^r) = \beta(0,F^r)$ is the normalized $(r-1)$-dimensional spherical volume of the $(r-1)$-face $\Omega \cap F^r$ of P. Thus φ_r is the spherical analogue of the intrinsic $(r-1)$-volume V_{r-1} for a euclidean polytope in $\mathbb{E}^{d-1}$. In fact, the functions $\varphi_1,\ldots,\varphi_d$ can also be obtained from a Steiner formula in Ω^{d-1} analogous to (3.9) (but with the powers of ρ replaced by other functions, see Allendoerfer [1948] in the smooth case). This approach can also be used to extend the definition of the φ_r to general spherically convex sets and to prove some of their properties. In particular, one gets rotation invariant continuous valuations. For polytopes, the valuation property can also be deduced directly from (3.31). The local Steiner formula (3.20) and the definition of the curvature measures also carry over. One could proceed similarly in hyperbolic space.

However, the analogy to euclidean space breaks down in several respects. While the euclidean quermassintegrals are monotonically increasing with respect to set inclusion, this is not generally true for the spherical φ_r. Clearly φ_d is increasing, and it can be shown that φ_{d-1} is increasing (e.g., Shephard [1968d], (32)). By duality (considering polar sets) it follows that φ_0 and φ_1 are decreasing. For $2 \leq r \leq d-2$, φ_r is neither increasing nor decreasing. To see this, let D_j denote a half j-space. If $2 \leq r \leq d-2$, we can arrange that $D_{r-1} \subset D_r \subset D_{r+1}$

$\subset D_{r+2}$ and φ_r successively takes the values 0, 1/2, 1/2, 0. Then we approximate the halfspaces by pointed d-dimensional polyhedral cones obeying the corresponding inclusion relations. The result follows by continuity.

Another difference to the euclidean case occurs when we consider the integral geometric approach (3.11). The spherical analogue to that formula is

$$(3.32) \quad \int \chi(P \cap L_r)\, dL_r = 2 \sum_{m \geq 0} \varphi_{d+1-r+2m}(P) = :2\psi_r(P)$$

for $r = 1,\ldots,d$ and spherically convex polytopes P (this can be generalized); χ is the Euler characteristic, and the integral is over the Grassmannian of all r-flats L_r through 0, with the invariant measure normalized to total measure 1 (see Santaló [1976], p. 310, with different terminology). Thus the ψ_r are also spherical analogues of the quermassintegrals, and perhaps the better analogues, since ψ_r is increasing for each r.

§4. The lattice point enumerator

Among the valuations derived from a measure, the next natural one after volume is perhaps the lattice point enumerator G defined by

$$G(K) := \operatorname{card}(K \cap \mathbb{Z}^d),$$

where $\mathbb{Z}^d$ is the integer lattice in $\mathbb{E}^d$. This function, of course, has its important place in geometry of numbers, and we refer the reader to the survey article of Gruber [1979]. An equally useful review of the known relations between the lattice point enumerator and other functionals on convex bodies has been given by Betke-Wills [1979]. Here we are only concerned with the valuation aspect of G. From this point of view, it turns out that most of the relations for G which are found in the literature are special cases of results which hold for more general valuations. Although these will be discussed in §§10, 12, we give a few references to the special results, since they added to the motivation for developing a general theory.

By $\mathscr{P}_L^d$ we denote the set of convex lattice polytopes in $\mathbb{E}^d$, and by $\mathbb{N}$ the set of positive integers. Ehrhart [1967a] proved the existence of a polynomial expansion

$$(4.1) \quad G(nP) = \sum_{i=0}^{d} n^i G_i(P) \quad \text{for } P \in \mathscr{P}_L^d, \quad n \in \mathbb{N},$$

where the coefficients G_i depend only on P, and he made applications of it to various counting problems. Ehrhart [1967b] also discovered and proved the so-called "reciprocity law"

$$(4.2) \quad G(\operatorname{relint} nP) = (-1)^{\dim P} \sum_{i=0}^{d} (-n)^i G_i(P) \quad \text{for} \quad P \in \mathscr{P}_L^d, \quad n \in \mathbb{N}.$$

Equality (4.1) has a generalization similar to the polynomial expansion (3.2), namely

$$(4.3) \quad G(n_1 P_1 + \cdots + n_k P_k) = \sum n_1^{r_1} \cdots n_k^{r_k} G(P_1,r_1;\ldots;P_k,r_k)$$

for $P_1,\dots,P_k \in \mathscr{P}^d_L$ and $n_1,\dots,n_k \in \mathbb{N}$; the sum extends over the nonnegative integers $r_1,\dots,r_k$ with $r_1 + \cdots + r_k \le d$. Expansion (4.3) for the lattice point enumerator was obtained by Bernštein [1976] at about the same time that it was discovered to hold for more general valuations, see §10.

Besides G, weighted lattice point numbers have been considered; these are simple valuations. Macdonald [1963], [1971] defined A(P) as the number which results when each lattice point in P is counted with weight $V(P \cap B)/V(B)$ for a sufficiently small ball B centred at the point. He proved an expansion similar to (4.1) and obtained some information on the coefficients. Hadwiger [1953], [1957], p. 69, took lattice-oriented cubes instead of balls and used the resulting simple valuation, applied to translates of lattice polytopes, in an equidissectability criterion with respect to lattice translations.

§5. Extension problems

Since the definition of a valuation involves unions and intersections of sets, it seems preferable that valuations be defined on set systems which are closed under (finite) unions and intersections. Thus for the valuations considered here which are, in the first instance, defined on the set $\mathscr{K}^d$ of convex bodies or the set $\mathscr{P}^d$ of convex polytopes, there arises the problem of extending them, as valuations, to the lattices $U(\mathscr{K}^d)$ or $U(\mathscr{P}^d)$, respectively. This extension problem is not only of formal interest. For instance, in integral geometry and its applications, one wants the formulae involving specific valuations to hold not only for convex sets, but at least for the elements of the convex-ring. A second reason for investigating extensions is technical: Even if only information on certain valuations on convex bodies is the aim, it may be necessary in the course of the proof first to extend the valuations to a broader class of sets.

For the classical valuations considered in the foregoing sections, one knows special constructions for extending them to $U(\mathscr{K}^d)$. Although the existence of such extensions would often also follow from more general theorems to be reviewed later, the explicit definitions are nevertheless of interest, since they exhibit the geometric meaning of the extended valuations, and sometimes permit the deduction of additional information.

There is, of course, no problem with volume or the lattice point enumerator, since they are measures, and hence valuations, on a ring containing $U(\mathscr{K}^d)$. For the Euler characteristic, the explicit construction of extensions was described in §2. Once the Euler characteristic χ is available on $U(\mathscr{K}^d)$, one can define (as Hadwiger [1957] does)

$$(5.1)\quad W_r(K) := b_{d,r} \int_{\mathscr{E}^d_r} \chi(K \cap E_r)\, d\mu_r(E_r)$$

for $K \in U(\mathscr{K}^d)$. For $K \in \mathscr{K}^d$ this gives, by (3.11), the r-th quermassintegral of K; and since the valuation property of χ carries over to W_r, this yields an additive extension of the quermassintegrals to the convex-ring.

In a similar way one can proceed with the quermassvectors. In analogy to (2.3) one first defines (following Mani [1971]) for $K \in U(\mathscr{K}^d)$, $u \in \Omega^{d-1}$, and $\lambda \in \mathbb{R}$ (with $H_{u,\lambda}$ as in (2.2))

$$(5.2)\quad h(K,u) := \sum_{\lambda \in \mathbb{R}} \lambda \left[\chi(K \cap H_{u,\lambda}) - \lim_{\mu \downarrow \lambda} \chi(K \cap H_{u,\mu}) \right],$$

where the limit exists and the sum is finite since K is a union of finitely many convex bodies. If K is convex and non-empty, then the right-hand side of (5.2) is the value of the support function of K at u, thus the definition is consistent. On the other hand, the valuation property of χ carries over to $h(\cdot,u)$, so that (5.2) extends the support function, as a valuation, to $U(\mathscr{K}^d)$. Observe that the support function of $K \in U(\mathscr{K}^d)$ no longer determines K uniquely, and that

$$(5.3) \quad h(K + t,u) = h(K,u) + \chi(K)\langle t,u\rangle$$

for $t \in \mathbb{E}^d$.

Now equation (3.28) may serve as definition of the Steiner point s(K) for $K \in U(\mathscr{K}^d)$; then s is a valuation on $U(\mathscr{K}^d)$ extending the classical Steiner point. Equation (5.3) implies the translation covariance property

$$(5.4) \quad s(K + t) = s(K) + \chi(K)t.$$

As proposed by Hadwiger-Schneider [1971], one may now define the quermassvectors $q_r(K)$ of $K \in U(\mathscr{K}^d)$ by equation (3.29). Then q_r is a valuation on $U(\mathscr{K}^d)$, and by (5.4) and (3.11) one has

$$(5.5) \quad q_r(K + t) = q_r(K) + W_r(K)t.$$

The Euler characteristic on $U(\mathscr{K}^d)$ can also be used to obtain an additive extension of the area functions and the curvature measures. We briefly explain this method of Schneider [1980]. For $z \in \mathbb{E}^d$ and $\rho > 0$, let $B(z,\rho) \subset \mathbb{E}^d$ denote the closed ball with centre z and radius ρ. Then for $K \in U(\mathscr{K}^d)$ and $q,x \in \mathbb{E}^d$, one defines the *index* of K at q with respect to x by

$$(5.6) \quad j(K,q,x) := \begin{cases} 1 - \lim\limits_{\delta \downarrow 0} \lim\limits_{\varepsilon \downarrow 0} \chi\big(K \cap B(x, \|x - q\| - \varepsilon) \cap B(q,\delta)\big) & \text{if} \quad q \in K \\ 0 \quad \text{if} \quad q \notin K. \end{cases}$$

If K is convex, then $j(K,q,x) = 1$ if $q = p(K,x)$ (where $p(K,\cdot)$ is the metric projection on to K), and $j(K,q,x) = 0$ otherwise. The additivity of the Euler characteristic implies that $j(\cdot, q,x)$ is a valuation on $U(\mathscr{K}^d)$. Now for $K \in U(\mathscr{K}^d)$, $\rho > 0$, a Borel set $\omega \subset \Omega$, and $x \in \mathbb{E}^d$, one defines

$$(5.7) \quad s_\rho(K,\omega,x) := \sum_{\substack{q \in \mathbb{E}^d \setminus (x) \\ (x-q)_0 \in \omega}} j(K \cap B(x,\rho),q,x),$$

where $(x - q)_0 := (x - q)/\|x - q\|$. If K is convex, then $s_\rho(K,\omega,\cdot)$ is the characteristic function of the local parallel set $B_\rho(K,\omega)$ defined by (3.17), hence the Lebesgue integral of this function satisfies the Steiner formula (3.18). Since $s_\rho(\cdot, \omega,x)$ is a valuation on $U(\mathscr{K}^d)$, it is easy to see that the polynomial expansion extends, and one has

$$(5.8) \quad \int_{\mathbb{E}^d} s_\rho(K,\omega,x)\,dx = \frac{1}{d} \sum_{i=0}^{d-1} \binom{d}{i} \rho^{d-i} S_i(K;\omega)$$

for $K \in U(\mathscr{K}^d)$. This defines signed Borel measures S_i on Ω which are the additive extensions to $U(\mathscr{K}^d)$ of the area functions.

Similarly, for a Borel set β in $\mathbb{E}^d$ one defines

$$(5.9) \quad c_\rho(K,\beta,x) := \sum_{q \in \beta \setminus \{x\}} j(K \cap B(x,\rho),q,x);$$

then $K \mapsto c_\rho(K,\beta,\cdot)$ is a valuation on $U(\mathscr{K}^d)$ which extends the indicator function of the local parallel set $A_\rho(K,\beta)$ defined by (3.19). The Steiner formula (3.20) extends to give

$$(5.10)\quad \int_{\mathbb{E}^d} c_\rho(K,\beta,x)\,dx = \frac{1}{d}\sum_{i=0}^{d-1}\binom{d}{i}\rho^{d-i}C_i(K,\beta)$$

for $K \in U(\mathscr{K}^d)$ and thus signed Borel measures C_i on $\mathbb{E}^d$ which additively extend the curvature measures.

If the point x in (5.6) "tends to infinity", one gets a different notion of index (see Schneider [1977a]). For $K \in U(\mathscr{K}^d)$, $q \in \mathbb{E}^d$, and $u \in \Omega^{d-1}$, we write

$$(5.11)\quad i(K,q,u) := \begin{cases} 1 - \lim\limits_{\delta \downarrow 0}\lim\limits_{\varepsilon \downarrow 0} \chi(K \cap H_{u,<q,u>+\varepsilon} \cap B(q,\delta)) & \text{if } q \in K \\ 0 \quad \text{if } q \notin K. \end{cases}$$

If K is convex, then $i(K,q,u) = 1$ if u is an exterior normal vector to K at q, and $i(K,q,u) = 0$ otherwise. Again, $i(\cdot,q,u)$ is a valuation on $U(\mathscr{K}^d)$. In analogy to a differential geometric notion, one may call q a *critical point* of K with respect to the height function $\langle\cdot,u\rangle$ if $i(K,p,u) \neq 0$. This analogy extends in so far as the "critical point theorem"

$$(5.12)\quad \sum_{q \in K} i(K,q,u) = \chi(K)$$

is valid for σ-almost all directions u.

For $K \in \mathscr{K}^d$ and a Borel set $\beta \subset \mathbb{E}^d$ it is easy to see that

$$(5.13)\quad C_0(K,\beta) = \int_\Omega \sum_{q \in \beta} i(K,q,u)\,d\sigma(u),$$

saying that $C_0(K,\beta)$ measures the area of the spherical image of $\partial K \cap \beta$. By additivity, (5.13) holds for arbitrary $K \in U(\mathscr{K}^d)$, thus yielding a geometric interpretation of C_0 as measuring the spherical image "with multiplicity". A particular consequence of (5.12) and (5.13) is the equality

$$(5.14)\quad C_0(K,\mathbb{E}^d) = \sigma(\Omega^{d-1})\chi(K).$$

Although this seems trivial, since it is evident for convex K and both sides are valuations in K, it has the nontrivial interpretation of expressing that the Euler characteristic can be obtained by adding up the local information provided by the curvature measure C_0. Thus (5.14) is an analogue of the Gauss-Bonnet theorem of differential geometry. For a polyhedron $P \in U(\mathscr{P}^d)$, Hadwiger [1969b] gave a different, though equivalent, definition for the curvature $C_0(P,\{p\})$ and proved (5.14).

If $P \in U(\mathscr{P}^d)$ is the point set of a cell complex of which Δ^k is the set of k-cells, then the index defined above satisfies

$$(5.15)\quad i(P,q,u) = \sum_{k=0}^{d}(-1)^k \sum_{Z \in \Delta^k} i(Z,q,-u).$$

For the special case of the boundary complex of a convex polytope P, this equality, which is related to the Euler-type theorems to be discussed in §12, was proved by Shephard [1968c], Lemma (13). Due to the valuation property of the index, (5.15) can be extended to the general case by means of an argument by Perles-Sallee [1970]. A definition equivalent to (5.15) was used by Banchoff [1967], [1970], who discussed critical point theory, curvature, and the Gauss-Bonnet theorem for polyhedra. Finally we mention that for polyhedral cell complexes a purely combinatorial (and very elementary) analogue of the Gauss-Bonnet theorem was proposed by Schneider [1977b].

As we have seen, some of the classical valuations derived from volume have natural additive extensions to the sets of the convex-ring $U(\mathscr{K}^d)$. There remains the question whether there exist more general sets to which these functions can be additively extended in a reasonable and useful way. Since, for instance, the integral geometric formulae involving quermassintegrals have applications in geometric probability theory and stereology and thus to practical problems, one would like to have those formulae available for fairly general sets which might serve as approximate models for material bodies occurring in reality. This motivation led Hadwiger [1959] to the introduction of his so-called normal bodies. Roughly speaking, their definition is chosen in such a way that Hadwiger's inductive definition (2.3) of the Euler characteristic can be carried over, and then the quermassintegrals of normal bodies can be defined by means of (5.1). As Lenz [1970] remarks, it seems difficult to prove that point sets which "normally" occur (he mentions the example of compact solution sets of finitely many analytic inequalities) are normal in Hadwiger's sense.

In a different direction, the quermassintegrals have been extended beyond the convex-ring by Groemer [1972]. He considers his vector space A^d of approximable functions, which was mentioned in §2. Since the Euler characteristic and projections of functions in A^d are available, it is possible to use an analogue of Kubota's recursion formula for the quermassintegrals to extend the latter from $\mathscr{K}^d$ to continuous linear functionals on A^d. In particular, this yields an additive extension W_i of the quermassintegral to the class $\mathscr{S}_A$ of subsets of $\mathbb{E}^d$ whose characteristic function belongs to A^d. Among other results, Groemer shows that

$$(5.16)\quad W_i(\text{relint } K) = (-1)^{d-i+\dim K} W_i(K)$$

for $K \in \mathscr{K}^d$. However, it seems difficult to describe the sets of $\mathscr{S}_A$ geometrically.

A class of point-sets with an easy intuitive definition was considered in this context by Federer [1959]. A subset K of $\mathbb{E}^d$ is called *of positive reach* if there exists a number $\varepsilon > 0$ such that, for each $x \in \mathbb{E}^d$ with distance less than ε from K, there is a unique point $p(K,x)$ in M nearest to x. For such sets, Federer was able to show that (3.20) holds (for $\rho < \varepsilon$) and yields signed measures $C_0(K,\cdot),\dots,C_{d-1}(K,\cdot)$, the curvature measures of K. These are valuations, for compact K the Gauss-Bonnet formula (5.14) (where the topological definition of χ is used) holds, and the curvature measures satisfy generalizations of the kinematic and Crofton formulae of integral geometry. A special case is the principal kinematic formula for compact sets K,L of positive reach,

$$(5.17)\quad \int \chi(K \cap gL)\, d\mu(g) = \sum_{k=0}^{d} c_{d,k} W_k(K) W_{d-k}(L),$$

where the integration is over the group of rigid motions of $\mathbb{E}^d$ with Haar measure μ, the $c_{d,k}$ are certain positive numbers, and $dW_k(K) = C_{d-k}(K,K)$ for $k = 1,\dots,d$, $W_0(K) = V(K)$. From the point of view explained above, a class of subsets of $\mathbb{E}^d$ containing $\mathcal{K}^d$ to which the Euler characteristic and the quermassintegrals can be additively extended, gains in interest if (5.17) can be proved for K,L in this class. For the class of normal bodies and for the class $\mathcal{S}_A$ of sets with approximable characteristic functions, such a result is not known. To be sure, neither of these classes nor the class of sets of positive reach is intersectional, but in (5.17), $\chi(K \cap gL)$ need only be defined μ-almost everywhere on the motion group. In fact, Federer [1959] shows that for K,L of positive reach, $K \cap gL$ is of positive reach for almost all g.

On the other hand, the class of sets of positive reach has the disadvantage that it does not contain the convex-ring $U(\mathcal{K}^d)$. Perhaps the system of all finite unions of (compact) sets of positive reach is a good class of point sets to consider in integral geometry. However, it seems unknown whether Federer's curvature measures admit additive extensions to this class.

We mention that certain generalizations of Federer's curvature measures to arbitrary closed sets have been constructed by Stachó [1979], but these do not have the valuation property. Kuiper [1971] has used singular homology to define a sequence of curvature measures including a generalization of C_0. In either case it seems difficult to work with these measures except in the special cases already known.

Let us now turn to the extension problem for the mixed volume. This is the function, also denoted by V, on $(\mathcal{K}^d)^d$ which sends the d-tuple $(K_1,\dots,K_d)$ of convex bodies into $V(K_1,\dots,K_d)$, the coefficient of $\lambda_1 \cdots \lambda_d$ in (3.1) for $k = d$. Since the mixed volume is Minkowski additive in each argument, only such extensions to more general domains are of interest for which this property is maintained in some sense. Now Minkowski or vector addition is defined for arbitrary subsets of $\mathbb{E}^d$. But, as Groemer [1977a] (p. 141) remarks, "this generalized concept of addition is not suitable as a basis for a theory of mixed volumes that applies to a reasonably large class of non-convex sets". In a strong sense, this was made clear by Weil [1975a] who showed the following. Suppose that $\mathcal{K}'$ is a class of compact subsets of $\mathbb{E}^d$ containing $\mathcal{K}^d$ and closed under vector addition. If there exists a function $V:(\mathcal{K}')^d \to \mathbb{R}$ which is Minkowski additive in each variable and for which $V(K,\dots,K)$ is the Lebesgue measure of $K \in \mathcal{K}'$, then $\mathcal{K}' = \mathcal{K}^d$. Thus it appears that a theory of mixed volumes of point sets which is satisfactory from a geometric point of view is only possible for convex sets. But extensions of the concept of mixed volumes beyond the field of convexity are possible if functions instead of point sets are considered.

The following such extension is particularly useful. Clearly the mixed volume V can be viewed as a function defined on a subset of $C(\Omega)^d$, where $C(\Omega)$ denotes the real vector space of continuous real functions (with the maximum norm) on Ω, just by identifying a convex body with its support function on Ω. Then V can be linearly extended to differences of support functions. By use of uniform continuity, in one argument a further linear extension to continuous functions is possible (as mentioned already in connexion with (3.14)). This procedure occurs, as a technical device, already in the important work of Aleksandrov [1937a,b]. Later the extension of mixed volumes (and a similar one for mixed area functions) was exploited in two papers by Weil [1974a,b]. One might ask (compare Weil [1974a], pp. 355–356) whether a further extension of the mixed volume to a continuous d-linear functional on $C(\Omega)^d$ is possible, but this has been answered in the negative by Meier [1982]. Yet further progress along these lines is possible, as shown by Weil [1981], whose starting point was the representation (3.14). That equality exhibits the linearity of V in its first argument, but it does not reflect the symmetry of V in its arguments. Therefore, one might ask whether there exists a completely symmetric analytic representation of the mixed volume involving each of the support functions linearly. Weil [1981] was able to show that there exists a distribution T on $(\Omega^{d-1})^d$ such that

$$V(K_1,\ldots,K_d) = T(h(K_1,\cdot) \otimes \cdots \otimes h(K_d,\cdot)),$$

where $\otimes$ denotes the tensor product. As a consequence, there exists a sequence $(f_j)_{j\in\mathbb{N}}$ of C^∞-functions on $(\Omega^{d-1})^d$ such that

$$V(K_1,\ldots,K_d) = \lim_{j\to\infty} \int_{\Omega^{d-1}} \cdots \int_{\Omega^{d-1}} h(K_1,u_1)\cdots h(K_d,u_d) f_j(u_1,\ldots,u_d)\, d\sigma(u_1)\cdots d\sigma(u_d)$$

uniformly for all $K_1,\ldots,K_d$ in some fixed ball.

Quite a different extension problem arises if convex bodies are identified with their characteristic functions. An interesting theory of mixed volumes on $V(\mathscr{K}^d)^d$ was developed by Groemer [1977a]. He first shows that there exists a unique bilinear map ψ of $V(\mathscr{K}^d) \oplus V(\mathscr{K}^d)$ into $V(\mathscr{K}^d)$ such that

$$\psi(K^*,L^*) = (K + L)^* \quad \text{for} \quad K,L \in \mathscr{K}^d,$$

and that the vector space $V(\mathscr{K}^d)$ together with the multiplication $\times$ defined by $f \times g := \psi(f,g)$ is a commutative algebra over $\mathbb{R}$ with unit element. The Euler characteristic $\bar{\chi}$ on $V(\mathscr{K}^d)$ (as defined in §2) is an algebra homomorphism. To every affine map $a: \mathbb{E}^d \to \mathbb{E}^d$ there exists a unique linear map $\alpha: V(\mathscr{K}^d) \to V(\mathscr{K}^d)$ such that $\alpha(K^*) = (aK)^*$ for $K \in \mathscr{K}^d$. In the special case $ax = \lambda x$ ($x \in \mathbb{E}^d$) with $\lambda \in \mathbb{R}^+$ one writes $\alpha(f) =: \lambda \circ f$ for $f \in V(\mathscr{K}^d)$. Clearly there is a unique linear functional $\bar{V}$ on $V(\mathscr{K}^d)$ such that $\bar{V}(K^*)$ is the volume of $K \in \mathscr{K}^d$. Groemer shows that (3.2) can be generalized as follows. Let $k_1,\ldots,k_d$ be nonnegative integers so that $k_1 + \cdots + k_d = d$ and write $k = (k_1,\ldots,k_d)$. There exists exactly one d-linear mapping v_k of $V(\mathscr{K}^d)^d$ into $\mathbb{R}$ so that

$$v_k(K_1^*,\ldots,K_d^*) = V(K_1,k_1;\ldots;K_d,k_d) \quad \text{for} \quad K_1,\ldots,K_d \in \mathscr{K}^d.$$

For $f_1,\ldots,f_d \in V(\mathscr{K}^d)$ and $\lambda_1,\ldots,\lambda_d \geq 0$ one has

$$\bar{V}((\lambda_1 \circ f_1) \times \cdots \times (\lambda_d \circ f_d)) = \sum \binom{d}{k_1 \cdots k_d} \lambda_1^{k_1} \cdots \lambda_d^{k_d}\, v_{(k_1,\ldots,k_d)}(f_1,\ldots,f_d).$$

The functions $v_{(k_1,\ldots,k_d)}$ are called mixed volumes, and Groemer investigates their properties carefully. We remark that mixed area functions could also be extended similarly and that some parts of the convex theory, for instance certain integral geometric formulae for mixed area functions, carry ever (see Schneider [1980], Bemerkung 9). Similar extensions of some integral geometric formulae for mixed volumes have been given by Groemer [1977b]. But, on the whole, the theory seems rather algebraic in character, and it loses much of its elegance if one tries to go back to point sets whose characteristic functions belong to $V(\mathcal{K}^d)$, and to the generalized mixed volumes of such sets.

So much for the extensions of special valuations derived from the notion of volume. Now we consider the general problem of extending an arbitrary valuation on a class $\mathcal{S}$ (of subsets of some set S) to a valuation on $U(\mathcal{S})$ or to a linear functional on $V(\mathcal{S})$ as described in §1.

Volland [1957] has proved that every valuation on the class $\mathcal{P}^d$ of convex polytopes admits a unique additive extension to the class $U(\mathcal{P}^d)$ of polyhedra (related but simpler extension results can be found in Hadwiger [1957], p. 81, Böhm-Hertel [1980], p. 47). Volland first uses induction to prove the following result.

(5.18) **Theorem.** *Every valuation on* $\mathcal{P}^d$ *satisfies the inclusion-exclusion principle.*

The rest of Volland's proof is not restricted to the special geometric situation. Together with the obvious remark made in §1 after (1.2), it shows the following.

(5.19) **Theorem.** *Let* φ *be a valuation on an intersectional class* $\mathcal{S}$. *Then* φ has an *additive extension to* $U(\mathcal{S})$ *if and only if* φ *satisfies the inclusion-exclusion principle. The extension is unique.*

Volland's theorem was rediscovered by Perles-Sallee [1970]. They prove (5.19) (in its abstract form) in essentially the same way, and for the assertion of (5.18) they refer to Sallee [1966], where the corresponding result was obtained for the Steiner point. The proof (which is similar to Volland's) holds for arbitrary valuations on $\mathcal{P}^d$. Sallee [1966] subsequently used the continuity of the Steiner point to show that the Steiner point satisfies the inclusion-exclusion principle on $\mathcal{K}^d$. Spiegel [1976b] noticed that here one does not really need the continuity; he proved that any Minkowski additive function from $\mathcal{K}^d$ into $\mathbb{E}^d$ which vanishes on centrally symmetric bodies satisfies the inclusion-exclusion principle. This can be generalized:

(5.20) **Theorem.** *Every Minkowski additive function on* $\mathcal{K}^d$ *(with values in an abelian group) satisfies the inclusion-exclusion principle, hence by* (5.19) *it can be extended to a valuation on the convex-ring* $U(\mathcal{K}^d)$.

The proof is very simple. By formula (5.2) the support function was extended, as a valuation, to $U(\mathcal{K}^d)$, hence it satisfies (1.2). Thus for $K = K_1 \cup \cdots \cup K_m$ with $K, K_i \in \mathcal{K}^d$ $(i = 1,\dots,m)$ we have

$$h(K,\cdot) = \sum_{r=1}^{m} (-1)^{r-1} \sum_{i_1<\cdots<i_r} h(K_{i_1} \cap \ldots \cap K_{i_r},\cdot).$$

Observing that $h(\emptyset,\cdot) = 0$ by (5.2), we can write this in the form

$$K + \sum_{r \text{ even}} \sum_{i_1<\cdots<i_r}' K_{i_1} \cap \cdots \cap K_{i_r} = \sum_{r \text{ odd}} \sum_{i_1<\cdots<i_r}' K_{i_1} \cap \cdots \cap K_{i_r},$$

where the sums $\sum'$ extend only over those $K_{i_1} \cap \cdots \cap K_{i_r}$ which are non-empty. If now φ is a Minkowski additive function on $\mathcal{K}^d$, we may apply φ to either side of this equality and deduce that φ satisfies (1.2). This completes the proof of (5.20).

We turn to the question of linear extensions to the real vector space $V(\mathcal{S})$ spanned by the characteristic functions A^* (defined on S) of the sets $A \in \mathcal{S}$. Here we assume that φ is a function on $\mathcal{S}$ with values in some real vector space $\mathcal{X}$. By a *linear extension* of φ to $V(\mathcal{S})$ we understand a linear map $\bar{\varphi}$ from $V(\mathcal{S})$ into $\mathcal{X}$ for which $\bar{\varphi}(A^*) = \varphi(A)$ for $A \in \mathcal{S}$. If φ admits such an extension, it must be a valuation (see §1). The following was proved by Groemer [1978]:

(5.21) **Theorem.** *Let φ be a function from the intersectional class $\mathcal{S}$ into a real vector space so that $\varphi(\emptyset) = 0$. Then the following statements are equivalent*:

(a) *φ has an additive extension* to $U(\mathcal{S})$,
(b) *φ has a linear extension* to $V(\mathcal{S})$,
(c) $\alpha_1 K_1^* + \cdots + \alpha_m K_m^* = 0$ *with* $K_i \in \mathcal{S}$, $\alpha_i \in \mathbb{R}(i = 1,\ldots,m)$

implies $\alpha_1\varphi(K_1) + \cdots + \alpha_m\varphi(K_m) = 0$.

The implications (b) ⇒ (a) and (c) ⇒ (b) are easy and were already discussed in §1, so the essentially new result is the fact that (a) (or, what is equivalent by (5.19), the inclusion-exclusion principle for φ on $\mathcal{S}$) implies (c). Using (5.21), Groemer also proved an extension theorem for valuations on $\mathcal{K}^d$. Let us say that the function φ from $\mathcal{K}^d$ into a topological (Hausdorff) vector space is *σ-continuous* if for every decreasing sequence $(K_i)_{i\in n}$ in $\mathcal{K}^d$ one has

$$\lim_{i\to\infty} \varphi(K_i) = \varphi\left(\bigcap_{i\in\mathbb{N}} K_i\right).$$

Clearly continuity with respect to the usual Hausdorff metric on $\mathcal{K}^d$ implies σ-continuity. Groemer [1978] proved:

(5.22) **Theorem.** *Every σ-continuous weak valuation on $\mathcal{K}^d$ (with values in some topological vector space) has an additive extension to* $U(\mathcal{K}^d)$.

Apparently it is not known whether every valuation on $\mathcal{K}^d$ has an additive extension to $U(\mathcal{K}^d)$. About valuations on classes which are not intersectional, only a recent result of Stein [1982] is known. He showed that every valuation on $\mathcal{P}_L^d$, the class of convex lattice polytopes in $\mathbb{E}^d$, satisfies the inclusion-exclusion principle if it is invariant under lattice translations. The latter restriction has recently been removed by Betke.

We conclude this section with a remark concerning valuations on polytopes. Let φ be a valuation on $\mathcal{P}^d$ with values in a real vector space (this could be generalized). By (5.18), (5.19), (5.21) there is a linear function $\bar{\varphi}$ on $V(\mathcal{P}^d)$ such that $\bar{\varphi}(P^*) = \varphi(P)$ for $P \in \mathcal{P}^d$. We extend the definition of φ by means of $\varphi(P) := \bar{\varphi}(P^*)$ for each set P for which $P^* \in V(\mathcal{P}^d)$. It is not difficult to see that these sets are precisely the finite unions of relatively open convex polytopes. Thus φ as just defined is a valuation on $U(\mathcal{P}_{ro}^d)$. The value of φ on a relatively open polytope is given by

$$(5.23)\quad \varphi^0(P):=\varphi(\operatorname{relint} P)=\sum_F(-1)^{\dim P-\dim F}\varphi(F)$$

for $P \in \mathscr{P}^d$, where the sum extends over the faces F of P (including P). The proof is easily done by induction with respect to dim P if one uses the Euler relation for faces $F \subseteq G$ of a polytope. Equivalently, (5.23) results from the obvious formula

$$\varphi(P)=\sum_F \varphi(\operatorname{relint} P)$$

(which holds since φ is a valuation on $U(\mathscr{P}^d_{ro})$) by applying the Möbius inversion formula (see Rota [1964], p. 344, and observe that $\mu(F,P)=(-1)^{\dim P-\dim F}$). For a given valuation φ on $\mathscr{P}^d$, Sallee [1968] defined

$$(5.24)\quad \varphi^*(P):=\sum_F(-1)^{\dim F}\varphi(F)\quad \text{for}\quad P\in\mathscr{P}^d.$$

Formula (5.23) exhibits the meaning of φ^*, namely

$$(5.25)\quad \varphi^*(P)=(-1)^{\dim P}\varphi(\operatorname{relint} P).$$

Using the additivity of φ on $U(\mathscr{P}^d_{ro})$ it is easy to see that φ^* is a valuation on $\mathscr{P}^d$. Sallee proved this in a different way.

The derived valuation φ^* will be taken up in §12.

II. Dissections and simple valuations

Valuations, and particularly simple valuations, which are invariant under a given group G of transformations acting on $\mathbb{E}^d$, are intimately connected with equidissectability of polytopes with respect to the same group G. We are going to review the more recent developments in this area. The reader will find useful introductions and extensive references to the older literature in the books of Hadwiger [1957] and Boltianskii [1978] (see also Boltianskii [1963]). For special aspects also the shorter survey articles by Hadwiger [1968c], [1975] and Hertel [1977] may be helpful.

We adopt the convention here that "valuation", without specification of the range, means real valued valuation.

§6. The algebra of polytopes

Let us define our terms here. Let P and Q be two d-polytopes, or, more generally, elements of $U(\mathscr{P}^d)$. A *dissection* of P is an expression of P in the form $P = P_1 \cup \cdots \cup P_k$, where $P_1,\ldots,P_k \in \mathscr{P}^d$ (or $U(\mathscr{P}^d)$) satisfy $\operatorname{int}(P_i \cap P_j)=\emptyset$ for $i \neq j$. We write this as

$$P=P_1\,\dot\cup\cdots\dot\cup\,P_k \quad\text{or}\quad P=\dot{\bigcup}_{i=1}^k P_i.$$

We say that P and Q are *equidissectable with respect to* G, or G-*equidissectable*, if there are dissections

$$P=P_1\,\dot\cup\cdots\dot\cup\,P_k,\quad Q=Q_1\,\dot\cup\cdots\dot\cup\,Q_k,$$

such that $Q_i = g_i(P_i)$ for some $g_i \in G$ ($i = 1,\ldots,k$); we write this as $P \approx_G Q$. We call P and Q *equicomplementable with respect to* G, or G-*equicomplementable*, if there are P', P'', Q', Q'', such that $P'' = P \dot\cup P'$, $Q'' = Q \dot\cup Q'$, and $P' \approx_G Q'$, $P'' \approx_G Q''$; we write this as $P \sim_G Q$. (The corresponding terms in Hadwiger [1957], where most of the early work on this topic is collected, are "zerlegungsgleich" and "ergänzungsgleich", respectively. A variety of terms have been used in English; we have chosen to use those above because they more exactly convey the meaning of their definitions, and, equally importantly, do not have any other meaning within convexity.)

As examples of groups G commonly encountered, we have the full group T of translations, the group TH of translations and reflexions in points (half turns in $\mathbb{E}^2$), the group D of isometries of $\mathbb{E}^d$, its subgroup SD of direct isometries or rigid motions, the full affine group A, and the group EA of equiaffine mappings (determinant $= \pm 1$). (We also refer later to the orthogonal group O and the rotation subgroup SO.) In what follows, we shall always assume G to have T at least as a subgroup.

Before we go any further, let us mention two results by Hadwiger [1957].

(6.1) **Lemma.** *Let* $P, Q \in \mathscr{P}^d$. *Then* $P \approx_G Q$ *if and only if* $P \sim_G Q$.

The validity of this result depends upon the field $\mathbb{R}$ being archimedean.

Let φ be a simple valuation on $\mathscr{P}^d$. We say that φ is G-*invariant* if $\varphi(gP) = \varphi(P)$ for all $g \in G$. Then we have

(6.2) **Theorem.** *Let* $P, Q \in \mathscr{P}^d$. *Then* $P \approx_G Q$ *if and only if* $\varphi(P) = \varphi(Q)$ *for every real valued* G-*invariant simple* φ.

Unfortunately, Hadwiger's proof is highly non-constructive, depending as it does on using the axiom of choice to pick a basis of the polytope group Π_G, which we are about to define below. In fact, the problem which will concern us here and in §§8 and 9 can be succinctly phrased as:

(6.3) **Problem.** *Given a group* G, *find a "nice" subfamily of* G-*invariant simple valuations which separates* Π_G.

In other words, we would wish to find an easily describable family $\mathscr{V}$ of such valuations φ, so that $P \approx_G Q$ if and only if $\varphi(P) = \varphi(Q)$ for all $\varphi \in \mathscr{V}$.

We define the polytope group Π_G as follows. Let $\mathbb{Z}\mathscr{P}^d$ be the free abelian group with the d-polytopes as basis elements; we write $\boldsymbol{+}$ and $\boldsymbol{-}$ for the group operations, to distinguish them from the usual $+$ and $-$ in $\mathbb{E}^d$. Let $\mathscr{T} = \mathscr{T}_G$ be the subgroup generated by all elements of the form $P \boldsymbol{-} P_1 \boldsymbol{-} P_2$ and $P \boldsymbol{-} gP$, where $P = P_1 \dot\cup P_2$ and $g \in G$. In other words, $\mathscr{T}$ is generated by all $P \boldsymbol{-} Q$, with $P \approx_G Q$. The quotient group

$$\Pi_G^d = \mathbb{Z}\mathscr{P}^d / \mathscr{T}_G$$

is called the *polytope group with respect to* G. We also use notations such as $\Pi_G(\mathbb{E}^d)$, as we shall wish later to write $\Pi(L) = \Pi_T(L)$ with the obvious meaning for linear subspaces L of $\mathbb{E}^d$. Until we wish to emphasize the dimension d, we shall write Π_G instead of Π_G^d.

We now make an obvious remark.

(6.4) *Let $\mathscr{X}$ be a group. The (group) homomorphisms $\varphi:\Pi_G \to \mathscr{X}$ are precisely those mappings induced by the* G*-invariant simple valuations $\varphi:\mathscr{P}^d \to \mathscr{X}$.*

In such cases, we shall often use the same letter to denote two such related mappings.

For a number of reasons, the basic polytope group Π_G we need to investigate is $\Pi = \Pi_T$. For one thing, the familiar group isomorphism theorem gives:

(6.5) *Let* G *be a group containing* T, *and let $\Sigma_G = \mathscr{T}_G/\mathscr{T}_T$ be the subgroup of* Π *corresponding to $\mathscr{T}_G$. Then $\Pi/\Sigma_G \cong \Pi_G$.*

That is, every polytope group Π_G is a suitable quotient of Π.

After these preliminaries, we are now ready to embark upon our investigation of Π. We shall give the main results, always without proof, following the treatments of Jessen-Thorup [1978] and Sah [1979]. Our notation will be closer to that of Jessen-Thorup; Sah's is more comprehensive, but also more complicated.

An important role is played in these investigations by the *dilatation operator* m, which is induced by the corresponding operator μ on $\mathscr{P}^d$, defined by $\mu(\lambda)P = \lambda P = \{\lambda x | x \in P\}$, for $\lambda > 0$. That is, $m(\lambda)[P] = [\mu(\lambda)P]$, where [P] is the equivalence class of P under $\approx\ = \ \approx_T$. For $\lambda < 0$, we define $m(\lambda) = (-1)^d m(-\lambda)$, and also $m(0) = 0$. (The choice of $(-1)^d$ is due to our considering unoriented polytopes. Further justification comes from Theorem (7.2); compare also (6.6) below.)

We call a polytope P a (basic) r-cylinder, if there are independent linear subspaces $L_1,\ldots,L_r$ of $\mathbb{E}^d$, whose dimensions $d_i = \dim L_i$ are positive and satisfy $d_1 + \cdots + d_r = d$, and d_i-polytopes P_i in L_i, with $P = P_1 + \cdots + P_r$. We denote by $\mathfrak{Z}_r$ the family of all r-cylinders in $\mathscr{P}^d$, and by

$$Z_r = (\mathbb{Z}\mathfrak{Z}_r + \mathscr{T}_T)/\mathscr{T}_T$$

the corresponding subgroup of Π. Thus we have

$$\Pi = Z_1 \supseteq Z_2 \supseteq \cdots \supseteq Z_d \supseteq Z_{d+1} = \{0\}.$$

In fact, as will be clearer later, all these inclusions are strict. We notice as well that we have a natural embedding

$$\Pi(L_1) \otimes \cdots \otimes \Pi(L_r) \mapsto \Pi;$$

we shall write $x_1 \times \cdots \times x_r$ for the image of $(x_1,\ldots,x_r)$ under this embedding (the tensor product is, as yet, only over $\mathbb{Z}$).

We write $[a_1,a_2,\ldots,a_d]$ to denote the equivalence class of the simplex with vertices $a_0, a_0 + a_1, a_0 + a_1 + a_2,\ldots,a_0 + a_1 + \cdots + a_d$, for $a_0 \in \mathbb{E}^d$, where $\{a_1,\ldots,a_d\}$ is linearly independent. We have the two canonical simplex dissections:

(6.6) **Theorem.** $m(\lambda + \mu)[a_1,\ldots,a_d]$

$$= \sum_{j=0}^{d} \{m(\lambda)[a_1,\ldots,a_j] \times m(\mu)[a_{j+1},\ldots,a_d]\}.$$

(6.7) **Theorem.** *Let* n *be a nonnegative integer. Then*

$$m(n)[a_1,\ldots,a_d] = \sum_{r=1}^{d} \binom{n}{r} z_r,$$

where $z_r = \sum_{1 \le j_1 \le \cdots \le j_{r-1} < d} [a_1,\ldots,a_{j_1}] \times \cdots \times [a_{j_{r-1}+1},\ldots,a_d] \in Z_r$.

The general term is $[a_{j_s+1},\ldots,a_{j_{s+1}}]$.

For future reference, it is helpful to take note of the actual translations involved in (6.7). In a different notation, let T be the d-simplex $T = \mathrm{conv}\{x_0,\ldots,x_d\}$, and write

$$T(j_1,\ldots,j_{r-1}) = \mathrm{conv}\{x_0,\ldots,x_{j_1}\} + \cdots + \mathrm{conv}\{x_{j_{r-1}},\ldots,x_d\}.$$

Here the general term is $\mathrm{conv}\{x_{j_s},\ldots,x_{j_{s+1}}\}$. Then we have (compare McMullen [1977])

$$\mu(n)T = \dot{\bigcup}_{r=1}^{d} \dot{\bigcup}_{0<j_1<\cdots<j_{r-1}<d} \dot{\bigcup}_{k_i \ge 0, \Sigma k_i = n-r} \left(T(j_1,\ldots,j_{r-1}) + \sum_{i=0}^{r} k_i x_{j_i} \right).$$

Next, we have a fundamental result, from which much else follows.

(6.8) **Theorem.** *Let* $\mathbb{E}^d = L_1 \oplus L_2$ *(direct sum), and let* $x_i \in \Pi(L_i)\,(i = 1,2)$. *Then, for all* $\lambda,\mu \in \mathbb{R}$,

$$x_1 * x_2 = m(\lambda)x_1 \times m(\mu)x_2 - m(\mu)x_1 \times m(\lambda)x_2 \in Z_3.$$

From (6.6) follows directly

$$\sum_{j=1}^{d-1} [a_1,\ldots,a_j] * [a_{j+1},\ldots,a_d] \in Z_3.$$

(Curiously, Sah fails to notice that it is a trivial consequence.) Jessen-Thorup then deduce (6.8) from this by a geometric lemma due to Thorup, while Sah uses a more algebraic argument to give the same deduction.

The next results concern various properties of the map $m: \mathbb{R} \to \mathrm{End}\,\Pi$. We denote by $E \subseteq \mathrm{End}\,\Pi$ the subring of endomorphisms of Π generated by the $m(\lambda)$. It is clear that m is multiplicative:

$$m(\lambda\mu) = m(\lambda)m(\mu).$$

Also $m(0) = 0, m(1) = 1$. Our aim is to show that $\mathbb{R}$ is embedded in E, so that Π is a real vector space. Our first step is to embed $\mathbb{Q}$ (the rationals) in E.

We define the operation $\Delta_\alpha (\alpha \in \mathbb{R})$ by

$$\Delta_\alpha m(\lambda) = m(\lambda + \alpha) - m(\lambda).$$

Then $\Delta_\alpha \Delta_\beta = \Delta_\beta \Delta_\alpha$. We write $\Delta_\alpha^s = \Delta_\alpha \cdots \Delta_\alpha$ (s times). It easily follows by an inductive argument that

$$m(q\alpha) = \sum_{p=0}^{d} \binom{q}{p} \Delta_\alpha^p m(0)$$

for every $\alpha \in \mathbb{R}$ and $q \in \mathbb{Z}$ (we must interpret $\binom{q}{p}$ here as $q(q-1)\cdots(q-p+1)/p!$). There follows:

(6.9) **Lemma.** *If* $k \in \mathbb{Z}\backslash\{0\}$, *then* $1/k \in E$. *Thus* E *is divisible, and* $\mathbb{Q} \subseteq E$.

For,

$$\frac{1}{k} = \frac{1}{k}m(1) = \sum_{p=1}^{d} \frac{d!}{p}\binom{kd!-1}{p-1}\Delta^p_{1/kd!}m(0).$$

Now, by induction from (6.6) follows

$$\Delta_{\alpha_p}\cdots\Delta_{\alpha_1}m(\lambda)x_{0d}$$
$$= \sum_{0<i_1\cdots<i_p\le d} m(\alpha_1)x_{0i_1} \times \cdots \times m(\alpha_p)x_{i_{p-1}i_p} \times m(\lambda)x_{i_pd},$$

where we write $x_{ij} = [a_{i+1},\ldots,a_j]$. In particular,

$$\Delta_{\alpha_d}\cdots\Delta_{\alpha_0}m(\lambda) = 0$$

for all $\lambda,\alpha_0,\ldots,\alpha_d$. This means that $m(\lambda)$ exhibits polynomial like behaviour (which, using (6.9), is already clear from (6.7) for positive integral λ). We now define

$$\tilde{m}_d(\alpha_1,\ldots,\alpha_d) = \frac{1}{d!}\Delta_{\alpha_1}\cdots\Delta_{\alpha_d}m(\lambda)$$

(this is independent of λ, by the above), and for $r < d$,

$$\tilde{m}_r(\alpha_1,\ldots,\alpha_r) = \frac{1}{r!}\Delta_{\alpha_1}\cdots\Delta_{\alpha_r}(m - m_d - \cdots - m_{r+1})(\lambda),$$

where $m_s(\lambda) = \tilde{m}_s(\lambda,\ldots,\lambda)$. We also write $\xi_s = m_s(1)$, $\Xi_s = \xi_s\Pi$ and

$$\iota(\lambda) = \tilde{m}_1(\lambda) + \tilde{m}_2(1,\lambda) + \cdots + \tilde{m}_d(1,\ldots,1,\lambda).$$

We can conclude the description of the polytope group Π as follows:

(6.10) **Theorem.** *The dilatation operator* m *has the following properties:*

(i) $\xi_1,\ldots,\xi_d$ *are orthogonal idempotents, with* $1 = \xi_1 + \cdots + \xi_d$;
(ii) $\tilde{m}_s(\lambda_1,\ldots,\lambda_s) = \iota(\lambda_1\ldots\lambda_s)\xi_s$;
(iii) $\iota: \mathbb{R} \to E$ *is a (non-trivial) ring homomorphism;*
(iv) E *is isomorphic to the product ring* $\mathbb{R}^d$, *the isomorphism being given by* $(\lambda_1,\ldots,\lambda_d) \mapsto \tilde{m}_1(\lambda_1)\xi_1 + \cdots + \tilde{m}_d(1,\ldots,1,\lambda_d)\xi_d$.

(6.11) **Theorem.** *The polytope group* Π *has the following properties:*

(i) $Z_r = \Xi_r \oplus \Xi_{r+1} \oplus \cdots \oplus \Xi_d$ $(r = 1,\ldots,d)$;
(ii) *each* Ξ_s *is a non-trivial real vector space, generated by the elements* $x_1 \times \cdots \times x_s$, *with* $x_i \in \Xi_1(L_i)$ $(i = 1,\ldots,s)$, *and* $\mathbb{E}^d = L_1 \oplus \cdots \oplus L_s$;
(iii) *the scalar multiplication in* Ξ_s *is given by*
$\lambda(x_1 \times \cdots \times x_s) = (m(\lambda)x_1) \times \cdots \times x_s = \iota(\lambda)(x_1 \times \cdots \times x_s)$.

(iv) *the dilatation operator* m *acts on* Ξ_s *by*
$m(\lambda)x = \iota(\lambda^s)x, \quad x \in \Xi_s.$

Of course, in (6.11) (iii), to which factor x_j we apply $m(\lambda)$ is immaterial. We may remark that Ξ_d is isomorphic to $\mathbb{R}$ under the volume map.

It is now relatively easy to find a suitable family of functionals which separates Π. Let $\mathscr{U}^k$ be the Stiefel manifold of k-frames $U = (u_1,\ldots,u_k)$ of orthonormal sets in $\mathbb{E}^d$. If P is a polytope, we denote by P_u the face of P in direction $u \in \Omega$; that is, the face of P lying in the supporting hyperplane of P with outer normal u. We define $P_U(U = (u_1,\ldots,u_k))$ inductively by

$$P_{(u_1,\ldots,u_k)} = (P_{(u_1,\ldots,u_{k-1})})_{u_k}.$$

A *basic Hadwiger functional of type* r is a function φ_U defined by

$$\varphi_U(P) = \sum_{\varepsilon_i = \pm 1} \varepsilon_1 \cdots \varepsilon_{d-r} V_r(P_{(\varepsilon_1 u_1,\ldots,\varepsilon_{d-r}u_{d-r})})$$

where $U = (u_1,\ldots,u_{d-r}) \in \mathscr{U}^{d-r}$, and V_r is ordinary r-dimensional volume. (In case $r = d$ and $U = \emptyset$, we have $\varphi_\emptyset = V$, ordinary volume.) We observe that $\varphi_U(P) = 0$ unless, for some choice $\varepsilon_i = \pm 1$,

$$\dim P_{(\varepsilon_1 u_1,\ldots,\varepsilon_j u_j)} = d - j$$

for $j = 1,\ldots,d - r$. Thus an unrestricted "linear" combination $\sum_U \varphi_U c_U$ ($c_U \in \mathscr{X}$, where $\mathscr{X}$ is now a real vector space) of basic Hadwiger functionals is always finite-valued on $\mathscr{P}^d$; such a linear combination is called a *Hadwiger functional*. We write $\mathscr{H}_r$ for the real linear space spanned by the basic Hadwiger functionals of type r.

It is easy to verify that a (basic) Hadwiger functional is a simple translation invariant valuation, so that it induces a linear mapping on Π, denoted by the same symbol.

(6.12) **Theorem.** *The Hadwiger functionals separate* Π.

The theorem is trivial for $d = 1$ (the only Hadwiger functionals are multiples of length). One now proceeds by induction. The proof is in two steps. Let H be a (linear) hyperplane, and $e \notin H$ a vector. Let Z be the subgroup of Π generated by the "prisms" $m(\lambda)[e] \times y$, where $y \in \Pi(H)$. If L is any hyperplane not containing e, we can construct homomorphisms $\rho_L : \Pi \to \Pi(L)$ and $\pi_L : \Pi(L) \to \Pi/Z$ as follows. We define ρ_L on $\mathscr{P}^d$ first. Let u be a fixed normal vector to L (say that lying on the same side of L as e); for $P \in \mathscr{P}^d$, define $\rho_L(P) = P_u - P_{-u}$. If $x \in \Pi$ is such that $\varphi(x) = 0$ for all Hadwiger functionals φ, then (by the induction assumption) $\rho_L(x) = 0$ for all L.

We again define π_L on $\mathscr{P}^{d-1}(L)$ first. Let $F \in \mathscr{P}^{d-1}(L)$. Translate F so that it lies in the open half space bounded by H containing e. Let $\bar{F}$ be the convex hull of F and its projection on to H in direction e. Then $\pi_L(F)$ is the equivalence class of $\bar{F}$ in Π modulo Z. It is easy to see that π_L is well defined.

For the equivalence class of $P \in \mathscr{P}^d$ modulo Z, we have

$$[P] = \sum_F \varepsilon\pi(F) \quad (\varepsilon = \pm 1),$$

where $\pi(F) = \pi_L(F)$ if F is parallel to L, and the sum is over the facets of P not parallel to e. We can write this as

$$[P] = \sum_L \pi_L(\rho_L(P)).$$

So, if $x \in \Pi$ is such that $\rho_L(x) = 0$ for all L, we have $x \in Z$ (since its equivalence class modulo Z vanishes).

Now Z is generated by all

$$m(\lambda)[e] \times y_1 \times \cdots \times y_t = [e] \times m(\lambda)y_1 \times \cdots \times y_t,$$

where $H = H_1 \oplus \cdots \oplus H_t$ and $y_i \in \Xi_1(H_i)(i = 1,\ldots,t)$. That is, if $x \in Z$, then $x = [e] \times x^*$ for some $x^* \in \Pi(H)$. But now $\varphi(x) = 0$ for all Hadwiger functionals φ on Π is easily seen to imply $\varphi^*(x^*) = 0$ for all Hadwiger functionals φ^* on $\Pi(H)$, and so, by our induction assumption, $x^* = 0$, and hence $x = 0$. This then proves Theorem (6.12).

Jessen and Thorup [1978] also give further results. For example:

(6.13) **Theorem.** *Let* $x \in \Pi$. *Then* $x \in Z_s$ *if and only if* $\varphi_U(x) = 0$ *for all basic Hadwiger functionals* φ_U *of type* r *with* $r < s$.

In fact, if $x \in \Xi_s$ and φ_U is of type r, then $\varphi_U(x) = 0$ unless $r = s$. In particular, this implies that each Ξ_s $(s = 1,\ldots,d)$ is non-trivial.

We can also show:

(6.14) **Theorem.** *Let* $\mathscr{X}$ *be any real vector space. Then all linear mappings* $\varphi:\Pi \to \mathscr{X}$ *are of the form*

$$\varphi = \textstyle\sum_U \varphi_U c_U$$

where $U \mapsto c_U$ *is an arbitrary function from frames into* $\mathscr{X}$.

Jessen and Thorup [1978] appeal to a "well-known theorem on vector spaces" to prove this, but it is not clear to us what this theorem is. Sah [1979] explicitly constructs bases in duality of Ξ_1 and $\mathscr{H}_1$, and then remarks that these can be used inductively to find bases of Ξ_r and $\mathscr{H}_r$, not now in duality, but nevertheless showing that every (non-trivial) linear function on Ξ_r is in $\mathscr{H}_r$.

To conclude this section, we mention some other results on equidissectability. If G is any group of transformations of $\mathbb{E}^d$ containing T, let $\mathscr{H}(G)$ denote the family of real valued G-invariant Hadwiger functions; that is, $\varphi \in \mathscr{H}(G)$ if $\varphi(P) = \varphi(g(P))$ for each $P \in \mathscr{P}^d$ and $g \in G$. Then we have (Jessen and Thorup [1978]):

(6.15) **Theorem.** *Let* $x \in \Pi_G$. *Then* $x = 0$ *if and only if* $\varphi(x) = 0$ *for each* $\varphi \in \mathscr{H}(G)$.

In most cases of interest, (6.15) is as useful as (6.2). However, in two cases, (6.15) can be applied.

(6.16) **Theorem.** (a) *Two polytopes* P *and* Q *are* TH-*equidissectable if and only if* $\varphi(P) = \varphi(Q)$ *for each* $\varphi \in \mathscr{H}_d + \mathscr{H}_{d-2} + \cdots$

(b) *Let* $\lambda \notin \{0,1,-1\}$, *and let* $G(\lambda)$ *be the subgroup generated by* T *and a dilatation of ratio* λ. *Then* $\Pi_{G(\lambda)} = \{0\}$.

In (a), TH is the group of translations and reflexions in points, and $\mathscr{H}_d + \mathscr{H}_{d-2} + \cdots$ is the sum of the $\mathscr{H}_r$ with $d - r$ even (compare Hadwiger [1952e], see also Harazišvili [1978]).

(b) is due for $\lambda > 0$ to Meier [1972]; it says that any two d-polytopes are $G(\lambda)$-equivalent (see also Zylev [1968], Debrunner [1969], Hadwiger [1974b] and Harazišvili [1977]).

(6.16) (b) clearly implies that any two d-polytopes are A-equidissectable. It is also easy to prove:

(6.17) **Theorem.** *Two* d-*polytopes* P *and* Q *are* EA-*equidissectable if and only if* $V(P) = V(Q)$.

Jessen and Thorup [1978] and Sah [1979] base their description of Π on the fact that a general polytope can be dissected into simplices. Tverberg [1974] gives a particularly nice way of doing this: if a polytope P is not a simplex, it can be cut into two polytopes P_1 and P_2 by a hyperplane spanned by a $(d-2)$-face and a vertex of P; P_1 and P_2 are then treated similarly, and after a finite number of such cuts, P is dissected into simplices. Such a dissection avoids the need to appeal to the inclusion-exclusion principle. Another method of finding a dissection into simplices is given immediately below.

Meier [1977] has a theory of "mixed polytopes", to generalize that of mixed volumes, but it appears that his proof is flawed at one point. We describe here an alternative approach, which in fact employs a more elementary method.

We use the "lifting theorem" of Walkup-Wets [1969], which (for our purposes) states that if a polyhedron Q is the image of a polyhedron P under an affine map Φ, then there is a subcomplex $\mathscr{C}$ of faces of P, such that Φ is one-to-one on set $\mathscr{C}$, and $\Phi(\text{set } \mathscr{C}) = Q$. This same lifting theorem can be used to show that, if X is a finite set in $\mathbb{E}^d$, then the polytope $P = \text{conv } X$ admits a dissection into a simplicial complex, whose 0-cells are just the points of X; this can be proved by more direct methods.

We now apply the lifting theorem to the affine map

$$P_1 \times \cdots \times P_k \mapsto \lambda_1 P_1 + \cdots + \lambda_k P_k,$$

where the term on the right is a Minkowski linear combination of the polytopes P_i with non-negative coefficients λ_i. Since the d-faces of $P_1 \times \cdots \times P_k$ are of the form $F_1 \times \cdots \times F_k$, with F_i a face of $P_i (i = 1,\ldots,k)$ and $\sum_{i=1}^k \dim F_i = d$, we see that $\lambda_1 P_1 + \cdots + \lambda_k P_k$ admits a dissection into cylinders $\lambda_1 F_1 + \cdots + \lambda_k F_k$ (note that some F_i may be vertices, so these are not necessarily k-cylinders). Passing to the polytope group Π, expanding the terms $[\lambda_i F_i] = m(\lambda_i)[F_i]$ into their homogeneous components (in the appropriate polytope groups), and collecting together terms of the same degree, we now see that we have an expression

$$[\lambda_1 P_1 + \cdots + \lambda_k P_k] = \sum_{r_i \geq 0,\, 1 \leq r_1 + \cdots + r_k \leq d} \lambda_1^{r_1} \cdots \lambda_k^{r_k} \binom{r_1 + \cdots + r_k}{r_1 \cdots r_k} \cdot p(P_1,r_1;\cdots;P_k,r_k).$$

The *mixed polytope* $p(P_1,r_1;\ldots;P_k,r_k)$ is positive homogeneous of degree r_i in P_i (and hence independent of P_i if $r_i = 0$); we have taken out the multinomial coefficient as a factor, so that if $P_1 = \cdots = P_k = P$ and $r_1 + \cdots + r_k = r$, then $p(P_1,r_1;\ldots;P_k,r_k)$ is just the r-th homogeneous component (in Ξ_r) of $[P]$. (The notation is again as in (3.3).)

However, it must be emphasized that the mapping

$$(P_1,\ldots,P_k) \mapsto p(P_1,r_1;\ldots;P_k,r_k)$$

is not particularly nice, in that it does not induce a multi-linear mapping on $\Pi \times \cdots \times \Pi$ (k-times). Indeed, if we replace P_j by $-P_j$, we obtain only Euler-type relations, of the kind we discuss in §12 below.

We saw in (6.14) that the linear functionals on Π are precisely the unrestricted linear combinations of the Hadwiger functionals φ_U. It is thus a natural question to ask about (linear) relations between the Hadwiger functionals, or, in other words, about *syzygies* between them. The complete picture is, as yet, unclear, but Sah [1979] gives some partial results.

We first dismiss, with the barest mention, the trivial relationships between φ_U's, when the U's are obtained from each other merely by changing signs.

For the next relations, we recall that if the d-polytope P has n facets with unit outer normal vectors u_i and areas A_i ($i = 1,\ldots,n$), then $\sum_{i=1}^n u_i A_i = 0$. Applying this to the facets of an $(r + 1)$-face, we deduce:

(6.18) **Theorem.** *Let* $U = (u_1,\ldots,u_{d-r-1})$ *be fixed. Then*

$$\sum_u u\, \varphi_{(U,u)} = 0,$$

where the sum extends over all u *orthogonal to* $u_1,\ldots,u_{d-r-1}$.

A $(d - 2)$-face of a d-polytope P is of the form $P_{(u,v)}$ for exactly two (u,v)'s, with u normal to a facet of P, and these pairs span the same 2-dimensional subspace, but have opposite orientation. There easily follows:

(6.19) **Theorem.** *Let* $(u_1,\ldots,u_{s-1},u_{s+2},\ldots,u_{d-r})$ *be fixed. Then*

$$\sum\nolimits^* \varphi_U = 0,$$

where the sum extends over all $U = (u_1,\ldots,u_{d-r})$, *such that* (u_s,u_{s+1}) *is a basis of given orientation of a given 2-dimensional subspace orthogonal to the remaining* u_i's.

There are no other known syzygies. Sah further discusses this topic in a more general (algebraic) context, but we shall refrain from following his example, and instead refer the interested reader to Sah [1979], §5.2–5.8.

§7. Simple valuations

Let φ be a simple valuation, taking values (for the moment) in a divisible abelian group ($\mathbb{Q}$-module) $\mathscr{X}$. There follows directly from (6.7) and the subsequent remarks:

(7.1) **Theorem.** (a) *If* φ *is translation invariant, then for* $P \in \mathscr{P}^d$ *and rational* $\lambda \geq 0$, *there is a polynomial expansion*

$$\varphi(\lambda P) = \sum_{r=1}^{d} \lambda^r \varphi_r(P).$$

(b) *If* φ *is translation covariant, then for such* P *and* λ, *there is a polynomial expansion of degree* $d + 1$.
In each case, the function φ_r *is a translation invariant or covariant valuation on* $\mathscr{P}^d$, *which is (non-negative, rational) homogeneous of degree* r.

Translation covariance will be discussed in §10, where the term is defined.

We need some kind of continuity condition to extend the range of validity of (7.1) beyond the rationals, as the following example illustrates. Let $\theta: \mathbb{R} \to \mathbb{R}$ be a Cauchy-Hamel function:

$$\theta(\alpha + \beta) = \theta(\alpha) + \theta(\beta), \quad \theta(1) = 0;$$

if the axiom of choice is used to construct a basis for $\mathbb{R}$ over $\mathbb{Q}$, non-trivial Cauchy-Hamel functions can be found. Then, with V being as usual volume, $\varphi = \theta \circ V\left(\varphi(P) = \theta(V(P))\right)$ is a rational homogeneous translation invariant valuation which is not real homogeneous.

In this section, we wish only to note one additional result, which justifies our definition in §6 of $m(\lambda)$ for $\lambda < 0$.

(7.2) **Theorem.** *Let* φ *be a translation invariant simple valuation on* $\mathscr{P}^d$, *which is homogeneous of degree* r. *Then*

$$\varphi(-P) = (-1)^{d-r}\varphi(P).$$

In view of the description of the polytope group Π in §6, it follows that the behaviour of φ is completely determined by its behaviour on r-cylinders: $\varphi \equiv 0$ if and only if $\varphi(P) = 0$ for all $P \in \mathfrak{Z}^d_r$. If $\mathbb{E}^d = L_1 \oplus \cdots \oplus L_r$, with $d_i = \dim L_i$ $(i = 1,\ldots,r)$, then for $P = P_1 + \cdots + P_r$, $\varphi(P) = \varphi(P_1 + \cdots + P_r) = \tilde{\varphi}(P_1,\ldots,P_r)$ induces a simple valuation on each $\mathscr{P}(L_i)$, which is homogeneous of degree 1. So, $\varphi(-P) = (-1)^{d-r}\varphi(P)$ will follow from $\tilde{\varphi}(\ldots,-P_i,\ldots) = (-1)^{d_i-1}\tilde{\varphi}(\ldots,P_i,\ldots)$.

The case $r = 1$ is established by induction on d. For $d = 1$ it is trivial ($-P$ is a translate of P). For $d \geq 2$, φ vanishes on $\mathfrak{Z}^d_2$. We now use an argument similar to that of (6.12). Let H be a linear hyperplane, and e a vector not in H. Let L be a general hyperplane not parallel to H. We define φ_L as follows. Translate $F \in \mathscr{P}^{d-1}(L)$ so that it lies on the positive side of H relative to e, let F' be the image of F under parallel projection onto H in direction e, and define $\bar{F} = \operatorname{conv}(F \cup F')$. Since two such $\bar{F}$ differ only by a prism with upright e, $\varphi_L(F) = \varphi(\bar{F})$ is well-defined. Then for $P \in \mathscr{P}^d$, $\varphi(P) = \sum \varepsilon\varphi_L(F)$, where $\varepsilon = \pm 1$ as F is a positive or negative face, and from

$$\begin{aligned}\varphi(-P) &= \sum(-\varepsilon)\varphi_L(-F) = (-1)(-1)^{d-2}\sum \varepsilon\varphi_L(F)\\ &= (-1)^{d-1}\varphi(P)\end{aligned}$$

(7.2) now follows.

Theorem (7.2) also turns out to be at the basis of the Euler-type theorems which we shall discuss in §12.

Theorems (7.1) (with non-negative integer coefficients) and (7.2) apply equally to valuations on lattice polytopes which are invariant under integer translations. In McMullen [1975b, 1977], where these generalizations were first proved, it seemed necessary to assume that the valuation φ satisfied the inclusion-exclusion principle, but Stein [1982] has recently shown how to remove this assumption.

Of course, (7.1) and (7.2) also apply directly to valuations on rational polytopes invariant under rational translations. The extension to such valuations invariant only under integer translations is discussed in McMullen [1978, 1982b].

Theorems (7.1) and (7.2) are due, as are (6.6) and (6.7) on which they depend, to a series of papers culminating in Hadwiger [1957]. However, Hadwiger usually also imposes certain continuity or monotoneity conditions; for a discussion of their implications, see §11 below.

§8. Spherical dissections

It is natural to pose the same questions about equidissectability of polytopes in any spaces in which the question is reasonable; in particular, the spherical (or elliptic) and hyperbolic spaces. What little we shall say about the hyperbolic case we postpone to the end of this section. However, it has become clear that an understanding of the spherical case is a necessary prerequisite for the understanding of the general euclidean case.

We write $\Omega = \Omega^{d-1}$ for the unit sphere in $\mathbb{E}^d$. According to §1, a *spherical polytope* in Ω is an intersection $P = \Omega \cap K$ of Ω with a polyhedral cone K with apex 0. The dimension of P is thus $\dim P = \dim K - 1$. We often identify P with the corresponding cone K; as we shall see, for many reasons, it is more convenient to work with the cones K.

The only groups acting on Ω which we should naturally wish to consider are the full orthogonal group $O = O_d$, and its subgroup $SO = SO_d$ of rotations.

The concepts of equidissectability and O- or SO-equivalence of spherical $(d-1)$-polytopes are defined in the obvious way, and lead immediately to the *spherical polytope group* Σ^d. For $d = 0$, it is convenient to define $\Sigma^0 = \mathbb{Z} = \mathbb{Z}.[\emptyset]$. Clearly, $\Sigma^1 \cong \mathbb{Z}$ also. In what follows, we shall sketch a description of what is currently known about Σ^d; this is largely taken from Sah [1979].

(8.1) **Theorem.** (a) Σ^2 *is divisible;*
(b) Σ^d *is 2-divisible for* $d \geq 3$.

Part (a) is obvious. It is enough to prove (b) for a simplex T. Let T have facets $F_1,\ldots,F_d$, and let $G_{ij} = F_i \cap F_j$ $(i \neq j)$. Let the insphere of T have centre p, and meet F_i in q_i $(i = 1,\ldots,d)$, and for $i \neq j$, define

$$T_{ij} = \operatorname{conv}\{G_{ij}, q_i, q_j, p\}.$$

Then $T = \dot{\bigcup}_{1 \leq i \leq j \leq d} T_{ij}$, and T_{ij} is symmetric in the plane spanned by G_{ij} and p. This proves firstly that T is equidissectable under SO with any of its images under

O (justifying our not mentioning which group we used to define Σ), and consequently that T is itself 2-divisible, since each T_{ij} is.

Whether Σ^d is divisible for $d \geq 4$ is an open and apparently rather deep question (we shall treat the case $d = 3$ shortly).

While the openness of this question clearly prevents our assuming even a $\mathbb{Q}$-linear structure for Σ^d, we do (in a certain sense) have a reduction from odd dimension d to the even dimension next below. This is a consequence of the first of two dissection theorems we now describe.

Let K be any polyhedral set in $\mathbb{E}^d$ (for example, a cone or a polytope), and F a non-empty face of K. Translate K so that $0 \in \text{relint } F$; then the positive hull pos K of K, which is the cone generated by K with apex 0, is a polyhedral cone, which we shall call the *angle cone* of K at F, and denote by A(F,K).

Associated with K is another convex cone, its *recession cone* or *characteristic cone* rec K, which is defined by

$$\text{rec } K = \{x \in \mathbb{E}^d \mid x + y \in K \quad \text{for all} \quad y \in K\}.$$

Thus if K is a polytope, rec K = $\{0\}$, while if K is itself a cone with apex a, rec $K = K - a$.

Using $[\cdot]$ to denote corresponding equivalence classes in Σ^d, we then have the following recently proved (McMullen [1982a]) generalization and abstraction of the theorems of Brianchon-Gram (see Brianchon [1837], Gram [1874], and, for a proof in the present spirit, Shephard [1967]) and Sommerville [1927]:

(8.2) **Theorem.** $\sum_F (-1)^{\dim F}[A(F, K)] = (-1)^d[\text{rec}(-K)]$.

We have kept rec $(-K)$ instead of rec K to emphasize the geometric nature of the dissection, which follows by considering the orthogonal projections of K onto arbitrary hyperplanes.

In the case of cones (Sommerville's theorem), we can apply (8.1) to (8.2) to obtain

(8.3) **Theorem.** *For* d *odd, if* K *is a pointed polyhedral cone in* $\mathbb{E}^d$,

$$[K] = \frac{1}{2} \sum_{F \neq \{0\}} (-1)^{\dim F - 1}[A(F,K)].$$

Let Σ be the direct sum of the Σ^d $(d \geq 0)$. Then Σ admits a product, induced by the cartesian product of cones lying in orthogonal subspaces. We shall denote this product by $*$, so that $[K_1] * [K_2] = [K_1 \times \sigma(K_2)]$, where σ is a suitable rotation taking K_2 into a subspace orthogonal to that carrying K_1. With this product (and the carefully chosen definition of Σ^0), Σ becomes a graded $\mathbb{Z}$-algebra, Σ^d being assigned the degree d.

An r-*fold* (*orthogonal*) *join* is just an r-fold product

$$[K_1] * \cdots * [K_r],$$

with each $\dim K_i \geq 1$; we write Σ_r^d for the subgroup of Σ^d generated by the r-fold joins, and $\Sigma_r = \bigcup_{d \geq 0} \Sigma_r^d$. We then have

$$\Sigma = \Sigma_1 \supseteq \Sigma_2 \supseteq \cdots, \quad \Sigma_{r_1}^{d_1} * \Sigma_{r_2}^{d_2} \subseteq \Sigma_{r_1+r_2}^{d_1+d_2}.$$

We may now observe that each term on the right of (8.3) is a join. Indeed, let us write (with the convention introduced above) $B(F,K) = A(F,K) \cap L$, where $L \subseteq \mathbb{E}^d$ is the orthogonal complement of lin F; we call B(F,K) the *intrinsic inner cone* of K at F. Then for dim $F \geq 1$,

$$A(F,K) = B(F,K) \times \text{lin } F$$

is a non-trivial cartesian product. Yet a third angle function is also useful:

$$\tilde{A}(F,K) = B(F,K) \times \mathbb{E}^{\dim F - 1}$$

for dim $F \geq 1$, which is the corresponding interior angle of the spherical polytope $K \cap \Omega$ at its face $F \cap \Omega$. The mapping $e: \Sigma^d \to \Sigma^{d-1}$ given by

$$e[K] = \sum_{F \neq \{0\}} (-1)^{\dim F - 1} [\tilde{A}(F,K)]$$

is called by Sah (for reasons which are not entirely convincing) the *Gauss-Bonnet map*. Thus, for d odd, (8.3) can be written

$$[K] = p * e([K]),$$

where (illogically) we write p = [point] for the class of a point of Ω, or a ray of some $\mathbb{E}^d$ (we here use the 2-divisibility of Σ). If d is even, Sah [1981] shows that $e([K]) = 0$. Since it subsequently assumes some importance, let us write $\Gamma^d = p * \Sigma^{d-1}$ ($d \geq 1$). Then the foregoing can be summarized as:

(8.4) **Theorem.** *For* $j \geq 0$, $\Sigma^{2j+1} = \Gamma^{2j+1} = p * \Sigma^{2j}$. *The map* $e: \Sigma^{2j+1} \to \Sigma^{2j}$ *is an isomorphism inverse to* $x \mapsto p * x$.

In particular, writing $\Gamma = p * \Sigma$ for the direct sum of the subrings Γ^d ($d \geq 1$), we see that Σ/Γ is evenly graded by degree.

In order to give these results a more concrete interpretation, we introduce the graded volume map. Sah follows Schläfli in normalizing so that the volume of Ω^{d-1} is 2^d. For reasons that will become clearer below, we prefer a different normalization. The easiest way to introduce this is to define

$$\text{vol } K = \int_K \exp(-\pi \|x\|^2)\, dx,$$

where dx is ordinary Lebesgue measure in the linear subspace lin K spanned by the polyhedral cone K with apex 0. We then define the graded volume by

$$\text{gr.vol } [K] = \text{vol } K . T^{\dim K},$$

where T is an indeterminate, and extend to Σ by linearity. The normalization is such that vol L $(= \text{vol}(\Omega \cap L)) = 1$ for all linear subspaces L.

Using volume, we see that (8.3) becomes

$$\text{vol } K = \frac{1}{2} \sum_{F \neq \{0\}} (-1)^{\dim F - 1} \beta(F,K),$$

where we write $\beta(F,K) = \text{vol } A(F,K) = \text{vol } B(F,K)$; the latter follows from the elementary observation

$$\text{vol } [K_1] * [K_2] = \text{vol } [K_1].\text{vol } [K_2].$$

In case $d = \dim K = 3$, this gives the well-known formula expressing the spherical area of a spherical polygon in terms of the angles at its vertices.

We now come to our dissection result. To describe this, we need to introduce another important concept. The *polar* K^0 of a polyhedral cone K with apex 0 in $\mathbb{E}^d$ is defined, as usual, by

$$K^0 = \{x \in \mathbb{E}^d | \langle x,y \rangle \leq 0 \quad \text{for all} \quad y \in K\}.$$

Then $K^{00} = K$. We shall discuss properties of the polarity correspondence (or, rather, a suitable modification of it) as it applies to the group Σ a little later. For the moment, let us define the *normal cone* N(F,K) of a polyhedral set K at its face F to be $(\text{pos } K)^0 = K^0$, where (as previously) we have taken $0 \in \text{relint } F$.

(8.5) **Theorem.** *Let* K *be a polyhedral cone with apex* 0 *in* $\mathbb{E}^d$. *Then the cones* F *and* N(F,K) (F *a face of* K) *are orthogonal, and* $\mathbb{E}^d$ *is dissected into the cones* $F \times N(F,K)$.

In terms of Σ, this gives

$$[\mathbb{E}^d] = \sum_F [F] * [N(F,K)].$$

The proof of (8.5) is immediate, on noticing that an arbitrary point $z \in \mathbb{E}^d$ can be uniquely represented in the form $z = x + y$, where, for some face F of K (possibly K itself) $x \in \text{relint } F$ and $y \in N(F,K)$; x is the (unique) nearest point of K to z.

There are two other equidissection results that we shall want to use later, at least on the level of (graded) volume. To prove these results, it is convenient to introduce a little more notation.

We shall write $b(F,K) = [B(F,K)]$, and adopt a similar convention for other such cones. The *intrinsic outer cone* C(F,K) to K at F is just $C(F,K) = N(F,K) \cap \text{lin } K$; then C(F,K) is the polar of B(F,K) with respect to its linear hull. We let Z(F,K) be the orthogonal complement of lin F in lin K, and (with $z(F,K) = [Z(F,K)]$ as above) we define

$$m(F,K) = (-1)^{\dim Z(F,K)} z(F,K).$$

As we shall shortly see, z and m are Σ-valued analogues of the zeta and Möbius functions of Rota [1964] and McMullen [1975b]. Indeed, $*$ induces a multiplication on the Σ-valued functions f on pairs of cones, which are such that $f(F,G) = 0$ unless F is a face of G, by

$$f * g(F,G) = \sum_J f(F,J) * g(J,G),$$

the sum extending over all faces J of G. We then have from Euler's theorem:

(8.6) **Theorem.** $m * z = i = z * m$, *where* i *is the identity function; that is,* $i(F,G) = 0$ *if* $F \neq G$, *and* $1 (\in \mathbb{Z})$ *if* $F = G$.

We can now rephrase (8.2) (for cones) and (8.5) as

(8.7) **Theorem.** (a) $m * b = \tilde{b}$, *where* $\tilde{b}(F,G) = (-1)^{\dim G - \dim F} b(F,G)$;
(b) $b * c = z$.

Next, we can paraphrase the arguments of McMullen [1975b], to prove new equidissectability results by purely algebraic means. For, let us similarly write $\tilde{c}(F,G) = (-1)^{\dim G - \dim F} c(F,G)$. Then we have:

(8.8) **Theorem.** (a) $\tilde{b} * c = i = b * \tilde{c}$;
(b) $c * \tilde{b} = i\ (= \tilde{c} * b)$.

For, (8.7)(a) and polarity yields $\tilde{c} = c * m$. (8.8)(a) follows at once from (8.7)(b), since $\tilde{b} * c = m * b * c = m * z = i$; the other relation is proved similarly, or by another appeal to polarity. (8.8)(b) needs a further remark. Confining attention to pairs of faces of a fixed polyhedral cone, we see that $\tilde{b}$ and c correspond to triangular matrices with values in the ring Σ, whose diagonal entries are 1. So, $\tilde{b}$ and c are both invertible. But since $\tilde{b} * c = i$, we have $c * \tilde{b} = i$ also. (The other relation of (8.8)(b) is redundant, since $\tilde{c} * b = c * m * b = c * \tilde{b}$.)

Note that (8.8) (a) implies

$$0 = \sum_F (-1)^{\dim F} [F] * [N(F,K)]$$

whenever K is not a subspace. By adding and subtracting this and the relation of (8.5), we obtain generalizations of Sah [1979], Proposition 6.3.5. In odd dimensions, we obtain another way of expressing a cone K in terms of product cones, while in even dimensions, we get a relation between K, K^0 and product cones, whose components are also even dimensional.

We shall need the metrical consequences of the relation of (8.5), which can be written $b * c = z$, and those of (8.8), in §10 below.

(8.9) **Theorem.** *Let* K *be a pointed polyhedral* d*-cone with apex* 0 ($d \geq 1$). Then,

(a) $\sum_F \beta(0,F)\gamma(F,K) = 1$;

(b) $\sum_F (-1)^{\dim F} \beta(0,F)\gamma(F,K) = 0$;

(c) $\sum_F (-1)^{\dim K - \dim F} \gamma(0,F)\beta(F,K) = 0$.

Here, $\gamma(F,G) = \operatorname{vol} c(F,G)$ is the normalized *external angle* of the polyhedral set G at its face F (as used in §3).

Polarity plays a further role in Σ. To begin our discussion, we remark:

(8.10) **Lemma.** *Let* K_1, K_2 *be polyhedral cones in* $\mathbb{E}^d$ *with apex* 0. *Then*

$$(K_1 \cap K_2)^0 = K_1^0 + K_2^0,$$

where the sum is Minkowski addition.

Now, if $K_1 \cup K_2$ is also convex, we have $K_1 \cup K_2 = K_1 + K_2$. Further, if dim $K < d$, then K^0 has the orthogonal complement $(\text{lin } K)^\perp$ as its non-trivial face of apices, and so $[K^0] \in \Gamma^d$; the converse also holds. In other words:

(8.11) **Theorem.** *The polarity correspondence induces an involutory automorphism* § of Σ^d/Γ^d, *defined by*

$$[K]^{\S} = [K^0].$$

This automorphism, which for reasons explained below we call the *antipodal map*, extends to Σ/Γ. The algebra structure on Σ, with multiplication $*$, induces an algebra structure on Σ/Γ. To complete the picture, we now describe a coalgebra structure on Σ/Γ.

In fact, we begin with something more general. For $x \in \Sigma$, write $\bar{x} \in \Sigma/\Gamma$ for its representative. The *total spherical Dehn invariant* of a pointed polyhedral cone K (or corresponding spherical polytope) is

$$\Psi_S(K) = \sum_F^* [F] \otimes \overline{[B(F,K)]} \in \Sigma \otimes (\Sigma/\Gamma),$$

where the sum extends over all faces $F \neq \{0\}$ of K with dim K − dim F even. (In fact, the terms with dim K − dim F odd automatically drop out, due to the even grading of Σ/Γ. The term $F = \{0\}$, in case dim K is even, is not needed, since the information contained in the first term $K \otimes \overline{[\emptyset]}$ includes that contained in $[\emptyset] \otimes \overline{[K]}$.)

The map Ψ_S then induces a map $\bar{\Psi}_S : \Sigma/\Gamma \to (\Sigma/\Gamma) \otimes (\Sigma/\Gamma)$; defined on a generator $\overline{[K]}$ of Σ/Γ, its effect is

$$\bar{\Psi}_S \overline{[K]} = \sum_F^* \overline{[F]} \otimes \overline{[B(F,K)]}.$$

This is the *comultiplication on* Σ/Γ. The *counit* or *augmentation* (which is dual to the unit in an algebra) is the natural mapping whose kernel is the set of elements of Σ/Γ of positive degree. With these algebra and coalgebra structures, and the antipodal map, Σ/Γ then becomes a Hopf algebra (see Sah [1979]).

We are now in a position to discuss equidissectability in general. As with the case of the group Π^d, we look for a suitable separating family of functions on Σ^d. The map Ψ_S does separate Σ^d, but in a rather trivial way, since it is obviously injective. More to the point is the *total classical spherical Dehn invariant*

$$\Phi_S = (\text{gr. vol} \otimes \text{id}) \circ \Psi_S,$$

so that

$$\Phi_S(K) = \sum_F^* (\text{vol } F \cdot T^{\dim F}) \otimes \overline{[B(F,K)]} \in \mathbb{R}[T] \otimes (\Sigma/\Gamma).$$

The summation convention is that used in defining Ψ_S above.

Unfortunately, the general equidissectability problem is far from solved.

(8.12) **Theorem.** Φ_S *separates* Σ^d *for* $d = 2$ *and* 3.

All this says is that arc length (in case d = 2) or area (for d = 3) characterize spherical polytopes up to equidissectability.

For $d \geq 4$, we encounter severe problems; the case d = 4 already illustrates most of these. For, we observe that, if K has rational dihedral angles, then $\Phi_S(K) = \text{vol}\, K.T^4$, and the lower degree terms vanish identically. So, if Φ_S does separate Σ^4, then such K should be equidissectable with product cones. However, this is far from obviously true in general, although Dupont and Sah [1982] have recently shown that it is true in one important case.

(8.13) **Theorem.** *A fundamental polyhedral cone for a finite orthogonal group in* $\mathbb{E}^d$ *is equidissectable with a* (d − 2)*-fold product cone over a planar cone.*

In fact, what Dupont and Sah prove is that fundamental polyhedral cones for two finite orthogonal groups in $\mathbb{E}^d$ of the same order are equidissectable. The result ultimately reduces to proving this for the special case of p-groups (Sylow subgroups) of the same order.

But the general problem remains unsolved; indeed, polyhedral cones with rational dihedral angles are a possible source of torsion in Σ^4 (Sah [1979]).

As we said at the beginning, there is also some interest in the hyperbolic case. As with the spherical case, in the hyperbolic line or plane, equality of length or area is a necessary and sufficient condition for equidissectability. Once again, the real problems begin in the next dimension.

In general, as with the spherical case, there is a difference between odd and even dimensional hyperbolic spaces H^d. The "Gauss-Bonnet" map can be defined equally well for hyperbolic polytopes, and associates with an element x of the polytope group in H^d (with respect, of course, to the group of all hyperbolic motions) an element e(x) of the polytope group in Ω^{d-1}. On the level of volume, we have

$$\text{vol}_{d-1}(e(x)) = (-1)^{d/2}\chi(\Omega^d)\text{vol}_d(x).$$

where $\chi(\Omega^d) = 1 + (-1)^d$ is the Euler characteristic of Ω^d (in the spherical case, the factor $(-1)^{d/2}$ is omitted). It is possible that there are deeper connexions, but these have not been much investigated.

Of course, if d is odd, the above formula yields no information. In recent years, much attention has been payed to the first non-trivial case d = 3, but even so, it seems not to be near complete solution.

One can further consider equidissectability in $\bar{H}^d = H^d \cup \partial H^d$, obtained by adding to H^d the ideal points (at infinity) forming ∂H^d. However, we are going a little far from our topic, so we shall merely refer the reader to Sah [1979; 1981], Dupont-Sah [1982], and the bibliographies contained therein.

§9. Hilbert's third problem

Discussion of the spherical problem was a prerequisite for consideration of the euclidean equidissectability problem, as we shall see. From an historical point of view, of course, we have things back to front, as the euclidean problem was investigated earlier. The case of dimension d = 1 is trivial, and the case d = 2 is nearly as easy. As Gerwien [1833a] and F. Bólyai observed, a triangle is D- (or

SD- or even TH-) equidissectable with a parallelogram of the same area. As a consequence, and using the results of §6:

(9.1) **Theorem.** *Two polygons in* $\mathbb{E}^2$ *are* D-*equidissectable if and only if they have the same area.*

In a later extension of this result, Hadwiger-Glur [1951] showed that, if G is a group such that two polygons with the same area are always G-equidissectable, then $G \supseteq TH$.

However, it was early recognized that the situation in $\mathbb{E}^3$ was likely to be different, and in 1900 Hilbert [1900] formally posed the problem of finding two 3-polytopes of the same volume (specifically, pyramids with the same height on the same base) which were not D-equidissectable. Modifying an earlier incorrect attempt of Bricard [1896], Dehn [1900], [1902] found a suitable pair of examples in the same year the problem was posed.

Before going further into Dehn's examples, it will be helpful to make some remarks about the appropriate polytope group Π_D^d. (Once more, we distinguish the dimension d.) Firstly, the fact that reflected polytopes are SD-equidissectable shows that $\Pi_D^d = \Pi_{SD}^d$. Secondly, dilatation commutes with isometry (at least, modulo translation), so that Π_D^d as a quotient space of Π_T^d also admits a grading

$$\Pi_D^d = \Xi_1^d \oplus \cdots \oplus \Xi_d^d.$$

However, the presence of scaling by -1 ensures that $\Xi_r \equiv 0$ unless $d - r \equiv 0 \pmod 2$.

Writing $\Pi_D = \Sigma_{d \geq 0} \Pi_D^d$ (with $\Pi_D^0 = \mathbb{R}$), we see that we have a natural product structure induced by orthogonal cartesian product, which we denote by $\times$.

We now describe what are conjectured to be a separating family of functionals on Π_D. These are the euclidean Dehn invariants, which are exact analogues of the spherical Dehn invariants. The *total euclidean Dehn invariant* of a polytope P is

$$\Psi_E(P) = \sum_F^{*} [F] \otimes \overline{[B(F,P)]} \in \Pi_D \otimes (\Sigma/\Gamma),$$

where the sum extends over all faces F of P with ($\dim F > 0$ and) $\dim P - \dim F$ even. Similarly, the *classical total euclidean Dehn invariant* is just

$$\Phi_E = (\mathrm{gr.vol} \otimes \overline{\mathrm{id}}) \circ \Psi_E,$$

so that

$$\Phi_E(P) = \sum_F^{*} \mathrm{vol}\, F . T^{\dim F} \otimes \overline{[B(F,P)]}.$$

Theorem (9.1) shows that Φ_E separates Π_D^d for $d \leq 2$. The considerable achievement of Sydler [1965] was to extend this to $d = 3$; Jessen [1972] was then able to use Sydler's result further to extend this to $d = 4$. In fact, Jessen [1968] was also able to simplify Sydler's original proof, essentially by using the language of the algebra of polytopes.

It would be inappropriate to give full details of these proofs here, but we can point out some of the salient features.

(9.2) **Theorem.** Φ_E *separates* Π_D^3.

It is easy to see that the group Π_D^d is generated by the (equivalence classes of) *orthogonal* simplices $[a_1,\ldots,a_d]$, which are such that $\{a_1,\ldots,a_d\}$ is an orthogonal set of vectors. In the particular case $d = 3$, Sydler's proof of Theorem (9.2) depends upon considering those particular orthogonal simplices $T(\xi,\eta) = [a_1,a_2,a_3]$, where

$$\|a_1\| = \left(\frac{1-\xi}{\xi}\right)^{1/2}, \quad \|a_3\| = \left(\frac{1-\eta}{\eta}\right)^{1/2}, \quad \|a_2\| = \|a_1\|\,\|a_3\|,$$

with $\xi,\eta \in\,]0,1[$. The more important and less obvious of Sydler's results is

(9.3) **Lemma.** *For* $\xi,\eta_1,\eta_2 \in\,]0,1[$,

$$T(\xi,\eta_1) + T(\xi\eta_1,\eta_2) = T(\xi,\eta_2) + T(\xi\eta_2,\eta_1).$$

The proof of this result relies on equicomplementability implying equidissectability. Another result, which follows from dissecting one tetrahedron in two ways, is much easier.

(9.4) **Lemma.** *For* $\xi,\eta,\zeta > 0$,

$$\mu(\xi)T\left(\frac{\xi+\eta}{\xi+\eta+\zeta},\frac{\xi}{\xi+\eta}\right) + \mu(\eta)T\left(\frac{\xi+\eta}{\xi+\eta+\zeta},\frac{\eta}{\xi+\eta}\right)$$
$$= \mu(\xi)T\left(\frac{\xi+\zeta}{\xi+\eta+\zeta},\frac{\xi}{\xi+\zeta}\right) + \mu(\xi)T\left(\frac{\xi+\zeta}{\xi+\eta+\zeta},\frac{\zeta}{\xi+\zeta}\right).$$

Finally, there is a further easy result.

(9.5) **Lemma.** *Let* $\alpha,\beta,\gamma \in\,]0,\tfrac{1}{2}\pi[$, *with* $\alpha+\beta+\gamma=\pi$. *Then there is a rectangular parallelepiped, which when dissected by the three planes containing a fixed diagonal and pairs of opposite edges of the box, yields orthogonal tetrahedra whose dihedral angles at the diagonal are* $\alpha,\beta,\gamma,\alpha,\beta,\gamma$.

Now Theorem (9.2) reduces to the following assertion.

(9.6) **Lemma.** *Let* $\varphi:\Xi_1^3 \to Y$ *be a linear map into a real linear space* Y. *Then there is a linear map* $\vartheta:\mathbb{R}\otimes(\mathbb{R}/\mathbb{Z}) \to Y$, *such that*

$$\varphi = \vartheta \circ \Phi^{(1)},$$

where $\Phi_E = \Phi^{(3)}T^3 + \Phi^{(1)}T$ *on* Π_D^3.

In particular, with $Y = \Xi_1^3$, this shows that $\Phi^{(1)}$ is a monomorphism. Note that the real linear structure of $\mathbb{R}\otimes(\mathbb{R}/\mathbb{Z})$ is inherited from its first component.

If we write $[T(\xi,\eta)] = t_3(\xi,\eta) + t_1(\xi,\eta)$ with $t_r(\xi,\eta) \in \Xi_r$ $(r = 1,3)$, and $\varphi(\xi,\eta)$ for $\varphi(t_1(\xi,\eta))$, then (9.3) and (9.4) yield corresponding functional relations for φ. From

(9.3), there is an $f:]0,1[\to Y$, such that

$$\varphi(\xi,\eta) = f(\xi) + f(\eta) - f(\xi\eta).$$

(Curiously, the proof of this fact, and that to follow, need the axiom of choice.) From f, one defines $G:\mathbb{R}_+ \times \mathbb{R}_+ \to Y$ by

$$G(\xi,\eta) = \xi f\left(\frac{\xi}{\xi+\eta}\right) + \eta f\left(\frac{\eta}{\xi+\eta}\right),$$

so that (9.4) and its definition yield

$$G(\xi,\eta) = G(\eta,\xi), \quad G(\lambda\xi,\lambda\eta) = \lambda G(\xi,\eta) \quad (\lambda > 0)$$

$$G(\xi,\eta) + G(\xi+\eta,\zeta) = G(\xi+\zeta,\eta) + G(\xi,\zeta).$$

G can be extended to $\mathbb{R} \times \mathbb{R}$ satisfying the same functional equations. Now we find $g_1:\mathbb{R}_+ \to Y$, such that

$$g_1(\xi\eta) = \eta g_1(\xi) + \xi g_1(\eta),$$

and

$$G(\xi,\eta) = g_1(\xi) + g_1(\eta) - g_1(\xi+\eta).$$

Writing $g_1(\xi) = \xi g(\xi)$ $(\xi > 0)$, we have $g:\mathbb{R}_+ \to Y$, such that

$$g(\xi\eta) = g(\xi) + g(\eta),$$

$$G(\xi,\eta) = \xi g(\xi) + \eta g(\eta) - (\xi+\eta)g(\xi+\eta).$$

We note that $g(1) = 0$. Now, if $\xi,\eta > 0$ with $\xi + \eta = 1$,

$$\xi f(\xi) + \eta f(\eta) = G(\xi,\eta) = \xi g(\xi) + \eta g(\eta).$$

If $h = f - g:]0,1[\to Y$, then

$$\varphi(\xi,\eta) = h(\xi) + h(\eta) - h(\xi\eta),$$

with

$$\xi h(\xi) + \eta h(\eta) = 0 \quad \text{if} \quad \xi,\eta > 0, \quad \xi + \eta = 1.$$

We complete the proof by defining $\sigma:\mathbb{R} \to Y$, where σ is periodic with period $\frac{1}{2}\pi$, by

$$\sigma(\alpha) = \tan\alpha\, h(\sin^2\alpha), \quad 0 < \alpha < \tfrac{1}{2}\pi$$

with $\sigma(\frac{1}{2}\pi) = 0$. That σ is additive follows from the above and (9.5), which gives

$$\sigma(\alpha) + \sigma(\beta) + \sigma(\gamma) = 0, \quad 0 < \alpha,\beta,\gamma < \tfrac{1}{2}\pi, \quad \alpha + \beta + \gamma = \pi.$$

From its definition follows

$$\varphi(\xi,\eta) = \cot\alpha\, \sigma(\alpha) + \cot\beta\, \sigma(\beta) - \cot(\alpha * \beta)\sigma(\alpha * \beta),$$

where

$$\xi = \sin^2\alpha, \quad \eta = \sin^2\beta$$

and

$$\sin(\alpha * \beta) = \sin\alpha \sin\beta, \quad 0 < \alpha,\beta,\alpha * \beta < \tfrac{1}{2}\pi.$$

Finally, if we define $\theta: \mathbb{R} \otimes (\mathbb{R}/\mathbb{Z}) \to \mathrm{Y}$ by

$$\theta(\xi \otimes \omega) = \xi\sigma(\tfrac{1}{2}\pi\omega),$$

then θ is the required linear map.

The extension by Jessen [1972] easily follows.

(9.7) **Theorem.** Φ_E *separates* Π_D^4.

We have $\Pi_D^4 = \Xi_2^4 \oplus \Xi_4^4$, so that every $x \in \Pi_D^4$ is equivalent to a prism $e \times y$, where e is a unit segment. (Jessen [1972] gives a direct proof of this). The result now follows directly from (9.2).

We end this section by making a few remarks, and stating some problems. A fact we have used above is that the Dehn invariant Φ_E is compatible with the product structure. Indeed,

$$\Psi_E(P \times Q) = \Psi_E(P)\Psi_E(Q),$$

since the angle $P \times Q$ at its face $F \times G$ is

$$B(F \times G, P \times Q) = B(F,P) \times B(G,Q).$$

A general question raised by the proof of (9.7) is:

(9.8) **Problem.** *Is every even-dimensional polytope equivalent to a direct sum of products of odd-dimensional components?*

For example, $\Pi_D^2 = \Xi_2^2 \cong \Xi_1^1 \otimes \Xi_1^1$. Further, $\Pi_D^4 = \Xi_2^4 \oplus \Xi_4^4$. Now $\Xi_4^4 \cong \otimes^4 \Xi_1^1$, and Ξ_2^4 is generated by $\Xi_1^1 \otimes \Xi_1^3$ and $\Xi_1^2 \otimes \Xi_1^2$, the latter term vanishing, so the result holds here also. Certainly, the space of indecomposable elements of Π_D is the sum of the spaces Ξ_1^{2s+1}. So, a variant of (9.8) is:

(9.9) **Problem.** *Is* Π_D *isomorphic to a symmetric algebra based on the space of indecomposable elements?*

In particular:

(9.10) **Problem.** *Is* Π_D *an integral domain? Is* Π_D *a Hopf algebra?*

Finally, related to (9.6), there is a question about Dehn invariants.

(9.11) **Problem.** *Is* $\Phi^{(1)}: \Xi_1^3 \to \mathbb{R} \otimes (\mathbb{R}/\mathbb{Z})$ *an isomorphism? More generally, what are the images of the Dehn invariants?*

Those readers who, after their excursion into dissection theory, might wish some recreation, are referred to the amusing book of Lindgren [1972].

III. General valuations

§10. Polynomial expansions for general valuations

While the polynomial expansion of $\varphi(\lambda P)$ for a simple (translation invariant) valuation φ dates back, as we have seen, to Hadwiger [1957], the question of the existence of such polynomial expansions for general valuations was settled much more recently. The question was posed (in the context of continuous or monotone valuations, with real coefficients) by McMullen at the Oberwolfach meeting on Convex Bodies in 1974, and settled by him the same year (see McMullen [1975a; 1977]). Somewhat later, but independently, Spiegel [1978] and Meier [1977] gave proofs using different approaches; in retrospect, their proofs should have been available earlier, since they involve fairly elementary modifications of the ideas of §7. However, McMullen's approach yields considerably more, and so we shall largely follow that here.

The basic idea is to use the angle-sum relations described in §8, to relate a given valuation φ (which for simplicity of exposition will always take values in a real vector space $\mathscr{X}$) to a family of simple valuations. Let $\mathscr{A}$ be the class of translates of a given flat A in $\mathbb{E}^d$ (possibly $A = \mathbb{E}^d$ itself), and write $\mathscr{P}(\mathscr{A})$ for the family of polytopes P such that aff P is a subflat of some flat in $\mathscr{A}$.

(10.1) **Lemma.** *Let φ be a translation invariant valuation on $\mathscr{P}^d$. Let the function $\psi_{\mathscr{A}}$ be defined by*

$$\psi_{\mathscr{A}}(P) = \begin{cases} \sum_F (-1)^{\dim P - \dim F} \beta(F,P)\varphi(F), & \textit{if} \quad \operatorname{aff} P \in \mathscr{A}, \\ 0, & \textit{otherwise.} \end{cases}$$

The $\psi_{\mathscr{A}}$ is a simple translation invariant valuation on $\mathscr{P}(\mathscr{A})$.

We can use (8.9) to invert (10.1); the converse result was first proved by Hadwiger [1953b; 1957].

(10.2) **Lemma.** *For each translation class $\mathscr{A}$ of flats in $\mathbb{E}^d$, let $\psi_{\mathscr{A}}$ be a simple translation invariant valuation on $\mathscr{P}(\mathscr{A})$. Define $\psi(P) = \psi_{[\operatorname{aff} P]}(P)$, where* [A] *is the translation class of* A, *and*

$$\varphi(P) = \sum_F \gamma(F,P)\psi(F).$$

Then $\psi_{\mathscr{A}}$ is a simple translation invariant valuation on $\mathscr{P}(\mathscr{A})$.

The proofs of (10.1) and (10.2) are straightforward but tedious.

From (7.1a), (10.1) and (10.2) lead at once to

(10.3) **Theorem.** *Let φ be a translation invariant valuation on $\mathscr{P}^d$. Then for rational $\lambda \geq 0$,*

$$\varphi(\lambda P) = \sum_{r=0}^{d} \lambda^r \varphi_r(P).$$

The coefficient $\varphi_r(P)$ *(which is independent of* λ*) is a translation invariant valuation on* $\mathscr{P}^d$*, which is homogeneous of degree* r.

In particular, $\varphi_0(P) = \varphi(\{x\})$ is the value taken by φ on a point.

We extend (10.3) to general linear combinations by means of (1.4). Let φ be a translation invariant valuation on $\mathscr{P}^d$, and let $Q \in \mathscr{P}^d$ be fixed. Then by (1.4) the function ψ defined by $\psi(P) = \varphi(P + Q)$ is also a translation invariant valuation on $\mathscr{P}^d$.

We deduce immediately the general result:

(10.4) **Theorem.** *Let* φ *be a translation invariant valuation on* $\mathscr{P}^d$*. Then for* $P_1,\ldots,P_k \in \mathscr{P}^d$ *and rational* $\lambda_1,\ldots,\lambda_k \geq 0$, $\varphi(\lambda_1 P_1 + \cdots + \lambda_k P_k)$ *is a polynomial in* $\lambda_1,\ldots,\lambda_k$ *of total degree at most* d. *The coefficient of* $\lambda_1^{r_1} \cdots \lambda_k^{r_k}$ *is a translation invariant valuation in* P_j *which is homogeneous of degree* r_j $(j = 1,\ldots,k)$.

We call these coefficients *mixed valuations.* If φ_r is the homogeneous valuation occurring in (10.3), so that $\varphi = \Sigma_{r=0}^{d} \varphi_r$, we may write

$$\varphi_r(\lambda_1 P_1 + \cdots + \lambda_k P_k) = \sum \binom{r}{r_1 \ldots r_k} \lambda_1^{r_1} \ldots \lambda_k^{r_k} \varphi_r(P_1,r_1;\ldots;P_k,r_k),$$

in analogy to (3.2).

Let us briefly survey the other two approaches. Spiegel [1978] uses the canonical simplex dissection, the inclusion-exclusion principle and induction on the dimension to obtain a direct proof. Meier [1977], which we have mentioned earlier, uses his mixed polytopes and the inclusion-exclusion principle to obtain the same result. (In fact, these approaches cover valuations taking values in a rational vector space, while McMullen's needs a real vector space; the modification of Meier's argument outlined in §6 possibly shows this most easily.)

The discussion of covariant valuations proceeds on very similar lines. We say that a valuation $\varphi: \mathscr{P}^d \to \mathscr{X}$ is *translation covariant* if there exists an *associated* function $\Phi: \mathscr{P}^d \to \mathrm{Hom}_{\mathbb{Z}}(\mathbb{E}^d,\mathscr{X}) = \mathrm{Hom}_{\mathbb{Q}}(\mathbb{E}^d,\mathscr{X})$, such that

$$\varphi(P + t) - \varphi(P) = \Phi(P)t,$$

for $t \in \mathbb{E}^d$ and $P \in \mathscr{P}^d$.

(10.5) **Lemma.** *Let* φ *be a translation covariant valuation on* $\mathscr{P}^d$*. Then the associated function* Φ *is a translation invariant valuation on* $\mathscr{P}^d$.

Lemmas (10.1) and (10.2) carry over at once to covariant valuations. So, all that is needed is to appeal to the expression after (6.7) for the specific translations involved in the canonical simplex dissection, in order to generalize (10.3) and (10.4), and show:

(10.6) **Theorem.** *Let* φ *be a translation covariant valuation on* $\mathscr{P}^d$*. Then for* $P_1,\ldots,P_k \in \mathscr{P}^d$ *and rational* $\lambda_1,\ldots,\lambda_k \geq 0$, $\varphi(\lambda_1 P_1 + \cdots + \lambda_k P_k)$ *is a poly-*

nomial in $\lambda_1,\ldots,\lambda_k$ *of total degree at most* d + 1. *The coefficient of* $\lambda_1^{r_1}\cdots\lambda_k^{r_k}$ *is a translation covariant valuation in* P_j *which is homogeneous of degree* r_j ($j = 1,\ldots,k$).

The degree d + 1 (rather than d, as in (10.4)) arises from the translations.

The results discussed above can be generalized in various ways. The main way is by restricting the translations which are allowed. For several reasons, the most interesting cases concern the translation subgroups $\mathbb{Q}^d$ and $\mathbb{Z}^d$ (or, of course, isomorphic subgroups). It might be thought that the case $\mathbb{F}^d$, for a general subfield $\mathbb{F}$ of $\mathbb{R}$, would also be of interest; however, analogues of the weak continuity condition to be discussed in §11 are needed to extend $\mathbb{Q}$-linearity to $\mathbb{F}$-linearity.

The case of the translation group $\mathbb{Q}^d$ and rational polytopes needs no further comment; the results above carry over with no change of language. When the translation group is $\mathbb{Z}^d$ and the polytopes are lattice polytopes, (10.4) and (10.6) remain valid with integers λ_i (McMullen [1977], and using Stein [1982]; see also Bernštein [1976] for the special case of the lattice point enumerator). More generally, though, some changes are needed.

Let P be a rational polytope. The r-*index* $\mathrm{ind}_r(P)$ is the smallest positive integer m, such that, for every r-face F of P, aff (mF) is spanned by lattice points (or, equivalently, contains a lattice point). Then we have:

(10.7) **Theorem.** *Let* φ *be a lattice translation invariant valuation on rational polytopes in* $\mathbb{E}^d$. *Then for rational polytopes* P *and integer* $n \geq 0$, *there is an expression*

$$\varphi(nP) = \sum_{r=0}^{d} n^r \varphi_r(P,n),$$

where $\varphi_r(P,n)$ *depends only on the congruence class of n modulo* $\mathrm{ind}_r(P)$.

Such an expression is called a *near-polynomial* in McMullen [1978], where this theorem is proved (and "polynome mixte" by Ehrhart [1967a] in the case of the lattice point enumerator G). The coefficient φ_r is *near-homogeneous* of degree r, in that

$$\varphi_r(mP,n) = m^r \varphi_r(P,mn)$$

for all integer $m,n \geq 0$. There are analogous expressions for integer combinations of rational polytopes; similarly, (10.7) and its generalizations are valid for lattice translation covariant valuations (McMullen [1982b]).

The proof of (10.7) depends upon a specific representation of the valuation, and so we defer further discussion until §17.

In view of (10.4) and (10.6), a translation invariant or covariant valuation on $\mathscr{P}^d$ which is homogeneous of degree 1 is also Minkowski additive:

$$\varphi(P + Q) = \varphi(P) + \varphi(Q);$$

this was earlier proved (for a special case of covariance) by Spiegel [1976a]. We have already remarked on the converse in (1.3).

§11. Additional properties

We now complete the discussion of the original question of McMullen, by considering various continuity conditions. As usual, continuity of functions on $\mathscr{P}^d$ or $\mathscr{K}^d$ will be with respect to the Hausdorff metric.

A different concept of continuity is due to Hadwiger [1952e]. Let $U = (u_1,\ldots,u_n)$ be a (for the moment) fixed set of unit outer normal vectors, and write $\mathscr{P}^d(U)$ for the family of polytopes of the form

$$P(y) = \{x \in \mathbb{E}^d \mid \langle x,u_i\rangle \leq \eta_i \quad (i = 1,\ldots,n)\},$$

where $y = (\eta_1,\ldots,\eta_n)$. We call a function φ on $\mathscr{P}^d$ *weakly continuous* if, for each such U, the function φ_U defined by $\varphi_U(y) = \varphi(P(y))$ is continuous. Clearly we have:

(11.1) **Lemma.** *A continuous function on* $\mathscr{P}^d$ *is weakly continuous.*

It turns out that, to extend (10.4) and (10.6) to real (rather than rational) coefficients, all that is needed is weak continuity. In fact, we have:

(11.2) **Theorem.** *The following conditions on a translation invariant or covariant valuation* φ *are equivalent:*

(a) φ *is weakly continuous;*
(b) *for all polytopes* $P_1,\ldots,P_k$ *and all real numbers* $\lambda_1,\ldots,\lambda_k \geq 0$, $\varphi(\lambda_1 P_1 + \cdots + \lambda_k P_k)$ *is a polynomial in* $\lambda_1,\ldots,\lambda_k$.
(c) *for each* U, *the one-sided partial derivatives of* φ_U *exist.*
(d) φ *is continuous under dilatations; that is, the mapping* Θ_P *on* $\mathbb{R}$ *defined by* $\Theta_P(\lambda) = \varphi(\lambda P)$ *is continuous for each* P.

In fact, condition (c) can be replaced by
(c′) *for each* U *and each* $a \in \mathbb{E}^d$, *the (one-sided) derivative of* φ_U *in direction* a *exists.*

The equivalence of (a), (b) and (c) is due to McMullen [1977]. The equivalence of (a) and (d), which was left as an open problem by Hadwiger [1952e], follows from the inversion formulae (10.1) and (10.2), and the fact that Π^d is a real vector space (see also (11.4) below).

As far as (weakly) continuous valuations are concerned, one general remark is often useful.

(11.4) **Theorem.** *The mixed valuations associated with a (weakly) continuous translation invariant or covariant valuation are (weakly) continuous in each of their arguments.*

To what extent this result extends to monotone valuations (which we shall next consider) is unknown. We call a function φ on $\mathscr{P}^d$ or $\mathscr{K}^d$ taking values in a partially ordered real linear space *monotone* if $\varphi(P) \leq \varphi(Q)$ whenever $P \subseteq Q$.

Then we have the following result of McMullen [1977] (for $d = 2$, Hadwiger [1951b], §5):

(11.5) **Theorem.** *A monotone translation invariant valuation φ is continuous.*

It is enough to prove this on $\mathscr{P}^d$; any extension to $\mathscr{K}^d$ is fairly easy. Let $(P_j) \subset \mathscr{P}^d$ be a sequence with $\lim P_j = P$. If $\dim P = d$, assume $0 \in \operatorname{int} P$; from $(1-\varepsilon)P \subseteq P_j \subseteq (1+\varepsilon)P$ for each rational $\varepsilon > 0$ and all large enough j, $\lim \varphi(P_j) = \varphi(P)$ follows from the polynomial expansion of φ. If $\dim P < d$, we consider instead $\varphi(P_j + nQ)$, where Q is a fixed d-polytope and $n > 0$ a positive integer; $\varphi(P_j)$ is the constant term in the expansion of $\varphi(P_j + nQ)$ as a polynomial in n, and the result follows from the previous case.

§12. Valuations and Euler-type relations

If φ is a function defined on $\mathscr{P}^d$ (with values in, for the moment, an abelian group), we can define a new function φ^* on $\mathscr{P}^d$ by means of

$$\varphi^*(P) = \sum_F (-1)^{\dim F} \varphi(F).$$

(Our usual convention for such sums prevails.) It is a consequence of Euler's theorem that $\varphi^{**}(=(\varphi^*)^*) = \varphi$. As shown by Sallee [1968],

(12.1) **Theorem.** *If φ is a valuation on $\mathscr{P}^d$, then so is φ^*.*

Sallee's proof is direct, and needs the same kind of considerations which prove (10.1) and (10.2); a different version of the proof was indicated at the end of §5, and an alternative one as a consequence of (10.1) and (10.2) is indicated below.

As an extension of the definition due to Sallee, let us say that φ satisfies an *Euler-type relation of kind* (ε,η) if, for all $P \in \mathscr{P}^d$,

$$\varphi^*(P) = \varepsilon\varphi(\eta P).$$

Evidently, from $\varphi^{**} = \varphi$ follows $\varphi(P) = \varepsilon^2\varphi(\eta^2 P)$. Sallee only considers the case $\eta = 1$, so that $\varepsilon = \pm 1$ (assuming φ non-trivial); more generally, we shall see that $\eta = \pm 1$ is usual, so that $\varepsilon = \pm 1$ also holds (we know of no valuation satisfying an Euler-type relation with $\eta^2 \neq 1$).

If φ is any function on $\mathscr{P}^d$, now taking values in a rational vector space, and we define $\varphi_\pm = \frac{1}{2}(\varphi \pm \varphi^*)$, then $\varphi = \varphi_+ + \varphi_-$, and $\varphi_+^* = \varphi_+$, $\varphi_-^* = -\varphi_-$. Thus φ is always the sum of functions satisfying Euler-type relations. This might seem to make the concept of an Euler-type relation of little significance, were it not for the rather deep connexions between them and valuations which we shall now describe.

We first discuss results of Sallee [1968].

(12.2) **Theorem.** *Let φ be a continuous function on $\mathscr{P}^d$ which satisfies an Euler-type relation $\varphi^* = (-1)^{d-1}\varphi$. Then φ is a valuation.*

(12.2) depends upon several results. Firstly

(12.3) **Lemma.** *If* φ *is a continuous function on* $\mathscr{P}^d$, *satisfying* $\varphi(P \cup Q) + \varphi(P \cap Q) = \varphi(P) + \varphi(Q)$ *whenever* P,Q, $P \cup Q$ *are* d-*polytopes, then* φ *is a valuation.*

Secondly, a function φ satisfying the condition of (12.2) yields expressions of the form

$$\varphi(P) = \frac{1}{2} \sum_{F, \dim F < d} (-1)^{d-1-\dim F} \varphi(F)$$

for d-polytopes P, enabling $\varphi(P)$ to be calculated from its values on lower dimensional faces.

Thirdly, using the continuity of φ, to verify that φ is a valuation, it is enough to consider P and Q of (12.3) satisfying $\dim(P \cap Q) = d - 1$, such that each face of $P \cup Q$ is a face of P or Q. Checking the valuation property in this case is straightforward.

Sallee appears to claim a more general result than (12.2), involving continuous functions satisfying $\varphi^* = \varepsilon\varphi$ for any ε ($= \pm 1$ by previous remarks). But his φ has domain $\mathscr{P} = \cup \mathscr{P}^d$ (presumably in $\mathbb{E}^\infty = \cup \mathbb{E}^d$). In fact, (12.2) does not extend to $\varepsilon = (-1)^d$; a counterexample is $\varphi(P) = V(P_{-\rho})$, where $P_{-\rho}$ is the inner parallel body of P at any fixed positive distance ρ. (The crucial feature of this example is that $\varphi(P) = 0$ if $\dim P < d$, the example can clearly be modified to give functions φ positive on all d-polytopes.)

The results in the other direction are more general and interesting. Various individual cases of translation invariant or covariant valuations satisfying Euler-type relations were shown by Shephard: the Steiner point [1966]; the mean width [1968a]; mixed volumes [1968c]. This last result, although its proof did not admit of generalization, gave a pointer to the general result, due to McMullen [1975a, 1977].

(12.4) **Theorem.** *Let* φ *be a translation invariant or covariant valuation on* $\mathscr{P}^d$ *which is homogeneous of degree* r. *Then* $\varphi^*(P) = (-1)^r \varphi(-P)$ *for all* $P \in \mathscr{P}^d$.

For translation invariant valuations, (12.4) reduces to (7.2) in view of

(12.5) **Lemma.** *For each translation class of flats* $\mathscr{A}$, *let* $\psi_{\mathscr{A}}$ *be the simple valuation corresponding to* φ. *Then* $\psi^*_{\mathscr{A}}$ *corresponds to* φ^*.

Since $\psi^*_{\mathscr{A}}(P) = (-1)^{\dim \mathscr{A}} \psi_{\mathscr{A}}(P) = (-1)^r \psi_{\mathscr{A}}(-P)$ for aff $P \in \mathscr{A}$, by (7.2), and $\gamma(F,P) = \gamma(-F,-P)$, (12.4) clearly follows.

For translation covariant valuations, we consider the valuation $\bar{\varphi}$ defined by $\bar{\varphi}(P) = \varphi(P) - (-1)^r \varphi^*(-P)$. It is easy to verify that the associated valuation $\bar{\Phi}$ vanishes (by the already proved cases of (12.4)); hence $\bar{\varphi}$ is translation invariant, so $\bar{\varphi}^*(P) = (-1)^r \bar{\varphi}(-P)$ (again, by the previously proved cases). Thus $\bar{\varphi} = 0$, as required.

Nothing we have said above needs to be altered if we confine our attention to lattice polytopes and valuations on them which are invariant or covariant under lattice translations. In the particularly interesting case of the lattice point enumerator G, the Euler-type relation

$$G_r^*(P) = (-1)^r G_r(-P) = (-1)^r G_r(P)$$

yields the *reciprocity law* of Ehrhart [1967b]:

(12.6) **Theorem.** *For a lattice polytope* P, *let* $G^0(P) = G(\text{relint } P)$. *Then*

$$G^0(nP) = (-1)^{\dim P} \sum_{r \geq 0} (-n)^r G_r(P),$$

for integer $n \geq 1$.

For,

$$G^0(P) = (-1)^{\dim P} G^*(P)$$

by (5.24).

Erhart [1967c], [1968] has also investigated G(P) for rational polytopes P (that is, vert $P \subseteq \mathbb{Q}^d$), and has proved an analogous formula to (12.6). More generally, we have (with the conventions and terminology of §10):

(12.7) **Theorem.** *Let* φ *be a lattice translation invariant or covariant valuation on rational polytopes, which is near-homogeneous of degree* r. *Then for all rational polytopes* P *and all integer* n,

$$\varphi^*(P,n) = (-1)^r \varphi(-P,-n).$$

The proof of (12.7) depends upon a specific characterization of such valuations φ, for which see §17 below, and McMullen [1978; 1982b].

For the extension of some Euler-type relations to systems larger than $\mathscr{P}^d$, see Perles-Sallee [1970], Groemer [1972], Hadwiger [1973].

IV. Characterization theorems

The purpose of this last chapter is a survey over the existing results, and some open problems, that concern the classification of valuations. For the most part, we shall deal with characterizations of the classical valuations described in §§3, 4 by some of their properties, where invariance under some group plays a crucial role. Further, a few results will be given on the explicit representation of more general classes of valuations.

§13. Minkowski additive functions

Before considering more general valuations, it seems appropriate to study those special valuations φ on $\mathscr{K}^d$ or $\mathscr{P}^d$ which are Minkowski additive, that is, satisfy

(13.1) $\varphi(K + L) = \varphi(K) + \varphi(L)$

for K,L in the domain of φ. Here $K + L$ is the Minkowski sum of K and L. These valuations deserve special interest at least for two reasons: Due to the strong assumptions, more specific results are available; on the other hand, some proofs for more general characterization results make essential use of certain information on Minkowski additive functions.

The Minkowski additive functions which we consider will have their values in either $\mathbb{R}$, $\mathbb{E}^d$, or $\mathscr{K}^d$. It is clear that (13.1) then implies $\varphi(\lambda K) = \lambda\varphi(K)$ for rational $\lambda \geq 0$, and if φ is continuous, then this holds for all real $\lambda \geq 0$.

A familiar example of a real-valued Minkowski additive function on $\mathscr{K}^d$ is the mean width $\bar{b}$,

$$\bar{b}(K) = \frac{2}{\sigma(\Omega)} \int_{\Omega} h(K,u)\, d\sigma(u) \quad \text{for} \quad K \in \mathscr{K}^d,$$

which is a constant multiple of the quermassintegral W_{d-1}. It is continuous and rigid motion invariant, and it is essentially the only function with all these properties:

(13.2) **Theorem.** *If $\varphi: \mathscr{K}^d \to \mathbb{R}$ is Minkowski additive, continuous, and invariant under rigid motions, then $\varphi = \alpha W_{d-1}$ with some real constant α.*

Hadwiger's proof ([1957], p. 213) uses a rotation averaging process and shows more. Suppose that $\varphi: \mathscr{K}^d \to \mathbb{R}$ is Minkowski additive and rigid motion invariant. Let $K \in \mathscr{K}^d$ be given. If $K' = \lambda_1 g_1 K + \cdots + \lambda_r g_r K$ with positive rational numbers $\lambda_1, \ldots, \lambda_r$ and rotations $g_1, \ldots, g_r \in SO_d$, then $\varphi(K') = (\lambda_1 + \cdots + \lambda_r)\varphi(K)$. Since the same holds for the mean width, we have $\varphi(K)/\bar{b}(K) = \varphi(K')/\bar{b}(K')$ assuming that $\bar{b}(K) \neq 0$. Now one knows that the λ's and g's can be chosen such that K' is arbitrarily close to the unit ball B. If φ is continuous at B, then it follows that $\varphi(K)/\bar{b}(K) = \varphi(B)/2$ is independent of K. Thus in (13.2) it suffices to assume continuity of φ only at the unit ball.

On the other hand, this proof does not work if φ is only defined on polytopes. Since (13.2) is a tool in the proof of later characterization theorems, this defect restricts the generality of those results in a similar way. Of course,if φ is locally uniformly continuous (with respect to the Hausdorff metric) on $\mathscr{P}^d$, then it has a unique continuous extension to $\mathscr{K}^d$, and the additivity and invariance properties would carry over. (We say that a function is *locally uniformly continuous* or *locally bounded*, if it is uniformly continuous, respectively bounded, on the elements of its domain inside any fixed ball.)

(13.3) **Problem.** *If $\varphi: \mathscr{P}^d \to \mathbb{R}$ is Minkowski additive, invariant under rigid motions, and either continuous or locally bounded, must it be a constant multiple of the mean width?*

As a vector valued counterpart to mean width we have the Steiner point s which can be defined by

$$s(K) = \frac{1}{\varkappa(d)} \int_{\Omega} h(K,u)u\, d\sigma(u)$$

and is obviously Minkowski additive. We say that a map f from $\mathscr{K}^d$ or $\mathscr{P}^d$ into $\mathbb{E}^d$ or $\mathscr{K}^d$ is *rigid motion equivariant* (*translation equivariant*) if $f(gK) = gf(K)$ for every rigid motion g of $\mathbb{E}^d$ ($f(K + t) = f(K) + t$ for $t \in \mathbb{E}^d$, respectively) and for all K. The following analogue of (13.2) holds:

(13.4) **Theorem.** *If* $f: \mathscr{K}^d \to \mathbb{E}^d$ *is Minkowski additive, continuous, and rigid motion equivariant, then* f(K) *is the Steiner point of* $K \in \mathscr{K}^d$.

Again, it suffices to assume that f is merely continuous at the unit ball. A proof for this case, which also uses rotation averaging, though in a more subtle way, was given by Posicel'skiĭ [1973]. The problem of characterizing the Steiner point by some of its properties was first posed by Grünbaum [1963], p. 239, who asked whether Minkowski additivity and similarity equivariance are sufficient to characterize the Steiner point. This is not the case, as can be shown by counter-examples. The first example to this effect was constructed by Sallee [1971]. A different example which is easier to describe, was mentioned by Schneider [1974a], p. 76. The first author to add the continuity assumption to Grünbaum's conditions was Shephard [1968b]. He proved (13.4) for $d = 2$, using Fourier series. His method was extended, though not in a straightforward way, to $d \geq 3$ by Schneider [1971], who proved (13.4) by making use of the fact that certain representations of the rotation group in spaces of spherical harmonics are irreducible. Although this method is less elementary and needs stronger assumptions than Posicel'skii's proof, the application of spherical harmonics seems to be a proper tool in this context, since it has proved useful in treating similar questions to be discussed below. Before Schneider's [1971] proof of (13.4), a slightly weaker version was obtained by Meyer [1970], who assumed uniform continuity. Two attempts to prove (13.4) (Schmitt [1968], Hadwiger [1969a]) contained errors. For the two-dimensional case, interesting elementary proofs were given by Hadwiger [1971] and Berg [1971]. The latter author obtained additional results for polytopes. To describe them, define

$$(13.5)\quad s_\varphi(P) := \sum_{i=1}^{f_0(P)} \varphi(N(v_i,P) \cap \Omega) v_i \quad \text{for} \quad P \in \mathscr{P}^d,$$

where the sum extends over the vertices $v_1, \ldots, v_{f_0(P)}$ of the convex polytope P, $N(v_i,P)$ denotes the cone of exterior normal vectors of P at v_i (see §8) and φ is a simple valuation on spherical polytopes. If $\varphi = \sigma/\sigma(\Omega)$, where σ is spherical volume, then s_φ is the Steiner point s; this representation is a special case of (3.26). The proof for the Minkowski additivity of s which uses this representation (see Grünbaum [1967], p. 309) extends to yield that s_φ as defined above is Minkowski additive. Clearly $s_\varphi(\lambda P) = \lambda s_\varphi(P)$ for $\lambda \geq 0$, and if $\varphi(\Omega) = 1$, then s_φ is translation equivariant. If $f: \mathscr{P}^d \to \mathbb{E}^d$ is Minkowski additive and commutes with all similarities (including improper ones), then Berg [1971] calls this map an *abstract Steiner point*. Formula (13.5) yields an abstract Steiner point if φ is invariant under rotations and reflections and is normalized to $\varphi(\Omega) = 1$. Whether every abstract Steiner point is obtained in this way is not known, but Berg [1971] showed that this is true for $d = 2.3$. In these dimensions one then also has $\varphi = \theta \circ (\sigma/\sigma(\Omega))$ with some function $\theta: [0,1] \to \mathbb{R}$ satisfying $\theta(1) = 1$ and $\theta(u + v) =$

$\theta(u) + \theta(v)$ for $u,v,u + v \in [0,1]$. Berg deduced that every abstract Steiner point on $\mathscr{P}^d$, $d = 2,3$, which is locally bounded, is the usual Steiner point.

When Minkowski additive maps on $\mathscr{K}^d$ are considered, it appears particularly natural to investigate those ones whose range is also $\mathscr{K}^d$. A Minkowski additive map $\Phi: \mathscr{K}^d \to \mathscr{K}^d$ which is continuous and rigid motion equivariant has been called an *endomorphism* of $\mathscr{K}^d$ in Schneider [1974a], since such a map is compatible with the most natural and geometrically important structures which one usually associates with $\mathscr{K}^d$. Special questions concerning endomorphisms of $\mathscr{K}^d$ were apparently first posed by Grünbaum ([1963], p. 239, [1967], p. 315; answers are given by Schneider [1974a], p. 54 and pp. 55–56). The investigation of endomorphisms of $\mathscr{K}^d$ shows different features in dimensions $d = 2$ and $d > 2$, due to the commutativity of the rotation group in dimension two. Let us first consider the two-dimensional case.

A particular example of an endomorphism $\Phi: \mathscr{K}^2 \to \mathscr{K}^2$ is given by

$$(13.6)\quad \Phi(K) := \lambda_1 g_1[K - s(K)] + \cdots + \lambda_r g_r[K - s(K)] + s(K)$$

for $K \in \mathscr{K}^2$, where $\lambda_1,\dots,\lambda_r \geq 0$ are real numbers and $g_1,\dots,g_r \in SO_2$ are rotations. It turns out that the general endomorphism is a limit of such rotation averages. To formulate a precise result, we choose an orthonormal basis e_1,e_2 of $\mathbb{E}^2$ and write $u(\alpha) := (\cos\alpha)e_1 + (\sin\alpha)e_2$ for $\alpha \in [0,2\pi)$; further, for $K \in \mathscr{K}^2$ we write $h(K,\alpha)$ instead of $h(K,u(\alpha))$.

(13.7) **Theorem.** *Let Φ be an endomorphism of $\mathscr{K}^2$. Then there exists a (positive) measure ν on the Borel subsets of $[0,2\pi)$ such that*

$$(13.8)\quad h(\Phi(K),\alpha) = \int_0^{2\pi} h(K - s(K),\alpha + \beta)\,d\nu(\beta) + \langle s(K),u(\alpha)\rangle$$

for $\alpha \in [0,2\pi)$ and all $K \in \mathscr{K}^2$.

This was proved by Schneider [1974b]. Conversely, any Borel measure ν on $[0,2\pi)$ defines an endomorphism Φ by means of (13.8). If Φ is given, the measure ν in (13.8) is unique up to the indefinite integral of $(a_1\cos\alpha + a_2\sin\alpha)$ with constants a_1,a_2. From (13.8) it can be deduced that any endomorphism Φ of $\mathscr{K}^2$ whose image contains a polygon (with more than one point) is of the form (13.6). Moreover, every endomorphism which maps $\mathscr{K}^2$ onto all of $\mathscr{K}^2$ is of the form $\Phi(K) = \lambda g[K - s(K)] + s(K)$ with $\lambda > 0$ and $g \in SO_2$. These are also precisely the extreme endomorphisms, if the set of all endomorphisms of $\mathscr{K}^2$ is made into a convex cone in a natural way (loc. cit., p. 310). Another consequence of (13.7) is the fact that some properties which Inzinger [1949] has proved for a certain very special class of endomorphisms, are shared by all endomorphisms, at least after suitable normalizations.

The proof of Theorem (13.7) uses a characterization, essentially due to Hadwiger [1951b], of the Minkowski additive, continuous, and translation invariant real functions on $\mathscr{K}^2$ (see §16). This result first yields an integral representation of the form

$$(13.9)\quad h(\Phi(K) - s(K),\alpha) = \int_0^{2\pi} g(\beta - \alpha)\,dS_1(K,\alpha)$$

for $\alpha \in [0,2\pi)$ and $K \in \mathscr{K}^2$, where g is a continuous real function determined by the endomorphism Φ and $S_1(K,\cdot)$ is the first order area function of K, considered as a measure on $[0,2\pi)$ instead of Ω^1. The passage from (13.9) to (13.8) uses mainly analytic arguments.

For $d \geq 3$, a comparatively complete description of the endomorphisms of $\mathscr{K}^d$ is not known, but a series of partial results were obtained by Schneider [1974a]. Non-trivial examples of endomorphisms can be obtained as follows. We consider the support function $h(K,u) := \max\{\langle x,u\rangle : x \in K\}$ of $K \in \mathscr{K}^d$ for arbitrary $u \in \mathbb{E}^d$. Let $q:[0,\infty) \to [0,\infty)$ be a function for which all the following integrals exist and are finite. Let $K \in \mathscr{K}^d$ be given. It can be shown that $x \mapsto \int_{\mathbb{E}^d} h(K, x - \|x\|z)q(\|z\|)dz$ is a support function, hence there exists a unique convex body $\Phi_q(K)$ for which

$$(13.10)\quad h(\Phi_q(K),x) = \int_{\mathbb{E}^d} h(K - s(K), x - \|x\|z)q(\|z\|)\,dz + \langle s(K),x\rangle$$

for $x \in \mathbb{E}^d$. It is then easy to see that the map $\Phi_q : \mathscr{K}^d \to \mathscr{K}^d$ defined by (13.10) is an endomorphism of $\mathscr{K}^d$. We remark that this map Φ_q, with special choices for q, is particularly useful in the treatment of certain approximation problems for convex bodies, see Berg [1969], Weil [1975b]. Further constructions for endomorphisms of $\mathscr{K}^d$ were described in Schneider [1974a]. The main theme of that paper was the investigation of additional assumptions by which, from the variety of endomorphisms, those with a simple geometric meaning could be singled out. The following results were obtained. Here we assume $d \geq 3$.

(13.11) **Theorem.** (a) *Every endomorphism of $\mathscr{K}^d$ is uniquely determined by the image of one suitably chosen convex body, for example, a triangle with at least one irrational* angle.

(b) *Let $\Phi : \mathscr{K}^d_d \to \mathscr{K}^d$ be a Minkowski additive and continuous map such that $\Phi(aK) = a\Phi(K)$ for every nonsingular affine transformation* a *of* $\mathbb{E}^d$. *Then*

$$\Phi(K) = K + \lambda[K + (-K)] \quad \textit{for} \quad K \in \mathscr{K}^d_d,$$

where $\lambda \geq 0$ is a real constant.

(c) *Let Φ be an endomorphism of $\mathscr{K}^d$. If the image under Φ of some at least one-dimensional convex body is a point, then $\Phi(K) = \{s(K)\}$ for $K \in \mathscr{K}^d$. If the image under Φ of some convex body is a segment, then*

$$\Phi(K) = \lambda[K - s(K)] + \mu[-K + s(K)] + s(K) \quad \textit{for} \quad K \in \mathscr{K}^d$$

with real numbers $\lambda,\mu \geq 0$, $\lambda + \mu > 0$.

(d) *The only surjective endomorphisms of $\mathscr{K}^d$ are given by*

$$\Phi(K) = \lambda[K - s(K)] + s(K) \quad \textit{for} \quad K \in \mathscr{K}^d$$

with $\lambda \neq 0$.

(e) *If Φ is an endomorphism of $\mathscr{K}^d$ satisfying $W_k(\Phi(K)) = W_k(K)$ for some $k \in \{0,1,\ldots,d-2\}$ and all $K \in \mathscr{K}^d$, then*

$$\Phi(K) = \varepsilon[K - s(K)] + s(K) \quad \textit{for} \quad K \in \mathscr{K}^d$$

where $\varepsilon \in \{1,-1\}$.

The proof uses a combination of elementary facts from harmonic analysis (for the rotation group acting on the sphere Ω) with convexity arguments.

So far the invariance or equivariance with respect to rigid motions has played an important role. It is clear that, in absence of such an assumption, particular Minkowski additive functions can only be characterized if fairly strong assumptions are imposed. Here is an example for such a result, due to Schneider [1974c]:

(13.2) **Theorem.** *Let* $T:\mathscr{K}^d \to \mathscr{K}^d$ $(d \geq 2)$ *be a Minkowski additive map satisfying* $V(T(K)) = V(K)$ *for all* $K \in \mathscr{K}^d$. *Then there exists a volume preserving affine map* $a:\mathbb{E}^d \to \mathbb{E}^d$ *such that* $T(K)$ *is a translate of* aK *for each* $K \in \mathscr{K}^d$.

Finally we mention that Valette [1974] has studied the continuous maps $F:\mathscr{K}^d \to \mathscr{K}^d$ which commute with affine maps and, instead of being Minkowski additive, only satisfy $F(K_1 + K_2) \supseteq F(K_1) + F(K_2)$ for $K_1, K_2 \in \mathscr{K}^d$.

§14. Volume and centroid

From now on, it seems convenient to define that "valuation", without further specification of the range, means real-valued valuation.

We shall now review the characterization theorems for volume. The essential uniqueness of Haar measure, applied to the additive group of $\mathbb{E}^d$, gives the following well-known uniqueness theorem: If φ is a translation invariant (positive) measure on the Borel sets of $\mathbb{E}^d$ with $\varphi(C) = 1$ for some fixed unit cube C, then φ is the volume (Lebesgue measure). From a geometric point of view one would, of course, not want to assume from the beginning that φ is defined on the Borel sets and is σ-additive, but rather that φ is a simple valuation defined on $\mathscr{P}^d$ or $\mathscr{K}^d$. We shall often consider only $\mathscr{P}^d$ as domain, since the assumptions which we have to impose in order to guarantee uniqueness will afterwards also give uniqueness if $\mathscr{K}^d$ was the original domain. If φ is a simple valuation on $\mathscr{P}^d$, then we know from (5.19) and (5.20) (or from the simpler version in Hadwiger [1957], p. 81) that φ has an additive (simple) extension to the set $U(\mathscr{P}^d)$ of polyhedra, and if φ was invariant under translations or rigid motions, then the extension also is. If φ is nonnegative, then the extension is obviously monotone, that is, it satisfies $\varphi(A) \leq \varphi(B)$ for $A \subseteq B$ $(A, B \in U(\mathscr{P}^d))$, and hence also nonnegative. Thus for simple valuations it makes no difference whether they are defined on $\mathscr{P}^d$ or on $U(\mathscr{P}^d)$, and whether they are assumed nonnegative or monotone. For these valuations we have:

(14.1) **Theorem.** *Let* φ *be a translation invariant, nonnegative, simple valuation on* $\mathscr{P}^d$. *Then* $\varphi = \alpha V$ *with some real constant* $\alpha \geq 0$.

The easy proof is well-known from analysis text books, see, e.g., Maak [1960], §46, Satz 8; see also Hadwiger [1955b], p. 47. These and similar proofs for the characterization of elementary volume cannot be considered as strictly elementary from a geometric point of view, since they use infinite processes for simple geometric figures, like "exhaustion" or polyhedral approximation. By Dehn's theorem, no proof can work with finite dissections alone, without some limit process, even if translation invariance is sharpened to rigid motion invariance. But one can give a proof where the unavoidable limit process is only used to establish the essential uniqueness of a monotone solution of Cauchy's functional equation, while the geometric part uses only finite dissections of polytopes. A proof which is elementary in this sense was given by Hadwiger [1950] (for $d = 3$ also [1949a]), [1957], 2.1.3.

We digress a bit and consider the corresponding problem in spherical and hyperbolic space. For the notion of convex polytope in these spaces the reader may consult Böhm-Hertel [1980]. These authors also discuss at length the problem of an elementary theory of volume in those spaces, and they give such a proof of existence and uniqueness in dimension two. The uniqueness problem in general is the following: Suppose that φ is a nonnegative simple valuation on the convex polytopes of R^d (this stands for either euclidean, spherical, or hyperbolic space of dimension d) which is invariant under the motion group of R^d. Is φ a constant multiple of the usual volume? All the methods of proof mentioned above for $R^d = \mathbb{E}^d$ make essential use of the vector space structure of $\mathbb{E}^d$ and hence do not extend to non-euclidean spaces. An affirmative answer for spherical space was given by Schneider [1978], Th. (6.2). The proof, which can be generalized to yield an abstract version of the result for compact homogeneous spaces (Schneider [1981]), relies heavily on compactness. For hyperbolic space, too, the answer is in the affirmative, although no proof seems to be known which is as directly geometric as Hadwiger's in the euclidean case. We sketch a less elementary proof which works for all three types of spaces.

Suppose that $\varphi:\mathscr{P}(R^d) \to \mathbb{R}$ (where $\mathscr{P}(R^d)$ denotes the set of convex polytopes in R^d) is a nonnegative simple valuation, which is invariant under the motions of R^d. For an arbitrary subset $A \subset R^d$, define $\varphi^*(A) := \inf \sum \varphi(P_n)$, where the infimum extends over all sequences $(P_n)_{n \in \mathbb{N}}$ of convex polytopes in R^d which cover A. Then φ^* is an outer measure on R^d (see, e.g., Munroe [1953], ch. II). If $A, B \subset R^d$ are such that their distance (with respect to the usual metric on R^d) is positive, then it is easy to see (using coverings by convex polytopes with sufficiently small diameters) that $\varphi^*(A \cup B) = \varphi^*(A) + \varphi^*(B)$. Thus φ^* is a metric outer measure and hence all Borel sets of R^d are φ^*-measurable (Munroe, loc. cit.). Since φ is a simple valuation, it has an additive extension to the ring generated by the convex polytopes. We assert that the extension, also denoted by φ, is a premeasure (i.e., countably additive). This can be shown as in the usual construction of Lebesgue measure on $\mathbb{E}^d$ (e.g., Bauer [1978], p. 29), as soon as the following has been proved: (*) To any convex polytope $P \subset R^d$ and any $\varepsilon > 0$, there exists a convex polytope $P' \subset R^d$ such that $P' \subset \operatorname{int} P$ and $\varphi(P) - \varphi(P') \leq \varepsilon$. If P and ε are given, let P_1 be obtained from P by "pushing a facet" of P towards the interior of P. For all P_1 sufficiently close to P we have $\varphi(P \backslash P_1) < \varepsilon/n$, where n is the number of facets of P. Otherwise, the nonnegativity and motion invariance of the extension φ would easily yield a contradiction (since a large number of congruent copies of a "sufficiently flat" $P \backslash P_1$ can be packed into some fixed polytope). Continuing in this way, (*) is proved. After that, it follows that the restriction $\bar{\varphi}$ of φ^* to the φ^*-measurable sets is a measure which extends φ (e.g., Bauer [1978], §5). Since by construction it is invariant, its restriction to the Borel sets must be a constant multiple of Haar measure, from which the assertion follows.

We return to euclidean space and consider characterizations of volume that use assumptions different from those of (14.1). Hadwiger [1957], p. 79, proved that a valuation (not necessarily simple or nonnegative) on $\mathscr{P}^d$ which is translation invariant and homogeneous of degree d, must be a constant multiple of volume. A different invariance property was supposed by Hadwiger [1970]. In the course of proving a more general theorem he showed: A nonnegative simple valuation on $\mathscr{P}^d$ which is invariant under volume preserving linear maps of $\mathbb{E}^d$ (keeping the origin fixed) must be a constant multiple of volume.

More important, with a view to the extension to quermassintegrals (see §15), is a characterization of volume where the nonnegativity is replaced by continuity with respect to the usual Hausdorff metric for convex bodies. The following theorem is due to Hadwiger [1952d], [1957], p. 221.

(14.2) **Theorem.** *Let φ be a rigid motion invariant, continuous, simple valuation on $\mathscr{K}^d$. Then φ is a constant multiple of volume.*

It seems to be unknown whether here $\mathscr{K}^d$ can be replaced by $\mathscr{P}^d$. Hadwiger's proof works with polytopes for a long while, but at the end it essentially uses (13.2). An affirmative answer to Problem (13.3) would, therefore, yield a corresponding generalization of (14.2). We remark that in (14.2) it would not be sufficient to assume translation invariance instead of rigid motion invariance, see §16.

An apparently hard problem is the analogue of (14.2) in non-euclidean spaces; evidently the proof does not carry over. We restate Problem 74 of Gruber-Schneider [1979]:

(14.3) **Problem.** *Let φ be a rotation invariant, continuous, simple valuation on the spherically convex polytopes or convex bodies in Ω^{d-1}. Is φ a constant multiple of spherical volume?*

Partial information can be obtained from §8. Instead of considering spherically convex polytopes, it is convenient to use the convex polyhedral cones with apex 0 which they generate. Theorem (8.8) and the remarks following it imply:

(14.4) **Theorem.** *If* d *is odd, then a rotation invariant simple valuation on polyhedral cones can be expressed as a linear combination of the valuation on proper product cones.*

In particular, this can be used to give an affirmative answer to Problem (14.3) for d = 3. But whether (14.4) is helpful for the general case remains open.

We conclude this paragraph with a view to vector valued valuations. Some of the characterizations of volume in euclidean space have counterparts for the centroid (centre of gravity). For a bounded measurable set $K \subset \mathbb{E}^d$ with $V(K) \neq 0$ the centroid is defined by

$$c(K) := \frac{1}{V(K)} \int_K x \, dV(x)$$

(so that, in the terminology of §3, $V(K)c(K) = z(K) = q_0(K)$ for $K \in \mathscr{K}^d$). Clearly Vc is a simple valuation with values in $\mathbb{E}^d$. We recall that $\mathscr{P}^d_d \subset \mathscr{P}^d$ ($\mathscr{K}^d_d \subset \mathscr{K}^d$) is the subset of polytopes (resp. convex bodies) with interior points. The following theorem of Schneider [1973] may be viewed as a counterpart to (14.1); the proof is also similar.

(14.5) **Theorem.** *Let* $f: U(\mathscr{P}^d_d) \to \mathbb{E}^d$ *be a translation equivariant function such that* $f(P)$ *lies in the convex hull of* P *and* Vf *is a simple valuation. Then* $f(P)$ *is the centroid of* P *for* $P \in U(\mathscr{P}^d_d)$.

And the following result of Schneider [1972b] (p. 211) is a counterpart to (14.2):

(14.6) **Theorem.** *Let* $f: \mathscr{K}^d_d \to \mathbb{E}^d$ *be a rigid motion equivariant continuous function such that* Vf *is a valuation. Then* $f(K)$ *is the centroid of* K *for* $K \in \mathscr{K}^d_d$.

§15. Quermassintegrals, mixed volumes, moment vectors, curvature measures

The following famous theorem of Hadwiger, which characterizes the linear combinations of quermassintegrals, is certainly the central result in a theory of valuations on convex bodies.

(15.1) **Theorem.** *If* φ *is a continuous and rigid motion invariant valuation on* $\mathscr{K}^d$, *then*

$$\varphi(K) = \sum_{i=0}^{d} c_i W_i(K) \quad \textit{for} \quad K \in \mathscr{K}^d \tag{15.2}$$

with real constants $c_0, \ldots, c_d$.

There is a companion to (15.1) with continuity replaced by monotoneity (with respect to set inclusion):

(15.3) **Theorem.** *If* φ *is an increasing and rigid motion invariant valuation on* $\mathscr{K}^d$, *then* (15.2) *holds with nonnegative real constants* $c_0, \ldots, c_d$.

By means of (11.5), which was proved by McMullen [1977], it has later become clear that (15.3) can be deduced from (15.1).

Results of this type were first considered by Blaschke [1937], §43. He investigated the rigid motion invariant, locally bounded valuations on $U(\mathscr{P}^3)$, but in order to obtain a representation of type (15.2), he had to impose an additional assumption, namely that the volume part be invariant under volume preserving affinities. Since the other quermassintegrals do not have this property, the assumption, which is dictated by the method of proof, seems artificial; moreover, it can only be formulated in the course of the proof, since a "volume part" of the valuation is not defined from the beginning. Hadwiger proved (15.1) for $d = 3$ in [1951a] (see also [1955b], §16), and for general d in [1952d]. The proof of (15.3) was then given in Hadwiger [1953a]. Both proofs were reproduced in Hadwiger [1957], 6.1.10 (see also Leichtweiß [1980], §17). It may be remarked that the proof needs only the weak valuation property (without first having to deduce the valuation property). Some (obvious) supplementary remarks to (15.1) were published by Müller [1967].

Hadwiger's proof of (15.1) uses induction with respect to the dimension and relies on (14.2) and thus on (13.2). It is, therefore, necessary to consider valuations on $\mathscr{K}^d$, while the proof would not work if $\mathscr{P}^d$ were the domain. This is also true for Hadwiger's proof of (15.3). The following seems to be open.

(15.4) **Problem.** *Let φ be a rigid motion invariant valuation on $\mathscr{P}^d$ which is either continuous, or increasing, or locally bounded, or nonnegative. Is it true that $\varphi(P) = \sum c_i W_i(P)$ for $P \in \mathscr{P}^d$ with constants $c_0,\dots,c_d$?*

Questions of this type were already suggested in the work of Hadwiger. The assumption of nonnegativity, known to imply monotoneity only in the simple case, was supposed by Spiegel. Both assumptions, local boundedness and nonnegativity, appear particularly natural in the case of polytopes. They are, however, not appropriate for general convex bodies: For $K \in \mathscr{K}^d$, let $\varphi(K)$ be the sum (finite or infinite, but obviously well defined) of the $(d-1)$-volumes of the $(d-1)$-dimensional faces of K (and twice the $(d-1)$-volume if K is of dimension at most $d-1$). Then φ is a rigid motion invariant valuation which is locally bounded and nonnegative, but not a linear combination of quermassintegrals.

The characterization theorem (15.1) has important applications in integral geometry. The principal idea also goes back to Blaschke [1937]. To demonstrate this method in a simple case, consider formula (3.11). The integral on the right-hand side, considered as a function of K, clearly defines a rigid motion invariant continuous valuation on $\mathscr{K}^d$, hence it can be expressed as a linear combination of quermassintegrals. Choosing for K a ball with variable radius, one then easily calculates the coefficients and thus proves (3.11). Hadwiger [1950d], [1955b], [1956], [1957] used this approach systematically for the derivation of several old and new integral geometric formulae. An interesting application of (15.1) of a different kind appears in Matheron's [1975] work on random sets.

We mention two variants of (15.1). As described in §5, Groemer [1972] has defined extensions of the quermassintegrals in the form of continuous linear functions on a certain vector space A^d of so-called approximable functions on $\mathbb{E}^d$. From Hadwiger's theorem he could then deduce a corresponding characterization theorem for the extended quermassintegrals. Baddeley [1980], motivated by requirements of stereology, developed an integral geometric theory of certain absolute curvature integrals, and he proved a characterization result analogous to, and motivated by, Hadwiger's theorem.

Results analogous to Theorems (15.1) and (15.3) should be expected to hold in noneuclidean spaces. Let us state this as a problem for spherical space. For spherically convex polytopes we defined the functionals φ_r by (3.31), and we mentioned that they have continuous extensions to general spherically convex sets. The functionals ψ_r defined by (3.32) are increasing.

(15.5) **Problem.** *Let φ be a rotation invariant valuation on the spherically convex sets (or polytopes) in the sphere Ω. If φ is continuous, is it a linear combination of the φ_r with constant coefficients? If φ is increasing, is it a linear combination of the ψ_r with nonnegative coefficients?*

The first part could be answered in the affirmative if (14.3) were true, since the induction part of Hadwiger's argument easily carries over.

We turn back to euclidean space and consider analogues of Hadwiger's theorem for the other particular valuations described in §3. A result which closely parallels (15.1) exists for the quermassvectors q_r defined by (3.23). These are continuous, rotation equivariant $\mathbb{E}^d$-valued valuations ($\mathbb{E}^d$-valuations, for short), as is clear from the definition. The behaviour under translations is exhibited by (3.24), namely

$$q_r(K + t) = q_r(K) + W_r(K)t \quad \text{for} \quad K \in \mathcal{K}^d \quad \text{and} \quad t \in \mathbb{E}^d.$$

From the characterization (13.4) of the Steiner point $s = q_d/\varkappa_d$ one can deduce the following stronger version, where Minkowski additivity is replaced by the valuation property (Schneider [1972b]). If one assumes that f commutes also with similarities, then a simpler reduction to (13.4) is possible, see Hadwiger [1971].

(15.6) **Theorem.** *Let* f *be a rigid motion equivariant continuous* $\mathbb{E}^d$*-valuation on* $\mathcal{K}^d$. *Then* f(K) *is the Steiner point of* K *for* $K \in \mathcal{K}^d$.

From this result and Hadwiger's theorem (15.1) it is not difficult to conclude the following.

(15.7) **Theorem.** *Let* f *be a rotation equivariant continuous* $\mathbb{E}^d$*-valuation on* $\mathcal{K}^d$ *such that* f(K + t) − f(K) *is always parallel to* t. *Then*

$$f(K) = \sum_{i=0}^{d} c_i q_i(K) \quad \textit{for} \quad K \in \mathcal{K}^d$$

with real constants $c_0,\dots,c_d$.

The proof may be found in Hadwiger-Schneider [1971] and Schneider [1972b]; these papers also contain applications to integral geometric formulae for the quermassvectors.

Also the area functions and curvature measures satisfy characterization theorems which may be compared with Hadwiger's theorem (15.1). These were proved by Schneider [1975a], [1975b], [1978] and were also applied to the derivation of integral geometric formulae. Since these results have already been reviewed in Schneider [1979] (§§6,7), we refer the reader to that survey. A new application of the characterization theorem for area functions to a problem on geometric probabilities was recently made by Vogiatzaki.

Since the quermassintegrals are special mixed volumes, one may ask whether more general mixed volumes can be characterized in a similar way. Only a very few special results in this direction are known. Fáry [1961] characterized, for a given convex body U, the functionals

$$\varphi: K \mapsto \sum_{i=0}^{d} c_i V(K,i;U,d-i)$$

with $c_0,\dots,c_d \in \mathbb{R}$ as the translation invariant, continuous valuations φ satisfying

$\varphi(K) = \varphi(L)$ whenever $V(K,i;U,d-i) = V(L,i;U,d-i)$ for $i = 0,\ldots,d$. Clearly an assumption of this kind cannot be omitted, but it makes the whole characterization theorem appear slightly artificial. Firey [1976] replaced the valuation property by Minkowski linearity and he showed that an increasing function $\varphi:\mathscr{K}^d \to \mathbb{R}$ with this property which is zero on one-pointed sets must be of the form

$$\varphi(K) = V(K,1;\bar{K},p-1;s_{p+1},\ldots,s_d)$$

with an essentially unique convex body $\bar{K}$ and pairwise orthogonal segments $s_{p+1},\ldots,s_d$ of unit length which span the orthogonal complement of the affine hull of $\bar{K}$. Obviously the assumptions of this theorem are quite strong, implying, in particular, that φ is translation invariant, nonnegative, and continuous; together with the linearity this opens the way to an application of the Riesz representation theorem.

Further characterization results related to mixed volumes appear in §16.

§16. Translation invariant valuations

Hadwiger's theorem (15.1) characterizes the (linear combinations of) quermassintegrals as those real valuations on $\mathscr{K}^d$ which are continuous and invariant under rigid motions. Among the various possibilities of relaxing the assumptions, the condition of translation instead of motion invariance seems both natural and important. Translation invariant or covariant valuations are the topic of the present section. Theorem (14.1) is an example where translation invariance, together with some other conditions, is still sufficient to characterize a particular valuation. But without such strong additional assumptions, translation invariant valuations abound, and a classification seems difficult.

Let us first consider some examples. Generally speaking, any representation of the quermassintegrals, suitably modified, will yield more general translation invariant valuations. For polytopes $P \in \mathscr{P}^d$ we have

$$(16.1)\quad V_r(P) = \sum_{F^r} \gamma(F^r,P)V_r(F^r)$$

by (3.30). Two modifications offer themselves. We may replace V_r on the right-hand side by a real function which, restricted to the polytopes in any r-dimensional affine subspace, yields a simple valuation in that space. The resulting function is a valuation on $\mathscr{P}^d$. In fact, every valuation on $\mathscr{P}^d$ (not necessarily enjoying an invariance property) can be represented as a finite sum of such valuations (Hadwiger [1953b]). In the translation invariant case, a special such representation was described and used in §10. On the other hand, the external angle $\gamma(F^r,P)$ in (16.1) was defined via the notion of spherical volume. We may replace this by other simple valuations on spherical polytopes and obtain a translation invariant valuation on $\mathscr{P}^d$. For the characterization of such valuations, see Theorem (16.6) below.

Now we consider general convex bodies. Definition (3.7) of the quermassintegrals as special mixed volumes generalizes to

$$(16.1)\quad \varphi(K) := V(K,p;\mathscr{C}) \quad \text{for} \quad K \in \mathscr{K}^d,$$

where $p \in \{1,\ldots,d-1\}$ and $\mathscr{C} = (K_{p+1},\ldots,K_d)$ is a fixed $(d-p)$-tuple of convex bodies. This yields a translation invariant valuation φ on $\mathscr{K}^d$ which is increasing and (hence) continuous. The integral geometric approach to quermassintegrals motivates the definitions (3.12) and (3.13) of valuations on $\mathscr{K}^d$ which are translation invariant if φ is. Finally, the mixed area functions defined by (3.14) can be used to generalize (16.2) by

(16.3) $$\varphi(K) := \int_\Omega g(u)\, dS(K,p;\mathscr{C};u) \quad \text{for} \quad K \in \mathscr{K}^d,$$

where $p \in \{1,\ldots,d-1\}, \mathscr{C} = (K_{p+1},\ldots,K_{d-1})$ is a fixed $(d-p-1)$-tuple of convex bodies, and g is a fixed continuous real function on Ω. It follows from the properties of the mixed area functions that φ is a continuous translation invariant valuation on $\mathscr{K}^d$ which is homogeneous of degree p. The special case

(16.4) $$\varphi(K) := \int_\Omega g(u)\, dS_{d-1}(K;u) \quad \text{for} \quad K \in \mathscr{K}^d$$

with an odd continuous function g yields a simple valuation. This shows that in (14.2) the rigid motion invariance is indispensable (and also that the last sentence on p. 387 of Hadwiger [1952e] is erroneous).

For each of these special classes of translation invariant valuations it would be interesting to have an axiomatic characterization. Let us now describe the known results.

First we consider the translation invariant valuations on $\mathscr{P}^d$ which are weakly continuous (see §11). Here a complete description is available. The case of simple valuations goes back to Hadwiger [1952e]. It extends, without change of proof, to valuations with values in a real topological vector space $\mathscr{X}$. To describe the result, recall that $\mathscr{U}^s$ is the Stiefel manifold of s-frames $U = (u_1,\ldots,u_s)$, that is, ordered orthonormal s-tuples of vectors, in $\mathbb{E}^d$. We call a function $\eta: \mathscr{U}^s \to \mathscr{X}$ *odd* if

$$\eta(\varepsilon_1 u_1,\ldots,\varepsilon_s u_s) = \varepsilon_1 \cdots \varepsilon_s \eta(u_1,\ldots,u_s)$$

whenever $\varepsilon_i = \pm 1$ $(i = 1,\ldots,s)$. For $p \in \mathscr{P}^d$ and $U \in \mathscr{U}^s$, the face P_U was defined in §6; by convention, $P_\varnothing = P$. Then we have (Hadwiger [1952e]):

(16.5) **Theorem.** *A function* $\varphi: \mathscr{P}^d \to \mathscr{X}$ *is a weakly continuous translation invariant simple valuation if and only if there is an expression*

$$\varphi(P) = \sum_{r=0}^{d} \sum_{U \in \mathscr{U}^{d-s}} V_r(P_U)\eta_r(U) \quad \text{for} \quad P \in \mathscr{P}^d,$$

where $\eta_r: \mathscr{U}^{d-r} \to \mathscr{X}$ *is an odd function* $(r = 0,\ldots,d-1;$ η_d *a constant*).

Clearly the sums occurring in (16.5) are finite. Note that the term for $r = 0$ vanishes identically when $d \geq 1$.

Essentially by employing the method used in §10, McMullen [1982b] extended Hadwiger's result to non-simple valuations:

(16.6) **Theorem.** *A function* $\varphi: \mathscr{P}^d \to \mathscr{X}$ *is a weakly continuous translation invariant valuation if and only if there is an expression*

$$\varphi(P) = \sum_{r=0}^{d} \sum_{F \in \mathscr{F}^r(P)} V_r(F)\lambda_r(F,P) \quad \textit{for} \quad P \in \mathscr{P}^d.$$

Here, $\mathcal{F}^r$ denotes the family of r*-faces of* P, λ_r *is a simple valuation on* $(d-r)$*-cones, and* $\lambda_r(F,P) = \lambda_r(N(F,P))$, *where* N(F,P) *is the cone of all outer normal vectors to* P *at* F.

Further extension is possible to translation covariant valuations (McMullen, loc. cit.). According to §11, a valuation $\varphi:\mathcal{P}^d \to \mathcal{X}$ is translation covariant if there exists a map $\Phi:\mathcal{P}^d \to \mathrm{Hom}_{\mathbb{Q}}(\mathbb{E}^d,\mathcal{X})$ such that $\varphi(P + t) - \varphi(P) = \Phi(P)t$ for $t \in \mathbb{E}^d$ and $P \in \mathcal{P}^d$. If φ is weakly continuous, then the rational linearity of Φ extends to real linearity, so that $\Phi(P) \in \mathrm{Hom}_{\mathbb{R}}(\mathbb{E}^d,\mathcal{X}) = \mathrm{Hom}(\mathbb{E}^d,\mathcal{X})$. Moreover, Φ is a weakly continuous translation invariant valuation, and this permits to deduce the following from (16.5) and (16.6). Here the moment vector $m_{r+1}(P)$ of an r-dimensional polytope P in $\mathbb{E}^d$ is defined by

$$m_{r+1}(P) := \int_P x \, dx,$$

where the integration is with respect to r-dimensional Lebesgue measure in aff P.

(16.7) **Theorem.** *A function* $\varphi:\mathcal{P}^d \to \mathcal{X}$ *is a weakly continuous translation covariant simple valuation if and only if there is an expression*

$$\varphi(P) = \sum_{r=0}^{d} \sum_{U \in \mathcal{U}^{d-r}} [H_r(U)m_{r+1}(P_U) + V_r(P_U)\eta_r(U)],$$

where $H_r:\mathcal{U}^{d-r} \to \mathrm{Hom}(\mathbb{E}^d,\mathcal{X})$ *and* $\eta_r:\mathcal{U}^{d-r} \to \mathcal{X}$ *are odd functions.*

(16.8) **Theorem.** *A function* $\varphi:\mathcal{P}^d \to \mathcal{X}$ *is a weakly continuous translation covariant valuation if and only if there is an expression*

$$\varphi(P) = \sum_{r=0}^{d} \sum_{F \in \mathcal{F}^r(P)} [\Lambda_r(F,P)m_{r+1}(F) + V_r(F)\lambda_r(F,P)],$$

where λ_r *is a simple* $\mathcal{X}$*-valued valuation on normal cones of dimension* $d - r$ *and* Λ_r *is a simple* $\mathrm{Hom}(\mathbb{E}^d,\mathcal{X})$*-valued such valuation.*

Let us now turn to translation invariant (real valued) valuations on $\mathcal{K}^d$. The problem of characterizing the translation invariant continuous valuations on $\mathcal{K}^d$ is open, but one has some partial results. A complete explicit representation is known in the two-dimensional case. We recall from §3 that $S_p(K;\cdot)$ is the p-th order area function of $K \in \mathcal{K}^d$, it is a positive measure on $\Omega = \Omega^{d-1}$. For $K \in \mathcal{P}^2$ or $\mathcal{K}^2$ let

(16.9) $$\varphi(K) = a + \int_{\Omega^1} g(u)\, dS_1(K;u) + bV_2(K),$$

where a,b are real constants and g is a real function on Ω^1 so that the integral exists for all K. Then φ is a translation invariant valuation.

(16.10) **Theorem.** *If* φ *is a locally bounded translation invariant valuation on* $\mathcal{P}^2$, *then constants* a,b *and a bounded function* g *exist so that* (16.9) *holds for* $K \in \mathcal{P}^2$.

If φ is a continuous translation invariant valuation on $\mathscr{K}^2$, then constants a,b *and a continuous function* g *exist so that* (16.9) *holds for* $K \in \mathscr{K}^2$.

This was proved by Hadwiger [1949b], [1951b]. Actually, he did not use the area function $S_1(K;\cdot)$, but his results are easily seen to be equivalent to the above. The function g is uniquely determined by the valuation φ up to a summand of the form $\langle v,\cdot\rangle$ with a constant vector v. If φ in the second part of (16.10) is even Minkowski additive, then (16.9) holds with $a = b = 0$. This result was used to obtain (13.9), in the course of the proof of Theorem (13.7).

For $d \geq 3$, no such explicit representation is known. If φ is a continuous translation invariant valuation on $\mathscr{K}^d$, then it follows from §§10, 11 that $\varphi = \sum_{r=0}^{d} \varphi_r$, where φ_r is a continuous translation invariant valuation on $\mathscr{K}^d$ which is homogeneous of degree r. By Hadwiger [1957], p. 79, φ_d is a constant multiple of volume, and clearly φ_0 is constant. Thus there remains the problem of determining the continuous translation invariant valuations on $\mathscr{K}^d$ which are homogeneous of degree $r \in \{1,\dots,d-1\}$. For $r = d-1$ the following solution was given by McMullen [1980].

(16.11) **Theorem.** *Let φ be a continuous translation invariant valuation on $\mathscr{K}^d$ which is homogeneous of degree* $d-1$. *Then there is a continuous function* g *on the unit sphere Ω such that*

$$\varphi(K) = \int_\Omega g(u)\, dS_{d-1}(K;u) \quad \textit{for} \quad K \in \mathscr{K}^d.$$

Thus (16.3), with $p = d-1$, describes the general translation invariant valuation which is continuous and homogeneous of degree $d-1$. If one prefers an expression in terms of mixed volumes, that is, valuations of type (16.2), one can deduce from (16.11) (see McMullen, loc. cit.) that to any such valuation φ correspond sequences $(L_i)_{i\in\mathbb{N}}$, $(M_i)_{i\in\mathbb{N}}$ in $\mathscr{K}^d$ such that

$$\varphi(K) = \lim_{i\to\infty} [V(K,d-1;L_i) - V(K,d-1;M_i)]$$

for $K \in \mathscr{K}^d$.

For degrees $p \in \{1,\dots,d-2\}$, no analogue of (16.11) is known. Clearly any finite linear combination of functions of type (16.3) leads to a continuous translation invariant valuation on $\mathscr{K}^d$ which is homogeneous of degree p, but one cannot obtain the general such valuation in this way. Also, it seems difficult to draw any further conclusions from Theorem (16.6) in case φ is continuous.

Only for $p = 1$ and under a stronger continuity assumption has one a result (McMullen [1980]; compare also Schneider [1974b], p. 306). If φ is a uniformly continuous translation invariant valuation on $\mathscr{K}^d$ which is homogeneous of degree 1, then one easily deduces from the Riesz representation theorem that there exists a signed Borel measure μ on Ω such that

$$\varphi(K) = \int_\Omega h(K,u)\, d\mu(u) \quad \text{for} \quad K \in \mathscr{K}^d.$$

Hence, there exist $L,M \in \mathscr{K}^d$ such that

$$\varphi(K) = V(K;L,d-1) - V(K;M,d-1).$$

But the assumption is fairly strong, and there exist continuous translation invariant valuations on $\mathscr{K}^d$, homogeneous of degree 1, which are not uniformly continuous; for an example, see Schneider [1974b], p. 306.

§17. Lattice invariant valuations

A function on subsets of $\mathbb{E}^d$ which is invariant under the translations of the integer lattice $\mathbb{Z}^d$, will briefly be called *lattice invariant*. The lattice point enumerator G, the functionals G_i appearing in (4.1), and the weighted lattice point numbers are natural examples of lattice invariant valuations. In this section we consider, under the aspects of characterization and representation, the lattice invariant valuations on $\mathscr{P}^d_L$, the class of lattice polytopes in $\mathbb{E}^d$, and on $\mathscr{P}^d_{\mathbb{Q}}$, the class of polytopes with vertices in $\mathbb{Q}$.

It is an interesting challenge to prove characterization theorems in the spirit of §§16, 17 for valuations on $\mathscr{P}^d_L$. The following result of Betke [1979; 1982] may be viewed as an analogue of Hadwiger's theorem (15.1).

(17.1) **Theorem.** *Let* φ *be a real valuation on* $\mathscr{P}^d_L$ *which satisfies the inclusion-exclusion principle and is invariant under unimodular transformations. Then*

$$\varphi(P) = \sum_{i=0}^{d} a_i G_i(P) \quad \textit{for} \quad P \in \mathscr{P}^d_L$$

with real constants $a_0,\ldots,a_d$.

A unimodular transformation of $\mathbb{E}^d$ is a volume preserving affine map of $\mathbb{E}^d$ into itself which leaves the lattice $\mathbb{Z}^d$ invariant. Due to Stein's [1982] result, the assumption that the valuation φ satisfy the inclusion-exclusion principle can be omitted.

We turn to lattice invariant valuations on $\mathscr{P}^d_{\mathbb{Q}}$. Some of the structural results of §16 have analogues for this case; these were used to prove Theorems (10.7) and (12.7). The following can be proved along the lines of Hadwiger's characterization of weakly continuous translation invariant simple valuations.

(17.2) **Theorem.** *A function* $\varphi:\mathscr{P}^d_{\mathbb{Q}} \to \mathbb{R}$ *is a lattice invariant simple valuation if and only if there is an expression*

$$\varphi(P) = \sum_{r=0}^{d} \sum_{U \in \mathscr{U}^{d-r}} V_r(P_U)\eta_r(U,P_U) \quad \textit{for} \quad P \in \mathscr{P}^d_{\mathbb{Q}},$$

where $\eta_r(U,P_U)$ *is odd in its first argument and depends only upon the translation class modulo* $\mathbb{Z}^d$ *of the translate of the subspace orthogonal to the frame* U *which contains the face* P_U.

This was proved by McMullen [1978]. As a consequence, one has (10.7) for simple valuations, and then the methods of §10 can be adapted to this case to yield (10.7) in general.

Inspection of the proof of (17.2) shows that the range $\mathbb{R}$ of the valuation φ may be replaced by an arbitrary $\mathbb{Z}$-module $\mathscr{X}$. As remarked in McMullen [1982b], it is further possible to extend Theorems (16.6), (16.7) and (16.8) to lattice invariant or covariant valuations on $\mathscr{P}^d_{\mathbb{Q}}$, as long as the coefficients λ_r,Λ_r,H_r and η_r exhibit the same additional dependence as the η_r in Theorem (17.2); further, Hom($\mathbb{E}^d$,$\mathscr{X}$) is replaced by $\mathrm{Hom}_{\mathbb{Z}}(\mathbb{Z}^d,\mathscr{X})$.

References

Numbers in square brackets at the end of a reference denote the sections of this report in which that reference is quoted. However, we have also included a number of references of which no specific mention is made in the text (we have attempted a certain degree of completeness in the references, but we make no claim to success).

Aleksandrov, A.D.

1937a Zur Theorie der gemischten Volumina von konvexen Körpern. I. Verallgemeinerung einiger Begriffe der Theorie der konvexen Körper. (Russian) Mat. Sbornik N.S. **2** (1937), 947–972. Zbl 17–426. [5]

1937b Zur Theorie der gemischten Volumina von konvexen Körpern. II: Neue Ungleichungen zwischen den gemischten Volumina und ihre Anwendungen. (Russian) Mat. Sbornik N.S. **2** (1937), 1205–1238. Zbl 18–276. [5]

Allendoerfer, C.B.

1948 Steiner's formulae on a general S^{n+1}. Bull. Amer. Math. Soc. **54** (1948), 128–135. MR 9–607. [3]

Baddeley, A.

1980 Absolute curvatures in integral geometry. Math. Proc. Camb. Phil. Soc. **88** (1980), 45–58. MR. 81k: 52011. [15]

Banchoff, T.

1967 Critical points and curvature for embedded polyhedra. J. Diff. Geom. **1** (1967), 245–256. MR 37 # 921. [5]

1970 Critical points and curvature for embedded polyhedral surfaces. Amer. Math. Monthly **77** (1970), 475–485. MR 41 # 4444. [5]

Bauer, H.

1978 Wahrscheinlichkeitstheorie und Grundzüge der Maßtheorie. 3rd ed., W. de Gruyter, Berlin-New York 1978. MR 39 # 983, MR 80b: 60001. [14].

Berg, C.

1969 Corps convexes et potentiels sphériques. Danske Vid. Selsk. Mat.-Fys. Medd. **37, 6** (1969). MR 40 # 7996. [13]

1971 Abstract Steiner points for convex polytopes. J. London Math. Soc. (2) **4** (1971), 176–180. MR 45 # 7593. [13]

Bernštein (Bernshtein), D.N.

1976 The number of integral points in integral polyhedra. (Russian) Funkcional. Anal. i Priložen. **10** (1976), no. 3, 72–73. English translation: Functional Anal. Appl. **10** (1976), 293–294. MR 54 # 8451. [4, 10]

Betke, U.

1979 Gitterpunkte und Gitterpunktfunktionale. Habilitationsschrift, Siegen 1979. [17]

1982 Ein Funktionalsatz für Gitterpolytope. (to appear) [17]

Betke, U., J.M. Wills

1979 Stetige und diskrete Funktionale konvexer Körper. In: Contributions to Geometry, ed. by J. Tölke and J.M. Wills. Birkhäuser Verlag, Basel etc. 1979, pp. 226–237. MR 81m: 10062. [4]

Blaschke, W.

1937 Vorlesungen über Integralgeometrie. 3rd ed., VEB Deutsch. Verl. d. Wiss., Berlin 1955 (First edition: 1937). [3, 15]

Böhm, J., E. Hertel

1980 Polyedergeometrie in n-dimensionalen Räumen konstanter Krümmung. VEB Deutsch. Verl. d. Wiss., Berlin 1980. Zbl 464.51009 + 466.52001. [3, 5, 14]

Boltianskii, V.G.
1956 Equivalent and equidecomposable figures. D.C. Heath Comp., Boston 1963 (Russian original: 1956).
MR 37 # 5761. [6]
1958 Zerlegungsgleichheit ebener Polygone. Bul. Inst. Politehnic Jaşi **4** (8), 33–38.
MR 21 # 2197.
1966 Equidecomposability of polygons and polyhedra. (Russian) Encyclopedia of Elementary Mathematics, vol. V, Moscow 1966, pp. 142–180.
1969 On Hilbert's third problem (Russian). Hilbert's problems (Russian). Izdat. "Nauka", Moscow, 1969, 92–94.
MR 40 # 7920.
1976 Decomposition equivalence of polyhedra and groups of motions. Dokl. Akad. Nauk SSSR **231** (1976), 788–790. English translation: Soviet Math. Dokl. **17** (1976), 1628–1631.
MR 54 # 13703.
1978 Hilbert's third problem. John Wiley & Sons, New York etc. 1978.
MR 57 # 7346, MR 58 # 18074. [3, 6]
Brianchon, C.J.
1837 Théorème nouveau sur les polyèdres. J. Ecole (Royale) Polytechnique **15** (1837), 317–319. [8]
Bricard, R.
1896 Sur une question de géométrie relative aux polyèdres. Nouv. Ann. Math. **15** (1896), 331–334. [9]
Busemann, H.
1958 Convex surfaces. Interscience Publishers, New York 1958.
MR 21 # 3900. [3]
Debrunner, H.
1952 Translative Zerlegungsgleichheit von Würfeln. Arch. Math. **3** (1952), 479–480.
MR 14–1115.
1969 Zerlegungsähnlichkeit von Polyedern. Elem. Math. **24** (1969), 1–6.
MR 38 # 6432. [6]
1978 Zerlegungsrelationen zwischen regulären Polyedern des E^d. Arch. Math. **30** (1978), 656–660.
MR 50 # 18089.
1980 Über Zerlegungsgleichheit von Pflasterpolytopen mit Würfeln. Arch. Math. **35** (1980), 583–587.
MR 82e: 52012.
Dehn, M.
1900 Über raumgleiche Polyeder. Nachr. Akad. Wiss. Göttingen, Math.-Phys. Kl. 1900, 345–354. [9]
1902 Über den Rauminhalt. Math. Ann. **55** (1902), 465–478. [9]
1905 Über den Inhalt sphärischer Dreiecke. Math. Ann. **60** (1905), 166–174.
Dupont, J.L., C.-H. Sah
1982 Scissors congruences, II. J. Pure Appl. Algebra **25** (1982), 159–195. [8]
Eckhoff, J.
1980 Die Euler-Charakteristik von Vereinigungen konvexer Mengen im $\mathbb{R}^d$. Abh. Math. Sem. Univ. Hamburg **50** (1980), 133–144.
MR 82d: 52008. [2]
Ehrhart, E.
1967a Sur un problème de géométrie diophantienne linéaire, I. J. reine angew. Math. **226** (1967), 1–29.
MR 35 # 4184. [4, 10]
1967b Démonstration de la loi de réciprocité pour un polyèdre entier. C.R. Acad. Sc. Paris, Sér. A **265** (1967), 5–7.
MR 39 # 6826. [4]
1967c Démonstration de la loi de réciprocité du polyèdre rationnel. C.R. Acad. Sc. Paris, Sér. A **265** (1967), 91–94.
MR 36 # 1393. [12]
1968 Sur la loi de réciprocité des polyèdres rationnels. C.R. Acad. Sc. Paris, Ser. A **266** (1968), 696–697.
MR 38 # 3233. [12]

Emch, A.
1946 Endlichgleiche Zerschneidung von Parallelotopen in gewöhnlichen und höheren euklidischen Räumen. Comment. Math. Helvet. **18** (1945), 224–231.
MR 8–83.

Fáry, I.
1961 Functionals related to mixed volumes. Illinois J. Math. **5** (1961), 425–430.
MR 24 # A811. [15]

Federer, H.
1959 Curvature measures. Trans. Amer. Math. Soc. **93** (1959), 418–491.
MR 22 # 961. [5]

Fenchel, W., B. Jessen
1938 Mengenfunktionen und konvexe Körper. Danske Vid. Selskab. Mat.-fys. Medd. **16,** 3 (1938).
Zbl 18, p. 424. [3]

Firey, W.J.
1976 A functional characterization of certain mixed volumes. Israel J. Math. **24** (1976), 274–281.
MR 58 # 30744. [15]

Gerwien, P.
1833a Zerschneidung jeder beliebigen Anzahl von gleichen geradlinigen Figuren in dieselben Stücke. J. reine angew. Math. **10** (1833), 228–234. [9]
1833b Zerschneidung jeder beliebigen Menge verschieden gestalteter Figuren von gleichem Inhalt auf der Kugelfläche in dieselben Stücke. J. reine angew. Math. **10** (1833), 235–240.

Gram, J.P.
1874 Om Rumvinklerne i et Polyeder. Tidsskr. Math. (Copenhagen) (3) **4** (1874), 161–163. [8]

Groemer, H.
1972 Eulersche Charakteristik, Projektionen und Quermaßintegrale. Math. Ann. **198** (1972), 23–56.
MR 48 # 7116. [2, 5, 12, 15]
1973 Über einige Invarianzeigenschaften der Eulerschen Charakteristik. Comment. Math. Helvet. **48** (1973), 87–99.
MR 49 # 7967. [2]
1974 On the Euler characteristic in spaces with a separability property. Math. Ann. **211** (1974), 315–321.
MR 50 # 5635. [2]
1975 The Euler characteristic and related functionals on convex surfaces. Geom. Dedicata **4** (1975), 91–104.
MR 56 # 6574. [2]
1977a Minkowski addition and mixed volumes. Geom. Dedicata **6** (1977), 141–163.
MR 57 # 1386. [3, 5]
1977b On translative integral geometry. Arch. Math. **29** (1977), 324–330.
MR 58 # 12872. [5]
1978 On the extension of additive functionals on classes of convex sets. Pacific J. Math. **75** (1978), 397–410.
MR 58 # 24003. [1, 5]

Gruber, P.M.
1979 Geometry of numbers. In: Contributions to Geometry, ed. by J. Tölke and J.M. Wills. Birkhäuser Verlag, Basel etc. 1979, pp. 186–225.
MR 81h: 10044. [4]

Gruber, P.M., R. Schneider
1979 Problems in geometric convexity. In: Contributions to Geometry, ed. by J. Tölke and J. M. Wills. Birkhäuser Verlag, Basel etc. 1979, pp. 255–278.
MR. 82d: 52001. [14]

Grünbaum, B.
1963 Measures of symmetry for convex bodies. Proc. Symp. Pure Math. **7** (1963), 223–270; Amer. Math. Soc., Providence, R.I.
MR 27 # 6187. [13]
1967 Convex polytopes. Interscience Publishers, London etc. 1967.
MR 37 # 2085. [3, 13]

Hadwiger, H.
1947 Über eine symbolisch-topologische Formel. Elem. Math. **2** (1947), 35–41. Portuguese translation: Gazeta Mat. **35** (1948), 6–9.
MR 8–484. [2]
1949a Bemerkung zur elementaren Inhaltslehre des Raumes. Elem. Math. **4** (1949), 3–7.
MR 10–564. [14]
1949b Über beschränkte additive Funktionale konvexer Polygone. Publ. Math. Debrecen **1** (1949), 104–108.
MR 12–124. [16]
1949c Zerlegungsgleichheit und additive Polyederfunktionale. Arch. Math. **1** (1949), 468–472.
MR 11–165.
1950a Zum Problem der Zerlegungsgleichheit der Polyeder. Arch. Math. **2** (1950), 441–444.
MR 13–377.
1950b Zerlegungsgleichheit und additive Polyederfunktionale. Comment. Math. Helvet. **24** (1950), 204–218.
MR 12–526.
1950c Zur Inhaltstheorie der Polyeder. Collect. Math **3** (1950), 137–158.
MR 13–154. [14]
1950d Einige Anwendungen eines Funktionalsatzes für konvexe Körper in der räumlichen Integralgeometrie. Monatsh. Math. **54** (1950), 345–353.
MR 12–734. [15]
1950e Translative Zerlegungsgleichheit k-dimensionaler Parallelotope. Collect. Math. **3** (1950), 11–23.
MR 13–768.
1951a Beweis eines Funktionalsatzes für konvexe Körper. Abh. Math. Sem. Univ. Hamburg **17** (1951), 69–76.
MR 12–851. [15]
1951b Translationsinvariante, additive und stetige Eibereichfunktionale. Publ. Math. Debrecen **2** (1951), 81–94.
MR 13–971. [11, 13, 16]
1952a Ergänzungsgleichheit k-dimensionaler Polyeder. Math. Z. **55** (1952), 292–298.
MR 14–309.
1952b Über addierbare Intervallfunktionale. Tôhoku Math. J. **4** (1952), 32–37.
MR 14–28.
1952c Mittelpunktspolyeder und translative Zerlegungsgleichheit. Math. Nachr. **8** (1952), 53–58.
MR 14–309.
1952d Additive Funktionale k-dimensionaler Eikörper. I. Arch. Math. **3** (1952), 470–478.
MR 14–1114. [14, 15]
1952e Translationsinvariante, additive und schwachstetige Polyederfunktionale. Arch. Math. **3** (1952), 387–394.
MR 14–964. [11, 15]
1953a Additive Funktionale k-dimensionaler Eikörper. II. Arch. Math. **4** (1953), 374–379.
MR 15–551.
1953b Über additive Funktionale k-dimensionaler Eipolyeder. Publ. Math. Debrecen **3** (1953), 87–94.
MR 16–345. [10, 16]
1953c Lineare additive Polyederfunktionale und Zerlegungsgleichheit. Math. Z. **58** (1953), 4–14.
MR 14–964.
1953d Über Gitter und Polyeder. Monatsh. Math. **57** (1953), 246–254.
MR 15–551. [4]
1954a Zum Problem der Zerlegungsgleichheit k-dimensionaler Polyeder. Math. Ann. **127** (1954), 170–174.
MR 15–691.
1954b Zur Zerlegungstheorie euklidischer Polyeder. Ann. Mat. pura appl. (IV) **36** (1954), 315–334.
MR 16–396.
1955a Eulers Charakteristik und kombinatorische Geometrie. J. reine angew. Math. **194** (1955), 101–110.
MR 17–402. [2]

1955b Altes und Neues über konvexe Körper. Birkhäuser Verlag, Basel und Stuttgart 1955. MR 17-401. [14, 15]

1956 Integralsätze im Konvexring. Abh. Math. Sem. Univ. Hamburg **20** (1956), 136–154. MR 18-503. [3, 15]

1957 Vorlesungen über Inhalt, Oberfläche und Isoperimetrie. Springer-Verlag, Berlin etc. 1957. MR 21 # 1561. [1–7, 10, 13–16]

1959 Normale Körper im euklidischen Raum und ihre topologischen und metrischen Eigenschaften. Math. Z. **71** (1959), 124–140. MR 23 # A1008. [2, 5]

1960 Zur Eulerschen Charakteristik euklidischer Polyeder. Monatsh. Math. **64** (1960), 49–60. MR 23 # A992. [2]

1963 Ungelöstes Problem Nr. 45. Elem. Math. **18** (1963), 29–31.

1968a Translative Zerlegungsgleichheit der Polyeder des gewöhnlichen Raumes. J. reine angew. Math. **233** (1968), 200–212. MR 38 # 6454.

1968b Eine Schnittrekursion für die Eulersche Charakteristik euklidischer Polyeder mit Anwendungen innerhalb der kombinatorischen Geometrie. Elem. Math. **23** (1968), 121–132. MR 38 # 5112. [2]

1968c Neuere Ergebnisse innerhalb der Zerlegungstheorie euklidischer Polyeder. Jber. Deutsche Math. Ver. **70** (1968), 167–176. MR 39 # 869. [6]

1969a Zur axiomatischen Charakterisierung des Steinerpunktes konvexer Körper. Israel J. Math. **7** (1969), 168–176. MR 40 # 4863. [13]

1969b Eckenkrümmung beliebiger kompakter euklidischer Polyeder und Charakteristik von Euler-Poincaré. L'Enseignement Math. **15** (1969), 147–151. MR 40 # 864. [5]

1969c Notiz zur Eulerschen Charakteristik offener und abgeschlossener Polyeder. Studia Sci. Math. Hung. **4** (1969), 385–387. MR 40 # 1888. [2]

1970 Zentralaffine Kennzeichnung des Jordanschen Inhaltes. Elem. Math. **25** (1970), 25–27. MR 55 # 8322. [14]

1971 Zur axiomatischen Charakterisierung des Steinerpunktes konvexer Körper; Berichtigung und Nachtrag. Israel J. Math. **9** (1971), 466–472. MR 43 # 3919. [1, 13, 15]

1972 Polytopes and translative equidecomposability. Amer. Math. Monthly **79** (1972), 275–276.

1973 Erweiterter Polyedersatz und Euler-Shephardsche Additionstheoreme. Abh. Math. Sem. Univ. Hamburg **39** (1973), 120–129. MR 48 # 9555. [2, 12]

1974a Begründung der Eulerschen Charakteristik innerhalb der ebenen Elementargeometrie. L'Enseignement Math. **20** (1974), 33–43. MR 50 # 14484. [2]

1974b Homothetieinvariante und additive Polyederfunktionen. Arch. Math. **25** (1974), 203–205. MR 49 # 11387. [6]

1975 Zerlegungsgleichheit euklidischer Polyeder bezüglich passender Abbildungsgruppen und invariante Funktionale. Math.-Phys. Sem. Ber. **22** (1975), 125–133. MR 51 # 6586. [6]

Hadwiger, H., P. Glur

1951 Zerlegungsgleichheit ebener Polygone. Elem. Math. **6** (1951), 97–106. MR 13-576. [9]

Hadwiger, H., P. Mani

1972 On the Euler characteristic of spherical polyhedra and the Euler relation. Mathematika **19** (1972), 139–143. MR 48 # 1040. [2]

1974 On polyhedra with extremal Euler characteristic. J. Combinat. Theory, Ser. A **17** (1974), 345–349. MR 50 # 11032. [2]

Hadwiger, H., R. Schneider
1971 Vektorielle Integralgeometrie. Elem. Math. **26** (1971), 49–57.
MR 44 # 967. [3, 5, 15]
Harazišvili, A.B.
1977 Equicomposition of polyhedra relative to the group of homotheties and translations. (Russian) Dokl. Akad. Nauk SSSR **236** (1977), 552–555. English translation: Soviet Math. Dokl. **18** (1977), 1246–1249.
MR 58 # 12726. [6]
1978 On an invariant defined on the class of three-dimensional polytopes. (Russian) Bull. Acad. Sci. Georgian SSR **91** (1978), 33–36.
MR 81b: 52006 [6]
Hertel, E.
1969 Über Intervallpolygone. Wiss. Z. Friedrich-Schiller-Univ. Jena, Math.-Nat. Reihe **18** (1969), 299–303.
MR 42 # 955.
1971 Über Intervallpolyeder im R_n. Beitr. Algebra Geom. **1** (1971), 77–83.
MR 46 # 2550.
1973 Zur translativen Zerlegungsgleichheit n-dimensionaler Polyeder. Publ. Math. Debrecen **20** (1973), 133–140.
MR 48 # 7117.
1974a Ein Subtraktionssatz der Polyederalgebra. Beitr. Algebra Geom. **2** (1974), 83–86.
MR 53 # 11498.
1974b Mittelpunktspolyeder im E^4. Elem. Math. **29** (1974), 59–64.
MR 49 # 7920.
1974c Polyederstrukturen. Math. Nachr. **62** (1974), 57–63.
MR 50 # 14494.
1977 Neuere Ergebnisse und Richtungen der Zerlegungstheorie von Polyedern. Mitt. Math. Ges. DDR 1977, Heft 4, 5–22.
Zbl 401. 52004. [6]
Hertel, E., H.E. Debrunner
1980 Zur Rolle von Subtraktionssatz und Divisionssatz in Zerlegungsstrukturen. Beitr. Algebra Geom. **10** (1980), 145–148.
Zbl 469. 52004.
Hilbert, D.
1899 Grundlagen der Geometrie. Teubner, Leipzig 1899. [3]
1900 Mathematische Probleme. Nachr. Königl. Ges. Wiss. Göttingen, Math.-Phys. Kl. 1900, 253–297. English translation: Bull. Amer. Math. Soc. **8** (1902), 437–479. [9]
Inzinger, R.
1949 Über eine lineare Transformation in den Mengen der konvexen und der stützbaren Bereiche einer Ebene. Monatsh. Math. **53** (1949), 227–250.
MR 11–455. [13]
Jessen, B.
1939 Om Polyedres Rumfang. Mat. Tidsskrift A 1939, 35–44.
MR 1–156.
1941 En Bemaerkning om Polyedres Volumen. Mat. Tidsskrift B 1941, 59–65.
MR 7–68.
1946 Om Aekvivalens af Aggregater af regulaere Polyedre. Mat. Tidsskrift B 1946, 145–148.
MR 7–386.
1967 Orthogonal icosahedra. Nordisk Mat. Tidsskrift **15** (1967), 90–96.
MR 37 # 2088.
1968 The algebra of polyhedra and the Dehn-Sydler theorem. Math. Scand. **22** (1968), 241–256.
MR 40 # 4860. [9]
1972 Zur Algebra der Polytope. Nachr. Akad. Wiss. Göttingen. II. Math.-phys. Kl. 1972, 47–53.
MR 50 # 5636. [9]
1978 Einige Bemerkungen zur Algebra der Polyeder in nicht-euklidischen Räumen. Comment. Math. Helvet. **53** (1978), 525–528.
MR 80b: 51010.

Jessen, B., J. Karpf, A. Thorup
1968 Some functional equations in groups and rings. Math. Scand. **22** (1968), 257–265.
MR 44 # 1963.
Jessen, B., A. Thorup
1978 The algebra of polytopes in affine spaces. Math. Scand. **43** (1978), 211–240.
MR 80i: 52008. [6]
Kagan, W.F.
1903 Über die Transformation der Polyeder. Math. Ann. **57** (1903), 421–424.
Kirsch, A.
1978 Polyederfunktionale, die nicht translationsinvariant, aber injektiv sind. Elem. Math. **33** (1978), 105–107.
MR 80a: 52009.
Klee, V.
1963 The Euler characteristic in combinatorial geometry. Amer. Math. Monthly **70** (1963), 119–127.
MR 26 # 3627. [2]
Kuiper, N.H.
1971 Morse relations for curvature and tightness. In: Proc. Liverpool Singularities Symp. II, ed. by C.T.C. Wall, pp. 77–89. Lecture Notes in Math. **209,** Springer, Berlin etc. 1971.
MR 49 # 8039. [5]
Kummer, H.
1956 Translative Zerlegungsgleichheit k-dimensionaler Parallelotope. Arch. Math. **7** (1956), 219–220.
MR 18–228.
Lebesgue, H.
1938 Sur l'équivalence des polyèdres, en particulier des polyèdres réguliers, et sur la dissection des polyèdres réguliers en polyèdres réguliers. Ann. Soc. Polon. Math. **17** (1938), 193–226.
1945 Sur l'équivalence des polyèdres. Ann. Soc. Polon. Math. **18** (1945), 1–3.
MR 8–218.
Leichtweiss, K.
1980 Konvexe Mengen. VEB Deutsch. Verl. d. Wiss., Berlin 1980.
MR 81b: 52001. [15]
Lenz, H.
1970 Mengenalgebra und Eulersche Charakteristik. Abh. Math. Sem. Univ. Hamburg **34** (1970), 135–147.
MR 42 # 3674. [2, 5]
Lindgren, H.
1972 Recreational problems in geometric dissections and how to solve them. Revised and enlarged by Greg Frederickson. Dover Publ., New York 1972.
Zbl 261. 50006.
Maak, W.
1960 Differential- und Integralrechnung. 2nd ed., Vandenhoeck und Ruprecht, Göttingen 1960.
MR 22 # 3764. [14]
Macdonald, I.G.
1963 The volume of a lattice polyhedron. Proc. Camb. Phil. Soc. **59** (1963), 719–726.
MR 27 # 4139. [4]
1971 Polynomials associated with finite cell-complexes. J. London Math. Soc. **4** (1971),181–192.
MR 45 # 7594. [4]
Mani, P.
1971 On angle sums and Steiner points of polyhedra. Israel J. Math. **9** (1971), 380–388.
MR 43 # 3913. [5]
Matheron, G.
1975 Random sets and integral geometry. Wiley, New York etc. 1975.
MR 52 # 6828. [15]
McMullen, P.
1975a Metrical and combinatorial properties of convex polytopes. Proc. Internat. Congr. Math. Vancouver 1974. (1975), pp. 491–495.
MR 55 # 3974. [10, 12]

1975b Non-linear angle-sum relations for polyhedral cones and polytopes. Math. Proc. Camb. Phil. Soc. **78** (1975), 247–261.
MR 52 # 15238. [7, 8]
1977 Valuations and Euler-type relations on certain classes of convex polytopes. Proc. London Math. Soc. **35** (1977), 113–135.
MR 56 # 6548. [6, 7, 10–12, 15]
1978 Lattice invariant valuations on rational polytopes. Arch. Math. **31** (1978), 509–516.
MR 80d: 52011. [10, 12, 17]
1980 Continuous translation invariant valuations on the space of compact convex sets. Arch. Math. **34** (1980), 377–384.
MR 81m: 52013. [16]
1982a The Brianchon-Gram theorem for polyhedral sets. (to appear) [8]
1982b Weakly continuous valuations on convex polytopes. (to appear) [6, 7, 10, 12, 16, 17]

Meier, C.
1972 Zerlegungsähnlichkeit von Polyedern. J. reine angew. Math. **253** (1972), 193–202.
MR 46 # 792. [6]
1977 Multilinearität bei Polyederaddition. Arch. Math. **29** (1977), 210–217.
MR 57 # 4005. [6, 10]
1982 Ein Skalarprodukt für konvexe Körper. (in preparation) [5]

Meyer, W.J.
1970 Characterization of the Steiner point. Pacific J. Math. **35** (1970), 717–725.
MR 43 # 3917. [13]

Minkowski, H.
1911 Theorie der konvexen Körper, insbesondere Begründung ihres Oberflächenbegriffs. Ges. Abh., vol. II, pp. 131–229. B.G. Teubner, Leipzig etc. 1911. [3]

Müller, H.R.
1967 Zur axiomatischen Begründung der Eikörperfunktionale. Monatsh. Math. **71** (1967), 338–343.
MR 36 # 5818. [15]

Munroe, M.E.
1953 Introduction to measure and integration. Addison-Wesley Publ. Comp., Reading, Mass. 1953.
MR 14–734. [14]

Mürner, P.
1974 Zwei Beispiele zur Zerlegungsgleichheit 4-dimensionaler Polytope. Elem. Math. **29** (1974), 132–135.
MR 50 # 8303.
1975 Translative Parkettierungspolyeder und Zerlegungsgleichheit. Elem. Math. **30** (1975), 25–27.
MR 51 # 11282.
1977 Translative Zerlegungsgleichheit von Polytopen. Arch. Math. **29** (1977), 218–224.
MR 57 # 4006.

Nicoletti, O.
1914 Sulla equivalenza dei poliedri. Rend. Circ. Mat. Palermo **37** (1914), 47–75.
1915 Sulla equivalenza dei poliedri. Rend. Circ. Mat. Palermo **40** (1915), 194–210.

Perles, M.A., Sallee, G.T.
1970 Cell complexes, valuations, and the Euler relation. Canad. J. Math. **22** (1970), 235–241.
MR 41 # 7530. [5, 12]

Posicel'skiĭ (Positsel'skii), E.D.
1973 Characterization of Steiner points. (Russian) Mat. Zametki **14** (1973), 243–247. English translation: Math. Notes **14** (1973), 698–700.
MR 48 # 4916. [13]

Rota, G.-C.
1964 On the foundations of combinatorial theory. I. Theory of Möbius functions. Z. Wahrscheinlichkeitstheorie verw. Geb. **2** (1964), 340–368.
MR 30 # 4688. [2, 5, 8]
1971 On the combinatorics of the Euler characteristic. Studies in Pure Mathematics (Papers presented to Richard Rado), pp. 221–233. Academic Press, London 1971.
MR 44 # 126. [2]

Sah, C.-H.
1979 Hilbert's third problem: Scissors congruence. Pitman Advanced Publishing Program, San Francisco etc. 1979.
MR 81g: 51011. [6, 8]
1981 Scissors congruences. I: The Gauss-Bonnet map. Math. Scand. **49** (1981), 181–210. [8]
Sallee, G.T.
1966 A valuation property of Steiner points. Mathematika **13** (1966), 76–82.
MR 33 # 7937. [1, 5]
1968 Polytopes, valuations, and the Euler relation. Canad. J. Math. **20** (1968), 1412–1424.
MR 38 # 605. [1, 5, 12]
1971 A non-continuous "Steiner point". Israel J. Math. **10** (1971), 1–5.
MR 45 # 5873. [13]
Santaló, L.A.
1976 Integral geometry and geometric probability. Addison-Wesley Publ. Comp., Reading, Mass. 1976.
MR 55 # 6340. [3]
Scherk, P.
1969 Über eine Klasse von Polyederfunktionalen. Comment. Math. Helvet. **44** (1969), 191–201.
MR 39 # 6189.
Schmitt, K.A.
1968 Kennzeichnung des Steinerpunktes konvexer Körper. Math. Z. **105** (1968), 387–392.
MR 37 # 6834. [13]
Schneider, R.
1971 On Steiner points of convex bodies. Israel J. Math. **9** (1971), 241–249.
MR 43 # 3918. [13]
1972a Krümmungsschwerpunkte konvexer Körper, I. Abh. Math. Sem. Univ. Hamburg **37** (1972), 112–132.
MR 46 # 6160. [3]
1972b Krümmungsschwerpunkte konvexer Körper, II. Abh. Math. Sem. Univ. Hamburg **37** (1972), 204–217.
MR 48 # 9554. [3, 14, 15]
1973 Volumen und Schwerpunkt von Polyedern. Elem. Math. **28** (1973), 137–141.
MR 48 # 12305. [14]
1974a Equivariant endomorphisms of the space of convex bodies. Trans. Amer. Math. Soc. **194** (1974), 53–78.
MR 50 # 5633. [13]
1974b Bewegungsäquivariante, additive und stetige Transformationen konvexer Bereiche. Arch. Math. **25** (1974), 303–312.
MR 49 # 9738. [13, 16]
1974c Additive Transformationen konvexer Körper. Geom. Dedicata **3** (1974), 221–228.
MR 50 # 8299. [13]
1975a Kinematische Berührmaβe für konvexe Körper. Abh. Math. Sem. Univ. Hamburg **44** (1975), 12–23.
MR 52 # 15249. [3, 15]
1975b Kinematische Berührmaβe für konvexe Körper und Integralrelationen für Oberflächenmaße. Math. Ann. **218** (1975), 253–267.
MR 52 # 806. [15]
1977a Kritische Punkte und Krümmung für die Mengen des Konvexringes. L'Enseignement Math. **23** (1977), 1–6.
MR 57 # 7474. [5]
1977b Ein kombinatorisches Analogon zum Satz von Gauss-Bonnet. Elem. Math. **32** (1977), 105–108.
MR 58 # 13057. [5]
1978 Curvature measures of convex bodies. Ann. Mat. pura appl. **116** (1978), 101–134.
MR 80d: 52012. [3, 14, 15]
1979 Boundary structure and curvature of convex bodies. In: Contributions to Geometry, ed. by J. Tölke and J.M. Wills. Birkhäuser Verlag, Basel etc. 1979, pp. 13–59.
MR 81i: 52001. [3, 15]

1980 Parallelmengen mit Vielfachheit und Steiner-Formeln. Geom. Dedicata **9** (1980), 111–127. MR 81f: 52006. [5]

1981 A uniqueness theorem for finitely additive invariant measures on a compact homogeneous space. Rend. Circ. Mat. Palermo (Ser. II) **30** (1981), 341–344. [14]

Shephard, G.C.

1966 The Steiner point of a convex polytope. Canad. J. Math. **18** (1966), 1294–1300. MR 35 # 4814. [12]

1967 An elementary proof of Gram's theorem for convex polytopes. Canad. J. Math. **19** (1967), 1214–1217. MR 37 # 822. [8]

1968a The mean width of a convex polytope. J. London Math. Soc. **43** (1968), 207–209. MR 37 # 2087. [12]

1968b A uniqueness theorem for the Steiner point of a convex region. J. London Math. Soc. **43** (1968), 439–444. MR 37 # 3447. [13]

1968c Euler-type relations for convex polytopes. Proc. London Math. Soc. (3) **18** (1968), 597–606. MR 38 # 606. [5, 12]

1968d Angle deficiencies of convex polytopes. J. London Math. Soc. **43** (1968), 325–336. MR 37 # 5784. [3]

Sommerville, D.M.Y.

1927 The relations connecting the angle-sum and volume of a polytope in space of *n* dimensions. Proc. Roy. Soc. London Ser. A **115** (1927), 103–119. [8]

Spiegel, W.

1976a Zur Minkowski-Additivität bestimmter Eikörperabbildungen. J. reine angew. Math. **286/287** (1976), 164–168. MR 55 # 3969. [10]

1976b Ein Zerlegungssatz für spezielle Eikörperabbildungen in den euklidischen Raum. J. reine angew. Math. **283/284** (1976), 282–286. MR 53 # 6422. [5]

1978 Ein Beitrag über additive, translationsinvariante, stetige Eikörperfunktionale. Geom. Dedicata **7** (1978), 9–19. MR 57 # 13699. [10]

Stachó, L.L.

1979 On curvature measures. Acta Sci. Math. **41** (1979), 191–207. MR 80j: 53065. [5]

Stein, R.

1982 Additivität und Einschließungs-Ausschließungsprinzip für Funktionale von Gitterpolytopen. Dissertation, Dortmund 1982. [7, 10, 17]

Strambach, K.

1966 Über die Zerlegungsgleichheit von Polygonen bezüglich Untergruppen nichteuklidischer Bewegungsgruppen. Math. Z. **93** (1966), 276–288. MR 34 # 660.

Sydler, J.-P.

1965 Conditions nécessaires et suffisantes pour l'équivalence des polyèdres de l'espace euclidien à trois dimensions. Comment. Math. Helvet. **40** (1965), 43–80. MR 33 # 632. [9]

Tverberg, H.

1974 How to cut a convex polytope into simplices. Geom. Dedicata **3** (1974), 239–240. MR 50 # 1127. [6]

Valette, G.

1974 Subadditive affine-invariant transformations of convex bodies. Geom. Dedicata **2** (1974), 461–465. MR 51 # 11279. [13]

Volland, W.

1957 Ein Fortsetzungssatz für additive Eipolyederfunktionale im euklidischen Raum. Arch. Math. **8** (1957), 144–149. MR 19–1074. [5]

Walkup, D.W., R.J. Wets
1969 Lifting projections of convex polyhedra. Pacific J. Math. **28** (1969), 465–475. MR 39 # 3390. [6]
Weil, W.
1974a Über den Vektorraum der Differenzen von Stützfunktionen konvexer Körper. Math. Nachr. **59** (1974), 353–369. MR 49 # 6033. [5]
1974b Decomposition of convex bodies. Mathematika **21** (1974), 19–25. MR 51 # 1611. [5]
1975a On mixed volumes of nonconvex sets. Proc. Amer. Math. Soc. **53** (1975), 191–194. MR 52 # 4130. [5]
1975b Einschachtelung konvexer Körper. Arch. Math. **26** (1975), 666–669. MR 53 # 9036. [13]
1981 Das gemischte Volumen als Distribution. Manuscripta math. **36** (1981), 1–18. [5]
Wieacker, J.A.
1982 Translative stochastische Geometrie der konvexen Körper. Dissertation, Freiburg 1982. [3]
Zylev, V.B.
1965 Equicomposability of equicomplementable polyhedra. (Russian) Dokl. Akad. Nauk. SSSR **161** (1965), 515–516. English translation: Soviet Math. Dokl. **6** (1965), 453–455. MR 31 # 1610.
1968 G-composedness and G-complementability (Russian) Dokl. Akad. Nauk. SSSR **179** (1968), 529–530. English translation: Soviet Math. Dokl. **9** (1968), 403–404. MR 37 # 2069. [6]

Remarks and further references (added in proof)

An application of the Euler characteristic of Hadwiger's normal bodies, which were mentioned in §5, to a question in probability theory may be found in

Adler, R. J., Hasofer, A. M., Level crossings for random fields. Ann. Prob. **4** (1976), 1–12.
Adler, R. J., The geometry of random fields. Wiley, Chichester, etc. 1981.

An additive extension of Federer's curvature measures to certain (but not all) finite unions of sets of positive reach was recently studied by

Zähle, M., Curvature measures and random sets I (to appear).

The expression of Weil [1981] in §5 has been extended (and the proof corrected) by

Goodey, P. R., W. Weil, Distributions and valuations (to appear).

They show that, if $\varphi:(\mathscr{K}^d)^r \to \mathbb{R}$ is any continuous multilinear function, then there is a distribution T on $(\Omega^{d-1})^r$, such that

$$\varphi(\mathrm{K}_1,\dots,\mathrm{K}_r) = T\big(h(K_1,\cdot) \times \cdots \times \mathrm{h}(\mathrm{K}_r,\cdot)\big)$$

The problem of finding the syzygies between the Hadwiger functionals, discussed at the end of §6, is in effect settled in Proposition 3.16 of

Dupont, J. L., Algebra of polytopes and homology of flag complexes. Osaka J. Math. **19** (1982), 599–641.

The recently published volume "Convexity and related combinatorial geometry", ed. by D. C. Kay and M. Breen, Marcel Dekker, New York etc. 1982, contains two articles which are concerned with valuations on polytopes:

Sallee, G. T., Euler's theorem and where it led; pp. 45–55.
Spiegel, W., Nonnegative, motion-invariant valuations of convex polytopes; pp. 67–72.

In the discussion in §10, it was necessary in following McMullen's approach to assume that $\mathscr{X}$ was a real vector space. It is worth remarking, though, that a rational vector space or abelian group (regarded as a $\mathbb{Z}$-module) $\mathscr{X}$ can be embedded in a real vector space $\mathscr{Y} = \mathbb{R}\mathscr{X}_{\mathbb{Z}}\,\mathscr{X}$; $\mathscr{Y}$ inherits its real vector space structure from the first component of the tensor product, and $\mathscr{X}$ itself can be identified with $\{1\} \times \mathscr{X}$. We can now pass from the general valuation φ to the simple valuations $\psi_{\mathscr{A}}$ and back, using the angle-sum relations. Analysis of the proof of (7.1) (compare (6.7)) shows that we obtain expressions of the form

$$\varphi(nP) = \sum_{r=0}^{d} \binom{n}{r} \tilde{\varphi}_r(P),$$

where each $\tilde{\varphi}_r\, \mathscr{P}^d \to \mathscr{X}$ is a continuous translation invariant valuation. If $\mathscr{X}$ is a rational vector space, then we obtain the required rational polynomial expansion of (10.3).

Considerable progress with some of the problems mentioned in §15, 16 has been achieved recently. In the work of Goodey and Weil quoted above (for $r = 1$) and by *U.* Betke and Goodey (in preparation, for $r \in \{2,\dots,d-2\}$), it is shown that, if φ is a continuous translation invariant valuation on $\mathscr{K}^d$, which is homogeneous of degree r, then there exist sequences $(L_i)_{i \in \mathbb{N}}$, $(M_i)_{i \in \mathbb{N}}$ in $\mathscr{K}^d$, such that

$$\varphi(K) = \lim\, [V(K,r;B,d-1-r;L_i) - V(K,r;B,d-1-r;M_i)],$$

uniformly for K in a compact subset of $\mathscr{K}^d$. This exactly generalizes the reformulation of (16.11) (immediately following), and proves a suitably modified conjecture of McMullen [1980]. Betke and Goodey make use of (16.6) to show that, if a polytope P is identified with its r-th order area function $S_r(P;\cdot)$, then α induces a continuous linear mapping on the space spanned by such functions. The method of Betke and Goodey also provides a new approach to Hadwiger's characterization theorem (15.1), as well as an affirmative solution to the continuous case of Problem (15.4).

P. McMullen
Department of Mathematics
University College London
Gower Street
London WCIE 6BT (England)

R. Schneider
Mathematisches Institut
Universität Freiburg
Albertstrasse 236
D–7800 Freiburg (FRG)

Pier Luigi Papini*

Minimal and Closest Points Nonexpansive and Quasi-Nonexpansive Retractions in Real Banach Spaces

Summary

The notions indicated in the title and related ones have been studied independently by various authors. Some characterizations of inner product spaces have been given in terms of properties of minimal and closest points, or by considering special retractions. Yet, the very strong connections existing among the above notions have not been considered in full. This present analysis shows that some results can be obtained from older ones, and other results can be restated in a sharper form using known facts and simple connections relating the various notions.

1. Introduction

Let us consider a real Banach space X. Recall the following theorem of Kakutani: the existence of a norm-one projection onto every closed subspace (or onto every closed hyperplane) characterizes inner product spaces (i.p.s.) among spaces of at least three dimensions. This is a now classical result, and a key tool to prove other characterizations of i.p.s. But norm-one projections are very special maps (they are linear, nonexpansive and idempotent). Thus it has some interest to indicate formally stronger formulations of Kakutani theorem by considering more general maps. To this end, C. Franchetti (in a private communication to the author) observed the following; it is not difficult to prove that for a hyperplane M these properties are equivalent (see section 2 for the definitions):

(a) M is the range of a norm-one projection from X;
(b) M is a nonexpansive retract of X;
(c) M is a quasi-nonexpansive retract of X;
(d) M is an optimal set.

Other characterizations of i.p.s. can be obtained by considering different sets and maps: e.g., closed convex and bounded sets (or balls) instead of linear sets (or respectively hyperplanes), and different kinds of retractions. The analogues to linear, norm-one projections (which are—in some sense—the best ones for linear sets) are the sunny nonexpansive retractions (for special sets, the radial map plays this role).

In a few years, weaker "contraction extension properties" have been used to obtain—at least under some restrictions—characterizations of i.p.s. for spaces whose dimension is not less than three. Results in this area are due mainly to Schönbeck, deFiguereido and Karlovitz (see [43]), and the successive improvements are very strong. We summarize some of the conditions used in the

* Work performed under the auspices of the G.N.A.F.A. of the C.N.R. (National Council of Research of Italy).

following list:

(e) any nonexpansive map from any set C into X can be extended to a nonexpansive map on X.

(f) any set C is a nonexpansive retraction of X (i.e., the identity on C can be extended to a nonexpansive map on X such that $T(X) \subset C$).

(e′),(f′) same as in (e), (f) respectively, but only for closed convex sets C.

(g) any bounded closed and convex set C is the fixed-point set for a nonexpansive map defined on X.

(h) there exists a retraction onto the unit ball of X (or onto some similar set) which is nonexpansive (in particular: the radial map onto such a set is nonexpansive).

Also, we could study similar conditions by considering e.g. quasi-nonexpansive (instead of nonexpansive) maps.

The notions we are dealing with are related to the characterizations of i.p.s., but they have also some independent interest. Note that in most cases the sets we shall consider (fixed-point sets for special maps, etc.) are convex.

Let us make a short history of the notions we are discussing. The notion of a closest point to a set was introduced by Fejér [17]. Later on Phelps [35] studied this concept and used it to give a characterization of i.p.s. More recently Beauzamy [1] introduced the notion of a minimal point with respect to a set, and obtained results concerning such points which were published in [3]. Here we point out that minimal and closest points are the same when the space is strictly convex. The center of a set, as defined and studied by Chebyshev and Garkavi (see [23, §31] or [42, II. §6.4]) can be related to the kind of points we are considering. In particular, it is of some interest to know whether a center of a set must be a closest or a minimal point for the set (see [42, p. 289]).

The notion of quasi-nonexpansive map is essentially due to Diaz and Metcalf [13]. These maps were studied mainly in the context of fixed point theory by Dotson, Massa, and many others (see e.g. [40]). In particular, the approximation of fixed points by means of iterative processes has been studied (see e.g. [13] and [34]).

Some kinds of retractions, considered as maps of approximation (and called R_C maps) were defined explicitly by Franchetti and Furi [18]; initially considered only for linear subspaces, the definition of R_C maps was extended by Bruck [7], then also in [32], so as to include any set C (see also [44] for other references on them).

We show here that quasi-nonexpansive maps onto a set C are related to R_C maps. We also try to show that also in the setting of quasi-nonexpansive maps, some questions already studied for nonexpansive maps are of some interest.

The existence of nonexpansive retractions onto the unit ball, or some other bounded sets, was studied by Bruck, deFiguereido, Karlovitz and others (see the references in [21]). We relate these questions to the nonexpansiveness of some quasi-nonexpansive maps.

Firmly nonexpansive mappings were defined by Bruck [7]. We recall some facts about these maps, since they have strong connections with sunny properties of other maps which play an important role in our context. When the dimension of the space is two, the situation is much simpler than the case of three or more dimensions. Some properties which in more dimensions rely on the existence of an inner product, are simply a consequence of strict convexity in two-dimensional spaces.

The paper is divided into seven parts. After the introduction (part 1), part 2 contains the definitions and notations used. In part 3 the main results appearing in the existing literature are listed. In part 4, general comments and the connections we alluded to are exposed. Part 5 deals with firmly nonexpansive maps. Part 6 deals with centers, while part 7 is concerned with two-dimensional spaces. Finally, an appendix (containing only some marginal comments or remarks) completes the paper. We consider only real Banach spaces, though a good deal of the results we are considering remains true in a general normed space. But under our assumption the overlaps with existing literature are larger. Also, we do not consider in detail related results concerning hyperplanes.

2. Definitions and notations

Throughout this paper X denotes a Banach space over the real field $\mathbb{R}$; X^* is the conjugate space of X and the dimension of X is denoted by dim(X). The following possible properties of X will be considered (and abbreviations used): reflexivity (r), strict convexity (SC), smoothness (S), the norm of X is generated by an inner product (H).

Let C be a proper subset of X; we denote by $\bar{C}$ its closure and by co(C) [$\overline{\text{co}}$(C)] its convex [respectively: closed convex] hull.

A *smooth set* is a convex body such that at each of its boundary points there is a unique supporting hyperplane.

If $x \in X$ and $r > 0$ we denote by $S(x,r)$ the set $\{y \in X; \|y - x\| < r\}$, and by $B(x,r)$ its closure. Given two points x,y in X, we set $\langle x,y \rangle = \{tx + (1-t)y; 0 \leq t \leq 1\}$; also, we set $\tau(x,y) = \lim_{t \to 0^+} (\|x + ty\| - \|x\|/t)$.

A point $x \in X$ is a *closest point* to C if no point is "pointwise closer" than x to C, that is no point $x' \in X$ satisfies $\|x' - y\| < \|x - y\|$ for every $y \in C$. We denote by m(C) the set of all closest point to C:

$$(1) \qquad m(C) = \left\{ x \in X; \bigcap_{y \in C} S(y, \|y - x\|) = \varnothing \right\}.$$

A point $x \in X$ is a *minimal point* with respect to C if no point $x' \in X$, $x' \neq x$ satisfies $\|x' - y\| \leq \|x - y\|$ for every $y \in C$. We denote by min(C) the set of all minimal points with respect to C:

$$(2) \qquad \min(C) = \left\{ x \in X; \bigcap_{y \in C} B(y, \|y - x\|) = \{x\} \right\}.$$

We call the set C *optimal* if C = min(C). We denote by I the identity map on X.

Given a map T from a subset of X to X, we denote by dom(T) its domain, by R(T) its range, and by F(T) the set of its fixed points (*fixed set* of T): i.e., $F(T) = \{x \in \text{dom}(T); Tx = x\}$. Given a map T, we say that:

T is *nonexpansive* if $\|Tx - Ty\| \leq \|x - y\|$ for every pair x,y in dom(T);
T is *quasi-nonexpansive* if $F(T) \neq \varnothing$, and $\|Tx - y\| \leq \|x - y\|$ whenever $x \in \text{dom}(T)$, $y \in F(T)$;

T is *firmly nonexpansive* if $\|Tx - Ty\| \leq \tau(Tx - Ty, x - y)$ for every pair x,y in dom(T);
T is a *retraction* from dom(T) onto R(T) if F(T) = R(T) (i.e., T is "idempotent": $T^2 = T$);
T is called *sunny* if $x \in \text{dom}(T)$ and $x' = Tx$ imply $T(x' + t(x - x')) = x'$ whenever $t > 0$ and $x' + t(x - x') \in \text{dom}(T)$.

We say that a subset C of some set D is a *sunny*, or a *nonexpansive*, or a *quasi-nonexpansive retract* of D when there exists a retraction from D onto C with the relative property.

We call R_C*-map* a set-valued map onto a set C such that any selection from it is a quasi-nonexpansive retraction onto C. As proved in [32, Theorem 2], one of those selection, R'_C, is sunny if and only if the elements $x' \in R'_C(x)$, for $x \in \text{dom}(R_C)$, can be characterized by the property:

(3) $\tau(x' - y, x - x') \geq 0$ for every $y \in C$.

Thus the set-valued maps R'_C, defined in [32] as those satisfying (3), are particular R_C maps.

We observe that condition (3) is easily seen to be equivalent to the following fact: x′ is a best approximation to every $y \in C$ from the segment $\langle x', x \rangle$. The equivalence between the presently defined maps and the best approximation maps from the subspaces of X holds only in (H) spaces, and in a few other two-dimensional spaces (see [18]).

We recall that in [7] sunny retractions, and retractions satisfying (3) (that is, sunny quasi-nonexpansive retractions) from convex subsets of X were considered, and called respectively "projections" and "orthogonal retractions". Sunny retractions were called "ray retractions" in [8]. R_C maps were called (metric) "coprojections" in [44] and [22]. Hetzelt, in his thesis, studied in detail R_C maps in $\mathbb{R}^n$: this thesis is surveyed in [22]. For the definition of "center" we refer to section 6.

In the following a theorem will always denote a known result; some new results are stated as propositions (or corollaries).

3. General known results

We collect here (without proof) the main known results dealing with the notions defined in section 2. These are stated in a unified terminology and are listed respecting more or less the chronological order in which they were published.

Theorem 1. ([35, Theorems 4 and 5]). *If* X *is* (H), *then* $m(C) = \overline{co}(C)$ *for every* $C \subset X$. *If* $\dim(X) \geq 3$ *and* $m(C) \subset \overline{co}(C)$ *for every set* $C \subset X$ (*or equivalently: for every hyperplane*), *then* X *is* (H).

Theorem 2. ([14, Theorem 1]). *Let the map* T *be quasi-nonexpansive. Then* T *is continuous in* F(T), *which is closed in* dom(T). *If in addition* X *is* (SC) *and* dom(T) *is convex, then* F(T) *is convex.*

Theorem 3. ([7, Theorem 1] or [36, Lemma 2.7]). *Let* X *be* (S). *Then there exists at most one sunny quasi-nonexpansive retraction* (*from any set* D) *onto a given* (*closed*) *set* C. *Moreover, such a retraction* (*whenever it exists*) *is nonexpansive.*

Theorem 4. ([21; Satz 2, p. 192 and Hilfssatz 9]). *Suppose that* X *is* (SC) *and* $3 \leq \dim(X) < \infty$, *and let* F *be the fixed set of a nonexpansive map from* X *and such that* $\theta \in F$; *then* F *is a sunny nonexpansive retract of* X.

Theorem 5. ([5]). *Suppose* $\dim(X) \geq 3$. *Then* X *is* (H) *if and only if there exists a bounded smooth nonexpansive retract of* X *with a nonempty interior.*

Theorem 6. ([3, Proposition III.6]). *If* X *is* (r), (SC), (S), *and* $\dim(X) \geq 3$, *then* $B(\theta,1)$ *is optimal if and only if* X *is* (H).

Theorem 7. ([3, p. 108 and Proposition 1.4]). *Let be* $C \subset X$; *then we have* $C \subset \min(C) = \min(\bar{C})$. *If* X *is* (*H*), *then* $\min(C) = \min(\mathrm{co}(C)) = \overline{\mathrm{co}}(C)$.

Theorem 8. ([3, Proposition III.1]). *Any quasi-nonexpansive retract from* X *is optimal.*

Theorem 9. ([3, Proposition I.2]). *Let* X *be the range of a norm-one projection from* X** *and let* C *be a subset of* X. *To each* $x \in X$ *we can associate a point* $x' \in m(C)$ *such that* $\|x' - y\| \leq \|x - y\|$ *for every* $y \in C$.

Theorem 10. ([2]). *Suppose* X *is* (r), (SC), (S) *and separable. Let* C *be a closed convex set with a nonempty interior such that* $\theta \in C$. *If* C *is optimal, then it is a nonexpansive retract of* X.

Theorem 11. ([38, Theorem 4.1]). *Suppose that* X *is a* (r) *space with a uniformly Gâteaux differentiable norm. Then any nonexpansive retract* C *of a closed convex set* D *is also a sunny nonexpansive retract of* D.

Comments and remarks

Theorem 1: by using the equivalence between the conditions (a) and (d) of introduction and Kakutani theorem, the second part of this theorem can be reformulated by only assuming $\min(C) \subset \overline{\mathrm{co}}(C)$ for every set $C \subset X$ (or equivalently: for every hyperplane). Such a formulation would be slightly stronger, but only when X is not (SC) (see (4) and Proposition 1 in section 4). Compare this theorem also with the Corollary of Proposition 4 in section 6 (see also [18, Theorem 1]).

Theorem 2: if the assumption of (SC) for X is dropped, F(T) is not necessarily convex; if T is nonexpansive this remark applies too (see [7, Example 2]).

Theorem 3: note that nonexpansive retractions are not unique in general, also when X is (H) (if π_C denotes the best approximation map onto C, let $C_1 \subset C_2$; then consider $\pi_{C_1} \circ \pi_{C_2}$ and π_{C_1}).

Theorem 6 (and also Theorem 5; see also the discussion after Proposition 3 in section 4): a special sunny retraction from X onto $B(\theta,1)$, as well as onto a "similar" set C, is the *radial map*, associating to every $x \in X$ its nearest point along the section between C and the straight line passing through x and θ. If X is (H), then such a map is nonexpansive. The same is not true in general when X is not (H) (see e.g. [21, §1 and the discussion preceding Zusatz 3 at p. 192]).

Theorem 9: in [3] this result is stated in a slightly different form. It springs out from a careful reading of that proof that our statement is true. Note that by Theorem 9 we can define a retraction from X onto m(C): but the inequalities we consider are not generally true for $y \in m(C)$ (thus, in general, such a retraction is not quasi-nonexpansive if $C \neq m(C)$).

Some of the above indicated results are related to the following possible implications (note that the set we denote by C will often be assumed to be closed and convex in our discussion):

(α) a nonexpansive retract C of X is a sunny nonexpansive retract of X;

(β) the fixed set of a nonexpansive map from C to C is a nonexpansive retract of C;

(γ) a quasi-nonexpansive retract C of X is a sunny quasi-nonexpansive retract of X;

(δ) the fixed set of a quasi-nonexpansive map from C to C is a quasi-nonexpansive retract of C.

In particular, Theorem 4 is concerned with (α) and (β); Theorem 11 is concerned with (α).

Other conditions related to the implication (α) have been considered in [7, Theorems 2 (b) and 3] and in [36, Lemma 2.9], where the smoothness assumptions made for X are weaker than in Theorem 11, but some compactness assumption on C is made (thus these results are somewhat similar to Theorem 4).

Conditions implying (β) have been considered e.g. in [6, Theorem 2 and Example 1] and, for maps defined in X, also in [36, Lemma 2.9 (b), (c)]. Concerning (γ) and (δ) we quote the following result:

Theorem 12. ([7, Corollary 2 to Theorem 4]). *If* X *is* (S), C *is closed and convex and* $T: C \to C$ *is continuous, compact and quasi-nonexpansive, then* F(T) *is a sunny nonexpansive retract of* C.

Note that in general, for C a nice set, the sunny retract (when exists) is not the radial map.

As noted in section 2, the implication (γ) is related to the implication: an R_C map has a selection satisfying (3). Some results concerning the latter have been given in [44]: they rely upon an existence and unicity assumption on R_C maps, which turns out to have very strong implications on the structure of the set C. Finally, the implication (δ) has been considered in [25, Corollary 2 and Theorems 5 and 6].

We quote another, rather old, characterization of (H). We denote by $\mathscr{C}$ the class of bounded, nonempty closed convex subsets of X.

Theorem 13. ([4, Theorem 3.1]). *Suppose* X *to be* (r) *and* $\dim(X) \geq 3$, *and let be* $F(T) \neq \emptyset$ *for any nonexpansive map* $T: C \to C$, *with* $C \in \mathscr{C}$. *Then* X *is* (H) *if and only if for any* $C \in \mathscr{C}$ *we have* $C = \mathrm{co}(F(T))$ *for some nonexpansive map* T *on* X.

A full comparison of this theorem with other (probably stronger) characterizations of (H), partly depends on the solution of the most famous open problem in the context of fixed point theory.

4. Some more results

We begin with some simple remarks. In [3, p. 110] also minimal points "in a wider sense" were considered. These are the points $x \in X$ such that

(2′) if $x' \in X$ satisfies $\|x' - y\| \leq \|x - y\|$ for every $y \in C$, then $\|x' - y\| = \|x - y\|$ for every $y \in C$.

Of course minimal points are also minimal in a wider sense, and these are in turn closest points. So in particular we have

(4) $\min(C) \subset m(C)$ for every set C.

In general the converse implications are not true. In fact, consider in $\mathbb{R}^2$, with the max norm, the points $x = (1,1)$ and $y = (1,-1)$: if $C_1 = \{x,y\}$, the origin is minimal in a wider sense but not minimal with respect to $C_1 = \min(C_1)$; for $C_2 = \langle -x,x \rangle$ the origin is a closest point to C_2, which is not minimal in a wider sense. However we have:

Proposition 1. *Given* X, *the following statements are equivalent:*

(i) X *is* (SC);
(ii) $m(C) = \min(C)$ *for every set* C;
(iii) $m(C) \subset \min(C)$ *for any 2-point set* C;
(iv) $co(C) \subset \min(C)$ *for every set* C.

Proof. (i) ⇒ (ii). Let X be (SC), and take any $C \subset X$. Because of (4), it is enough to prove that $m(C) \subset \min(C)$. Take $x \notin \min(C)$, and suppose that $x' \neq x$, $x' \in \bigcap_{y \in C} B(y, \|x - y\|)$. For $0 < \lambda < 1$, set $z = \lambda x + (1 - \lambda)x'$. Then, given $y \in C$, we have $\|z - y\| \leq \|x - y\|$. Moreover, either $x' - y = \mu(x - y)$ for some $\mu \geq 0$, which implies $\|z - y\| < \|x - y\|$ since $z \neq x$, or $\|z - y\| = \|\lambda(x - y) + (1 - \lambda)(x' - y)\| < \|x - y\|$ because of (SC). Thus we conclude that $z \in \bigcap_{y \in C} S(y, \|x - y\|)$, so $x \notin m(C)$: thus $m(C) \subset \min(C)$.

(ii) ⇒ (iii). Trivial.

(iii) ⇒ (iv). It is enough to prove that the following is true in every space X:

(5) $co(C) \subset m(C)$ for every set C.

Let x,y belong to C; then the segment joining them is contained in m(C). Thus

$$co(C) = \left\{ \sum_{i=1}^{n} \lambda_i x_i; \quad x_i \in C; \quad n \in N; \quad \sum_{i=1}^{n} \lambda_i = 1; \quad \lambda_i \geq 0 \right\} \subset m(C).$$

(iv) ⇒ (i). If X is not (SC), then we can find two points x,y such that the segment joining them $\langle y,z \rangle = co\{y,z\}$ is not contained in $\min(\{x,y\})$ (see [3, Proposition I.3]).

Remark. If $C = \{x,y\}$, then we necessarily have either $\min(C) = C$ or $\min(C) = \langle x,y \rangle$ (see e.g. [19, Theorem 1]).

Concerning the inclusion (5), see also the comments to Theorem 19 in section 7. If X is (r) and (SC), then m(C) is closed whenever C is compact, but not in general (see [3, p. 115] and the comments to Theorem 19). In fact, the structure of m(C) is not clear in general. For that reason, in [3, section III] a new operation was considered, by reiteration of the operation "minimal", in order to overcome these difficulties; the new set so constructed there (and containing the original one) has—at least in good spaces—nice properties.

Of course, any nonexpansive map T with $F(T) \neq \emptyset$ is quasi-nonexpansive. Optimal sets and quasi-nonexpansive retracts are related in the following simple way.

Proposition 2. (a) *A set* C *is optimal if and only if it is the fixed set of a quasi-nonexpansive map from* X.

(b) *If* X *is* (r) *and* (SC), *then a set* C *is optimal if and only if it is a quasi-nonexpansive retract of* X.

Proof. Part (a). The "if" statement (which is slightly stronger than Theorem 8) is easily seen to be true. Consider the "only if" statement. Given an optimal set C, associate to every $x \notin \min(C)$ a point $x' \neq x$ such that $\|x' - y\| \leq \|x - y\|$ for every $y \in C$. By extending this map to X with the identity in min(C), we obtain a quasi-nonexpansive map T such that $F(T) = \min(C) = C$.

Part (b). The "if" statement is contained in Theorem 8. Consider the "only if" statement. It is enough to note that under the present assumptions the map T constructed in proving the "only if" part of (a), because of Theorem 9 and Proposition 1, can be chosen so as to be retraction onto $m(C) = C$.

Note that a finite-dimensional version of Proposition 2 appears in [22, Propositions 2.2 and 2.3].

Proposition 2 can be useful in proving some known results (by older ones) in a simpler way.

In view of Proposition 2, it could be of some interest to find conditions (if possible, weaker than (r) and (SC)) implying (δ) of section 3. Also, we could try to consider for a quasi-nonexpansive map T some iterative process (e.g., a "normal Mann process": see [15]) hoping the process to be such that any sequence (starting from a given $x \in X \setminus F(T)$) should converge to a point in F(T). But if no assumption of continuity on T is made, such a convergence is known to hold only when the assumptions made on X are still stronger than (r) and (SC) (in general: (H), or at least uniform convexity: see [11], [12], [15], [28], [29], [30], [34], [39], [40], [41]).

Let C be an optimal set. Because of Proposition 2 and Theorem 2, C is closed. C is also convex if X is (SC), but in general it is not (see the Remark following Proposition 1). Thus there occurs a misprint in [2, p. 43]: see also [3, Proposition III.3].

Consider now the optimality of sets like $B(\theta,1)$. In general the condition "$B(\theta,1)$ is optimal" is very strong, as Theorem 6 shows. The proof of that theorem in [3] consists of several steps, which we analyse next. Assume X to be (r), (SC) and (S), and let $B(\theta,1)$ be optimal. Then every halfspace is optimal, since:

(j) if X is (S), then every halfspace is the union of balls ([3, Lemma III.3]) and

(jj) let X be (r) and (SC); if for every $n \in N$ the set C_n is optimal and $C_n \subset C_{n+1}$, then $\bigcup_{n=1}^{\infty} C_n$ is optimal ([3, Corollaire 1 to Proposition III.4 bis]). This implication is not true in general: see [3, Remarques at p. 123 and p. 132].

Therefore, every hyperplane is optimal since by [3, Proposition III.2]:

(jjj) the intersection of optimal sets is optimal.

Thus every hyperplane is optimal. This implies (H) if $\dim(X) \geq 3$. In fact, this last implication is true for any X (see section 1), while in [3] a weaker fact was used (see [3, Proposition III.5]). The use of (r), (SC) and (S) could be thus avoided in the last step. Therefore a weakening of the conditions (r) and (SC) in (jj) could imply an improvement of Theorem 6. Note that the analogues of the implications in (jj) and (jjj), for nonexpansive retracts, were considered in [9, Lemma 4 and Theorem 5], and in [8, Theorem 2] (see also [6, Lemma 1]).

Of course, better results concerning (γ) would be useful to obtain new versions of Theorems 5 and 6, or to revisite Theorem 10 (by using Theorem 5); in fact, any optimal set is a nonexpansive retract of X if X is (r), (SC) and (S), and (γ) holds. We summarize the above discussion by the following proposition:

Proposition 3. *Let the space* X *of three or more dimensions be* (r), (SC) *and* (S). *Let* C *be a bounded subset of* X *with a nonempty interior, and let* T *denote the map onto* min(C) *that we can define by Theorem* 9. *Then* X *is* (H) *if the following conditions hold:*

(i) min(C) *is a smooth set and* T *is quasi-nonexpansive;*

(ii) *the map* T *is sunny, or* T *is continuous and compact.*

Proof. By (i), min(C) is a quasi-nonexpansive retract of X, and it is a smooth set. Then by (ii) and Theorem 12 it is also a sunny quasi-nonexpansive retract of X. Thus (by Theorem 3) it is a nonexpansive retract of X. The conclusion follows from Theorem 5.

Still supposing $\dim(X) \geq 3$, consider the following properties:

(a) X is (H);
(b) $B(\theta,1)$ is a nonexpansive retract of X;
(c) $B(\theta,1)$ is a quasi-nonexpansive retract of X;
(d) $B(\theta,1)$ is optimal.

The implications (a) $\Rightarrow$ (b) and (b) $\Rightarrow$ (c) are trivial, while by Theorem 8 (c) $\Rightarrow$ (d). By Theorem 5, (b) $\Rightarrow$ (a) if X is (S); but this is not true in general, as seen by considering $\mathbb{R}^3$, with the max norm (see [16, Remarques]). By Theorem 3, if X is (S) and (γ) holds, then (c) $\Rightarrow$ (b) (see the above discussion). If X is (r) and (SC), then (d) $\Rightarrow$ (c) by Proposition 2. Finally, (d) $\Rightarrow$ (a) if X is (r), (SC) and (S) (by Theorem 6), but not in general: consider e.g. the infinite-dimensional space L^∞ [0,1] (see [3, p. 126]).

In this context, it seems to be plausible to raise the following **Question**. For what spaces (possibly, for any) are the conditions (b) through (d) equivalent?

Note that, by considering the radial map, we obtain that $B(\theta,1)$ always possess a weaker contractive property (it is a "1-set contraction": see [33, Theorem 1]). We observe also that if $B(\theta,1)$ is optimal and X is (r) and (SC), then X has the "Ball Intersection Property" (BIP) defined by Nussbaum in [31]: this can be easily proved, via Proposition 2, by adapting the proof of [31, Proposition 1]. But (BIP) does not imply that $B(\theta,1)$ is optimal: for example, l^1 has the (BIP) but its unit ball is not optimal (see [31, p. 932] and [3, p. 136]).

5. Firmly nonexpansive maps

The condition we assumed to define firmly nonexpansive maps can be restated in several equivalent ways, as it was shown in [20] (where these maps were called "regularly nonexpansive". A firmly nonexpansive map is obviously nonexpansive.

In Hilbert spaces these maps had already been considered (and called "firmly contractive") by Browder (see ref. [3, p. 212] of [7]). In (H) spaces, these maps are related to projection maps, and T is firmly nonexpansive if and only if $2T - I$ is nonexpansive (see [45, I, §1], where a detailed study of these maps was made; see also [24]). More generally, the following was shown to be true in [10, §3]: in some Banach spaces, every firmly nonexpansive map T is an *averaged mapping*, i.e., there exist a nonexpansive map S and a real c with $0 < c < 1$ such that $T = (1 - c)S + cS$, and conversely: if S is nonexpansive, then $(1 - c)S + cS$ is firmly nonexpansive for sufficiently small c.

Concerning firmly nonexpansive maps, we recall only two more results.

(R_1) ([7, Theorem 1]). Nonexpansive sunny retractions are firmly nonexpansive retractions, and these are in turn sunny quasi-nonexpansive (recall also Theorem 3 in section 3).

(R_2) ([7, Lemma 3]). The fixed set of a nonexpansive map T is also the fixed set of a firmly nonexpansive map T′ of the type $T' = \lambda I + (1 - \lambda)T$ T′ for some $\lambda \in (0,1)$, and such that (3) is satisfied for $x' = T'x$.

Note that we cannot say that any nonexpansive retraction is a firmly nonexpansive retraction: otherwise, by (R_1), (α) in section 3 would be always true. The extension of a firmly nonexpansive map defined on any set C to a map of the same type defined on X has been discussed in [37, Lemmas 2.1 and 7.1]. More precisely, the following properties were studied:

(P_1) whenever D is a subset of X, every firmly nonexpansive mapping $T:D \to X$ can be extended to a firmly nonexpansive mapping $\tilde{T}:X \to X$.

(P_2) same as in (P_1), but with $\tilde{T}(X) \subset \overline{co}(T(D))$.

By assuming $\dim(X) \geq 3$, the following facts were proved in [37]:

(P_2) ⇒ (H) (Proposition 2.2: this result is contained in our modified version of Kakutani theorem, see section 1).

(P_1) ⇒ (H) if X is (r) and (S), but not in general (see Corollary 7.3 and p. 392).

6. Centers

Throughout this section, C denotes a bounded nonempty set. Given C, we set (for $x \in X$)

$$(6) \qquad r_x(C) = \sup_{y \in C} \|x - y\| \quad \text{and} \quad r(C) = \inf_{x \in X} r_x(C).$$

An element $\bar{x} \in X$ such that $r_{\bar{x}}(C) = r(C)$ is called a *center* of C. We denote by *E(C)* the (possibly empty) set consisting of the centers of C:

$$(7) \qquad E(C) = E(\overline{co}(C)) = \{\bar{x} \in X; \ \sup_{y \in C} \|\bar{x} - y\| \leq \sup_{y \in C} \|x - y\| \quad \text{for every } x \in X\}.$$

The main results about centers are the following.

Theorem 14. ([23, p. 184]). *If* X *is constrained in* X**, *then* $E(C) \neq \emptyset$ *for every* C.

Theorem 15. ([23, p. 187–188]). *If* X *is* (SC) *and* C *is compact, then* E(C) *is a singleton.* X *is uniformly convex in every direction (UCED) if and only if for each subset* C *of* X, *the set* E(C) *contains at most one element.* For the notion of (UCED) we refer to [42, p. 288].

Theorem 16. ([23, p. 190] or [27, §§2 and 3]). *Assume* $\dim(X) \geq 3$. *Then the following statements are equivalent:*

(i) X *is* (H);

(ii) $E(C) \cap \overline{co}(C) \neq \emptyset$ *for every set* C;

(iii) $E(C) \subset \overline{co}(C)$ *for every set* C.

Moreover, if X *is not* (H), *then there exists at least one 3-point set* C *for which* (ii) *fails.*

It was noted in [42, p. 289] that in general E(C) is not contained in m(C); but we can prove the following

Proposition 4. (a) *For every compact set* C *we have* $E(C) \subset m(C)$.

(b) *Let* C *be any subset for which* E(C) *contains at most one element. Then* $E(C) \subset \min(C)$.

(c) *If* X *is* (r), *then* $E(C) \cap m(C) \neq \emptyset$ *for every set* C.

Proof. (a). Assume $x \notin m(C)$. Take $\bar{x} \neq x$; $\|\bar{x} - y\| < \|x - y\|$ for every $y \in C$. So for C compact we obtain $r_{\bar{x}}(C) < r_x(C)$. Thus $x \notin E(C)$.

(b) Assume that $x \notin \min(C)$, and let $\bar{x} \neq x$; $\|\bar{x} - y\| \leq \|x - y\|$ for every $x \in C$.

Hence $r_{\bar{x}}(C) = \sup_{y \in C} \|\bar{x} - y\| \leq \sup_{y \in C} \|x - y\| = r_x(C)$. Our assumption then implies $x \notin E(C)$. This proves (b).

(c) Let X be (r) and let $\|x - y\| \leq r(C)$ for every $y \in C$. Because of Theorem 9 there exists $x' \in m(C)$ such that $\sup_{y \in C} \|x' - y\| \leq \sup_{y \in C} \|x - y\| \leq r(C)$. This proves (c).

Remark. In Proposition 4, if X is (SC) then (b) implies (a). Moreover, (c) implies (a) if X is (r) and (SC). Note that in (a) and (c) we cannot replace m(C) with min(C). Consider e.g. the set C_1 defined at the beginning of section 4.

Corollary. (cf. Theorem 1 and the comments to it). *Suppose* $\dim(X) \geq 3$. *Then* X *is* (H) *if and only if* $m(C) \subset \overline{co}(C)$ *for every compact set* C.

Proof. The "only if" part is known (see Theorem 1), so we must prove the "if" part. If X is not (H), then (by Theorem 16) there exists a set $C = \{x,y,z\}$ such that $E(C) \cap \overline{co}(C) = \emptyset$, while by Proposition 4 our assumption would imply $E(C) \subset m(C) \subset \overline{co}(C)$. This is a contradiction, proving our corollary.

7. Two-dimensional spaces

In what follows, the space X is assumed to be two-dimensional. We recall some special results concerned with this case which turn out to be very special. For example, the analogue of Kakutani theorem is trivially false; but also many facts concerning non linear sets are completely different. For a detailed study of R_C maps in two-dimensional spaces, we refer to [22].

Theorem 17. ([26, Theorem 1], or [7, Theorem 5], or [21, Satz 1]). *Every closed and convex set is a sunny nonexpansive retract of* X.

Theorem 18. ([27, §§2 and 3]). *For every bounded set* C *we have* $E(C) \cap \overline{co}(C) \neq \emptyset$, *while* X *is* (SC) *if and only if* $E(C) \subset \overline{co}(C)$ *for every set* C (*or, equivalently, for every compact set* C).

Theorem 19. ([35, Theorems 7 and 8]). *If* $m(C) \subset \overline{co}(C)$ *for every set* C (*or, equivalently, for every compact set* C), *then* X *is* (SC). *Conversely, if* X *is* (SC), *then we have* $\overline{co}(C) = m(C)$ *for every bounded set* C.

Comments and remarks

Theorem 17: for results on the nonexpansiveness of the radial map, we refer again to [21].

Theorem 19. If X is not (SC), then (by Theorem 18) for some compact set C we have $E(C) \not\subset co(C)$. Thus, by Proposition 4, we obtain $m(C) \not\subset \overline{co}(C)$, that is the first part of theorem 19. Note how in that statement we cannot replace m(C) by min(C), since the inclusion $\min(C) \subset \overline{co}(C)$ is always true.

Consider again a space X, of dimension ≥ 2, which is (SC). By using (5), Proposition 1 and Theorem 7, we have:

$$co(C) \subset m(\bar{C}) = \min(\bar{C}) = \min(C) = m(C).$$

Therefore, we obtain $\overline{co}(C) \subset m(C)$ whenever m(C) is closed or $\overline{co}(C) = co(\bar{C})$ (for example: when C is convex, or when C is bounded and $\dim(X) < \infty$). As noted in [35, p. 872], also if X is a two-dimensional (SC) space, for C unbounded we may have $\overline{\text{co}}(C) \not\subset m(C)$ (thus $m(C) = m(\bar{C})$ not closed; see also the Remark following Proposition 1).

Appendix

We collect here some more comments on the terminology in use, as well as some (less important) complementary results appearing in the literature. A supplementary list of references (figures preceded by A) is given for this section.

Concerning minimal points, we recall that an extension of such notion was considered and studied in [19]. The notion of "Pareto optimum", though used in general for C a finite set, generalizes the notion of minimal point (see e.g. [A1, p. 15]) or—more frequently—that of minimal point in a wider sense (see e.g. [A6]). A generalized framework for Pareto optimization is given in [A3]. In general the notion of Pareto optimum is related to economic problems: usually the setting of finite-dimensional spaces fit them.

The notion of minimal (or "admissible", or "not dominated") point, as well as that of min(C) (or "efficient set" for C), is used in location problems (see [A16], or [A6] for the 2-dimensional case). Note that the terms indicated now are those used in the works of this kind, where some simple results concerning these notions had already been given (sometimes in journals which are not mathematical in a strict sense).

A notion which—in some sense—forerun in $\mathbb{R}^n$ the definition of quasi-nonexpansive map was that of "point of concentration", due to Kirszbraun (see [A8]).

In the terminology used here, a quasi-nonexpansive map T has at least a fixed point (thus not every nonexpansive map is quasi-nonexpansive). Note that for the results on these maps given in [15] and in [29] such a condition is not assumed, while in [34, p. 468] the maps obtained by dropping the condition $F(T) \neq \emptyset$ are called *conditionally quasi-nonexpansive*. If in the definition of quasi-nonexpansiveness we require inequalities in the strict sense for all $x \in F(T)$, the maps obtained are called in general *quasi-contractive* (the set F(T) is then not only *non-repelling*, but also *attracting*). Note that the term quasi-nonexpansive was also used by some authors to denote different types of maps which are not nonexpansive (but satisfy some weaker condition).

In [7, §2] multi-valued maps such that any selection from them is sunny and quasi-nonexpansive were studied in terms of "approximation regions" (see also [44]). Multi-valued quasi-nonexpansive maps were defined in [A14].

Several generalizations of quasi-nonexpansive or quasi-contractive maps have been considered. In [A11] maps which are contractive with respect to a nonempty T-invariant set A were considered; maps of that kind had been considered, for $A = F(T)$, also in [13] (see condition (ii*) at page 462 there, and also condition (ii_w) at page 480 for the case of weak inequalities); see also [A10]. Still more general (and similar) maps have been considered elsewhere: e.g. in [A13], where the non-repelling set may vary with the point x.

"Strongly quasi-nonexpansive" maps and "uniformly quasi-nonexpansive" families of quasi-nonexpansive maps were considered in [A2], in metric spaces too. In more general spaces, quasi-nonexpansive maps were studied also in [A7] and in [A9], and a generalization of them in [A15]. "Asymptotically quasi-nonexpansive "families of mappings, also in more general spaces, were considered in [A12].

The following question was studied mainly by Reich: given an accretive set A, when is A a sunny nonexpansive retract of X (or of a closed convex set)? For results of this kind, see [37] and [38].

Results concerning the possibility to extend any nonexpansive map $T:C \to X$ to a nonexpansive map $\tilde{T}:X \to X$ (eventually: with $\tilde{T}(X) \subset \overline{co}(T(C))$ are listed in [43]. Note that the last condition is equivalent to C being a nonexpansive retract of X whenever C is closed and convex (see [43, p. 63]). Two of the main results in this context are Theorems 12.11 and 13.5 in [43], but see also [A5, Theorem 3]; note how the separability assumption in Theorem 12.11 is not essential, as proved in [A5, Theorem 2].

The following proposition is contained in [A4; Corollary 1 of Theorem 2]:

(P) If X is (SC) and $3 \leq \dim(X) < \infty$, then X is (H) if and only if $B(\theta,1)$ is a nonexpansive retract of X. If $\dim(X) = 3$, the assumption (SC) on X may be omitted.

In fact, the proof of (P) seems to work only with some smoothness assumption on X (see [43, Theorem 13.4], and also [16, Theorem 2.3] where an infinite-dimensional version of it is given). In this sense, a nicer version of such result is given by Theorem 5 here.

References

[1] Beauzamy, B.: Points minimaux dans les espaces de Banach. C.R. Acad. Sci. Paris Sér. A–B **280** (1975), A-717–720.

[2] Beauzamy, B.: Projections contractantes dans les espaces de Banach. Bull. Sci. Math. (2) **102** (1978), 43–47.

[3] Beauzamy, B., and B. Maurey: Points minimaux et ensembles optimaux dans les espaces de Banach. J. Funct. Anal. **24** (1977), 107–139.

[4] Belluce, L.P., and W.A. Kirk: Nonexpansive mappings and fixed points in Banach spaces. Illinois J. Math. **11** (1967), 474–479.

[5] Bruck Jr., R.E.: A characterization of Hilbert space. Proc. Amer. Math. Soc. **43** (1974), 173–175.

[6] Bruck Jr., R.E.: A common fixed point theorem for a commuting family of nonexpansive mappings. Pacific J. Math. **53** (1974), 59–71.

[7] Bruck Jr., R.E.: Nonexpansive projections on subsets of Banach spaces. Pacific J. Math. **47** (1973), 341–355.

[8] Bruck Jr., R.E.: Nonexpansive retracts of Banach spaces. Bull. Amer. Math. Soc. **76** (1970), 384–386.

[9] Bruck Jr., R.E.: Properties of fixed-point sets of nonexpansive mappings in Banach spaces. Trans. Amer. Math. Soc. **179** (1973), 251–262.

[10] Bruck Jr., R.E., and S. Reich: Nonexpansive projections and resolvents of accretive operators in Banach spaces. Houston J. Math. **3** (1977), 459–470.

[11] Cheng, F.C.: On the Mann iteration process in a uniformly convex Banach space. Kyungpook Math. J. **18** (1978), 189–193.

[12] Das, K.M., S.P. Singh and B. Watson: A note on Mann iteration for quasi-nonexpansive mappings. Nonlinear Anal. **6** (1981), 675–676.

[13] Diaz, J.B., and F.T. Metcalf: On the set of subsequential limit points of successive approximations. Trans. Amer. Math. Soc. **135** (1969), 459–485.

[14] Dotson Jr., W.G.: Fixed points of quasi-nonexpansive mappings. J. Austral. Math. Soc. **13** (1972), 167–170.

[15] Dotson Jr., W.G.: On the Mann iterative process. Trans. Amer. Math. Soc. **149** (1970), 65–73.
[16] Fakhoury, H.: Deux caractérisations des espaces de Banach hilbertisables. 1973–74, unpublished (partially exposed at the Sémin. "G. Choquet" d'Initiation à l'Analyse).
[17] Fejér, A.: Über die Lage der Nullstellen von Polynomen, die aus Minimumforderungen gewisser Art entspringen. Math. Ann. **85** (1922), 41–48.
[18] Franchetti, C., and M. Furi: Some characteristic properties of real Hilbert spaces. Rev. Roumaine Math. Pures Appl. **17** (1972), 1045–1048.
[19] Godini, G.: On minimal points. Comment. Math. Univ. Carolinae **21** (1980), 407–419.
[20] Goebel, K., and M. Koter: Regularly nonexpansive maps. An Ştiinţ. Univ. "Al. I. Cuza" Iaşi Secţ, I a Mat. (N.S.) **24** (1978), 265–269.
[21] Gruber, P.M.: Fixpunktmengen von Kontraktionen in endlichdimensionalen normierten Räumen. Geom. Dedicata **4** (1975), 179–198.
[22] Hetzelt, L.: On suns and cosuns in finite dimensionalen normed real vector spaces. Submitted.
[23] Holmes, R.C.: A course on optimization and best approximation. Lecture Notes in Mathematics **257**. Berlin: Springer-Verlag, 1972.
[24] Havrda, J.: Generalization of the notion of projector. Časopis Pěst. Mat. **98** (1973), 265–268. Czech., english summary.
[25] Itoh, S., and W. Takahashi: The common fixed point theory of single-valued mappings and multivalued mappings. Pacific J. Math. **79** (1978), 493–508.
[26] Karlovitz, L.A.: The construction and application of contractive retractions in 2-dimensional normed linear spaces. Indiana Univ. Math. J. **22** (1972), 473–481.
[27] Klee, V.: Circumspheres and inner products. Math. Scand. **8** (1960), 363–370.
[28] Kuhfittig, P.K.F.: The mean value iteration for set-valued mappings. Proc. Amer. Math. Soc. **80** (1980), 401–405.
[29] Massabó, I.: On the construction of fixed points for a class of nonlinear mappings. Boll. Un. Mat. Ital. (4) **10** (1974), 512–528.
[30] Miyazaki, K., T. Kawatani and S. Miyaura: Some remarks on the Mann iterative process for nonexpansive mappings. Bull. Kyushu Inst. Tech. Math. Natur. Sci. **29** (1982), 17–25.
[31] Nussbaum, R.D.: The ball intersection property for Banach spaces. Bull. Acad. Polon. Sci. Sér. Sci. Math. Astronom. Phys. **19** (1971), 931–936.
[32] Papini, P.L.: Approximation and strong approximation in normed spaces via tangent functionals. J. Approx. Theory **22** (1978), 111–118.
[33] Petryshin, W.V.: Structure of the fixed points sets of k-set contractions. Arch. Rational Mech. Anal. **40** (1971), 312–328.
[34] Petryshin, W.V., and T.E. Williamson Jr.: Strong and weak convergence of the sequence of successive approximations for quasi-nonexpansive mappings. J. Math. Anal. Appl. **43** (1973), 459–497.
[35] Phelps, R.R.: Convex sets and nearest points. II. Proc. Amer. Math. Soc. **9** (1958), 867–873.
[36] Reich, S.: Asymptotic behavior of contractions in Banach spaces. J. Math. Anal. Appl. **44** (1973), 57–70.
[37] Reich, S.: Extension problems for accretive sets in Banach spaces. J. Funct. Anal. **26** (1977), 378–395.
[38] Reich, S.: Product formulas, nonlinear semigroups, and accretive operators. J. Funct. Anal. **36** (1980), 147–168.
[39] Rhoades, B.E.: Fixed point iterations using infinite matrices. Trans. Amer. Math. Soc. **196** (1974), 161–176.
[40] Roux, D.: Applicazioni quasi non espansive: approssimazione dei punti fissi. Rend. Mat. (6) **10** (1977), 597–605.
[41] Senter, H.F., and W.G. Dotson Jr.: Approximating fixed points of nonexpansive mappings. Proc. Amer. Math. Soc. **44** (1974), 375–380.
[42] Singer, I.: Best approximation in normed linear spaces by elements of linear subspaces. Berlin: Springer-Verlag, 1970.
[43] Wells, J.H., and L.R. Williams: Embeddings and extensions in analysis. Berlin: Springer-Verlag, 1975.
[44] Westphal, U.: Über die Existenz-und Eindeutigkeitsmenge bei der besten Ko-approximation. In "Linear spaces and approximation", P. Butzer, E. Görlich and B.-Sz.-Nagy, ed., ISMN 60. Basel: Birkhäuser Verlag, 1982; 255–264.

[45] Zarantonello, E.H.: Projections on convex sets in Hilbert space and spectral theory. In "Contributions to Nonlinear Functional Analysis", E.H. Zarantonello ed., New York: Academic Press, 1971; 237–424.

References for the appendix

[A1] Beauzamy, B., and P. Enflo: Théorèmes de point fixe et d'approximation. Centre Math. Ecole. Polytech., Palaiseau (1976), pp. 28. Communicated.

[A2] Bruck, R.E.: Random products of contractions in metric and Banach spaces. J. Math. Anal. Appl. **88** (1982), 319–332.

[A3] Censor, Y.: Necessary conditions for Pareto optimality in simultaneous Chebyshev best approximation. J. Approx. Theory **27** (1979), 127–134.

[A4] DeFiguereido, D.G., and L.A. Karlovitz: On the extension of contractions on normed spaces. Proc. Symp. Pure Math., Vol. 18, Part I. Providence, RI: Amer. Math. Soc., 1970; 95–104.

[A5] DeFiguereido, D.G., and L.A. Karlovitz: The extension of contractions and the intersection of balls in Banach spaces. J. Funct. Anal. **11** (1972), 168–178.

[A6] Hansen, P., and J.-F. Thisse: Recent advances in continuous location theory. Atti Giornate AIRO 81, Vol. I. Torino: 1981; 1–27.

[A7] Huffmann, E.: Strict convexity in locally convex spaces and fixed point theorems. Math. Japon. **22** (1977), 323–333.

[A8] Kirszbraun, M.D.: Über die zusammenziehende und Lipschitzsche Transformationen. Fund. Math. **22** (1934), 77–108.

[A9] Lee, C.M.: A development of contraction mapping principles in Hausdorff uniform spaces. Trans. Amer. Math. Soc. **226** (1977), 147–159.

[A10] Maitu, M., and A.C. Babu: On subsequential limit points of a sequence of iterates. Proc. Amer. Math. Soc. **82** (1981), 377–381.

[A11] Margolis, B.: Attractive operators and fixed points. Notas de Matemática, Univ. Nacional de La Plata **17** (1970), pp. 13.

[A12] Naimpally, S.A., K.L. Singh and J.H.M. Whitfield: Fixed points and sequences of iterates in locally convex spaces. Abstracts Amer. Math. Soc. **3** (1982), 355.

[A13] Schöneberg, R.: Some fixed point theorems for mappings satisfying Frum-Ketkov conditions. Comment. Math. Univ. Carolinae **17** (1976), 399–411.

[A14] Shiau, C., K. Tan and C.S. Wong: Quasi-nonexpansive multi-valued maps and selections. Fund. Math. **87** (1975), 109–119.

[A15] Tarafdar, E.: An approach to fixed point theorems on uniform spaces. Trans. Amer. Math. Soc. **191** (1974), 209–225.

[A16] Wendell, R.E., and A.P. Hurter Jr.: Location theory, dominance and convexity. Oper. Res. **21** (1973), 314–320.

P.L. Papini
Istituto Matematico "Salvatore Pincherle"
Università di Bologna
Piazza di Porta San Donato 5
I-40127 Bologna (Italy)

Clinton M. Petty

Ellipsoids

1. Introduction

Ellipsoids, which are the affine transforms of Euclidean balls, have many captivating properties and have been studied for their own intrinsic interest. Many extremal problems of an affine nature have ellipsoids as extremal bodies and this has motivated additional investigations. In studies of Minkowski geometries, Hilbert geometries, and affine differential geometry, ellipsoids play an important role. Known results on ellipsoids have been interpreted in the context of these geometries. On the other hand, problems formulated in these geometries have led to new results on ellipsoids.

Section 3 gives a reasonably complete account of work done during the past decade on characterizations of ellipsoids by sections or projections. Sections 4 and 5 are far from a complete account of results in the areas indicated by their titles. However, a study of the papers cited and their references will provide a good account. Section 6 is concerned with classes of convex bodies having some given ellipsoidal property. Many such classes have been studied. However, in some an explicit determination of their members has not yet been obtained and in others the defining property is the only way they are known to arise. We have restricted attention to those classes whose members have been determined and which arise in at least two important but apparently unrelated ways. The latter provides evidence of the importance of the class.

The standard text Bonnesen-Fenchel [1934], chapter 16, provides a good survey of the work done prior to the date of its publication. All papers cited here have appeared since that date. A much more recent survey was made by Gruber-Höbinger [1976]. We have tried to keep the overlap with the latter at a minimum in both content and references.

2. Preliminaries

Here a convex body is a compact, convex set having interior points in d-dimensional Euclidean space E^d. We complete E^d to the d-dimensional projective space P^d by adding the hyperplane at infinity. A projective transformation on P^d is said to be permissible for a convex body K if the image of K remains in E^d.

A knowledge of a few basic results is needed, in order to read the papers cited, since they are frequently used. The first of these will be called a Blaschke type result. Let $K \subset E^d$ $(d \geq 3)$ be a convex body and let $p \in P^d \setminus K$. The supporting lines to K through p form a cone (or cylinder) denoted by C_p. The intersection $C_p \cap K$ is called the graze of p with respect to K. Restrictions on the graze can be of two types:

(1) The graze is contained in some hyperplane.

(2) The graze contains some section of C_p made by a hyperplane.

The restriction (1) is stronger than the restriction (2). A Blaschke type result specifies an $S_i \subset P^d \setminus K$ (i = 1 or 2) such that restriction (i) is satisfied for every $p \in S_i$ and this hypothesis is strong enough to force K to be an ellipsoid. In the classical Blaschke theorem, the hyperplane at infinity is shown to be an S_1. It is also an S_2. In fact, Gruber [1974c] shows that a certain subset of the hyperplane at infinity is an S_2. Burton [1977b] shows that for $\delta > 0$, the set of points in $E^d \setminus K$ whose distance from K is less than δ is an S_1. See Gruber-Höbinger [1976] for additional references.

The second basic result is the following theorem:

(2.1) **Theorem.** *Let* $K \subset E^d$ $(d \geq 2)$ *be a convex body. Suppose that for each family of parallel chords of* K *there is a number* $t \in [0,1]$ *such that the points, which divide the (directed) chords in the proportion* $t:1-t$, *lie in a hyperplane. Then if* $d \geq 3$, K *is an ellipsoid and if* $d = 2$, K *is a triangle or an ellipse.*

See Gruber-Höbinger [1976] for references.

Finally, we mention the Loewner ellipsoid which is frequently used in proofs. This is the unique ellipsoid of minimal volume which contains a given convex body (sometimes the center of the ellipsoid is specified). References to this ellipsoid and related results can be found in Grünbaum [1963], p. 241, and Gruber-Höbinger [1976]. To these references we add Firey [1964] where some interesting properties of Loewner ellipsoids are established.

3. Sections and projections

There is a considerable body of literature in the theory of convexity on the problem of determining the properties of a convex body from information about its lower dimensional sections or projections. First we will discuss some recent characterizations of ellipsoids from information about the sections through a fixed point. The basic result in this direction is:

(3.1) **Theorem.** *Let* $K \subset E^d$ $(d \geq 3)$ *be a convex body and let* $p \in P^d$. *If for some* k, $2 \leq k < d$, *every* k*-dimensional section of* K *through* p *is an ellipsoid then* K *is an ellipsoid.*

Clearly the case $k = 2$ implies that case $k > 2$. This result was proved by Busemann [1955], p. 91, for $p \in \operatorname{int} K$ and was extended to any $p \in E^d$ by Burton [1975]. The proof when p is a point at infinity follows from the latter by use of a permissible projective transformation. The following results involve a substantial reduction in the properties of the concurrent sections of the convex body given in Theorem (3.1).

(3.2) **Theorem.** (False Center Theorem). *Let* $K \subset E^d$ $(d \geq 3)$ *be a convex body and let* $p \in E^d$. *If for some* k, $2 \leq k < d$, *every* k*-dimensional section of* K *through* p *is centrally symmetric, then either* p *is the center of* K *or* K *is an ellipsoid.*

Whereas Theorem (3.1) may be viewed as a projective characterization of an ellipsoid, the False Center Theorem is an affine characterization. The latter is not true if p is a point at infinity. For example, let K be a right cylinder and let p be the point at infinity in the direction of a generator. Then every planar section of K through p is centrally symmetric but p is neither the center of K nor is K an ellipsoid.

The False Center Problem was conjectured by Rogers [1965]. The first proof was obtained by Aitchison-Petty-Rogers [1971] for the case when p is an interior point of K. Larman [1974] extended the result to any point in E^d. More recently, Burton-Mani [1978] reproved the False Center Theorem as a consequence of their investigation of sections of a convex body (see Theorem (3.8)).

A corollary of the False Center Theorem is the following useful result:

(3.3) **Corollary.** *Let* $K \subset E^d$ $(d \geq 3)$ *be a convex body and let* $p \in E^d$. *If for some* k, $2 \leq k < d$, *every* k*-dimensional section of* K *through* p *is centrally symmetric, then* K *is centrally symmetric.*

Actually, this corollary served as a lemma in proving the False Center Theorem and therefore required a separate proof. This was first proved by Rogers [1965] when p is an interior point of K and $k = 2$. Burton [1975] extended this to any point $p \in E^d$. Finally, Burton-Mani [1978] gave a proof for $k > 2$ and $p \in E^d$.

We now turn to a different type of reduction on the properties of the concurrent sections in Theorem (3.1). We state this as an assertion since it has not been established in all cases.

(3.4) **Assertion.** *Let* $K \subset E^d$ $(d \geq 3)$ *be a convex body and let* $p \in \mathit{int}K$. *If for some* k, $2 \leq k < d$, *all* k*-dimensional sections of* K *through* p *are affinely equivalent then* K *is an ellipsoid.*

This assertion has been proved for the following cases. First Süss [1974] proved the assertion for $d = 3$, $k = 2$. Gromov [1967] gave a proof for k even or k odd and $k \leq d - 2$. Burton [1979] observed that the results of Mani [1970] proved the assertion when k is even. Consequently, the only remaining case for which the assertion has not been established is d even and $k = d - 1$.

There is an interesting connection between this assertion and a problem in Minkowski geometry. Now, two Minkowski metrics in a common affine space are isometric if and only if the Minkowski spheres are affinely equivalent (see Busemann [1955], p. 98). Consequently, the assertion would imply that a Minkowski space is Eucledean if any two k-flat are isometric. Actually, this connection was the motavation for the work of Gromov [1967].

The next two theorems involve centrally symmetric sections which are not concurrent. Let $H_K(0,u)$ be the support hyperplane to K with outer normal u and let $H_K(\alpha,u)$ be the hyperplane parallel to $H_K(0,u)$ at distance α and on the same side of $H_K(0,u)$ as K.

(3.5) **Theorem.** *Let* $K \subset E^d$ $(d \geq 3)$ *be a convex body. Then for each direction* u *there exists* $\varepsilon(u) > 0$ *such that* $H_K(\alpha,u) \cap K$ *is centrally symmetric for*

$0 < \alpha < \varepsilon(u)$ *if and only if* K *is the sum of a (perhaps degenerate) zonotope and a* d*-dimensional ellipsoid.*

This result is due to Burton [1977a]. It follows at once that if K is strictly convex then this condition implies that K is an ellipsoid. This characterization of an ellipsoid had been obtained earlier by Aitchison [1974] for $d = 3$.

Burton [1977b] has also proved the following interesting characterization:

(3.6) **Theorem.** *Let* $K \subset E^d$ $(d \geq 3)$ *be a convex body. If for some* k, $2 \leq k < d$, *and for some* $\delta > 0$, *every* k*-dimensional section of* K *having diameter less then* δ *is centrally symmetric, then* K *is an ellipsoid.*

The next two theorems involve equivalence relations on certain parallel sections. Let $w_K(u)$ be the width of K in the direction u. Two compact convex sets C and D are said to be width-equivalent if there is a positive constant λ such that $w_C(u) = \lambda w_D(u)$ for all directions u. Two bodies of constant width or two homothetic bodies are width-equivalent.

(3.7) **Theorem.** *Let* $K \subset E^d$ $(d \geq 3)$ *be a convex body. If there exists* $\varepsilon > 0$ *such that for every direction* u *and* $0 < \alpha < \beta < \varepsilon$ *the sections* $H_K(\alpha,u) \cap K$ *and* $H_K(\beta,u) \cap K$ *are width-equivalent, then* K *is an ellipsoid.*

This characterization was proved by Aitchison [1970] for $d = 3$ and extended (by induction) to higher dimensions by Burton [1977b]. For $d = 3$ and K strictly convex, Aitchison [1974] weakened the hypothesis in this theorem by replacing ε by $\varepsilon(u)$ in the spirit of Theorem (3.5). However, in this case, he was only able to conclude that the boundary of K is composed of finitely many pieces of ellipsoids.

The following theorem was conjectured by Peter Gruber at Oberwolfach in 1974 and proved by Burton-Mani [1978].

(3.8) **Theorem.** *Let* $K \subset E^d$ $(d \geq 3)$ *be a convex body and let* a_1 *and* a_2 *be distinct points of* E^d. *If for some* k, $2 \leq k < d$, *and every* k*-flat* L *through the origin* $(a_1 + L) \cap K$ *is directly homothetic to* $(a_2 + L) \cap K$, *then* K *is an ellipsoid.*

We now turn to a recent characterization of an ellipsoid by projections. The convex hull of two parallel hyperplanes normal to a direction u is called a slab with normal u. A doublet with normal u is a pair (A,B) where A is slab or a hyperplane with normal u and B is a hyperplane with normal u and such that $B \not\subset A$. The doublet is said to be degenerate if A is a hyperplane. A set of doublets is called independent if their normals are linearly independent. Burton-Larman [1976] have proved the following result:

(3.9) **Theorem.** *Let* $K \subset E^d$ $(d \geq 3)$ *be a convex body and let* $(A_i,B_i)_{i=1}^{d-1}$ *be* $d - 1$ *independent doublets at least one of which is nondegenerate. If* K *lies strictly between* A_i *and* B_i *such that the projection of* K *from each point of* A_i *into* B_i *is centrally symmetric for* $i = 1,\ldots,d - 1$, *then* K *is an ellipsoid.*

We close this section with a theorem on almost ellipsoidal sections and projections due to Larman and Mani [1975] where a history of this problem and references to earlier work may be found. Let K be a centrally symmetric convex body and let $\alpha(K)$ be asphericity of K, that is, $1 - \alpha(K)$ is the ratio of the distances from the center of K to the nearest and farthest points on the boundary of K. Let $M_{n,k}$ be the Grassmann manifold of all k-dimensional subspaces of E^n and let $\mu_{n,k}$ be the normalized Haar measure in $M_{n,k}$. Let $K|E$ be the orthogonal projection of K on the subspace E.

(3.10) **Theorem.** *Given* $\varepsilon, 0 < \varepsilon < 1$, *and* $\delta, 0 < \delta < 1$ *and an integer* $k > 1$, *there exists an integer* $N = N(\varepsilon,\delta,k)$ *with the following property. If* $K \subset E^n$ $(n \geq N)$ *is a centrally symmetric convex body with center at the origin, then there exists a transform* K^* *of* K *under a non-singular central affinity such that*

$$\mu_{n,k}\{E: E \in M_{n,k}, \alpha(K^* \cap E) < \varepsilon, \alpha(K^*|E) < \varepsilon\} > 1 - \delta.$$

This theorem may be interpreted as saying that all but δ of the k-dimensional orthogonal projections and the corresponding central sections of K^* are within ε of being spherical. The corresponding sections and projections of K are therefore within ε of being ellipsoidal. However their measure need not be greater than $1 - \delta$ since this measure is not an affine invariant.

4. Extremal problems

A large number of papers prove results which yield characterizations of ellipsoids as solutions to extremal problems. In addition to those referenced in Gruber-Höbinger [1976], we add Rogers and Shephard [1958], Petty [1967], McKinney [1974], Lutwak [1975], Lenz [1957]. Here, we will discuss those extremal problems which involve multiple integrations over a convex body. If k points are randomly and independently selected from a convex body $K \subset E^d$, then $\operatorname{conv}\{x_1,\ldots,x_k\}$ may be interpreted as a random polytope in K with at most k vertices. Let $T(x_1,\ldots,x_k)$ be the volume of $\operatorname{conv}\{x_1,\ldots,x_k\}$. Then

$$m(k,K) = (1/|K|)^k \int_K \cdots \int_K T(x_1,\ldots,x_k)\, dx_1 \cdots dx_k$$

is defined as the expected volume of this polytope, where $|K|$ is the volume of K. Groemer [1974] proves the following result:

(4.1) **Theorem.** *Let* $K \subset E^d$ *be a convex body. Given an integer* $k > d$ *and a real number* $r \geq 1$, *let*

$$m_r(k,K) = (1/|K|)^k \int_K \cdots \int_K (T(x_1,\ldots,x_k))^r\, dx_1 \cdots dx_k.$$

Then, if B *is a ball with the same volume as* K, $m_r(k,B) \leq m_r(k,K)$ *with equality if and only if* K *is an ellipsoid.*

Thus, for $r = 1$, we see that the expected volume of a random polytope with at most k vertices is minimal among all convex bodies K of given volume if and only if K is an ellipsoid. For extensions, history, and other references see Klee [1969], Groemer [1973], Schöpp [1977], Gigena [1978].

Other integrals of the above type have interesting geometric interpretations. Let K be a centrally symmetric convex body with center at the origin z. An oriented hyperplane through z normal to the direction u bounds a closed halfspace lying on its positive side. Let C(u) be the intersection of this halfspace with K. The set of all centroids of C(u) as u varies is called the centroid surface of K. It is a smooth closed convex surface and bounds the centroid body of K. For the definition of the centroid body for arbitrary K, see Petty [1961]. Now, let $K_1,\ldots,K_d$ be arbitary convex bodies with corresponding centroid bodies $C_1,\ldots,C_d$. Then

$$V(C_1,\ldots,C_d) = \frac{2^d}{|K_1| \cdots |K_d|} \int_{K_1} \cdots \int_{K_d} T(z,x_1,\ldots,x_d)\, dx_1 \cdots dx_d$$

where the left-hand side is a mixed volume. See Petty [1961]. Busemann [1953] proves that

$$\int_{K_1} \cdots \int_{K_d} T(z,x_1,\ldots,x_d)\, dx_1 \cdots dx_d \geq \frac{2}{(d+1)!} \frac{\pi_{d+1}^{d-1}}{\pi_d^{d+1}} |K_1|^{\frac{d+1}{d}} \cdots |K_d|^{\frac{d+1}{d}}$$

where π_r is the volume of the r-dimensional unit ball. The equality holds only for homothetic ellipsoids with center z. Using the identity $n!\, \pi_n \pi_{n-1} = 2^n \pi^{n-1}$ we obtain the following result:

(4.2) **Theorem.** *Let* $K \subset E^d$ *be a convex body and let* C *be the centroid body of* K *with respect to the origin* z. *Then*

$$\left[\frac{(d+1)\pi_d}{2\pi_{d-1}}\right]^d |C| \geq |K|$$

with equality if and only if K *is an ellipsoid with center* z.

It may be noted that if we dilate the centroid body of an ellipsoid with center z by the factor $(d+1)\pi_d/2\pi_{d-1}$ then it will coincide with the ellipsoid. Other integrals of the above type and other characterizations of ellipsoids can be found in Busemann [1953], Busemann-Straus [1960], and Petty [1961].

5. Minkowski geometry, Hilbert geometry, Affine differential geometry

A (symmetric) Minkowski metrization of affine d-space is Euclidean if and only if the balls are ellipsoids. Consequently, properties of the space which imply that it is Euclidean necessarily involve characterizations of the ellipsoid among all centrally symmetric convex bodies. The recent work of Gruber [1974a,b,c] is in this direction although the setting is more general than Minkowski spaces. Busemann-Straus [1960] characterize Euclidean spaces among Minkowski spaces by properties of normality and area functions. All of these papers use the basic results of section 2.

Ellipsoids play a similar role in characterizations of hyperbolic geometry among Hilbert geometries. A Hilbert geometry arises from a special metrization on the interior of a convex body K such that the open chords of K which pass through an interior point are geodesics (straight lines) and they are the only geodesics if any plane which intersects the interior of K intersects the boundary of K in a curve which does not have two noncollinear segments. A Hilbert geometry is hyperbolic if and only if K is an ellipsoid. See Busemann [1955], section 18.

Kelly-Straus [1958] prove the following theorem:

(5.1) **Theorem.** *If a* d-*dimensional Hilbert geometry has negative curvature at each point, then the geometry is hyperbolic.*

The characterization of the ellipsoid used in the proof is of a different type than those mentioned so far. An interior point p of a convex body K is called a projective center of K if there exists a projective transformation, permissible for K, such that the image of p is the affine center of the image of K. To prove this result, they first show that if the curvature at a point p is negative then p must be a projective center. Then they use the result that the ellipsoid is the only convex body all of whose interior points are projective centers. Actually, the hypothesis could be reduced since it is known that all the interior points are projective centers if a suitable subset consists of projective centers. See Marchaud [1948], Kelly-Straus [1960].

Kelly-Paige [1952] give a characterization of hyperbolic geometry among Hilbert geometries from a hypothesis on symmetric perpendicularity of lines. In their proof, basic Theorem (2.1) is used.

If $K \subset E^d$ is a convex body with affine surface area $A_a(K)$, the affine isoperimetric inequality is

(5.2) $$(A_a(K))^{d+1} \leq (d\pi_d)^2(d|K|)^{d-1}$$

with equality if and only if K is an ellipsoid. See Santaló [1949]. The affine surface area is invariant under volume preserving affine transformations but does not satisfy the usual properties of a surface area. For example, if K_1, K_2 are convex bodies with $K_1 \subset K_2$ it need not follow that $A_a(K_1) \leq A_a(K_2)$. Petty [1974] introduces a surface area G(K), called the geominimal surface area, which is also invariant under volume preserving affine transformations and does satisfy the usual properties of a surface area. The inequality (5.2) is then replaced by

(5.3) $$G^d(K) \leq d^d \pi_d |K|^{d-1}$$

with equality if and only if K is an ellipsoid. The inequality between these two surface areas is given by

(5.4) $$(A_a(K))^{d+1} \leq d\pi_d G^d(K)$$

with equality for a certain class of convex bodies described in Petty [1974].

Other characterizations of ellipsoids in this setting are given in Petty [1972], [1974].

6. Classes of convex bodies with an ellipsoidal property

If the hypothesis in some characterization of ellipsoids is replaced by a weaker condition it is possible that the class of convex bodies satisfying this condition contains bodies other than ellipsoids. The next step would be to determine all members in this class. Finally, if this class is to be of some importance we would expect it to share other interesting properties with ellipsoids. For example, a natural weakening of the hypothesis in Assertion (3.4) is to impose the condition that the sections are projectively equivalent. The class of convex bodies satisfying this condition is unknown, but since the hypothesis is still strong we would expect it to be a rather exclusive but interesting class.

A classic example of this program is the class of Radon curves. A Radon curve is the boundary of a centrally symmetric planar convex body with the following property. The line through the center z parallel to a supporting line at a point p intersects the boundary in a point at which there exists a supporting line parallel to the line zp. Thus, a Radon curve shares with the ellipse the property of having conjugate diameters.

Constructions for all Radon curves are given by Day [1947] and Petty [1955]. In a Cartesian coordinate system, draw any convex curve S_1 (turning its concave side toward the origin) from (0,1) to (1,0) and remaining within or on the unit square in the first quadrant. The polar reciprocal S_2 of S_1 with respect to the unit circle shares these same properties and the polar reciprocal of S_2 is again S_1. Now rotate S_2 through 90° and complete the curve by reflection through the origin. Every Radon curve is one of these or an affine transform of one.

Now, there is another characterization of Radon curves which is not at all evident from their definition. Busemann [1947] shows that the Minkowski circles solve the isoperimetric problem in the Minkowski plane if and only if they are Radon curves.

We now turn to a more recent problem which also leads to an interesting class of convex bodies. Let $K \subset E^d$ $(d \geq 2)$ be a convex body and let Γ_K be the group of projectivities in P^d which map K onto itself. If Γ_K is transitive on the boundary of K, the situation is quite simple.

(6.1) **Theorem.** *If Γ_K is transitive on the boundary of* K, *then* K *is an ellipsoid.*

See Auerbach [1940], Busemann [1967]. In E^3, Burton [1978] describes those convex bodies K for which Γ_K acts transitively on the set of extreme points.

Now, the problem we have referred to is this: Find all convex bodies $K \subset E^d$ such that Γ_K is transitive on the interior of K. First, we state an interesting result of Busemann [1967] is this direction.

(6.2) **Theorem.** *If the boundary of* K *possesses an Euler point with non-vanishing Gauss curvature and Γ_K is transitive on the interior of* K, *then* K *is an ellipsoid.*

Dropping this harmless looking smoothness assumption completely alters the situation. D.G. Larman, P. Mani, and C.A. Rogers determined all convex bodies $K \subset E^d$ for which Γ_K acts transitively on the interior of K. For an abstract see Larman-Rogers [1976], p. 20. In higher dimensions, the variety is surprisingly large.

There remains the question of whether there is a different property which characterizes exactly the same class of convex bodies. Two convex bodies in E^d

are said to be projectively equivalent if there exists a projectivity of P^d which maps one onto the other. In general, the projective equivalence class of a convex body is much larger than its affine equivalence class. For example, the affine equivalence class of a square is the class of parallelograms whereas its projective equivalence class is the class of all convex quadrilaterals. We will provide a proof of the following theorem.

(6.3) **Theorem.** *Let* $K \subset E^d$ $(d \geq 2)$ *be a convex body. Then* Γ_K *is transitive on the interior of* K *if and only if the projective equivalence class of* K *is identical to its affine equivalence class.*

Let K_p be the polar reciprocal of K with respect to the unit sphere centered at $p \in \operatorname{int} K$. We first prove:

(6.4) *If* z *and* p *are interior points of* K, *then* K_z *may be mapped onto* K_p *with* $z \to p$ *by a projectivity.*

Proof. Let z be the origin and let $H = \{x | \langle u,x \rangle = h\}$ be the supporting plane to K with unit outer normal u. Then $x = h^{-1}u$ and $y = p + (h - \langle u,p \rangle)^{-1}u$ are the poles of H with respect to unit spheres centered at z and p and therefore they are boundary points of K_z and K_p respectively. Thus, the projectivity $\pi(x) = (1 - \langle p,x \rangle)^{-1}x + p$ maps K_z onto K_p and sends z into p.

(6.5) *If a projectivity* π *maps* K *onto* $K' \subset E^d$ *and sends* $z \in \operatorname{int} K$ *into* $p \in \operatorname{int} K'$, *then there exists an affinity* α *which maps* K_z *onto* K'_p.

Proof. If z is the origin, π can be expressed by

$$(6.6) \quad \pi(x) = \frac{Ax}{cx + 1} + p$$

where x and p are column vectors, c is a row vector and A is a $d \times d$ nonsingular matrix. We will show the affinity

$$(6.7) \quad \alpha(x) = A^{-T}(x + c^T) + p$$

maps K_z onto K'_p. We may assume $p = z$, since the general result follows from the validity of this special case.

We introduce homogeneous coordinates in projective space P^d so that $x_{d+1} = 0$ is the equation of the hyperplane at infinity. Points are denoted by column vectors and hyperplanes by row vectors. Then the projectivity π is expressed by $x' = Px$ where

$$P = \begin{bmatrix} A & 0 \\ c & 1 \end{bmatrix}$$

and the polarity determined by the unit sphere centered at z is expressed by $\xi^T = Bx$ where B is a diagonal matrix with $b_{ii} = 1$ for $i = 1,\ldots,d$ and $b_{d+1,d+1} = -1$. This polarity maps the interior of K_z onto the set of hyperplanes in P^d which do not intersect K. The projectivity π, with $(\xi')^T = P^{-T}\xi^T$, maps this set of hyperplanes onto the set of hyperplanes in P^d which do not intersect K'. Finally, the polarity, with $x' = B(\xi')^T$, maps the latter set of hyperplanes onto the interior of K'_p. Thus $x' = BP^{-T}Bx$ maps K_z onto K'_p. The last row of $BP^{-T}B$ is $(0,\ldots,0,1)$ and therefore this projectivity is an affinity which in nonhomogeneous coordinates is given by (6.7) where in this special case p is the zero column vector. This completes the proof of (6.5).

Let c(K) be the centroid of K and let s(K) be the unique interior point of K for which the volume $|K_p|$ is minimal. Both c(K) and s(K) are affine-invariant points and if s = s(K) then $c(K_s) = s$. See Petty [1972], pp. 26–27, [1974], pp. 77–80.

(6.8) *If a projectivity π maps* K *onto* $K' \subset E^d$ *and sends* s(K) *into* s(K'), *then π is an affinity.*

Proof. We may assume that z = s(K) is the origin and π is given by (6.6) with p = s(K'). Now $\alpha(x)$ in (6.7) maps K_z onto K'_p and must send the centroid $c(K_z) = z$ into the centroid $c(K'_p) = p$. Hence c in (6.7) is the zero row vector and therefore π is an affinity.

(6.9) *For each* $z \in$ *int* K *there exists a permissible projectivity π for* K *such that* $\pi(z) = c(\pi(K))$.

Proof. Let $p = s(K_z)$. Then by (6.4) there is a projectivity which maps $K = (K_z)_z$ onto $(K_z)_p$ and sends z into $p = c((K_z)_p)$.

We can now prove Theorem (6.3). Suppose Γ_K is transitive on the interior of K. Let π_1 be a projectivity which maps K onto $K' \subset E^d$. Let $z \in \text{int } K$ such that $\pi_1(z) = s(K')$. Let $\pi_2 \in \Gamma_K$ send s(K) into z. Then the composition of π_2 followed by π_1 sends s(K) into s(K') and maps K onto K'. By (6.8) this is an affinity and therefore K and K' are affinely equivalent.

Now suppose the projective and affine equivalence classes of K are identical. Let $z \in \text{int} K$. By (6.9), there is a projectivity π such that $\pi(z)$ is the centroid of $\pi(K)$ and by hypothesis there is an affinity which maps $\pi(K)$ onto K and therefore sends $\pi(z)$ into c(K). The composition gives an element of Γ_K which sends z into c(K) and it follows that the group Γ_K is transitive on the interior of K.

References

Aitchison, P.W.

1970 A characterization of the ellipsoid. J. Austral. Math. Soc. **11** (1970) 385–394. MR 43 # 1042.

1974 The determination of convex bodies by some local conditions. Duke Math. J. **41** (1974), 193–209. MR 48 # 12293.

Aitchison, P.W., C.M. Petty, C.A. Rogers

1971 A convex body with a false centre is an ellipsoid. Mathematika **18** (1971), 50–59. MR 45 # 2572.

Auerbach, H.

1940 Sur une propriété caractéristique de l'ellipsoide. Studia Math. **9** (1940), 17–22. MR 3, p. 89.

Bonnesen, T., W. Fenchel

1934 Theorie der konvexen Körper, Berlin: Springer 1934, New York: Chelsea 1948.

Burton, G.R.
1975 Sections of convex bodies. J. London Math. Soc. (2) **12** (1975/76), 331–336.
MR 53 # 6421.
1977a On the sum of a zonotope and an ellipsoid. Comment. Math. Helv. **51** (1976), 369–387.
MR 55 # 11145.
1977b Some characterisations of the ellipsoid. Israel J. Math. **28** (1977), 339–349.
MR 58 # 24006.
1978 The transitive action of projective transformations on the extreme points of a three-dimensional convex body. J. London Math. Soc. (2) **18** (1978), 166–172.
MR 58 # 18141.
1979 Congruent sections of a convex body. Pacific J. Math. **81** (1979), 303–316.
MR 80i: 52002.
Burton, G.R., D.G. Larman
1976 On a problem of J. Höbiinger. Geometriae Dedicata **5** (1976), 31–42.
MR 57 # 1278.
Burton, G.R., P. Mani
1978 A characterisation of the ellipsoid in terms of concurrent sections. Comment. Math. Helv. **53** (1978), 485–507.
MR 80c: 52003.
Busemann, H.
1947 The isoperimetric problem in the Minkowski plane. Amer. J. Math. **69** (1947), 863–871.
MR 9, p. 372.
1953 Volume in terms of concurrent cross-sections. Pacific J. Math. **3** (1953), 1–12.
MR 14, p. 1115.
1955 The geometry of geodesics. New York: Academic Press 1955.
MR 17, p. 779.
1967 Timelike spaces. Dissertationes Math. Rozprawy Mat. **53** (1967), pp. 52.
MR 36 # 3304.
Busemann, H., E.G. Strauss
1960 Area and normality. Pacific J. Math. **10** (1960), 35–72.
MR 22 # 12501.
Day, M.M.
1947 Some characterizations of inner product spaces. Trans. Amer. Math. Soc. **62** (1947), 320–337.
Firey, W.J.
1964 Some applications of means of convex bodies. Pacific J. Math. **14** (1964), 53–60.
MR 28 # 4428.
Gigena, S.
1978 Integral invariants of convex cones. J. Differential Geom. **13** (1978), 191–222.
MR 80h: 53007.
Groemer, H.
1973 On some mean values associated with a randomly selected simplex in a convex set. Pacific J. Math **45** (1973), 525–533.
MR 47 # 5916.
1974 On the mean value of the volume of a random polytope in a convex set. Arch. Math. (Basel) **25** (1974), 86–90.
MR 49 # 6036.
Gromov, M.L.
1967 On a geometric hypothesis of Banach (Russian). Izv. Akad. Nauk. SSSR, Ser Mat. **31** (1967), 1105–1114.
MR 36 # 655.
Gruber, P.
1974a Über kennzeichnende Eigenschaften von euklidischen Raumen und Ellipsoiden. I.J. Reine Angew. Math. **265** (1974), 61–83.
MR 49 # 3694.
1974b Über kennzeichnende Eigenschaften von euklidischen Raumen und Ellipsoiden. II. J. Reine Angew. Math. **270** (1974), 123–142.
MR 54 # 5974.

1974c Über kennzeichnende Eigenschaften von euklidischen Raumen und Ellipsoiden. III. Monatsh. Math. **78** (1974), 311–340.
MR 57 # 7381.

Gruber, P., J. Höbinger
1976 Kennzeichnungen von Ellipsoiden mit Anwendungen. Jahrbuch Uberblicke Mathematik, 1976, pp. 9–29. Bibliographisches Inst. Mannheim, 1976.
MR 53 # 11494.

Grünbaum, B.
1963 Measures of symmetry for convex sets. Proc. Symp. Pure Math. 7 (Convexity), Amer. Math. Soc. (1963), 233–270.
MR 27 # 6187.

Kelly, P.J., L.J. Paige
1952 Symmetric perpendicularity in Hilbert geometries. Pacific J. Math. **2** (1952), 319–322.
MR 14, p. 308.

Kelly, P.J., E.G. Straus
1958 Curvature in Hilbert geometries. Pacific J. Math. **8** (1958), 119–125.
MR 20 # 2748.
1960 On the projective centres of convex curves. Canad. J. Math. **12** (1960), 568–581.
MR 22 # 11307.

Klee, V.
1969 What is the expected volume of a simplex whose vertices are chosen at random from a given convex body? Amer. Math. Monthly, **76** (1969), 286–288.

Larman, D.G.
1974 A note on the false centre problem. Mathematika **21** (1974), 216–227.
MR 50 # 14490.

Larman, D.G., P. Mani
1975 Almost ellipsoidal sections and projections of convex bodies. Math. Proc. Cambridge Philos. Soc. **77** (1975), 529–546.
MR 51 # 13866.

Larman, D.G., C.A. Rogers
1976 Durham symposium on the relation between infinite-dimensional and finite-dimensional convexity. Bull. London Math. Soc. **8** (1976), 1–33.

Lenz, H.
1957 Eine Kennzeichnung des Ellipsoids. Arch. Math. **8** (1957), 209–211.
MR 19, p. 977.

Lutwak, E.
1975 A general Bieberbach inequality. Math. Proc. Cambridge Philos. Soc. **78** (1975), 493–495.
MR 52 # 4134.

Mani, P.
1970 Fields of planar bodies tangent to spheres. Monatsh. Math. **74** (1970), 145–149.
MR 41 # 4386.

Marchaud, A.
1948 Sur les ovales. Ann. Soc. Polon. Math. **21** (1948), 324–331.
MR 11, p. 127.

McKinney, J.R.
1974 On maximal simplices inscribed in a central convex set. Mathematika **21** (1974), 38–44.
MR 51 # 6582.

Petty, C.M.
1955 On the geometry of the Minkowski plane. Riv. Mat. Univ. Parma. **6** (1955), 269–292.
MR 18, p. 760.
1961 Centroid surfaces. Pacific J. Math. **11** (1961), 1535–1547.
MR 24 # A3558.
1967 Projection bodies. Proc. Colloq. on Convexity, (Copenhagen, 1965), Kobenhavns Univ. Mat. Inst., Copenhagen, 1967, 234–241.
MR 35 # 7203.
1972 Isoperimetric problems. Proceedings of the Conference on Convexity and Combinatorial Geometry, Univ. of Oklahoma, June 1971, (1972), 26–41.
MR 50 # 14499.

1974 Geominimal surface area. Geometriae Dedicata **3** (1974), 77–97.
MR 48 # 12313.

Rogers, C.A.
1965 Sections and projections of convex bodies. Portugal. Math. **24** (1965), 99–103.
MR 33 # 6502.

Rogers, C.A., G.C. Shephard
1958 Some extremal problems for convex bodies. Mathematika **5** (1958), 93–102.
MR 21 # 2960.

Santaló, L.A.
1949 An affine invariant for convex bodies of n-dimensional space. Portugaliae Math. **8** (1949), 155–161. (Spanish)
MR 12, p. 526.

Schöpf, P.
1977 Gewichtete Volumsmittelwerte von Simplices, welche zufallig in einem konvexen Körper des R^n gewahlt werden. Monatsh. Math. **83** (1977), 331–337.
MR 57 # 1285.

Süss, W.
1947 Kennzeichnende Eigenschaften der Kugel als Folgerung eines Brouwerschen Fixpunktsatzes. Comment. Math. Helvet. **20** (1947), 61–64.
MR 9, p. 55.

R.R. Phelps

Convexity in Banach spaces: some recent results

1. Introduction

We all recognize that some of the most fruitful ideas in a particular area of mathematics may well have originated elsewhere. In the case of convexity in Banach spaces, this occurred in 1967, when Marc Rieffel wanted to do a thorough classroom presentation of the Radon-Nikodým theorem for Banach space-valued measures. In formulating a condition on the range of such a measure which would be sufficient for the validity of the Radon-Nikodým theorem, yet would avoid earlier compactness hypotheses, he introduced the notion of a "dentable" set [41]. This was the start of a remarkable chain of results connecting vector-valued integration, the extremal structure of bounded convex sets and generic Fréchet differentiability of convex continuous functions. The details of this story through 1976 have been related in a lively and thorough manner by J. Diestel and J. Uhl in their monograph [14], while the more recent results are covered in the forthcoming lecture notes by R. Bourgin [9]. We will present a portion of the chain, centering our attention on the extremal structure of bounded closed convex subsets of Banach spaces (and differentiability properties of some closely related convex functions). This first section contains a review of the notions of exposed points, strongly exposed points, dentable sets and the Radon-Nikodým property (RNP). In Section 2 there is a discussion of the differentiability of convex functions on Banach spaces and the duality between Fréchet differentiability and strongly exposed points, followed by one of the main recent results in the subject (Theorem 2.8): A bounded closed convex set with the RNP is the closed convex hull of its strongly exposed points. The concluding Section 3 describes the Krein-Milman property and its relation to the RNP, as well as the duality between Asplund spaces and the RNP.[1]

While much of this survey offers bibliographic references in place of proofs, Section 2 is an exception. There, one will find a largely self-contained exposition of all the fundamental material on convex functions needed to prove Theorem 2.8. Most of this is taken from our more extensive unpublished lecture notes [39], and constitutes a brief introduction to differentiability of convex continuous functions on Banach spaces.

We start by establishing some notation.

Throughout, E will denote a real Banach space, E^* its dual and $\langle x^*,x\rangle$ the natural pairing between elements x of E and x^* of E^*. We will let C denote a nonempty, bounded closed convex subset of E. For x^* in E^* let

$$M(x^*,C) = \sup\{\langle x^*,x\rangle : x \in C\}.$$

Recall the following well-known sharpening of the notion of *extreme point*:

Definition. *A point* $x \in C$ *is called an* exposed point *of* C *provided there exists*

[1] The author is grateful to Richard Bourgin and Isaac Namioka for their helpful criticism and suggestions.

$x^* \in E^*$ *such that* x^* *attains its supremum on* C *precisely at* x, *that is,*

$$\{y \in C : \langle x^*, y\rangle = M(x^*, C)\} = \{x\}.$$

In this circumstance one says that x *is exposed by* x^*, or that x^* *exposes* x. Clearly, if x^* exposes x then so does any positive multiple of x^*.

In the plane, the two tangency points of the convex hull C of a circle and a point outside the circle are extreme points which are not exposed, but these are the only ones. Generally, in finite dimensional spaces E our set C will have many exposed points; Klee [28] has shown that they are always dense in the extreme points. When C is infinite dimensional one is led to a proper subclass of the exposed points by following a familiar procedure: Consider those points defined by a property which characterizes exposed points in finite dimensional spaces but which fails to do so in the infinite dimensional situation. Such a property can be described in terms of *slices.*

Definition. *If* $x^* \in E^*$, $\|x^*\| = 1$, *and if* $\alpha > 0$ *let*

$$S(x^*, C, \alpha) = \{y \in C : \langle x^*, y\rangle > M(x^*, C) - \alpha\}.$$

Such a set is called a *slice of* C *defined by* x^*.

It is clear that if x^* has norm 1, say, then it exposes $x \in C$ if and only if the intersection of all the slices of C defined by x^* is precisely $\{x\}$. In finite dimensional spaces, C will necessarily be compact and these slices are easily seen to have diameters converging to zero as α converges to zero. In infinite dimensional spaces this need not happen, as shown by the following well-known example.

1.1 **Example.** *In the sequence space* l_2 *let* C *be the closed convex hull of the sequence* $\{e_n\}$ *of orthogonal basis vectors. Then* 0 *is an exposed point of the weakly compact convex set* C, *but every slice of* C *which contains* 0 *has diameter at least* $\sqrt{2}$.

Proof. Since $\{e_n\}$ converges weakly to 0 and C is weakly closed, we know that $0 \in C$. Moreover, *any* element $x = (x_n)$ of l_2 such that $x_n < 0$ for all n will expose 0. Indeed, if $y \in C$, then $y = (y_n)$ is a limit of convex combinations of basis vectors, hence $y_n \geq 0$ for all n. Thus, $\langle x, y\rangle = \sum x_n y_n \leq 0$ and will equal 0 only if $y = 0$. On the other hand, any slice of C which contains 0 is a weak neighborhood of 0, hence contains all but finitely many basis vectors, and $\|e_n - e_m\| = \sqrt{2}$ whenever $m \neq n$.

This example leads us to the following definition.

Definition. *A point* x *of* C *is* strongly exposed *by* x^* *in* E^*, $\|x^*\| = 1$, *provided* $\langle x^*, x\rangle = M(x^*, C)$ *and the diameter of the slices* $S(x^*, C, \alpha)$ *converge to* 0 *as* α *decreases to* 0. *We also say that* x^* strongly exposes x. *A convenient reformulation is this: The functional* $x^* \in E^*$, $\|x^*\| = 1$, *strongly exposes* $x \in C$ *provided* $\|x_n - x\| \to 0$ *whenever* $\{x_n\} \subseteq C$ *and* $\langle x^*, x_n\rangle \to M(x^*, C)$.

It is obvious from Example 1.1 above that such points can form a proper subclass of the exposed points (in infinite dimensional spaces).

Rieffel's concept of a dentable set can also be formulated in terms of small slices.

Definition. *A bounded nonempty set* A *is said to be* dentable *if for each* $\varepsilon > 0$ *there exists a slice of* A *of diameter less than* ε.

Note that our definition of a slice did not utilize the convexity of C, so it makes sense for any nonempty bounded set. Rieffel's original definition of dentable is more complicated. Let co B denote the convex hull of a set B and $\overline{\text{co}}$ B its norm closure. *A nonempty bounded set* A *is dentable if and only if for each* $\varepsilon > 0$ *there exists* $x \in A$ *such that* $x \notin \overline{\text{co}}[A \backslash N_\varepsilon(x)]$, where $N_\varepsilon(x)$ is the open ball of radius ε centered at x. The equivalence of this with our definition is easily established using the separation theorem. [We cite the original version partly because, after sketching a few examples, the reader might be led to discover the motivation for the term "dentable" (something which has eluded the author to date)]. An elementary argument using the separation theorem shows that a bounded set is dentable if the same is true of its convex hull. The converse may be found in [9].

It is clear that if C has a strongly exposed point, then C is dentable. The following well-known example [9] shows that the converse is false:

Example. *There exists a nonempty bounded closed convex subset* C *of the sequence space* c_0 *which is dentable but which has no strongly exposed points.*

Proof. We take advantage of the fact that the closed unit ball B of c_0 has no extreme points, by letting C be the closed convex hull of the union A of a widely scattered sequence of balls of decreasing diameter. Specifically, for $n = 1,2,3,\ldots$ let δ_n denote the usual n-th basis vector in c_0 and take balls of radius $(n+2)^{-1}$ centered at δ_n, that is,

$$A = \cup\{\delta_n + (n+2)^{-1}B\}.$$

The distance between any two balls in this union is greater than 1/3, so A is closed. Moreover, A is dentable, since for any $n \geq 1$, the closed convex hull of $A \backslash (\delta_n + (n+2)^{-1}B)$ does not contain δ_n. (Indeed, the continuous linear functional $x \to x_n$ has supremum on this set at most 1/3, while its value at δ_n is 1.) Thus, the set $C = \overline{\text{co}}\, A$ is dentable. On the other hand, since A is closed, any strongly exposed point of C would be in A and therefore would be a strongly exposed point of one of the balls which make up A, contradicting the fact that B has no extreme points.

We come now to the concept which Rieffel found to be so useful in proving his vector-valued Radon-Nikodým theorem.

Definition. *The set* C *is said to be* subset dentable *if every nonempty closed convex subset of* C *is dentable.*

Before examining the geometric implications of this property, let us see how it is related to the Radon-Nikodým theorem. Suppose that (Ω,Σ,μ) is a finite measure space and that $F:\Sigma \to E$ is a countably additive (in the norm topology) vector measure which is of bounded variation ($\sup \Sigma \|F(A_i)\| < \infty$, where the supremum is taken over all finite partitions of Ω into measurable sets) and μ-continuous ($F(A) = 0$ whenever $\mu(A) = 0$). Let AR(F) denote the *average range* of F, namely, the set of all points $\mu(A)^{-1}F(A)$, where $A \in \Sigma$ and $\mu(A) > 0$. The *Radon-Nikodým theorem* for F is the assertion that there exists a Bochner integrable function $f:\Omega \to E$ such that $F(A) = \int_A f\,d\mu$ for each $A \in \Sigma$.

Definition. *A bounded closed convex subset* C *of* E *is said to have the Radon-Nikodým property* (*RNP*) *provided the Radon-Nikodým theorem is valid for each such measure space and each such* F *for which* $AR(F) \subseteq C$. *The Banach space* E *is said to have the RNP if each of its bounded closed convex subsets has the RNP.*

Theorem. *A bounded closed convex subset* C *of a Banach space has the Radon-Nikodým property if and only if it is subset dentable.*

The proof of this remarkable relationship between measure theory and geometry extended over nearly seven years. In 1967 Rieffel [41] proved the sufficiency of his dentability condition (although it was formulated in terms of the space rather than a single subset C). In 1973 H. Maynard [31] introduced an intermediate geometric property (subset σ-dentability) which he proved equivalent to the RNP, and in 1974 W. Davis and R. Phelps [12] showed that his property was equivalent to dentability. At about the same time, R. Huff [19] independently modified Maynard's proof to show directly that the RNP implied subset dentability. Expositions of this may be found in Diestel-Uhl [14] or Bourgin [9].

Here are some concrete examples of spaces or sets with the RNP (see [14] for details): Every weakly compact convex set has the RNP (a nontrivial fact), hence every reflexive Banach space has the RNP. Separable dual spaces have the RNP and subspaces of spaces with the RNP have the property. A space will have the RNP if each of its separable subspaces has the RNP. The class of spaces *without* the RNP includes all infinite dimensional C(X) spaces and all $L_1(\mu)$ spaces for which μ is not purely atomic.

2. Differentiability of convex functions and strongly exposed points

The key to our approach to showing the existence of strongly exposed points in sets C with the Radon-Nikodým property is the duality between such points and points of Fréchet differentiability of a certain convex continuous gauge functional on E* (the *support* functional) associated with C. In this section we will first give an essentially self-contained exposition of this duality, developing all the necessary properties of convex continuous functions and their derivatives. Recall that a real-valued function φ on E is *convex* provided.

$$\varphi(\lambda x + (1 - \lambda)y) \leq \lambda\varphi(x) + (1 - \lambda)\varphi(y)$$

whenever, $x,y \in E$ and $0 \leq \lambda \leq 1$. If such a function is continuous, then it is *very* continuous, since it is *locally Lipschitzian* (see, e.g. [42 p. 93]): Given $x \in E$ there exists a neighbourhood U of x and $M > 0$ such that

$$|\varphi(y) - \varphi(z)| \leq M\|y - z\| \quad \text{for} \quad y,z \text{ in } U.$$

The simplest such function is the *norm*, which is a special case of a *gauge functional.*

Definition. *A real-valued function* p *on* E *is called a* gauge (*or* Minkowski) functional *provided it is continuous, nonnegative, subadditive and positive homogeneous. These last two mean that*

$$p(x + y) \leq p(x) + p(y) \quad \textit{and} \quad p(\lambda x) = \lambda p(x)$$

whenever $x,y \in E$ *and* $\lambda \geq 0$.

It is easy to produce gauge functionals which are not norms; for instance, let D be any bounded subset of E^* containing 0 and let

$$p(x) = \sup\{\langle x^*,x\rangle : x^* \in D\}, \quad x \in E.$$

The study of differentiability properties of convex continuous functions is simplified by the introduction of the notion of the *subdifferential.*

Definition. *Suppose that* φ *is convex and continuous on* E *and that* $x \in E$. *The subdifferential of* φ at x, *denoted* $\partial\varphi(x)$, *is the set of all* x^* *in* E^* *satisfying*

$$\langle x^*,y - x\rangle \leq \varphi(y) - \varphi(x), \quad y \in E.$$

The fact that $\partial\varphi(x)$ is always nonempty is a consequence of the separation theorem (applied to the closed convex *epigraph* of φ in $E \times R$). For our purposes it is enough to know that $\partial\varphi(x)$ is nonempty whenever φ is a gauge functional, and this is an immediate consequence of the usual form of the Hahn-Banach theorem, aided by the following alternative description of the subdifferential for this special case.

2.1 **Lemma.** *Let* p *be a gauge functional on* E *and suppose* $x \in E$. *Then* $x^* \in \partial\varphi(x)$ *if and only if*

$$\langle x^*,x\rangle = p(x) \quad \textit{and} \quad \langle x^*,y\rangle \leq p(y), \quad y \in E.$$

Proof. Suppose that $x^* \in \partial p(x)$, so that $\langle x^*,y - x\rangle \leq p(y) - p(x)$ for all y. By applying this to the point ry where $r > 0$ and y is fixed, one can divide both sides by r and take $r \to \infty$ to obtain $\langle x^*,y\rangle \leq p(y)$. Taking y (above) equal to 0 yields the reverse inequality for the point x. The converse is even easier.

Thus, to show that $\partial p(x)$ is nonempty, we need only apply the Hahn-Banach theorem to produce a functional x^* in E^* which is dominated by p and equals p(x) at x.

The set-valued mapping $x \to \partial\varphi(x)$ is called the *subdifferential map*; it has the following fundamental properties.

2.2 **Proposition.** *Suppose that* φ *is convex and continuous on* E. *For each* $x \in E$ *the nonempty subdifferential* $\partial\varphi(x)$ *is convex and weak* compact. The subdifferential map is both locally bounded and norm-weak* upper semi-continuous, that is:*

(i) *For any* $x \in E$ *there exist* $M > 0$ *and a neighborhood* U *of* x *such that* $\|y^*\| \leq M$ *whenever* $y^* \in \partial\varphi(y)$ *and* $y \in U$, *and*
(ii) *If* $\{x_n\} \subseteq E$ *with* $\|x_n - x\| \to 0$ *and if* W *is weak* open in* E^* *with* $\partial\varphi(x) \subseteq W$, *then* $\partial\varphi(x_n) \subseteq W$ *for all sufficiently large* n.

Proof. The convexity and weak* closure of $\partial\varphi(x)$ are immediate from the definition of $\partial\varphi$. Once we prove (i), we'll know that $\partial\varphi(x)$ is bounded, hence weak* compact. Local boundedness, however, is itself almost immediate from the defining inequality and the fact that φ is locally Lipschitzian. Finally, we prove (ii) by contradiction. Suppose it were false; we could then find a sequence $\{x_n\}$ converging to x, a weak* open set W containing $\partial\varphi(x)$ and $x_n^* \in \partial\varphi(x_n)$ with $x_n^* \in E^*\backslash W$. By local boundedness we can assume that $\{x_n^*\}$ is bounded, hence admits a weak* cluster point in the weak* closed set $E^*\backslash W$. On the other hand, for all $y \in E$ we have

$$\begin{aligned}\langle x_n^*, y - x\rangle &= \langle x_n^*, y - x_n\rangle + \langle x_n^*, x_n - x\rangle \\ &\leq \varphi(y) - \varphi(x_n) + \|x_n^*\| \cdot \|x_n - x\|.\end{aligned}$$

It follows that $\langle x^*, y - x\rangle \leq \varphi(y) - \varphi(x)$, that is, $x^* \in \partial\varphi(x) \subseteq W$, a contradiction.

Definition. *A continuous real-valued function* φ *on* E *is said to be* Fréchet differentiable *at the point* $x \in E$ *provided there exists an element* $\varphi'(x)$ *in* E^* *(called the* Fréchet differential *of* φ *at* x*) satisfying the following: For each* $\varepsilon > 0$ *there exists* $\delta > 0$ *such that*

$$|\varphi(x + y) - \varphi(x) - \langle\varphi'(x), y\rangle| \leq \varepsilon\|y\|$$

whenever $y \in E$, $\|y\| \leq \delta$.

The first connection of this with the subdifferential is simple.

2.3 **Proposition.** *If* φ *is convex, continuous and Fréchet differentiable at* $x \in E$, *then* $\partial\varphi(x) = \{\varphi'(x)\}$.

Proof. Suppose that $x^* \in \partial\varphi(x)$, so that for all $y \in E$,

$$\langle x^*, y\rangle \leq \varphi(x + y) - \varphi(x).$$

Since $\varphi'(x)$ exists, for any $\varepsilon > 0$ there exists $\delta > 0$ such that

$$\varphi(x + y) - \varphi(x) \leq \langle\varphi'(x), y\rangle + \varepsilon\|y\|$$

whenever $y \in E$ and $\|y\| \leq \delta$. Thus,

$$\langle x^*, y\rangle \leq \langle\varphi'(x), y\rangle + \varepsilon\|y\|$$

whenever $\|y\| \leq \delta$. By homogeneity, this inequality is valid for all $y \in E$. Thus, for a given $y \in E$ it holds for all $\varepsilon > 0$, so that

$$\langle x^*,y\rangle \leq \langle \varphi'(x),y\rangle, \quad y \in E.$$

This implies that $x^* = \varphi'(x)$. We conclude, therefore, that $\varphi'(x) \in \partial\varphi(x)$ and is the unique member of this set.

The converse to this proposition is not valid; in fact, $\partial\varphi(x)$ contains a unique functional x^* precisely when φ is *Gateaux* differentiable at x, that is, when

$$\lim_{t\to 0^+} \frac{\varphi(x + ty) - \varphi(x)}{t}$$

exists for each $y \in E$ (and equals $\langle x^*,y\rangle$). It is known that there exists a separable Banach space whose norm is Gateaux differentiable at each $x \neq 0$ yet is *nowhere* Fréchet differentiable [35, p. 981]. There is an elegant and useful characterization of Fréchet differentiability of φ in terms of the behavior of the subdifferential map. Before presenting this we will prove a slightly technical but important lemma which does not require convexity.

2.4 **Lemma.** *Suppose that* $x^* = \varphi'(x)$ *exists and that* $\{x_n^*\} \subseteq E^*$ *satisfies the following: There exists a real sequence* $\{\beta_n\}$ *with* $\beta_n \to 0$ *such that, for all* y *in some neighborhood of* 0,

(*) $$\langle x_n^*,y\rangle \leq \varphi(x + y) - \varphi(x) + \beta_n.$$

(*This says that* x_n^* *is "almost" a subdifferential of* φ *at* x.) *Then* $\|x_n^* - x^*\| \to 0$.

Proof. If the conclusion fails there exists a subsequence (which we call $\{x_n^*\}$) and $\varepsilon > 0$ such that $\|x_n^* - x^*\| > 2\varepsilon$ for all n. Hence there exists $\{z_n\} \subseteq E$ with $\|z_n\| = 1$ and

$$\langle x_n^* - x^*,z_n\rangle > 2\varepsilon$$

for all n. By definition of $x^* = \varphi'(x)$ there is a $\delta > 0$ such that

$$\varphi(x + y) - \varphi(x) - \langle x^*,y\rangle \leq \varepsilon\|y\|$$

whenever $\|y\| \leq \delta$. We can assume that δ is small enough that (*) is satisfied for $\|y\| \leq \delta$. Let $y_n = \delta z_n$; then $\|y_n\| = \delta$, hence by (*)

$$\begin{aligned} 2\varepsilon\delta < \langle x_n^* - x^*,y_n\rangle &\leq \varphi(x + y_n) - \varphi(x) + \beta_n - \langle x^*,y_n\rangle \\ &\leq \varepsilon\|y_n\| + \beta_n = \varepsilon\delta + \beta_n. \end{aligned}$$

Since $\beta_n \to 0$, this leads to a contradiction.

We saw in Proposition 2.2 that the subdifferential map is always norm-weak* upper semicontinuous. *Norm-norm upper semicontinuity* at a point x (that is, the property defined in Proposition 2.2(ii) but with the set W assumed to be norm open instead of weak* open and $\partial\varphi(x)$ assumed to be a single point) turns out to characterize Fréchet differentiability at x. This is part of the next proposition, which also contains the fundamental fact that *the set of points of Fréchet differentiability of a convex continuous function is always a* G_δ *subset.* To prove this we define a useful sequence of open sets $\{U_n\}$ as follows:

Definition. *For each* $n \geq 1$ *let* U_n *denote the set of all points* x *in* E *such that, for some* $\delta > 0$,

$$\|x_1 - x\| < \delta \quad \textit{and} \quad \|x_2 - x\| < \delta \quad \textit{imply}$$
$$\sup\{\|x_1^* - x_2^*\|: \ x_i^* \in \partial\varphi(x_i), \ i = 1,2\} < 1/n.$$

(*It is immediate that each of these sets is open.*)

2.5 **Proposition.** *Let* φ *be a continuous convex function on* E *and suppose that* $x \in E$. *The following assertions are equivalent.*

(i) *The subdifferential map for* φ *is norm-norm upper semicontinuous at* x.
(ii) $x \in \cap U_n$
(iii) *The Fréchet derivative* $\varphi'(x)$ *exists.*

Proof. Suppose that (i) holds. By definition, $\partial\varphi(x)$ consists of a single point, call it x^*. By applying the definition of norm-norm upper semicontinuity with W the ball centered at x^* of radius 1/2n, we see that $x \in U_n$, $n = 1,2,3,\ldots$.

Suppose, next, that (ii) holds. Obviously, diam $\partial\varphi(x) < 1/n$ for each n, so that $\partial\varphi(x)$ consists of a single point x^*, say. Given $n \geq 1$, choose a δ-neighborhood of x as in the definition of U_n. If $\|y\| < \delta$, choose $y^* \in \partial\varphi(x + y)$; then $\|y^* - x^*\| < 1/n$ and

$$-\langle y^*,y\rangle = \langle y^*,x - (x + y)\rangle \leq \varphi(x) - \varphi(x + y)$$

so that

$$\varphi(x + y) - \varphi(x) \leq \langle y^*,y\rangle \leq \langle x^*,y\rangle + n^{-1}\|y\|.$$

Thus, for each n there exists $\delta > 0$ such that

$$0 \leq \varphi(x + y) - \varphi(x) - \langle x^*,y\rangle \leq n^{-1}\|y\|$$

whenever $\|y\| < \delta$, so that $\varphi'(x)$ exists (and equals x^*).

Finally, assume that $\varphi'(x)$ exists. To see that (i) holds, it suffices to show that if $\|x_n - x\| \to 0$ and $x_n^* \in \partial\varphi(x_n)$, then $\|x_n^* - \varphi'(x)\| \to 0$. Now, for all $y \in E$ we have

$$\langle x_n^*,(x + y) - x_n\rangle \leq \varphi(x + y) - \varphi(x_n),$$

and consequently,

$$\langle x_n^*,y\rangle \leq \varphi(x + y) - \varphi(x) + \beta_n,$$

where

$$\beta_n = \langle x_n^*,x_n - x\rangle + \varphi(x) - \varphi(x_n).$$

By local boundedness of the subdifferential map we can assume that the sequence $\{x_n^*\}$ is bounded. Since $\|x_n - x\| \to 0$ and φ is continuous, we must have $\beta_n \to 0$; from Lemma 2.4 we can conclude that $\|x_n^* - \varphi'(x)\| \to 0$.

2.6 **Corollary.** *For any convex continuous φ the set of points* D *where the Fréchet derivative exists is a* G_δ *(possibly empty!) and the map* $x \to \varphi'(x)$ *is norm-norm continuous on* D. *In particular, any Fréchet differentiable convex function necessarily has a continuous derivative.*

This last fact, that Fréchet differentiable convex functions are always C^1, is a rather remarkable property of convex functions.

The next proposition is of central importance to us. It establishes the duality between Fréchet differentiability and strongly exposed points. This was first proved (for norms and dual unit balls) by Šmul'yan [44]. We prove it for bounded closed convex sets C and their support functionals. The most general formulation for such results (involving pairs of locally convex spaces in duality) was investigated by Asplund and Rockafellar [4].

Definition. *Let* C *be a bounded closed convex subset of* E, *with* $O \in C$. *Denote by* s_C*(or, more simply, by* s) *the* support functional for C, *that is, the function on* E^* *defined by*

$$s_C(x^*) = M(x^*,C) = \sup\{\langle x^*,x\rangle : x \in C\}, \quad x^* \in E^*.$$

This is clearly a continuous gauge functional on E^*. If C is the unit ball of E, for instance, then s_C is the dual norm on E^*. Note that since s_C is defined on E^*, its derivative (when it exists) is an element of E^{**}. In what follows we identify E with its canonical isometric embedding in E^{**}; this should not cause any major difficulty.

2.7 **Proposition.** *Let* C *be a bounded closed convex subset of* E, *with* $O \in C$. *Then the Fréchet derivative of* s_C *exists at* $x^* \in E^*$ *if and only if* $s'_C(x^*) \in C$ *and* x^* *strongly exposes* C *at the point* $s'_C(x^*)$.

Proof. Suppose that $s'(x^*)$ exists; call it x^{**}. Since s is a gauge functional and $\{x^{**}\} = \partial s(x^*)$, Lemma 2.1 implies that $\langle x^{**},x^*\rangle = s(x^*)$ and $\langle x^{**},y^*\rangle \le s(y^*)$ for all $y^* \in E^*$. Suppose that $\{x_n\} \subseteq C$ is any maximizing sequence for $s(x^*)$, that is, $\langle x^*,x_n\rangle \to s(x^*)$. Clearly, for any $y^* \in E^*$,

$$s(y^* + x^*) \ge \langle y^* + x^*,x_n\rangle, \quad \text{so}$$

$$\langle y^*,x_n\rangle \le s(y^* + x^*) - s(x^*) + \beta_n,$$

where $\beta_n = s(x^*) - \langle x^*,x_n\rangle \to 0$ as $n \to \infty$. By Lemma 2.4, we conclude that (in E^{**}) $\|x_n - x^{**}\| \to 0$. This shows that x^{**} is in the closed set C and also that x^* strongly exposes C at x^{**}.

Suppose, now, that $x \in C$ and that $x^* \in E^*$ strongly exposes C at x. From the boundedness of C we know that there is a constant $M > 0$ such that $s(y^*) \le M\|y^*\|$ for all $y^* \in E^*$. The fact that the slices of C defined by x^* become smaller as they become thinner implies that given $\varepsilon > 0$ there exists $\delta > 0$ such that

(*) $\quad y \in C \quad \text{and} \quad \langle x^*,y\rangle \ge s(x^*) - 3M\delta \quad \text{imply} \quad \|x - y\| \le \varepsilon.$

We will show that for all $\varepsilon > 0$ there exists $\delta > 0$ such that

$$(**) \qquad s(x^* + y^*) - s(x^*) - \langle y^*,x\rangle \leq \varepsilon(1 + M)\|y^*\|$$

whenever $\|y^*\| < \delta$. Since $s(x)^* = \langle x^*,x\rangle$, the left side of the inequality is nonnegative, so this will suffice to prove that $x = s'(x^*)$. Given $0 < \varepsilon < 1$, choose $\delta > 0$ as in (*) and suppose that $0 < \|y^*\| < \delta$. Choose $y \in C$ such that $\langle x^* + y^*,y\rangle > s(x^* + y^*) - \varepsilon M\|y^*\|$. Thus

$$\begin{aligned}\langle x^*,y\rangle &> s(x^* + y^*) - \langle y^*,y\rangle - \varepsilon M\|y^*\| \\ &\geq s(x^*) - s(-y^*) - \langle y^*,y\rangle - \varepsilon M\|y^*\| \\ &\geq s(x^*) - M\|y^*\| - s(y^*) - \varepsilon M\|y^*\| \\ &\geq s(x^*) - 3M\|y^*\| \geq s(x^*) - 3M\delta\end{aligned}$$

hence $\|x - y\| \leq \varepsilon$. It follows that

$$\begin{aligned}s(x^* + y^*) - \varepsilon M\|y^*\| &\leq \langle x^* + y^*,y\rangle \\ &\leq s(x^*) + \langle y^*,y\rangle = s(x^*) + \langle y^*,x\rangle + \langle y^*,y - x\rangle \\ &\leq s(x^*) + \langle y^*,x\rangle + \|y^*\|\cdot\|x - y\| \\ &\leq s(x^*) + \langle y^*,x\rangle + \varepsilon\|y^*\|\end{aligned}$$

which yields (**) and completes the proof.

Our goal for the remainder of this section is to prove the following result.

2.8 **Theorem.** *Let* C *be a bounded closed convex subset of a Banach space* E. *If every nonempty subset of* C *is dentable* (*that is, if* C *has the RNP*), *then* C *is the closed convex hull of its strongly exposed points.*

This theorem has a substantial history. It was first proved in [38] (under the stronger hypothesis that E has the RNP), using geometric techniques which make no mention of differentiability. J. Bourgain [6] then proved it in full generality, using somewhat different (but still very geometric) methods. C.A. Rogers subsequently showed that a simple modification of the original proof in [38] will yield the full theorem (see [30, p. 119]). K. Kunen and H.P. Rosenthal have produced a proof recently [29] which utilizes vector-valued martingales. All of these proofs have a substantial intersection of geometric ideas. The proof we present below is entirely different from these. Its key idea originated with Kenderov [26], who introduced it in his proof of a generic continuity theorem for set-valued maximal monotone operators on certain Banach spaces. (The subdifferential map is an important example of such an operator.)

One of our tools is the following well-known density theorem.

2.9 **Lemma.** (Bishop-Phelps [5]). *If* C *is a nonempty bounded closed convex subset of the Banach space* E, *then the functionals which attain their supremum on* C *are norm dense in* E^*. *That is, the set of all* x^* *in* E^* *such that* $\langle x^*,x\rangle = M(x^*,C)$ (*for some* $x \in C$) *is norm dense in* E^*.

It is also true that the corresponding *support points* x are dense in the boundary of C. Proofs and extensions of this result may be found in [5], [37], [14] and [22,

p. 112]. Far-reaching generalizations of the *proof* have been fruitfully applied to a number of optimization problems; results of this kind have been beautifully surveyed by Ekeland [17].

The next key result exhibits the duality between the RNP in C and generic Fréchet differentiability of s_C. It is a specialization, to the subdifferential map, of Fitzpatrick's generalization [18] of Kenderov's fundamental lemma on monotone operators [27]. Modulo Proposition 2.7, the same result was proved by Bourgain [6], using (as we have said above) different methods.

2.10 **Lemma.** *If* $0 \in C$ *and if* C *has the RNP, then the points where* s_C *is Fréchet differentiable form a dense* G_δ *subset of* E^*.

Proof. As in the proof of Proposition 2.7 we will write s for s_C and will frequently treat E as a subspace of E**. For each n let V_n be the set of all x* in E* with the property that, for some $y \in E$, there exists an open neighborhood V of x* such that

$$\partial s(y^*) \cap E \subseteq y + n^{-1}B.$$

for all $y^* \in V$; here, B is the closed unit ball of E. It is immediate from the definition that each of the sets V_n is open; we will show that they are each dense. To this end, fix $u^* \in E^*$ and suppose that U is any open set containing u*. Let

$$\begin{aligned} A &= \{x \in C : x \in \partial s(x^*) \quad \text{for some} \quad x^* \in U\} \\ &= \{x \in C : \text{Some } x^* \text{ in U supports C at } x\}. \end{aligned}$$

To see that A is nonempty, first note that by Lemma 2.9 we know there exist $x^* \in U$ and $x \in C$ such that $\langle x^*,x\rangle = s(x^*)$. Moreover, for any $y^* \in E^*$ we have $\langle y^*,x\rangle \leq s(y^*)$, so from Lemma 2.1 it follows that $x \in \partial s(x^*)$ and hence $x \in A$. By hypothesis, A is dentable, so there exist $z^* \in E^*$ and $\alpha > 0$ such that slice $S = S(z^*,A,\alpha)$ has diameter less than 1/2n. Choose $y \in S$; we know that $y \in \partial s(v^*)$ for some $v^* \in U$. Choose $r > 0$ sufficiently small so that the point $w^* = v^* + rz^*$ is in the open set U. Observe that $\partial s(w^*)$ is contained in the weak* open half-space

$$H = \{z^{**} \in E^{**} : \langle z^{**},z^*\rangle > M(z^*,A) - \alpha\};$$

indeed, if $z^{**} \in \partial s(w^*)$, then

$$\begin{aligned} &\langle z^{**},v^* - w^*\rangle \leq s(v^*) - s(w^*), \quad \text{while} \\ &\langle y,w^* - v^*\rangle \leq s(w^*) - s(v^*). \end{aligned}$$

If we add these two inequalities and transpose we get

$$0 \leq \langle z^{**} - y,w^* - v^*\rangle = r\langle z^{**} - y,z^*\rangle$$

so that

$$\langle z^{**},z^*\rangle \geq \langle y,z^*\rangle > M(z^*,A) - \alpha.$$

By norm-weak* upper semicontinuity of the subdifferential map, there is an open neighborhood V of w* in U such that $x^* \in V$ implies $\partial s(x^*) \subseteq H$ and hence

$$\partial s(x^*) \cap E \subseteq H \cap A = S.$$

Since S has diameter less that $1/2n$ and $y \in S$, this shows that $\partial s(x^*) \cap E \subseteq y + n^{-1}B$ for all $x^* \in V$; we conclude that $w^* \in V_n$ and hence the latter is dense in E^*. By the Baire category theorem, $\cap V_n$ is a dense G_δ subset of E^*. We now show that s is Fréchet differentiable at each point of this set. Indeed, suppose that $x^* \in \cap V_n$. We first show that $\partial s(x^*) \cap E$ is nonempty. We know that for each n there exist $y_n \in E$ and $\delta_n > 0$ such that

$$(1) \qquad \partial s(x^* + y^*) \cap E \subseteq y_n + n^{-1}B \quad \text{provided} \quad \|y^*\| < \delta_n.$$

We can assume that $\delta_n \to 0$. Let J_n denote the union, over all y^* with $\|y^*\| < \delta_n$, of the sets on the left side of (1). By Lemma 2.9, each J_n is nonempty, and since the δ_n are decreasing, we have $J_{n+1} \subseteq J_n$ for all n. Moreover, diam $J_n \leq 2/n$, so by completeness there exists $x \in \cap \bar{J}_n$. Thus, for each n we can find $\|y_n^*\| < \delta_n$ and $x_n \in \partial s(x^* + y_n^*) \cap E$ such that $\|x_n - x\| \to 0$. For any $z^* \in E^*$ and each n we have

$$\langle x_n, z^* - (x^* + y_n^*)\rangle \leq s(z^*) - s(x^* + y_n^*).$$

Since $\|y_n^*\| \to 0$ and $\|x_n - x\| \to 0$ we conclude that $x \in \partial s(x^*)$. We will now show that $x = s'(x^*)$. Fix n for the moment and suppose that $\|y^*\| < \delta_n$ and that the left side of (1) is nonempty. If $z \in \partial s(x^* + y^*) \cap E$, then $\|z - x\| \leq 2/n$ and

$$-\langle z, y^*\rangle = \langle z, x^* - (x^* + y^*)\rangle \leq s(x^*) - s(x^* + y^*)$$

so

$$(2) \qquad 0 \leq s(x^* + y^*) - s(x^*) - \langle x, y^*\rangle \leq \langle z - x, y^*\rangle \leq (2/n)\|y^*\|.$$

This inequality holds for all y^* for which $\|y^*\| < \delta_n$ and for which the left side of (1) is nonempty. By Lemma 2.9, $\partial s(u^*) \cap E$ is nonempty for a dense set of elements u^* of E^*, so the translate (by $-x^*$) of this dense set intersects the ball $\{y^* : \|y^*\| < \delta_n\}$ in a dense set. From the continuity of the functions in (2) we conclude that

$$0 \leq s(x^* + y^*) - s(x^*) - \langle x, y^*\rangle \leq 2n^{-1}\|y^*\|$$

for *all* y^* satisfying $\|y^*\| < \delta_n$. It is immediate from this and the definition of the Fréchet derivative that $x = s'(x^*)$. The foregoing argument shows that s is differentiable at the points of some dense G_δ set. Proposition 2.5, on the other hand, asserts that the set of all points of differentiability always forms a G_δ, so it must, in this case, be a *dense* G_δ set.

Proof of Theorem 2.8

It is relatively easy now to prove the main result. There is clearly no loss in generality in assuming that $0 \in C$, so by Lemma 2.10 we know that s_C is Fréchet differentiable at the points of a dense subset of E^*. Suppose that $S = (y^*, C, \alpha)$ is any slice of C; it suffices to produce a strongly exposed point of C which lies in S. By Proposition 2.7, we need only find a point $x^* \in E^*$ for which $s'(x^*)$ exists and is contained in S; this will be the requisite strongly exposed point. Since s_C is Fréchet differentiable on a dense set, we can approximate y^* as closely as we wish by such points x^*. Suppose that $M = \sup\{\|x\| : x \in C\}$ and choose $\varepsilon > 0$ and $\beta > 0$ such that $2M\varepsilon + \beta \leq \alpha$. It is straightforward to verify that if $\|x^* - y^*\| < \varepsilon$, then

$$S(x^*,C,\beta) \subseteq S(y^*,C,\alpha) = S.$$

Choose such a point x^*; as we have noted, Proposition 2.7 implies that $s'(x^*) \in C$ and is strongly exposed by x^*. In particular, every slice of C defined by x^* contains $s'(x^*)$, hence the latter is contained in S and the proof is complete.

3. The Krein-Milman property and related topics

A Banach space E is said to have the *Krein-Milman property* (KMP) provided every bounded closed convex subset C of E is the closed convex hull of its extreme points. For instance, if E is reflexive, then each such set C is necessarily weakly compact and it follows readily from the Krein-Milman theorem that E has the KMP. Theorem 2.8 trivially yields a more general result: *If a Banach space has the Radon-Nikodým property, then it has the Krein-Milman property.* This was originally proved in an elegant manner by J. Lindenstrauss, who also reformulated the KMP in the following simple way. (See [14, p. 190].)

3.1 **Proposition.** *A Banach space* E *has the Krein-Milman property provided each nonempty bounded closed convex subset* C *of* E *has at least one extreme point.*

Lindenstrauss' proof of this is quickly described: Let $D = \overline{\text{co}}(\text{ext } C)$ and suppose there exists $x \in C \setminus D$. Then $\{x^* \in E^*: \langle x^*,x \rangle > M(x^*,D)\}$ is open and nonempty, hence by Lemma 2.9 contains an element x^* for which the bounded closed convex set

$$F = \{y \in C: \langle x^*,y \rangle = M(x^*,C)\}$$

is nonempty (and clearly disjoint from D). By hypothesis, ext F is nonempty, which contradicts the fact that $\text{ext } F \subseteq \text{ext } C \subseteq D$. (An elementary geometrical proof has been given by R. Bourgin [9], using the idea that an extreme point of a slice of C is either an extreme point of C or lies on an extremal segment, one endpoint of which is an extreme point of C.)

The validity of the converse to Lindenstrauss' theorem is perhaps the outstanding unsolved problem in this area:

Problem. *If a Banach space has the Krein-Milman property, must it have the Radon-Nikodým property?*

There have been some striking affirmative partial solutions to this problem. For instance, R. Huff and P.D. Morris [20] (see also [14]) have obtained the following result.

Theorem. (Huff-Morris). *In a dual Banach space, the Krein-Milman property implies the Radon-Nikodým property.*

Another special case has been proved by J. Bourgain and M. Talagrand [8]. Recall that a Banach lattice E is *solid* if $\|x\| \leq \|y\|$ whenever $|x| \leq |y|$.

Theorem. (Bourgain-Talagrand). *In a solid Banach lattice the Krein-Milman property implies the Radon-Nikodým property.*

Huff and Morris also obtained a partial affirmative solution (without the duality hypothesis) by assuming a stronger form of the Krein-Milman property.

Definition. *Say that a closed convex subset* B *of a Banach space* E *has the* strong Krein-Milman *property if for every nonempty closed and bounded subset* A *of* B, *there is a point of* A *which is extreme in the convex hull* co A *of* A.

It is obvious that if E has the strong KMP, then it has the KMP. On the other hand, if E has the RNP and if A is a nonempty bounded closed subset of E, then by Theorem 2.8 the set $C = \overline{co}\, A$ has a strongly exposed point. It follows from the definition (and the fact that A is closed) that x is in A and is (clearly) extreme in co A. Thus, the RNP implies the strong KMP. Huff and Morris [21] showed the converse: *If* E *has the strong* KMP, *then it has the* RNP. This was subsequently improved by Bourgain [7], who obtained the following version; it extends the preceding result, since weakly closed sets are norm closed.

Theorem. (Bourgain-Huff-Morris). *A Banach space* E *will have the Radon-Nikodým property provided every weakly closed nonempty and bounded subset* A *of* E *contains a point which is extreme in* co A.

Morris [32] (see also [9]) has shown that the "local" version of this is valid, that is, one can replace the space E in the statement by a bounded closed convex subset G of a Banach space. R.C. James [25] obtains this result in the course of a determined attack on the main problem of whether the KMP implies the RNP.

For separable spaces, at least, Lindenstrauss' theorem that the RNP implies the KMP is also a byproduct of a different line of investigation, which looks into the extent to which the Choquet representation theorem holds for noncompact sets. Recall that for a metrizable (say) compact convex subset K of a locally convex space E, the Krein-Milman theorem can be sharpened to Choquet's theorem: Not only is a given point x of K the limit of a sequence of convex combinations of extreme points, but there exists a regular Borel probability measure μ on ext K with *barycenter* (or *resultant*) equal to x, in the sense that

$$f(x) = \int_{\mathrm{ext}\, K} f(y)\, d\mu(y)$$

for each continuous linear functional f on E. This latter equality is equivalent to saying that

$$x = \int_{\mathrm{ext}\, K} y\, d\mu(y)$$

where the integral is interpreted as the weak (or *Pettis*) integral. In other words, x is an infinite (even uncountable) "convex combination" of extreme points. G.A. Edgar [15] has extended this theorem to certain noncompact sets, as follows.

Theorem. (Edgar). *Suppose that* C *is a separable bounded closed convex subset of a Banach space and that* C *has the Radon-Nikodým property. Then for each* x *in* C *there exists a probability measure on the universally measurable subsets of* C *such that* $\mu(\text{ext C}) = 1$ *and*

$$x = \int y \, d\mu(y),$$

where the integral is understood to be the strong (*or* Bochner) *integral.*

Implicit in this theorem is the important assertion that the set ext C is universally measurable, that is, it is Carathéodory measurable with respect to every regular Borel measure on C. While this is relatively easy to prove (ext C is coanalytic), it is about all that can be claimed; J. Jayne and C.A. Rogers [26] have shown that, even in the RNP space l_1, there is a bounded closed convex set C for which ext C is not a Borel set. The improvement of the above theorem over the classical Choquet theorem lies in the fact that C need not be compact, and also in the fact that the integral converges in the strong topology. (A drawback is that it applies only to subsets of Banach spaces.) Note that it trivially implies that *if* C *is separable and has the* RNP, *then* ext C (having positive measure) *is nonempty*, which is equivalent to the separable version of Lindenstrauss' theorem. (The proof, which utilizes vector-valued martingales, does not depend on the latter theorem.)

By reformulating the Choquet theorem in terms of "maximal" measures, it is possible to obtain satisfactory versions which are valid for nonmetrizable compact convex sets. In an analogous way, Edgar [16] has proved a version of his theorem for nonseparable sets with the RNP. R. Bourgin and Edgar [10] have also proved some very nice results characterizing those sets C for which representing measures are unique; as in the compact case, they are the (noncompact) simplexes. (For the separable case this uniqueness result was obtained independently by J. Saint-Raymond [43].) A converse to Edgar's theorem would assert that the validity of the representation theorem for every bounded closed convex subset of C implies that C has the RNP. While this remains open, recent work by E.G.F. Thomas [49] has led him to a partial converse (which has been improved in a joint effort with B. Fuchssteiner). These ideas have been applied to noncompact convex sets of probability measures by H. v. Weizsäcker and G. Winkler [53]. For detailed treatments of classical Choquet theory see E. Alfsen [1], L. Asimow and A.J. Ellis [3], G. Choquet [51] or Phelps [36]. There is an introductory exposition of the metrizable case in [40] and an exposition of the noncompact case in [9].

We recall the statement of Lemma 2.10: If C has the RNP, then its support functional s_C is "generically" Fréchet differentiable in E*, that is, it is differentiable at the points of a dense G_δ subset of E*. This leads easily to the following fact: If E has the RNP, then every weak* lower semicontinuous gauge functional p on E* is generically differentiable. (All such functionals have the form s_C, where

$$C = \{x \in E : \langle x^*, x \rangle \leq p(x^*) \quad \text{for all} \quad x^* \in E^*\}.)$$

This comes close to—and in fact can be used to prove [39]—the assertion that *if* E *has the* RNP, *then* E* *is a weak* Asplund space*, that is, every convex continuous function on E* which is also weak* lower semicontinuous is generically Fréchet differentiable. This is one half of an interesting duality result due to J. Collier [11].

Theorem. (Collier). *A Banach space* E *has the Radon-Nikodým property if and only if* E* *is a weak* Asplund space.*

Recall that an E is an *Asplund space* provided every continuous convex function on E is generically Fréchet differentiable. The RNP in E does *not* imply that E* is an Asplund space; for instance l_1 has the RNP but its dual l_∞ admits a continuous convex function (namely, $x = (x_n) \to \lim\sup|x_n|$) which is nowhere Fréchet differentiable. (It is not even Gateaux differentiable.) On the other hand, there is a pleasing duality between Asplund spaces and dual spaces with the RNP. The following theorem combines results of I. Namioka and Phelps [34] and C. Stegall [45], [46].

Theorem. (Namioka-Phelps-Stegall). *The following assertions about a Banach space* E *are equivalent:*

(i) E *is an Asplund space.*

(ii) E* *has the Radon-Nikodým property.*

(iii) *Every separable subspace of* E *has a separable dual.*

Another proof of the equivalence between (i) and (ii) has been given by R. Anantharaman, T. Lewis and J.H.M. Whitfield [2] and this has been greatly shortened by F. Sullivan [48].

We conclude this survey with a comparison between weak compactness and the RNP for a bounded closed convex nonempty set C. As we noted earlier, if such a set is weakly compact, then it has the RNP. (One way to see this is via Lindenstrauss' result (see [13]) that a weakly compact convex set is the closed convex hull of its strongly exposed points. Namioka has given an elementary proof in [33].) An illuminating way of distinguishing between weak compactness and the RNP is to investigate the set S(C) of *support functionals* of C.

Definition. *For any closed bounded convex and nonempty subset* C *of a Banach space* E *let*

$$S(C) = \{x^* \in E^* : s(x^*) = \langle x^*, x\rangle \quad \text{for some} \quad x \in C\},$$

that is, let S(C) *be all those functionals which actually attain their supremum on* C.

First, recall R.C. James' classical result [23] (see also [24], [13]): *The set* C *is weakly compact if and only if* S(C) = E*. At the other extreme, the Bishop-Phelps theorem (Lemma 2.9) tells us that S(C) *is always norm dense* in E*. Now, it follows from Proposition 2.7 and Lemma 2.10 that *if* C *has the* RNP, *then the set of all those functionals, each of which strongly exposes some point of* C, *is a dense* G_δ *subset of* E*. In particular, this trivially implies that *if* C *has the* RNP, *then* S(C) *contains a dense* G_δ *subset of* E*, a conclusion lying between those of the James and Bishop-Phelps theorems, for a class of closed convex sets lying between the weakly compact sets and the bounded sets. It is striking that one can actually *characterize* the RNP in these terms, a result whose origins are a bit obscure. (It was related orally in 1976 to I. Namioka by J. Bourgain, with C. Stegall present, and Namioka has given an excellent exposition in [33]. It is cited by Stegall in [47] without attribution or proof. An exposition appears in Bourgain's course notes [50] and will appear in Bourgin's notes [9].) The heart of the matter is the following result.

Theorem. *If* C *is a separable bounded closed convex nonempty subset of a Banach space* E, *and if* S(C) *contains a dense* G_δ *subset of* E*, *then* C *is dentable.*

(It is an open question whether this result is valid without the separability hypothesis.) Next, it is shown that if C does not have the RNP, then it contains a separable subset which is not dentable (see, for instance, [14]). Combining this with the foregoing remarks and results one obtains another characterization of the RNP.

Theorem. *A bounded closed convex subset* C *has the RNP if and only if* S(B) *contains a dense* G_δ *set, for each nonempty closed convex subset* B *of* C.

J. Bourgain's course notes [50] contain a somewhat condensed exposition of the RNP, while J. Giles' lecture notes [52] contain a detailed and leisurely treatment of differentiability of convex functions, expanding considerably on [39].

References

1. Erik Alfsen, Compact convex sets and boundary integrals, Ergeb. der Math., vol 57, Springer-Verlag, Berlin, Heidelburg, New York (1971).
2. Anantharaman, R., T. Lewis and J.H.M. Whitfield, Smoothability and dentability in Banach spaces, Canad. Math. Bull. **24(1),** (1981), 59–68.
3. Asimow, L., and A.J. Ellis, "Convexity Theory and its Applications in Functional Analysis" London Math. Soc. Monograph No. 16, Academic Press, N.Y. 1980.
4. Edgar Asplund and R.T. Rockafellar, Gradients of convex functions, Trans. Amer. Math. Soc. **139** (1969), 433–467.
5. Bishop, E., and R.R. Phelps, The support functionals of a convex set, Proc. Sympos. Pure Math., vol 7, Amer. Math. Soc., Providence, R.I. (1963), 27–35.
6. Bourgain, J., Dentability and the Bishop-Phelps property, Israel J. Math. **28** (1977), 265–271.
7. ———, A geometric characterization of the Radon-Nikodým property in Banach spaces, Compositio Math. **36** (1978), 3–6.
8. ———, and M. Talagrand, Dans un espace de Banach reticulé solide, la propriété de Radon-Nikodým et celle de Krein-Milman sont equivalents, Proc. Amer. Math. Soc.; **81** (1981), 93–96.
9. Bourgin, R., Geometric aspects of convex sets with the Radon-Nikodým property, Springer Lecture Notes (to appear).
10. ———, and G.A. Edgar, Noncompact simplexes in Banach spaces with the Radon-Nikodým property, J. Funct. Anal. **23** (1976), 162–176.
11. Collier, J.B., The dual of a space with the Radon-Nikodým property, Pacific J. Math. **64** (1976), 103–106.
12. Davis, W.J., and R.R. Phelps, The Radon-Nikodým property and dentable sets in Banach spaces, Proc. Amer. Math. Soc. **45** (1974), 119–122.

13. Joseph Diestel, Geometry of Banach Spaces—Selected Topics, Lecture Notes in Math. 485, Springer-Verlag, Berlin-Heidelberg-New York, (1975).
14. Diestel, J., and J.J. Uhl, Jr., Vector Measures, Math. Surveys Nr. 15, Amer. Math. Soc., Providence, R.I. (1977).
15. Edgar, G.A., A noncompact Choquet theorem, Proc. Amer. Math. Soc. **49** (1975), 354–358.
16. ———, Extremal integral representations, J. Functional Anal. **23** (1976), 145–161.
17. Ivar Ekeland, Nonconvex minimization problems, Bull. Amer. Math. Soc. (N.S.) **1,** (1979), 443–474.
18. Fitzpatrick, S.P., Monotone operators and dentability, Bull. Australian Math. Soc. **18** (1978), 77–82.
19. Huff, R.E., Dentability and the Radon-Nikodým property, Duke Math. J. **41** (1974), 111–114.
20. ———, and P.D. Morris, Dual spaces with the Krein-Milman property have the Radon-Nikodým property, Proc. Amer. Math. Soc. **49** (1975), 104–108.
21. ———, Geometric characterizations of the Radon-Nikodým property in Banach spaces, Studia Math. **56** (1976), 157–164.
22. R. Israel, Convexity in the Theory of Lattice Gases, Princeton Series in Physics, Princeton Univ. Press, Princeton, N.J. (1979).
23. James, R.C., Weakly compact sets, Trans. Amer. Math. Soc. **113** (1964), 129–140.
24. ———, Reflexivity and the sup of linear functionals, Israel J. Math. **13** (1972), 289–300.
25. ———, Subbushes and extreme points in Banach spaces, Abstracts Amer. Math. Soc. **2** (1981), 599.
26. Jayne, J., and C.A. Rogers, The extremal structure of convex sets, J. Functional Anal. **26** (1977), 251–288.
27. Kenderov, P.S., Monotone operators in Asplund spaces, C.R. Acad. Bulgare Sci. **30** (1977), 963–964.
28. Klee, V.L., Jr., Extremal structure of convex sets, Archiv der Math. **8** (1957), 234–240.
29. Kunen, K., and H. Rosenthal, Martingale proofs of some geometrical results in Banach theory, preprint, 33pp.
30. Larman, D.G., and R.R. Phelps, Gateaux differentiability of convex functions on Banach spaces, J. Lond. Math. Soc. (2) **20** (1979), 115–127.
31. Maynard, H.B., A geometrical characterization of Banach spaces having the Radon-Nikodým property, Trans. Amer. Math. Soc. **185** (1973), 493–500.
32. Morris, P.D., Characterizations of closed convex sets with the RNP, due to J. Bourgain, John Rainwater Seminar Notes, Univ. of Wash., Jan. 1977.
33. Isaac Namioka, Characterizations of closed convex sets with the RNP, following Bourgain and Stegall, Rainwater Seminar, Univ. of Wash., Oct. 1976, 11pp.
34. ———, and R.R. Phelps, Banach spaces which are Asplund spaces, Duke Math. J. **42** (1975), 735–750.
35. Phelps, R.R., A representation theorem for bounded convex sets, Proc. Amer. Math. Soc. **11** (1960), 976–983.
36. ———, Lectures on Choquet's theorem, D. Van Nostrand, Princeton, N.J. (1966).
37. ———, Support cones in Banach spaces and their applications, Advances Math. **13** (1974), 1–19.
38. ———, Dentability and extreme points in Banach spaces, J. Funct. Anal. **16** (1974), 78–90.
39. ———, Differentiability of convex functions on Banach spaces, Lecture notes (unpublished), University College London, January, 1978, 65pp.
40. ———, Integral representation for elements of convex sets, Studies in Functional Analysis, MAA Studies in Math., vol. 21, R.G. Bartle, Ed., Math. Assoc. Amer. (1980), 115–157.
41. Rieffel, M.A., Dentable subsets of Banach spaces, with application to a Radon-Nikodým theorem, Functional Analysis (Proc. Conf., Irvine, Calif. (1966), B.R. Gelbaum, Ed., Academic Press, London, (1967), 71–77.
42. Wayne, A., Roberts and Dale E. Varberg, Convex Functions, Academic Press, New York-London (1973).
43. Jean Saint-Raymond, Représentation intégrale dans certains convexes, Sém. Choquet, Initiation à l'Analyse, 14e année, 1974/75, 11pp.
44. Smul'yan, V.L. Sur la structure de la sphère unitaire dans l'espace de Banach, Mat. Sbornik (N.S.) **9,** (51) (1941), 545–561.

45. Charles Stegall, The Radon-Nikodým property in conjugate Banach spaces, Trans. Amer. Math. Soc. **206** (1975), 213–223.
46. ———, The duality between Asplund spaces and spaces with the Radon-Nikodým property, Israel J. Math. **29** (1978), 408–412.
47. ———, Optimization of functions on certain subsets of Banach spaces, Math. Annalen **236** (1978), 171–176.
48. Francis Sullivan, On the duality between Asplund spaces and spaces with the Radon-Nikodým property, Proc. Amer. Math. Soc. **71** (1978), 155–156.
49. Thomas, E.G.F., A converse to Edgar's theorem, Measure theory, Oberwolfach 1979 (Proc. Conf., Oberwolfach, 1979) Lecture Notes in Math. 794, Springer-Verlag, Berlin (1980), 497–512.
50. Bourgain, J., La propriété de Radon-Nikodým, Publ. Math. de l'Université Pierre et Marie Curie, No. 36, (1979), 104.
51. Choquet, G., Lectures on Analysis, vol. II, J. Marsden, T. Lance, S. Gelbart (eds.), W.A. Benjamin (1969).
52. Giles, J.R., Convex analysis with application in differentiation of convex functions, Res. Notes in Math. 58, Pitman, Boston-London-Melbourne, 1982.
53. Weizsäcker, H.v., and G. Winkler, Non-compact extremal integral representations: Some probabilistic aspects, Functional Analysis: Surveys and Recent Results II, K.-D. Bierstedt, B. Fuchssteiner (eds.), North-Holland, (1980), 115–148.

R.R. Phelps
Department of Mathematics
University of Washington
Seattle WA 98195 (USA)

Rolf Schneider and Wolfgang Weil

Zonoids and Related Topics

A *zonoid* is a convex body in euclidean space which can be approximated, in sense of the Hausdorff metric, by finite vector sums of line segments. Several equivalent definitions are known, and zonoids appear in some surprisingly different contexts. The purpose of the present survey article is twofold. After giving some basic definitions and properties of zonotopes and zonoids, we describe the various ways in which zonoids enter the discussion of a number of seemingly unrelated topics. The stress is here on interrelations between the geometry of convex bodies and other fields. The second part treats those results and problems concerning zonoids and their generalizations which are of interest within the theory of convex bodies.

An excellent survey article on zonoids was written by Bolker [1969]. In the meantime, some of the open problems he mentioned have been solved, new problems have arisen, and some generalizations have been considered. We hope, therefore, that the reader will agree that the present survey is not superfluous, the more so since the point of view is considerably different.

Essentially, this survey is restricted to finite-dimensional convexity. But in those cases where infinite-dimensional generalizations of the described notions and results have been considered, we try to give appropriate references.

As far as Part I is concerned, our list of references does not aim at completeness. In particular, we have not duplicated the bibliographies of those quoted articles which themselves give surveys.

§1. Basic definitions and properties

First some notation. As a rule, we assume that the reader is familiar with the common notions from the theory of convex bodies. By $\mathbb{R}^d$ we denote d-dimensional euclidean vector space ($d \geq 2$) with scalar product $\langle \cdot,\cdot \rangle$. Its unit sphere is

$$\Omega^{d-1} := \{u \in \mathbb{R}^d : \langle u,u \rangle = 1\};$$

ω is the usual spherical Lebesgue measure on Ω^{d-1}, so that $\omega(\Omega^{d-1}) = d\varkappa_d$, where $\varkappa_d = \pi^{d/2}/\Gamma(1 + d/2)$ is the volume of the d-dimensional unit ball. The space of convex bodies (nonempty, compact, convex subsets) of $\mathbb{R}^d$ is equipped with the Blaschke-Hausdorff metric δ. By $h(K,x) := \max\{\langle x,y \rangle : y \in K\}$ the support function $h(K,\cdot): \mathbb{R}^d \to \mathbb{R}$ of K is defined. Throughout this article, we will mainly consider convex bodies K which have a centre of symmetry, and this may be assumed to be the origin of $\mathbb{R}^d$. We then say, shortly, that K is *centred*. Let $\mathscr{K}^d$ denote the space of centred convex bodies in $\mathbb{R}^d$ and $\mathscr{P}^d$ the subspace of centred polytopes.

The simplest zonoids are the finite vector sums of segments themselves, they are called *zonotopes*. Since the segment with endpoints $\pm\alpha u$, where $u \in \Omega^{d-1}$ and $\alpha \in \mathbb{R}^+$, has support function $\alpha|\langle u,\cdot\rangle|$, a convex body $Z \in \mathcal{K}^d$ is a zonotope if and only if its support function is of the form

$$(1.1)\quad h(Z,\cdot) = \sum_{i=1}^{n} \alpha_i|\langle u_i,\cdot\rangle|$$

with $n \in \mathbb{N}$, unit vectors $u_1,\ldots,u_n \in \Omega^{d-1}$, and positive real numbers $\alpha_1,\ldots,\alpha_n$.

Since faces of convex bodies behave additively under Minkowski addition, every face of a zonotope is itself a zonotope, in particular it is centrally symmetric. Vice versa: A convex polytope all of whose 2-dimensional faces are centrally symmetric is a zonotope. Explicit formulations and proofs occur in Bolker [1969] and Schneider [1970a], but certainly the result was known before to many people; for $d = 3$, compare Blaschke [1923], p. 250, and Coxeter [1963], section 2.8. In any case, it seems not justified to ascribe this criterion for zonotopes to Aleksandrov, as Bolker [1971], Witsenhausen [1973], and Assouad [1980a] do. Aleksandrov [1933] only proved that a convex d-polytope ($d \geq 3$) all of whose facets have a centre of symmetry is itself centrally symmetric (see also Burckhardt [1940]). This leads one to consider, as a generalization of the zonotopes, the class $\mathcal{P}_k^d$ of d-dimensional polytopes all of whose k-faces are centrally symmetric ($k \in \{2,\ldots,d\}$). Aleksandrov's theorem implies that $\mathcal{P}_k^d \subseteq \mathcal{P}_{k+1}^d \subseteq \cdots \subseteq \mathcal{P}_d^d$ for $k \geq 2$; see also Shephard [1967] for further related results and McMullen [1976] for a simpler proof. McMullen [1970] showed by an example that $\mathcal{P}_{d-2}^d \neq \mathcal{P}_{d-1}^d$ for $d \geq 4$, and he proved, somewhat surprisingly, that $\mathcal{P}_k^d$ for $2 \leq k \leq d-2$ is equal to $\mathcal{P}_2^d$, the class of d-dimensional zonotopes.

A convex body Z is, by definition, a *zonoid* if it is the limit, in the Hausdorff metric, of a sequence of zonotopes. This is equivalent to the property that the support function of a suitable translate of Z can be approximated, uniformly on Ω^{d-1}, by support functions of the form (1.1). From this one deduces:

(1.2) **Theorem.** *$Z \in \mathcal{K}^d$ is a zonoid if and only if its support function is of the form*

$$(1.3)\quad h(Z,x) = \int_{\Omega^{d-1}} |\langle u,x\rangle|\, d\rho(u) \quad \textit{for} \quad x \in \mathbb{R}^d$$

with some even measure ρ on Ω^{d-1}.

Here by a *measure* on Ω^{d-1} we understand a nonnegative, σ-additive, real valued function on the Borel subsets of Ω^{d-1}. ρ is called *even* if $\rho(\beta) = \rho(-\beta)$ for all Borel sets $\beta \subset \Omega^{d-1}$.

That a convex body Z with support function (1.3) is a zonoid, is easily seen by writing ρ as the weak limit of a sequence of measures with finite supports (more explicit arguments in Schneider [1970a] and Lindquist [1975a]). For the other direction of Theorem (1.2), it suffices to show that the set of convex bodies in $\mathcal{K}^d$ whose support function has a representation (1.3), is closed. This follows from a standard compactness argument and may be found, e.g., in Bolker [1969], p. 336, Matheron [1975], p. 94 (in a less direct way proved in Schneider [1970a] and Lindquist [1975a]).

The following theorem shows that the measure ρ in (1.3) is unique:

(1.4) **Theorem.** *If ρ is an even signed measure on Ω^{d-1} satisfying*

$$\int_{\Omega^{d-1}} |\langle u,v\rangle| \, d\rho(u) = 0 \quad \textit{for each} \quad v \in \Omega^{d-1},$$

then $\rho = 0$.

Two proofs for $d = 3$ and special measures appear in Blaschke [1916], pp. 152 and 154–155. The general case was first proved by Aleksandrov [1937], §8, although in a geometric disguise. It seems that the theorem has been rediscovered several times, at least by Petty [1961], Theorem (4.3), Rickert [1967a,b], and Matheron [1974a] (reproduced in [1975], ch. 4.5). Aleksandrov, Petty, and Rickert use spherical harmonics (see Schneider [1970b] for a survey over this method applied to similar uniqueness theorems). Matheron uses an integral representation for conditionally positive definite functions, which stems from the theory of infinitely divisible probability distributions. Clearly the assertion of (1.4) is equivalent to the fact that the real vector space spanned by the functions $u \mapsto |\langle u,v\rangle|$, $v \in \Omega^{d-1}$, is dense in the space $C_e(\Omega^{d-1})$ of even continuous real functions on Ω^{d-1} with the sup-norm. An elementary proof of this fact was given by Choquet [1969a], p. 53, [1969b], p. 171.

An interesting possible generalization of the uniqueness result (1.4), with the sphere replaced by Grassmannians, was conjectured by Matheron [1975], p. 73. As far as we know, this conjecture is still open.

In the following we denote the set of centred zonoids in $\mathbb{R}^d$ by $\mathscr{Z}^d$.

Part I. The various appearances of zonoids

§2. Zonotopes in geometry

Let us first mention a few instances where zonotopes occur in the treatment of particular geometric problems. From the combinatorial point of view, a d-dimensional zonotope is the same as an arrangement of hyperplanes (see Grünbaum [1967], ch. 18) in $(d - 1)$-dimensional projective space. This connection, which (for $d = 3$) apparently first appeared in Kelly-Moser [1958] and became wider known by an interesting application made by Coxeter [1962], is explained and utilized in the survey articles of Grünbaum [1971] and McMullen [1979], §7; to these we refer the reader for more information. More results on arrangements are found in Grünbaum [1972], Zaslavsky [1975], Shannon [1976].

The association of a 3-dimensional zonotope to a finite system of points in the projective plane was also applied by Burton-Purdy [1979]. By consideration of zonohedra they were able to determine a lower bound for the number of directions determined by n points in $\mathbb{R}^2$.

Various other results on zonotopes, including an interesting duality, were obtained by McMullen [1971] and Shephard [1974a], in part by using a variant of the Gale diagram technique. For a review and additional references, the reader should consult McMullen [1979], §7.

A special aspect of zonotopes deserves further mention, namely their relation to space-filling polytopes, that is, polytopes which tile space by translation. In $\mathbb{R}^3$ every space-filling polytope is a zonotope, with the additional property that every zone contains four or six facets (this fact was used by Coxeter [1962] in classifying in a simple way the space-filling polyhedra). The higher-dimensional generalization is the following: A convex body K tiles $\mathbb{R}^d$ by translation if and only if K is a centrally symmetric polytope with centrally symmetric facets, such that each belt of K (consisting of those of its facets which contain a translate of a given (d − 2)-face) contains four or six facets. This formulation is taken from McMullen [1980] (note [1981]), who rediscovered a result of Venkov [1954]. For $d \geq 4$, the polytopes described in this theorem need not be zonotopes. The space-filling zonotopes have been investigated by Shephard [1974b], McMullen [1975], Doignon [1980]. Groemer [1978] treated multiple space subdivisions by zonotopes.

The following connection with decomposition theory (scissors congruence) is also of interest. Every space-filling polytope is T-equidecomposable to a cube, where T denotes the group of translations. This was proved for $d = 3$ by Hadwiger [1952] and for general d by Mürner [1975]. In three-space, this property leads to a characterization of zonotopes, more generally: A d-polytope is T-equidecomposable to a cube if and only if it belongs to the class $\mathscr{P}^d_{d-1}$. This was conjectured (together with some other equivalences) by Hadwiger [1972] and proved by Mürner [1977].

§3. Ranges of vector measures

In analysis, zonoids apparently first came up in the form of ranges of vector measures. Liapounoff [1940] proved that the range of a nonatomic $\mathbb{R}^d$-valued measure (i.e., a countably additive function from a σ-algebra of subsets of some set into $\mathbb{R}^d$) is compact and convex. It is an elementary matter to show that such a convex body Z, after suitable translation, has a support function of the form (1.3), and vice versa, see Rickert [1967b], Bolker [1969]. Shorter proofs of Liapounoff's theorem were given, e.g., by Halmos [1948], Blackwell [1951], Karlin-Studden [1966], and Lindenstrauss [1966]. The latter proof is of particular interest from the point of view of convexity theory, since it uses the Krein-Milman theorem together with the determination of the extreme points of a certain (infinite-dimensional) compact convex set.

Liapounoff's theorem has applications in the theory of optimal control, for which we refer the reader to Hermes-LaSalle [1969] and Kluvánek-Knowles [1975]. For applications in statistics and game theory, connections with results of a topological nature, and generalizations, see Dvoretzky-Wald-Wolfowitz [1951], Dubins-Spanier [1961], Karlin-Studden [1966], and the literature cited there.

The investigation of the ranges of $\mathbb{R}^d$-valued measures can be reduced to the study of measures on Ω^{d-1} (Bolker [1969], Theorem 2.5). The uniqueness theorem (1.4) then implies that two nonatomic $\mathbb{R}^d$-valued measures φ,ψ on Ω^{d-1}, the ranges of which are translates of each other, yield the same symmetrized measures $\varphi + \varphi^* = \psi + \psi^*$ (the star indicates the measure obtained by reflection at 0). A related question, asking which ranges determine the vector measure on Ω^{d-1} uniquely, was treated by Neyman [1981].

Although the present survey restricts itself to finite-dimensional convexity, a few hints to literature on infinite-dimensional analogs may be useful. The range of a vector measure with values in a Banach space need neither be closed nor convex. The infinite-dimensional generalizations of Liapounoff's theorem that do hold are treated in Diestel-Uhl [1977], ch. IX. This book contains extensive references, also for the finite-dimensional case (p. 273). A certain infinite-dimensional generalization of the Rickert-Bolker characterization of the ranges of vector measures was proved by Kluvánek [1976]. He made essential use of Choquet's [1969a,b] work on conical measures.

§4. Subspaces of L_1-spaces

It is an elementary fact that a zonotope can be viewed as a projection of a (high-dimensional) cube, and every such projection is a zonotope. Equivalently, a centred polytope is the polar of a zonotope if and only if it is a central section of a higher-dimensional cross-polytope. This can be generalized to zonoids in the following way. For a centred d-dimensional convex body $Z \subset \mathbb{R}^d$, write $h(Z,\cdot) =: \|\cdot\|$; then $\|\cdot\|$ is a norm on $\mathbb{R}^d$ with Z^*, the polar of Z, as its unit ball. Now Z is a zonoid if and only if the normed space $(\mathbb{R}^d, \|\cdot\|)$ is isometric to a subspace of $L_1 = L_1([0,1])$ (see Bolker [1969], section 6, for a proof and further discussion). In this way, questions about the isometric embedding of finite-dimensional Banach spaces in L_1 reduce to problems concerning zonoids. In this context, Bolker formulated two conjectures which in the meantime have been decided. The first conjecture concerns the space ℓ_p^d, which is $\mathbb{R}^d$ with norm defined by $\|(x_1,\dots,x_d)\| = (\sum_{i=1}^d |x_i|^p)^{1/p}$. Bolker [1969], p. 339, pointed out that ℓ_p^d embeds in L_1 isometrically if $1 \le p \le 2$, and he conjectured that ℓ_p^d does not embed isometrically in L_1 if $d \ge 3$ and $p > 2$. This was proved for $p > 2.7$ by Witsenhausen [1973] and for all $p > 2$ by Dor [1976]. The second conjecture was motivated by a result and a question of Grothendieck [1953], p. 66. He proved that if a Banach space E has the property that E and its dual space E^* embed isomorphically in L_1, then E is isomorphic to a Hilbert space. Then he asked whether "isomorphic" can be replaced by "isometric". In particular, if E is a d-dimensional Banach space ($d \ge 3$) such that E and E^* embed isometrically in L_1, is E isometric to ℓ_2^d? In other words: If a centred convex body of dimension $d \ge 3$ and its polar are both zonoids, must the body be an ellipsoid? Bolker [1969], [1971] conjectured this to be true, but Schneider [1975] constructed counter-examples.

§5. Norm inequalities

Centred zonoids can be characterized by the fact that the norm defined by their support function satisfies certain sets of inequalities. Let us first recall some definitions. A function $f: \mathbb{R}^d \to \mathbb{R}$ is called *positive definite* if

$$(5.1) \quad \sum_{i,j=1}^{n} f(x_i - x_j) w_i w_j \ge 0$$

for $n \in \mathbb{N}$, all $x_1,\ldots,x_n \in \mathbb{R}^d$ and all real numbers $w_1,\ldots,w_n$. If (5.1) is merely assumed to hold under the additional assumption that $\sum w_i = 0$, then f is *conditionally positive definite.* The function f is said to be *of negative type,* if for every $t > 0$ the function e^{-tf} is positive definite. A real normed vector space $(V,\|\cdot\|)$ is called *hypermetric* if

$$\sum_{i,j=1}^{n} w_i w_j \|x_i - x_j\| \leq 0$$

for $n \in \mathbb{N}$, all $x_1,\ldots,x_n \in V$ and all integers $w_1,\ldots,w_n$ satisfying $\sum w_i = 1$. (See Kelly [1975] for a survey describing the origin and applications of this notion.) The following theorem collects some known results in a suitably specialized form.

(5.2) **Theorem.** *For a centred convex body $Z \in \mathscr{K}^d$ the following conditions are equivalent:*

(a) *Z is a zonoid,*
(b) *$h(Z,\cdot)$ is of negative type,*
(c) *$e^{-h(Z,\cdot)}$ is positive definite,*
(d) *$(\mathbb{R}^d,\|\cdot\|)$ with $\|\cdot\| = h(Z,\cdot)$ is hypermetric,*
(e) *$-h(Z,\cdot)$ is conditionally positive definite.*

The equivalence of the existence of an integral representation (1.3), hence of (a), with (b) is a theorem of P. Lévy [1937], §63; see also Choquet [1969b], p. 173, for a proof of (b)$\Rightarrow$(a) which uses the Lévy-Khintchine formula. (b)$\Leftrightarrow$(c) holds by definition because of $th(Z,u) = h(Z,tu)$. The equivalence of (c) and (d) is due to Witsenhausen [1973], who also showed that this is equivalent to $\sum^n w_i w_j \|x_i - x_j\| \leq 0$ for $w_j \in \mathbb{R}$ with $\sum w_i = 0$. Hence (d) is equivalent to (e). (The equivalence of (b) and (e) is a special case of a more general theorem due to Schoenberg, see, e.g., Choquet [1969a], pp. 55–56). Some of the results can also be obtained in the more general setting of metric spaces if condition (a) is replaced by embeddability in L_1 (see Assouad [1980b] for a survey).

We remark that the equivalence of (e) and (a) admits infinite-dimensional extensions, if the condition that Z be a zonoid is either interpreted by a generalization of the integral representation (1.3) involving a conical measure (Choquet [1968], [1969a], Th. (40.14)), or by the assumption that Z be the closed convex hull of the range of a vector measure (Kluvánek [1975], [1976]).

We remark also that some equivalences of this and the foregoing section generalize in the following way:

(5.3) **Theorem.** *Let $K \in \mathscr{K}^d$ be d-dimensional, and let $1 \leq p \leq 2$. Then the following conditions are equivalent:*

(a) *$h(K,\cdot)^p$ is of negative type,*
(b) *There is a positive measure ρ on Ω^{d-1} such that*

$$h(K,u) = \left[\int_{\Omega^{d-1}} |\langle u,v\rangle|^p d\rho(v)\right]^{1/p} \quad \textit{for} \quad u \in \mathbb{R}^d,$$

(c) *$(\mathbb{R}^d,\|\cdot\|)$ with $\|\cdot\| = h(K,\cdot)$ is isometric to a subspace of L_p.*

One may find these and more general results in Lévy [1937], Herz [1963], Choquet [1969a], §41, Bretagnolle-Dacunha Castelle-Krivine [1966]. However, from the viewpoint of geometric convexity it appears that the convex bodies having the properties of (5.3) are, for $p > 1$, far less interesting than for $p = 1$.

Theorem (5.2) characterizes zonoids among centred convex bodies by certain systems of inequalities which are satisfied by their support functions. For polytopes one has characterizations by much simpler inequalities:

(5.4) **Theorem.** *For a polytope* $Z \in \mathscr{P}^d$ *the following conditions are equivalent*:

(a) Z *is a zonotope*,

(b) *The norm* $\|\cdot\| = h(Z,\cdot)$ *satisfies Hlawka's inequality*

$$\|x\| + \|y\| + \|z\| + \|x + y + z\| \geq \|x + y\| + \|y + z\| + \|z + x\|$$

for all $x,y,z \in \mathbb{R}^d$,

(c) *The norm* $\|\cdot\| = h(Z,\cdot)$ *satisfies the 7-polygonal inequality*

$$\begin{aligned} &\|a - b\| + \|a - c\| + \|b - c\| + \|x - y\| + \|x - z\| \\ &+ \|x - t\| + \|y - z\| + \|y - t\| + \|z - t\| \\ &\leq \|a - x\| + \|a - y\| + \|a - z\| + \|a - t\| \\ &+ \|b - x\| + \|b - y\| + \|b - z\| + \|b - t\| \\ &+ \|c - x\| + \|c - y\| + \|c - z\| + \|c - t\| \end{aligned}$$

for all $a,b,c,x,y,z,t \in \mathbb{R}^d$.

The equivalence (a) $\Leftrightarrow$ (b) is due to Witsenhausen [1978]. In any normed space, the 7-polygonal inequality implies Hlawka's inequality (see Assouad [1980a], Prop. 4.4). The implication (b) $\Rightarrow$ (c) was proved by Assouad [1980a]. He also gave a simple proof for (b) $\Rightarrow$ (a). While Witsenhausen used properties of the polar polytope Z^*, Assouad showed that the Hlawka inequality for $h(Z,\cdot)$ implies the Hlawka inequality for the directional derivatives $h_u(Z,\cdot)$, $u \in \Omega^{d-1}$, and hence the central symmetry of the support sets of Z. He finally used, as did Witsenhausen, the characterization of the zonotopes as the polytopes with centrally symmetric faces. If Z is not a polytope but an arbitrary centred convex body, then (b) or (c) does not imply that Z is a zonoid. This follows from the fact that there are convex bodies in $\mathbb{R}^d$ (for $d > 3$) which are not zonoids, although all their projections are zonoids (Weil [1982]), a result which will be considered in more detail in §8. More generally, this result implies that zonoids cannot be characterized, in all dimensions d, by a finite set of inequalities for the support function involving only a fixed number k (independent of d) of variable points $x_1,\ldots,x_k \in \mathbb{R}^d$.

§6. Positive linear functionals

The set $\mathscr{Z}^d$ of centred zonoids is a closed convex cone. The uniqueness result (1.4) together with (1.2) implies that this cone is simplicial in the sense of Choquet (see, e.g. Alfsen [1971]) and that its extremal rays are made up by the centred line segments. The map $Z \mapsto h(Z,\cdot)$ (where we now restrict support functions to Ω^{d-1})

is an isometric isomorphism from the cone $\mathscr{Z}^d$ on to the cone $\mathscr{S} \subset C_e(\Omega^{d-1})$ of functions $h \in C_e(\Omega^{d-1})$ of the form (1.3). While §5 treated analytic characterizations of (the elements of) $\mathscr{S}$, geometric characterizations of zonoids will be discussed in the second part. Here we want to mention a third aspect, of a functional analytic or measure theoretic type, in considering the dual cone $\mathscr{S}^*$ of $\mathscr{S}$,

$$\mathscr{S}^* := \{\mu \in C'_e(\Omega^{d-1}) : \mu(h) \geq 0 \quad \text{for all} \quad h \in \mathscr{S}\}.$$

Since the dual space $C'_e(\Omega^{d-1})$ of $C_e(\Omega^{d-1})$ is (identified with) the space of even signed measures on Ω^{d-1} and because of the extremal structure of $\mathscr{S}$, we have

$$\mathscr{S}^* = \left\{\mu \in C'_e(\Omega^{d-1}) : \int_{\Omega^{d-1}} |\langle u,x\rangle| \, d\mu(u) \geq 0 \quad \text{for all} \quad x \in \mathbb{R}^d\right\}.$$

No simple characterization of $\mathscr{S}^*$ seems to be known.

Characterizations of $\mathscr{S}^*$, $\mathscr{S}$, and $\mathscr{Z}^d$ are closely connected. By the Hahn-Banach theorem, a body $Z \in \mathscr{K}^d$ is a zonoid (equivalently, $h(Z,\cdot)$ is in $\mathscr{S}$) if and only if $\mu(h(Z,\cdot)) \geq 0$ for all $\mu \in \mathscr{S}^*$. This criterion can be modified by considering a subset $\mathscr{N} \subset \mathscr{S}^*$ such that the closed convex cone cl co $\mathscr{N}$ generated by $\mathscr{N}$ is equal to $\mathscr{S}^*$.

(6.1) **Theorem.** *Let $\mathscr{N}$ be a set of signed even measures on Ω^{d-1} such that $\mathscr{S}^* = \operatorname{cl} \operatorname{co} \mathscr{N}$. Then a body $Z \in \mathscr{K}^d$ is a zonoid if and only if*

$$\int_{\Omega^{d-1}} h(Z,u) \, d\mu(u) \geq 0 \quad \textit{for all} \quad \mu \in \mathscr{N}.$$

By suitable choices of $\mathscr{N}$, several characterizations of zonoids can be deduced. These will be discussed in §9.

§7. Zonoids as auxiliary bodies

Talking of the ways in which zonoids appear in various branches of mathematics, we must also mention that sometimes the introduction of specially constructed zonoids as auxiliary bodies proves useful. Matheron [1974b], [1975], in his important work on stochastic geometry, associated a unique zonoid (a "Steiner compact", in his terminology) with every stationary Poisson hyperplane process, and he found that the projections and the Minkowski quermassintegrals of such a zonoid are linked with certain other processes of flats derived from the original hyperplane process. In a similar way, Schneider [1982a,b] associated zonoids with certain finitely many random hyperplanes meeting a given convex body, and he obtained extremal properties of isotropic random hyperplanes by applying classical inequalities from the theory of convex bodies to these zonoids and other bodies. Auxiliary zonoids are also implicit in the work of Wieacker [1982].

In quite a different field, the introduction of zonoids was found useful by Stanley [1981]. He discovered and utilized that certain numbers of combinatorial interest, pertaining to regular matroids, could be interpreted as mixed volumes of certain especially constructed zonotopes, to which then the Aleksandrov-Fenchel inequalities could be applied.

Part II. Special problems and generalizations

Let us now consider some problems related to zonoids which are of interest from the viewpoint of the geometry of convex bodies. The special role of the class of zonoids in $\mathbb{R}^d$ is already exhibited by the fact that it is the smallest subclass of the space of convex bodies which is closed under affine maps, Minkowski addition and limits, and which is nontrivial (i.e., does not only contain points). Characterization problems, the integral representation of the support function, and the interpretation as projection bodies are the main sources of the questions to be treated and the generalizations to be considered.

§8. Characterization problems

According to §2, a convex polytope is a zonotope if and only if all its two-dimensional faces are centrally symmetric. In this section we consider results which are somehow related to this criterion, but are not restricted to polytopes.

The problem of characterizing general zonoids was repeatedly posed by Blaschke. In [1923], p. 250, and in Blaschke-Reidemeister [1922], pp. 81–82, he asked for a characterization of three-dimensional zonoids "im unendlich Kleinen". Apparently he had in mind, possibly under smoothness assumptions, a local differential-geometric condition which, when satisfied globally on the boundary of $K \in \mathscr{K}^3$, would characterize K as a zonoid. He mentioned that the problem can also be formulated as an analytic one for support functions. The problem of giving intrinsic geometric criteria for deciding when a body is a zonoid, was also propagated by Bolker [1971]. However, it was shown by Weil [1977] that such a local characterization of zonoids cannot exist. He proved the existence of a convex body $K \in \mathscr{K}^d$ $(d \geq 3)$, arbitrarily smooth, which is not a zonoid, but has the following property: For every $u \in \Omega^{d-1}$ there exists a centred zonoid Z and a neighborhood U of u in Ω^{d-1} such that the boundaries of K and Z coincide at all points where the exterior unit normal vector falls in U. This clearly shows that zonoids cannot be characterized, in the class of centred convex bodies, by properties which involve only arbitrarily small neighborhoods of boundary points.

Although zonoids cannot be characterized by strictly local conditions, we conjecture that a "semi-local" characterization is possible (Weil [1977]):

(8.1) **Conjecture.** *Let* $K \in \mathscr{K}^d$ *be a convex body. Suppose that to any great sphere* $\sigma \in \Omega^{d-1}$ *there exists a zonoid* Z *and a neighborhood* U *of* σ *in* Ω^{d-1} *such that the boundaries of* K *and* Z *coincide at all points where the exterior normal vector falls in* U. *Then* K *is a zonoid.*

Zonotopes, as the polytopes with centrally symmetric faces, also come into play in an interesting paper of Burton [1976]. Aiming at an improvement of earlier results, he asks for the convex bodies in $\mathbb{R}^d$ $(d \geq 3)$ which have all their $(d-1)$-dimensional sections sufficiently close to the boundary (in a sense made precise) centrally symmetric. He proves the deep result that K is such a body if and only if it is the sum of a (not necessarily d-dimensional) zonotope and a (d-dimensional) ellipsoid.

Let $P \in \mathscr{P}^d$ $(d \geq 3)$ be a polytope with the property that all its three-dimensional projections are zonotopes. Then each 2-face of P, being a 2-face of a suitable three-dimensional projection of P, is centrally symmetric; thus P is a zonotope. This observation, used by Witsenhausen [1978], motivates the following question: If a centred convex body in $\mathbb{R}^d$ $(d > 3)$ has the property that all its three-dimensional projections are zonoids, must it itself be a zonoid? A more general question, for compact convex sets in a topological vector space, was posed by Choquet [1969b], p. 178. The answer is, however, in the negative. More generally, let $\mathscr{K}_j^d \subset \mathscr{K}^d$ denote the class of centred convex bodies for which all j-dimensional projections are zonoids. (It is sufficient to consider only orthogonal projections, since the class of zonoids is stable under affine maps.) Weil [1982] proved that each of the inclusions $\mathscr{K}_{j+1}^d \subset \mathscr{K}_j^d$, $j = 2,\ldots,d-1$, is strict. The essential point here is the construction of a centred (and, in fact, arbitrarily smooth) convex body $K \in \mathscr{K}^d$ for $d \geq 3$ which is not a zonoid, while all its orthogonal projections onto hyperplanes are zonoids. There arises the new problem of investigating the class $\mathscr{K}_j^d$ for $3 \leq j \leq d-1$. Clearly $\mathscr{K}_j^d$ is a closed convex cone which is stable under linear maps. The following conjecture was ventured in Weil [1982]. Let $\mathscr{E}_k^d$ be the Grassmannian of all k-dimensional linear subspaces of $\mathbb{R}^d$, topologized as usual. For $E,F \in \mathscr{E}_k^d$, the absolute value of the determinant of the (orthogonal) projection from E on to F is denoted by $|\langle E,F\rangle|$, and $v^{(k)}(K;E)$ is the k-dimensional volume of the projection of the convex body K on to E.

(8.2) **Conjecture.** *Let $K \in \mathscr{K}^d$, and let $j \in \{2,\ldots,d\}$. Then $K \in \mathscr{K}_j^d$ if and only if there exists a finite positive Borel measure ρ on $\mathscr{E}_{d-j+1}^d$ such that*

$$(8.3)\quad v^{(d-j+1)}(K;E) = \int_{\mathscr{E}_{d-j+1}^d} |\langle E,F\rangle| \, d\rho(F)$$

for all $E \in \mathscr{E}_{d-j+1}^d$.

The corresponding uniqueness question, that is, whether

$$\int_{\mathscr{E}_{d-j+1}^d} |\langle E,F\rangle| d\rho(F) = 0 \quad \text{for all} \quad E \in \mathscr{E}_{d-j+1}^d$$

for a signed measure ρ on $\mathscr{E}_{d-j+1}^d$ would imply $\rho = 0$, is Matheron's problem mentioned at the end of §2.

Let $\hat{\mathscr{K}}_j^d \subset \mathscr{K}^d$ denote the class of convex bodies K which have the property introduced in (8.2), namely, that there exists a positive Borel measure ρ on $\mathscr{E}_{d-j+1}^d$ such that (8.3) holds for all $E \in \mathscr{E}_{d-j+1}^d$. Conjecture (8.2) then states that $\hat{\mathscr{K}}_j^d = \mathscr{K}_j^d$ for $j = 2,\ldots,d$. Clearly this is true in the extreme cases $j = 2$ and $j = d$.

The above is closely related to the question of extending a well-known integral representation of the quermassintegrals to more general mixed volumes. Let $K \in \hat{\mathscr{K}}_j^d$ and $L_1,\ldots,L_{j-1} \in \mathscr{K}^d$ be given. If ρ is a measure satisfying (8.3), one can deduce from the formulae of Weil [1979] that

$$(8.4)\quad \binom{d}{j-1} V(\underbrace{K,\ldots,K}_{d-j+1},L_1,\ldots,L_{j-1}) = \int_{\mathscr{E}_{d-j+1}^d} v^{(j-1)}(L_1,\ldots,L_{j-1};F^\perp) \, d\rho(F).$$

Here, V on the left-hand side is a mixed volume, $F^\perp$ denotes the orthogonal complement of F, and the integrand is the $(j-1)$-dimensional mixed volume of the orthogonal projections of $L_1,\ldots,L_{j-1}$ on to $F^\perp$. On the other hand, if (8.4) holds and if we choose $L_1 = \cdots = L_{j-1} = L$ and $L \subset E^\perp$ with $(j-1)$-volume equal to one ($E \in \mathscr{E}^d_{d-j+1}$ fixed), then we have

$$v^{(j-1)}(L_1,\ldots,L_{j-1}; F^\perp) = v^{(j-1)}(L;F^\perp) = |\langle E,F\rangle|$$

and hence get (8.3), since it is easy to see that

$$\binom{d}{j-1} V(\underbrace{K,\ldots,K}_{d-j+1},\underbrace{L,\ldots,L}_{j-1}) = v^{(d-j+1)}(K;E).$$

Thus (8.3) can be interpreted as a special case of (8.4).

If K is a zonoid, then (8.4) (with suitable ρ) holds for arbitrary convex bodies $L_1,\ldots,L_{j-1}$ (not necessarily centrally symmetric), see Matheron [1975], p. 103. Whether this is true for $K \in \hat{\mathscr{K}}^d_j$, is an open question which is closely connected with the uniqueness problem mentioned above. It would be interesting to further investigate these classes and their relations.

§9. Generalized zonoids

The construction of the counter-examples discussed in the preceding section relies on the following fact. If the support function of a convex body $K \in \mathscr{K}^d$ has a representation

$$(9.1) \quad h(K,u) = \int_{\Omega^{d-1}} |\langle u,v\rangle| \, d\rho(v) \quad \text{for} \quad u \in \mathbb{R}^d$$

with an even signed Borel measure ρ on Ω^{d-1} which is not a positive measure, then K cannot be a zonoid. This is an immediate consequence of the representation theorem (1.2) and the uniqueness theorem (1.4). In particular, if f is a continuous real function on Ω^{d-1} attaining negative values but such that

$$(9.2) \quad h(u) = \int_{\Omega^{d-1}} |\langle u,v\rangle| f(v) \, d\omega(v) \quad \text{for} \quad u \in \Omega^{d-1}$$

defines a convex function h, then h is the support function of a body in $\mathscr{K}^d \backslash \mathscr{Z}^d$. By suitable choices of f the existence of the counter-examples in Weil [1977], [1982] was shown. With the same method and formula (9.4) below, one can also show the existence of a smooth body $K \in \mathscr{K}^d$ which has all its principal radii of curvature arbitrarily near to one, but which is not a zonoid. This provides a negative answer to a question of Choquet [1969b], p. 178.

For particular f (e.g., spherical harmonics) an explicit computation of the integral in (9.2) may be possible. In this way one can obtain simple explicit examples of smooth centred convex bodies which are not zonoids, for instance the body $K \in \mathscr{K}^3$ with support function given by

$$h(K,u) = 1 + \alpha P_2(\langle e,u\rangle) \quad \text{for} \quad u \in \Omega^2,$$

where $e \in \Omega^2$ is fixed, $P_2(t) = (3t^2 - 1)/2$ denotes the second Legendre polynomial, and $\alpha \in [-2/5, -1/4]$ is a real constant (see Schneider [1970a], p. 69).

It is often useful to know whether the integral equation (9.2), where h is a given even function on Ω^{d-1}, has a solution f. (Geometric applications of such a solution are made, for instance, in Chakerian [1967], Firey [1967], Petty [1967], Schneider [1967].) As remarked by Blaschke [1916], p. 154, spherical harmonics can be used to give an affirmative answer (under suitable smoothness assumptions, to which Blaschke does not pay attention). It is sufficient that h be of differentiability class C^k with $k = d + 2$ if d is even and $k = d + 3$ if d is odd; in this case the even solution f is continuous. One also has a stability result: $\|f\| \leq \|h\|_{2k}$, where $\|\cdot\|$ denotes the sup-norm and $\|\cdot\|_{2k}$ is a certain norm which involves derivatives up to order 2k. This was shown in Schneider [1975], where it was used to prove the existence of centred zonoids whose polars are also zonoids, but which are not ellipsoids.

The convex bodies $K \in \mathscr{K}^d$ whose support function has a representation (9.1) with an even signed Borel measure ρ, are called *generalized zonoids*, and ρ is then called the *generating measure* of K. Let $\mathscr{Z}^d_g$ denote the class of these bodies. Thus every sufficiently smooth centred convex body is a generalized zonoid, but $\mathscr{Z}^d_g$ does not exhaust $\mathscr{K}^d$. As shown by Schneider [1970a], §6, every face of a generalized zonoid is a translate of a generalized zonoid, hence centrally symmetric. This implies that a polytope is a generalized zonoid only if it is a zonotope. Of course, using the Jordan decomposition of a signed measure, we note that K is a generalized zonoid if and only if there exist centred zonoids Z_+, Z_- such that $Z_+ = K + Z_-$.

The integral representation (9.1) offers two main problems: To investigate the convex bodies which have such a representation with a signed measure ρ, and to characterize those signed measures ρ for which (9.1) defines a support function. A detailed study was begun in Weil [1976a]. This paper first develops a formula for the mixed surface area measure of generalized zonoids. For $u_1,\dots,u_j \in \Omega^{d-1}$ let $D(u_1,\dots,u_j)$ denote the (absolute) j-volume of the parallelepiped spanned by $u_1,\dots,u_j$. Further, define a partial map $T:(\Omega^{d-1})^{d-1} \to \Omega^{d-1}$ by the normalized vector product. Then if $K_1,\dots,K_{d-1}$ are generalized zonoids represented by

$$(9.3) \quad h(K_i,u) = \int_{\Omega^{d-1}} |\langle u,v\rangle| \, d\rho_i(v),$$

their mixed surface area measure is given by

$$(9.4) \quad S(K_1,\dots,K_{d-1};\cdot) = \frac{2^d}{(d-1)!} T\left[\int_{(\cdot)} D d(\rho_1 \times \cdots \times \rho_{d-1})\right],$$

where $T\mu$ denotes the image measure of μ under T (observe that, in the present case, T is defined μ-almost everywhere) and $\int_{(\cdot)} D d(\rho_1 \times \cdots \times \rho_{d-1})$ is the indefinite integral. Consequently, the mixed volume of the generalized zonoids $K_1,\dots,K_d$ can be computed from

$$(9.5) \quad V(K_1,\dots,K_d) = \frac{2^d}{d!} \int_{\Omega^{d-1}} \cdots \int_{\Omega^{d-1}} D(u_1,\dots,u_d) \, d\rho_1(u_1) \cdots d\rho_d(u_d).$$

From this it is deduced that the j-th quermassintegral of a generalized zonoid K with generating measure ρ is given by

$$(9.6) \quad W_j(K) = \frac{2^{d-j} j! \varkappa_j}{d!} \int_{\Omega^{d-1}} \cdots \int_{\Omega^{d-1}} D(u_1,\dots,u_{d-j}) \, d\rho(u_1) \cdots d\rho(u_{d-j})$$

for $j = 0,\ldots,d-1$. In a slightly different way this formula (for zonoids) is proved by Matheron [1975] (Corollary 4 on p. 104). From his argument (or by extending the argument used in Weil [1976a]) the following more general formula for the mixed volume of $d-j$ generalized zonoids $K_1,\ldots,K_{d-j}$ and j arbitrary convex bodies $L_1,\ldots,L_j \subset \mathbb{R}^d$ can be deduced:

$$\binom{d}{j} V(K_1,\ldots,K_{d-j},L_1,\ldots,L_j)$$

$$(9.7) \qquad = \frac{2^{d-j}}{(d-j)!} \int_{\Omega^{d-1}} \cdots \int_{\Omega^{d-1}} v^{(j)}(L_1,\ldots,L_j;F) D(u_1,\ldots,u_{d-j}) \times d\rho_1(u_1) \cdots d\rho_{d-j}(u_{d-j}).$$

Here F is the j-dimensional space orthogonal to $u_1,\ldots,u_{d-j}$ (provided they are linearly independent).

Special cases of (9.4) and (9.6) (for sufficiently smooth zonoids in $\mathbb{R}^3$) occur already in Blaschke [1916], p. 156. Formula (9.5) for generalized zonoids was also proved by Lindquist [1975b]. Related computations in a special case were made by Petty [1961]. Applications of formula (9.6) to problems in stochastic geometry appear in Matheron [1975], ch. 6, and Schneider [1982a], [1982b].

Now from (9.4) and the results of Weil [1974] it is possible to deduce a characterization of those signed measures ρ for which (9.1) defines a support function. According to Weil [1976a], this is the case if and only if

$$T\left[\int_{(\cdot)} D d(\rho^j \times \omega^{d-j-1})\right] \geq 0$$

for $j = 1,\ldots,d-1$. Specializing (9.4) to measures with densities (see Weil [1976a]) one also obtains that the function h defined by (9.2) is convex if and only if

$$\int_{u^\perp} \cdots \int_{u^\perp} D^2(u_1,\ldots,u_j) f(u_1) \cdots f(u_j)\, d\omega_u(u_1) \cdots d\omega_u(u_j) \geq 0$$

for $j = 1,\ldots,d-1$ and all $u \in \Omega^{d-1}$; here $u^\perp$ is the great sphere with pole u and ω_u is spherical Lebesgue measure on that sphere.

A different characterization of the signed measures ρ which generate convex bodies by means of (9.1) can be obtained by considering projections. For a signed measure ρ on Ω^{d-1} and for a j-dimensional great subsphere Ω_j of Ω^{d-1}, a projection of ρ onto Ω_j can be defined (see Weil [1982]) such that the following holds. If K is a generalized zonoid with generating measure ρ and if E_{j+1} is a $(j+1)$-dimensional linear subspace of $\mathbb{R}^d$, then the projection of K onto E_{j+1} has the projection of ρ on $E_{j+1} \cap \Omega^{d-1}$ as its generating measure. From this it was deduced in Weil [1982] that a signed measure ρ generates a convex body if and only if all its projections onto great circles are nonnegative. It also follows that h given by (9.2) is convex if and only if

$$\int_{\Omega_{d-2}} |\langle u,v\rangle|^2 f(v)\, d\omega_{\Omega_{d-2}}(v) \geq 0$$

for all $(d-2)$-dimensional great spheres $\Omega_{d-2} \subset \Omega^{d-1}$ (with spherical Lebesgue measure $\omega_{\Omega_{d-2}}$) and arbitrary $u \in \Omega_{d-2}$. This result is due to Lindquist [1975b].

The usefulness of the representation (9.1) on one hand and the fact that the generalized zonoids are only a proper subclass of $\mathscr{K}^d$ on the other hand, leads one to ask for a generalization of (9.1) to arbitrary centred convex bodies K. Such a generalization is possible if distributions are used instead of signed measures. It was shown by Weil [1976b] that to every convex body $K \in \mathscr{K}^d$ there exists an even distribution T_K on Ω^{d-1}, whose domain can be extended to include the functions $|\langle u,\cdot\rangle|$, $u \in \Omega^{d-1}$, such that $T_K(|\langle u,\cdot\rangle|) = h(K,u)$. The properties of the correspondence $K \mapsto T_K$ are further studied in Weil [1976b] and [1979]. Using this correspondence, the formulae (9.4)–(9.7) were generalized to arbitrary centred convex bodies. As another application, one can obtain characterizations of zonoids and generalized zonoids, making use of the known conditions for a distribution to be a measure or a signed measure. Thus one obtains:

(9.8) **Theorem.** *A body* $K \in \mathscr{K}^d$ *is a zonoid if and only if*

$$V(K,L,\ldots,L) \leq V(K,M,\ldots,M)$$

for all $L,M \in \mathscr{K}^d$ *which satisfy* $v_{d-1}(L,u) \leq v_{d-1}(K,u)$ *for* $u \in \Omega^{d-1}$.

Here $v_{d-1}(L,u) = v^{(d-1)}(L;\{\text{lin } u\}^{\perp})$ is the $(d-1)$-volume of the orthogonal projection of L in direction u.

(9.9) **Theorem.** *A body* $K \in \mathscr{K}^d$ *is a generalized zonoid if and only if there is a constant* c_K *such that*

$$|V(K,L,\ldots,L) - V(K,M,\ldots,M)| \leq c_K \max_{u \in \Omega^{d-1}} |v_{d-1}(L,u) - v_{d-1}(M,u)|$$

for all $L,M \in \mathscr{K}^d$.

Theorem (9.8) was proved in Weil [1976b]. Theorem (9.9) is due to Goodey [1977], who used a different method of proof, and it was proved again by Weil [1979] with the aid of distributions. These papers also contain generalizations and other results in the same spirit.

Now, having formulated Theorem (6.1), we observe that it allows a simple proof of Theorem (9.8) by using the fact that each measure $\mu \in \mathscr{S}^*$ is a difference of $(d-1)$-th order surface area measures of centred convex bodies (for the notion of surface area measures, or area functions, see Busemann [1958], p. 62). By an appropriate choice of $\mathscr{N} \subset \mathscr{S}^*$, (9.8) can be improved. Let

$$\mathscr{N} = \{\mu \in \mathscr{S}^* : \mu = c\omega - S_j(L,\cdot), \quad c \geq 0, \quad L \in \mathscr{K}^d\},$$

where $S_j(L,\cdot)$ is the j-th order surface area measure of L. By a suitable extension of a result of Weil [1980], $\mathscr{N}$ is dense in $\mathscr{S}^*$. Therefore, Theorem (6.1) implies:

(9.10) **Theorem.** *Let* $j \in \{1,\ldots,d-1\}$ *and* $K \in \mathscr{K}^d$ *be of mean width one. Then* K *is a zonoid if and only if*

$$V(K,\underbrace{L,\ldots,L}_{j},\underbrace{B,\ldots,B}_{d-1-j}) \leq \frac{\varkappa_d}{\varkappa_{d-1}} \max_{u \in \Omega^{d-1}} v_j(L,u)$$

for all $L \in \mathscr{K}^d$.

Here $v_j(L,u)$ is a quermassintegral of the orthogonal projection of L on to the hyperplane E through O orthogonal to u:

$$v_j(L,u) := v^{(d-1)}(\underbrace{L,\ldots,L}_{j},\underbrace{B,\ldots,B}_{d-1-j};E).$$

Using a different dense subset $\mathcal{N}$ of $\mathcal{S}^*$, one gets the following result.

(9.11) **Theorem.** *A body* $K \in \mathcal{K}^d$ *of mean width one is a zonoid if and only if*

$$\sum_{i=1}^{n} h(K,x_i) \leq \frac{\varkappa_d}{\varkappa_{d-1}} \max_{u \in \Omega^{d-1}} \sum_{i=1}^{n} |\langle x,u_i \rangle|$$

for all $x_1,\ldots,x_n \in \Omega^{d-1}$ *and all integers* $n > 0$.

Similarly as in §8, these characterization results suggest the investigation of more general classes of convex bodies. For instance, Theorem (9.8) leads one to consider, for fixed $j \in \{1,\ldots,d-1\}$, the set of those $K \in \mathcal{K}^d$ which satisfy

$$V(\underbrace{K,\ldots,K}_{j},\underbrace{L,\ldots,L}_{d-j}) \leq V(\underbrace{K,\ldots,K}_{j},\underbrace{M,\ldots,M}_{d-j})$$

for all $L,M \in \mathcal{K}^d$ with $v^{(d-j)}(L;E) \leq v^{(d-j)}(M;E)$ for $E \in \mathcal{E}^d_{d-j}$. For $j = 1$ this class coincides with $\mathcal{Z}^d$, and for $j = d-1$ with $\mathcal{K}^d$. For the intermediate cases, relations to the classes $\mathcal{K}^d_j$ and $\hat{\mathcal{K}}^d_j$ of §8 probably exist, but are unknown. Similar definitions are motivated by Theorem (9.10).

§10. Projection bodies

In geometry, zonoids first appeared as so-called projection bodies. Let $K \subset \mathbb{R}^d$ be a convex body and let $v_{d-1}(K,\cdot)$ be the projection function defined in §9. As observed by Minkowski [1911], §26 (see also Bonnesen-Fenchel [1934], p. 45), $v_{d-1}(K,\cdot)$ is the restriction to Ω^{d-1} of a support function and hence defines a centred convex body $\Pi_{d-1}K$, which is called the *projection body* of K. Since

$$v_{d-1}(K,u) = \frac{1}{2}\int_{\Omega^{d-1}} |\langle u,v \rangle| \, dS_{d-1}(K,v) = \int_{\Omega^{d-1}} |\langle u,v \rangle| \, d\rho(v)$$

with $\rho(\beta) = \frac{1}{4}[S_{d-1}(K,\beta) + S_{d-1}(K,-\beta)]$, every projection body is a zonoid. Vice versa, every d-dimensional centred zonoid is the projection body of a unique centred convex body. The existence follows from the solution of the generalized Minkowski problem (e.g., Busemann [1958], p. 64) and the uniqueness from the corresponding uniqueness result and from (1.4). (It was in proving a generalization of this result that Aleksandrov [1937] was led to show (1.4)).

The above construction can be generalized in the following way. For $K \in \mathcal{K}^d$ and $u \in \Omega^{d-1}$, we have

$$(10.1) \quad v_j(K,u) = \frac{1}{2}\int_{\Omega^{d-1}} |\langle u,v \rangle| \, dS_j(K,v), \quad j \in \{0,\ldots,d-1\}.$$

Thus $v_j(K,\cdot)$ is the support function of a centred zonoid which we denote by $\Pi_j K$; we call it the *projection body of order* j of the body K. It would be interesting to know more about the range of the transformation $K \mapsto \Pi_j K$ ($K \in \mathscr{K}^d$), in other words, to investigate the class $\mathscr{Z}^d_{(j)}$ of those zonoids which are projection bodies of order j. While $\mathscr{Z}^d_{(0)}$ contains only a ball and $\mathscr{Z}^d_{(d-1)}$ is the class of centred zonoids, the classes $\mathscr{Z}^d_{(j)}$ for $0 < j < d-1$ lie strictly between these two cases, and any nontrivial characterization is probably very difficult.

Since the transformation Π_j leads from general convex bodies to zonoids and thus to especial bodies, one may expect that the iterations Π_j^k of this transformation yield ever more restricted classes. This led Weil [1971] to consider the following problem. What can be said about the convex bodies $K \in \mathscr{K}^d$ which, for every $m \in \mathbb{N}$, satisfy $K = \Pi_{d-1}^m K_m$ for some $K_m \in \mathscr{K}^d$? He proved that a d-polytope has this property if and only if it is a direct sum of symmetric polygons. These polytopes P are also characterized by the fact that $\Pi_{d-1}^2 P$ is homothetic to P. On the contrary, Schneider [1977] proved that a convex body K for which some $\Pi_1^m K$ is homothetic to K, must be a ball.

The rest of this section will be concerned with the question in how far (partial) information on $\Pi_j K$ yields information on K.

First we have the uniqueness theorem of Aleksandrov [1937] (for a different proof and a generalization, see Chakerian [1967]).

(10.2) **Theorem.** *Let* $j \in \{1,\ldots,d-1\}$, *and let* $K,L \in \mathscr{K}^d$ *be centred convex bodies of dimension* $\geq j+1$. If $\Pi_j K = \Pi_j L$, *then* $K = L$.

That the assumption of central symmetry for K and L cannot be omitted, is plausible since the values of the integral (10.1) are independent of the odd part of the measure $S_j(K,\cdot)$. It is, for instance, possible to construct nonspherical bodies K for which $\Pi_j K$ is a ball (bodies of constant brightness, see Blaschke [1916] and Firey [1965]). We shall, therefore, only consider centred bodies in the following.

Theorem (10.2) says that the transformation $\Pi_j : \mathscr{K}^d \to \mathscr{Z}^d_{(j)}$ has an inverse, and it is not difficult to show that this inverse is continuous (with respect to the Hausdorff metric). One of the major open problems in this area is a corresponding stability result: If $K,L \in \mathscr{K}^d$ are such that $\Pi_j K$ and $\Pi_j L$ are close to each other, are K and L close in a sense which can be made precise in a quantitative manner? Even more desirable would be an effective way of reconstructing $K \in \mathscr{K}^d$ from $\Pi_j K$, but there seems to be little hope for such a result.

While the transformation Π_j is clearly monotone with respect to set inclusion, its inverse is not. It is even not possible to deduce from $\Pi_{d-1} K \supset \Pi_{d-1} L$ for $K,L \in \mathscr{K}^d$ that $V(K) \geq V(L)$, where V denotes the volume. More precisely, if $K \in \mathscr{K}^d \setminus \mathscr{Z}^d$ is sufficiently smooth and of positive curvature, then there exists a body $L \in \mathscr{K}^d$ such that $v_{d-1}(K,u) > v_{d-1}(L,u)$ for all $u \in \Omega^{d-1}$, but $V(K) < V(L)$. On the other hand, if $Z \in \mathscr{Z}^d$ and $L \in \mathscr{K}^d$, then $v_{d-1}(Z,u) > v_{d-1}(L,u)$ implies $V(Z) > V(L)$. For these and related results, see Petty [1967] and Schneider [1967].

Considering the uniqueness result (10.2), one might think of improving it as follows. The assumption $\Pi_j K = \Pi_j L$ demands that

(10.3) $v_j(K,u) = v_j(L,u)$

for all $u \in \Omega^{d-1}$. Would it be sufficient to assume (10.3) only for $u \in A$, where $A \subset \Omega^{d-1}$ is a proper closed symmetric subset? At least for $j = d-1$, the answer is "no" for any such set A, as shown by Schneider-Weil [1970]. However, they proved the following slightly curious result. Suppose that $K,L \in \mathscr{K}^d$ are such that, for some $z \in \Omega^{d-1}$, the support planes of K and L with normals in a neighborhood of z contain only singular points of the bodies. Suppose further that $v_{d-1}(K,u) = v_{d-1}(L,u)$ for all u in some neighborhood of the great sphere with pole z. Then if $d \geq 3$ is odd, K is equal to L. This result is deduced from the following supplement to Theorem (1.4). Let d be odd. If ρ is an even signed measure on Ω^{d-1} satisfying

$$\int_{\Omega^{d-1}} |\langle u,v \rangle| \, d\rho(u) = 0$$

for each u in a neighborhood of some great sphere with pole z and if $|\rho|$ is zero on a neighborhood of z, then $\rho = 0$. In even dimensions this conclusion would be false.

The following result in a similar spirit was proved by Schneider [1970c]. Let $K \in \mathscr{K}^d$ be a convex body and $P \in \mathscr{P}^d$ a polytope. Let $j \in \{1,\ldots,d-1\}$ and let $A \subset \Omega^{d-1}$ be an open set which contains to each j-face of P a vector parallel to it. If $v_j(K,u) = v_j(P,u)$ for all $u \in A$, then $K = P$. For $j \geq 2$ the set A could be, for instance, a neighborhood of an arbitrary great sphere.

In a different direction, conclusions from partial information on $\Pi_j K$ to information on K have been drawn by Betke-McMullen [1982]. Clearly $\Pi_j K$ determines the intrinsic j-volume $V_j(K)$ of K ($\varkappa_{d-j} V_j(K) = \binom{d}{j} W_{d-j}(K)$, where W_{d-j} is the usual quermassintegral), since $V_j(K)$ is, up to a constant factor, the mean value of $v_j(K,u)$ over Ω^{d-1}. Now let $\mathscr{L} = (u_1,\alpha_1;\ldots;u_n,\alpha_n)$ be a family of vectors $u_i \in \Omega^{d-1}$ with associated positive weights α_i. Betke and McMullen define

$$V_j(K,\mathscr{L}) := \sum_{i=1}^{n} \alpha_i v_j(K,u_i)$$

and ask for the range of the ratio

$$\pi_j(K,\mathscr{L}) = V_j(K,\mathscr{L})/V_j(K)$$

for convex bodies K of dimension $\geq j$. (They use different notation and treat more general problems.) Introducing the zonotope

$$Z(\mathscr{L}) := \sum_{i=1}^{n} \operatorname{conv}\{-\alpha_i u_i, \alpha_i u_i\},$$

they give sharp bounds (together with a criterion for the extremal bodies) for $\pi_{d-1}(K,\mathscr{L})$ in terms of the inradius and circumradius of $Z(\mathscr{L})$. For $1 \leq j \leq d-2$ they have a weaker result and an interesting conjecture.

A number of apparently difficult problems arises if our partial information on $\Pi_j K$ consists only in the knowledge of certain functionals. To give a specific example: What is the range of the affine-invariant ratio $V(\Pi_{d-1}K)/V(K)^{d-1}$ for d-dimensional centred bodies $K \in \mathscr{K}^d$ ($d \geq 3$)? The question for the minimum was posed by Petty [1967]. Considering a certain question on geometric probabilities, Schneider [1982a] was led to consider the maximum and conjectured that $V(\Pi_{d-1}K)/V(K)^{d-1} \leq 2^d$, with equality precisely when K is a direct sum of centred convex bodies of dimension ≤ 2. Some interesting inequalities for projection bodies were proved by Petty [1967].

Besides the projection bodies, a different way of associating with a convex body (or a more general set) a certain zonoid, was studied by Petty [1961]. He introduced the *centroid bodies*, which are special zonoids, and made interesting applications.

§11. Approximation problems for zonoids

In their investigation on estimating the intrinsic volumes of convex bodies from those of projections, Betke and McMullen [1982] were also led to consider the following problem on approximating the unit ball $B \subset \mathbb{R}^d$ by zonotopes. Let $\zeta(n,d)$ denote the infimum of the Hausdorff distance $\delta(B,Z)$, taken over all zonotopes Z which are the sums of n line segments. Betke and McMullen proved the existence of constants β_d and γ_d, independent of n, such that

$$\beta_d n^{-2} \leq \zeta(n,d) \leq \gamma_d n^{-2/(d-1)}.$$

It is an intriguing question to determine the precise order of $\zeta(n,d)$, which Betke and McMullen conjecture to be $n^{-2/(d-1)}$.

Approximation problems of a different kind arise for those bodies which are not zonoids. For instance, for which bodies $K \in \mathscr{K}^d$ of given mean width (say) is $\inf\{\delta(K,Z): Z \in \mathscr{Z}^d\}$ maximal? But perhaps it is more natural to measure the distance of K from $\mathscr{Z}^d$ in an affine-invariant way. As an example, for $K \in \mathscr{K}^d$ let

$$\lambda(K) := \inf\{\lambda \geq 1 : K \subseteq Z \subseteq \lambda K \quad \text{for some} \quad Z \in \mathscr{Z}^d\}.$$

For which bodies $K \in \mathscr{K}^d$ is λ maximal? We think that the cross-polytope X_d is a good candidate. One can prove that

$$\lambda(X_d) = 2^{-d+1} d \binom{d-1}{\left\lceil \frac{d-1}{2} \right\rceil}.$$

We mention that for $d = 3$, there are different zonotopes Z for which $X_3 \subset Z \subset \lambda(X_3)X_3$. One of these is the rhombic dodecahedron, and another one the truncated octahedron. Thus these two classical polytopes (two of the five parallelohedra in $\mathbb{R}^3$) share the extremum property that they serve best, among all zonohedra, to approximate (in this sense) the octahedron.

References

Aleksandrov, A.D.
1933 A theorem on convex polyhedra. (Russian) Trudy Fiz.-Mat. Inst. Im Steklova **4** (1933), 87.
1937 Zur Theorie der gemischten Volumina von konvexen Körpern. II. Neue Ungleichungen zwischen den gemischten Volumina und ihre Anwendungen. (Russian) Mat. Sbornik N.S. **2** (1937), 1205–1238.
Alfsen, E.M.
1971 Compact convex sets and boundary integrals. Springer, Berlin etc. 1971.
Assouad, P.
1980a Charactérisations de sous-espaces normés de L^1 de dimension finie. Séminaire d'analyse fonctionnelle (Ecole Polytechnique Palaiseau) 1979–1980, exposé No **19**.
1980b Plongements isométriques dans L^1: aspect analytique. Séminaire Choquet (Initiation à l'analyse) 1979–1980.

Betke, U., P. McMullen
1982 Estimating the sizes of convex bodies from projections. (to appear)
Blackwell, D.
1951 The range of certain vector integrals. Proc. Amer. Math. Soc. **2** (1951), 390–395.
Blaschke, W.
1916 Kreis und Kugel. Veit, Leipzig 1916. 2nd edition: de Gruyter, Berlin 1956.
1923 Vorlesungen über Differentialgeometrie. II. Affine Differentialgeometrie. Springer-Verlag, Berlin 1923.
Blaschke, W., K. Reidemeister
1922 Über die Entwicklung der Affingeometrie. Jber. DMV **31** (1922), 63–82.
Bolker, E.D.
1969 A class of convex bodies. Trans. Amer. Math. Soc. **145** (1969), 323–346.
1971 The zonoid problem. Amer. Math. Monthly **78** (1971), 529–531.
Bonnesen, T., W. Fenchel
1934 Theorie der konvexen Körper. Springer-Verlag, Berlin 1934.
Bretagnolle, J., D. Dacunha Castelle, J.L. Krivine
1966 Lois stables et espaces L^p. Ann. Inst. H. Poincaré N.S. **2** (1966), 231–259.
Burckhardt, J.J.
1940 Über konvexe Körper mit Mittelpunkt. Vierteljahresschr. Naturf. Ges. Zürich **85** (1940), 149–154.
Burton, G.R.
1976 On the sum of a zonotope and an ellipsoid. Comment. Math. Helvet. **51** (1976), 369–387.
Burton, G.R., G.B. Purdy
1979 The directions determined by n points in the plane. J. London Math. Soc. (2) **20** (1979), 109–114.
Busemann, H.
1958 Convex surfaces. Interscience Publishers, New York 1958.
Chakerian, G.D.
1967 Sets of constant relative width and constant relative brightness. Trans. Amer. Math. Soc. **129** (1967), 26–37.
Choquet, G.
1968 Mesures coniques et affines invariantes par isométries. Zonoformes, zonoèdres et fonctions de type négatif. C.R. Acad. Sc. Paris **266** (1968), 619–621.
1969a Lectures on analysis. Vol. III. W.A. Benjamin, Reading, Mass. etc. 1969.
1969b Mesures coniques, affines et cylindriques. Symposia Math., vol. II (INDAM, Rome 1968). Academic Press, London and New York 1969.
Coxeter, H.S.M.
1962 The classification of zonohedra by means of projective diagrams. J. Math. pures appl. (9) **41** (1962), 137–156.
1963 Regular polytopes. 2nd edition Macmillan, New York 1963.
Diestel, J., J.J. Uhl
1977 Vector measures. Math. Surveys No **15**. Amer. Math. Soc., Providence, R.I. 1977.
Doignon, J.-P.
1980 On space-filling zonotopes. Lecture at Oberwolfach, July 1980.
Dor, L.E.
1976 Potentials and isometric embeddings in L_1. Israel J. Math. **24** (1976), 260–268.
Dubins, L.E., E.H. Spanier
1961 How to cut a cake fairly. Amer. Math. Monthly **68** (1961), 1–17.
Dvoretzky, A., A. Wald, J. Wolfowitz
1951 Elimination of randomization in certain statistical decision procedures and zero-sum two-person games. Ann. Math. Stat. **22** (1951), 1–21.
Firey, W.J.
1965 The brightness of convex bodies. Tech. Report No. **19**, Oregon State University, Corvallis 1965.
1967 Blaschke sums of convex bodies and mixed bodies. Proc. Colloquium Convexity (Copenhagen 1965), Københavns Univ. Mat. Inst. 1967, 94–101.
Goodey, P.R.
1977 Centrally symmetric convex sets and mixed volumes. Mathematika **24** (1977), 193–198.

Groemer, H.
1978 On multiple space subdivisions by zonotopes. Monatsh. Math. **86** (1978), 185–188.
Grothendieck, A.
1953 Résumé de la théorie métrique des produits tensoriels topologiques. Bol. Soc. Mat. São Paulo **8** (1953), 1–79.
Grünbaum, B.
1967 Convex polytopes. Interscience Publishers, London etc. 1967.
1971 Arrangements of hyperplanes. Proc. 2nd Lousiana Conf. Combin. and Graph Theory. Baton Rouge (1971), 41–106.
1972 Arrangements and spreads. Amer. Math. Soc., Providence, R.I. 1972.
Hadwiger, H.
1952 Mittelpunktspolyeder und translative Zerlegungsgleichheit. Math. Nachr. **8** (1952), 53–58.
1972 Polytopes and translative equidecomposability. Amer. Math. Monthly **79** (1972), 275–276.
Halmos, P.R.
1948 The range of a vector measure. Bull. Amer. Math. Soc. **54** (1948), 416–421.
Hermes, H., J.P. LaSalle
1969 Functional analysis and time optimal control. Academic Press, New York 1969.
Herz, C.S.
1963 A class of negative definite functions. Proc. Amer. Math. Soc. **14** (1963), 670–676.
Karlin, S., W.J. Studden
1966 Tchebycheff systems: With applications in analysis and statistics. Interscience Publishers, New York etc. 1966.
Kelly, J.B.
1975 Hypermetric spaces. In: The geometry of metric and linear spaces. Lecture Notes in Mathematics 490, Springer, Berlin etc. 1975, 17–31.
Kelly, L.M., W.O.J. Moser
1958 On the number of ordinary lines determined by n points. Can. J. Math. **10** (1958), 210–219.
Kluvánek, I.
1975 The range of a vector measure. Bull. Amer. Math. Soc. **81** (1975), 609–611.
1976 Characterization of the closed convex hull of the range of a vector-valued measure. J. Funct. Anal. **21** (1976), 316–329.
Kluvánek, I., G. Knowles
1975 Vector measures and control systems. North-Holland, Amsterdam 1975.
Lévy, P.
1937 Théorie de l'addition des variables aléatoires. Gauthier-Villars, Paris 1937.
Lyapunov (Liapounoff) A.A.
1940 On completely additive vector functions. (Russian) Izv. Akad. Nauk. SSSR **4** (1940), 465–478.
Lindenstrauss, J.
1966 A short proof of Liapounoff's convexity theorem. J. Math. Mech. **15** (1966), 971–972.
Lindquist, N.F.
1975a Approximation of convex bodies by sums of line segments. Portugaliae Math. **34** (1975), 233–240.
1975b Support functions of central convex bodies. Portugaliae Math. **34** (1975), 241–252.
Matheron, G.
1974a Un théorème d'unicité pour les hyperplans poissoniens. J. Appl. Prob. **11** (1974), 184–189.
1974b Hyperplans poissoniens et compacts de Steiner. Adv. Appl. Prob. **6** (1974), 563–579.
1975 Random sets and integral geometry. Wiley, New York etc. 1975.
McMullen, P.
1970 Polytopes with centrally symmetric faces. Israel J. Math. **8** (1970), 194–196.
1971 On zonotopes. Trans. Amer. Math. Soc. **159** (1971), 91–110.
1975 Space tiling zonotopes. Mathematika **22** (1975), 202–211.
1976 Polytopes with centrally symmetric facets. Israel J. Math. **23** (1976), 337–338.
1979 Transforms, diagrams and representations. In: Contributions to Geometry, Proc. Geom. Symp. Siegen 1978, ed. by J. Tölke and J.M. Wills, Birkhäuser, Basel etc. 1979, 92–130.
1980 Convex bodies which tile space by translation. Mathematika **27** (1980), 113–121.
1981 Convex bodies which tile space by translation. Acknowledgement of priority. Mathematika **28** (1981), 191.

Minkowski, H.
1911 Theorie der konvexen Körper, insbesondere Begründung ihres Oberflächenbegriffs. Ges. Abh., vol. II. B.G. Teubner, Leipzig etc. 1911, 131–229.
Mürner, P.
1975 Translative Parkettierungspolyeder und Zerlegungsgleichheit. Elem. Math. **30** (1975), 25–27.
1977 Translative Zerlegungsgleichheit von Polytopen. Arch. Math. **29** (1977), 218–224.
Neyman, A.
1981 Decomposition of ranges of vector measures. Israel J. Math. **40** (1981), 54–64.
Petty, C.M.
1961 Centroid surfaces. Pacific J. Math. **11** (1961), 1535–1547.
1967 Projection bodies. Proc. Coll. Convexity (Copenhagen 1965), Københavns Univ. Mat. Inst. 1967, 234–241.
Rickert, N.W.
1967a Measures whose range is a ball. Pacific J. Math. **23** (1967), 361–371.
1967b The range of a measure. Bull. Amer. Math. Soc. **73** (1967), 560–563.
Schneider, R.
1967 Zu einem Problem von Shephard über die Projektionen konvexer Körper. Math. Z. **101** (1967), 71–82.
1970a Über eine Integralgleichung in der Theorie der konvexen Körper. Math. Nachr. **44** (1970), 55–75.
1970b Functional equations connected with rotations and their geometric applications. L'Enseignement Math. **16** (1970), 297–305.
1970c On the projections of a convex polytope. Pacific. J. Math. **32** (1970), 799–803.
1975 Zonoids whose polars are zonoids. Proc. Amer. Math. Soc. **50** (1975), 365–368.
1977 Rekonstruktion eines konvexen Körpers aus seinen Projektionen. Math. Nachr. **79** (1977), 325–329.
1982a Random hyperplanes meeting a convex body. Z. Wahrscheinlichkeitsth. verw. Geb. **61** (1982), 379–387.
1982b Random polytopes generated by anisotropic hyperplanes. Bull. London Math. Soc. **14** (1982), 549–553.
Schneider, R., W. Weil
1970 Über die Bestimmung eines konvexen Körpers durch die Inhalte seiner Projektionen. Math. Z. **116** (1970), 338–348.
Shannon, R.W.
1976 A lower bound on the number of cells in arrangements of hyperplanes. J. Combinatorial Th. A **20** (1976), 327–335.
Shephard, G.C.
1967 Polytopes with centrally symmetric faces. Canad. J. Math. **19** (1967), 1206–1213.
1974a Combinatorial properties of associated zonotopes. Canad. J. Math. **26** (1974), 302–321.
1974b Space-filling zonotopes. Mathematika **21** (1974), 261–269.
Stanley, R.P.
1981 Two combinatorial applications of the Aleksandrov-Fenchel inequalities. J. Combinatorial Th. A **31** (1981), 56–65.
Venkov, B.A.
1954 On a class of euclidean polytopes. (Russian) Vestnik Leningr. Univ. (Ser. Mat. Fiz. Him.) **9** (1954), 11–31.
Weil, W.
1971 Über die Projektionenkörper konvexer Polytope. Arch. Math. **22** (1971), 664–672.
1974 Über den Vektorraum der Differenzen von Stützfunktionen konvexer Körper. Math. Nachr. **59** (1974), 353–369.
1976a Kontinuierliche Linearkombination von Strecken. Math. Z. **148** (1976), 71–84.
1976b Centrally symmetric convex bodies and distributions. Israel J. Math. **24** (1976), 352–367.
1977 Blaschkes Problem der lokalen Charakterisierung von Zonoiden. Arch. Math. **29** (1977), 655–659.
1979 Centrally symmetric convex bodies and distributions. II. Israel J. Math. **32** (1972), 173–182.
1980 On surface area measures of convex bodies. Geom. Dedicata **9** (1980), 299–306.
1982 Zonoide und verwandte Klassen konvexer Körper. Monatsh. Math. **94** (1982), 73–84.

Wieacker, J.A.
1982 Translative stochastische Geometrie der konvexen Körper. Dissertation, Freiburg 1982.
Witsenhausen, H.S.
1973 Metric inequalities and the zonoid problem. Proc. Amer. Math. Soc. **40** (1973), 517–520.
1978 A support characterization of zonotopes. Mathematika **25** (1978), 13–16.
Zaslavsky, T.
1975 Facing up to arrangements: Face-count formulas for partitions of space by hyperplanes. Memoirs Amer. Math. Soc. **154** (1975), 102 p.

Added in proof.

For a recent application of zonoids in Mathematical Economics, see

W. Hildenbrand, Short-run production functions based on microdata. Econometrica **49** (1981), 1095–1125.

Generalizations of zonoids are considered in

W. Hildenbrand, A. Neyman, Integrals of production sets with restricted substitution. J. Math. Economics **9** (1982), 71–82.
W. Ricker, A new class of convex bodies. Contemporary Math. **9** (1982), 333–340.

These authors investigate, for fixed $k \in \{1,\dots,d-1\}$, the class of convex bodies in $\mathbb{R}^d$ which can be approximated by sums of k-dimensional convex sets or, respectively, of k-dimensional simplices.

The connection between zonoids and stationary Poisson hyperplane processes (§7) is further exploited by

C. Thomas, Extremum properties of the intersection densities of stationary Poisson hyperplane processes. (to appear).

R. Schneider
Mathematisches Institut
Universität Freiburg
Albertstrasse 236
D–7800 Freiburg (FRG)

W. Weil
Mathematisches Institut II
Universität Karlsruhe
Englerstrasse 2
D–7500 Karlsruhe (FRG)

G. Fejes Tóth

New Results in the Theory of Packing and Covering

Let $\mathscr{S}$ be a system of sets. If all members of $\mathscr{S}$ are contained in a given set C and each point of C belongs to at most one member of $\mathscr{S}$ then $\mathscr{S}$ is said to be a *packing into* C. If, on the other hand, each point of C belongs to at least one member of $\mathscr{S}$ then we say that $\mathscr{S}$ is a *covering of* C. Packings and coverings have been considered in various spaces and on various combinatorial structures. Here we are interested in problems concerning packings and coverings consisting of convex bodies in spaces of constant curvature, i.e. in Euclidean, spherical and hyperbolic space. Instead of saying that $\mathscr{S}$ is a packing into the whole space or $\mathscr{S}$ is a covering of the whole space we shall simply use the terms *$\mathscr{S}$ is a packing* and *$\mathscr{S}$ is a covering.*

While in the plane a great variety of problems has been studied and several of them completely solved, in the higher dimensional analogues we are either totally uninformed or we must be satisfied with more or less rough estimates. This difference between two and more dimensional spaces is revealed also by the two basic books on packing and covering. On the one hand, the book of L. Fejes Tóth (1972a) deals mainly with problems in two dimensions, on the other hand, Rogers' book (1964) is devoted to problems in n-dimensional space, with *n* large. The aim of the present paper is to survey the main results of the theory obtained after the publication of these books.

The subject of the paper is divided into two parts: Part 1 which is meant to supplement Rogers' book deals with problems in general n-dimensional spaces, while problems in the plane are treated in Part 2.

For some basic notions as well as for earlier literature the reader is referred to the above mentioned books. Part of our subject is treated also in the survey papers of Bambah (1970, 1971), Gruber (1979), L. Fejes Tóth (1977a), (1983c), Florian (1975a, 1980a, 1980b, 1980c), Krötenheerdt (1975), Ryskov (1977) and Saaty and Alexander (1975). A special class of packings and coverings are tilings. These have been excluded from our considerations as they are the subject of the recent book of Grünbaum and Shephard (1983).

1. Problems in *n*-dimensional space

1.1 *Densest packing and thinnest covering in Euclidean space*

First we recall one of the basic notions of the theory of packing and covering, the notion of density. The *density* of a system $\mathscr{S}$ of measurable sets with respect to a set C with finite and positive Lebesgue measure is defined by the ratio of the total volume of the sets of $\mathscr{S}$ intersecting C to the volume of C:

$$d(\mathscr{S},C) = \frac{\sum_{S\in\mathscr{S}-S\cap C\neq\emptyset} |S|}{|C|}.$$

(Here and in what follows we denote the volume of a set X by $|X|$.) The density $d(\mathscr{S})$ of $\mathscr{S}$ with respect to the whole space can be defined by a limiting process as follows: if C(r) denotes a sphere of radius r centred at a fixed point 0 then

$$d(\mathscr{S}) = \lim_{r\to\infty} d(\mathscr{S}, C(r)).$$

If the limit on the right hand side does not exist then we consider the *lower density* and the *upper density* of $\mathscr{S}$ defined by

$$d_-(\mathscr{S}) = \lim_{r\to\infty} \inf d(\mathscr{S}, C(r))$$

and

$$d_+(\mathscr{S}) = \lim_{r\to\infty} \sup d(\mathscr{S}, C(r)).$$

It can be easily seen that these limiting values do not depend on the choice of the point 0.

For a given set S we define the *packing density* $\delta(S)$ and the *covering density* $\vartheta(S)$ of S by

$$\delta(S) = \sup d_+(\mathscr{S})$$

and

$$\vartheta(S) = \inf d_-(\mathscr{S}),$$

the supremum being taken over all systems of congruent copies of S forming a packing and the infimum being taken over all systems of congruent copies of S forming a covering. Groemer (1963), (1968a) established the existence of packings and coverings with density $\delta(S)$ and $\vartheta(S)$, respectively, under very general conditions on S.

The basic problem in the theory of packing and covering is to determine the quantities $\delta(S)$ and $\vartheta(S)$, or at least to give good lower and upper bounds for them. The book of Rogers is devoted solely to the discussion of this problem. In the preface to his book Rogers writes: "The present seems to me an opportune time for the publication of such a book, since I have the impression that most of the simplest general results of the subject have already been discovered, and further progress may be rather slow, depending on detailed and complicated technical developments. If I am proved wrong in this I shall be happy". Unfortunately, regarding problems in higher-dimensional spaces, he was certainly right. New results since then have been attained only concerning packing of balls. Before discussing these results we recall some of the earlier ones.

We begin with results concerning the existence of reasonably dense packings and thin coverings. For packings we have the famous theorem of Minkowski-Hlawka which was stated, without proof, by Minkowski and first proved by Hlawka (1944). According to the Minkowski-Hlawka theorem we have

$$\delta(S) \geq \zeta(n)/2^{n-1} \quad \left(\zeta(n) = \sum_{k=1}^{\infty} k^{-n}\right)$$

for all n-dimensional centrally symmetric convex sets. This implies, together with an argument of Minkowski and an inequality of Rogers and Shephard (1957) for the volume of the difference set, that

$$\delta(S) \geq 2\zeta(n)(n!)^2/(2n)!.$$

for all n-dimensional convex sets. From among the various refinements of the Minkowski-Hlawka theorem the best result is due to Schmidt (1963). It states that

$$\delta(S) \geq (\log 2 + o(1))n/2^n$$

for symmetrical convex n-dimensional sets and

$$\delta(S) \geq (\pi \log 2 + o(1))n^{3/2}/4^n$$

for all convex n-dimensional sets.

Concerning coverings with congruent convex sets we have a result of Rogers (1957) according to which

$$\vartheta(S) \leq n \log n + n \log\log n + 5n$$

for all convex n-dimensional sets.

As to upper bounds for $\delta(S)$ and lower bounds for $\vartheta(S)$, it is clear that we cannot expect general results valid for all convex sets. For spacefillers the packing and the covering densities are equal to 1. The seemingly trivial conjecture that the packing and covering density of a smooth convex body cannot be 1 was first proved by Schmidt (1961). However his method does not give explicit bounds for any special set. Thus the main interest lies in giving estimates for the packing and covering densities of certain special sets such as the n-dimensional ball B_n.

The best lower bound for $\vartheta(B_n)$ is

$$\vartheta(B_n) \geq \tau_n \sim n/e\sqrt{e} \quad (n \to \infty)$$

due to Coxeter, Few and Rogers (1959). The geometric meaning of τ_n is the following: We consider a regular simplex of sidelength $\frac{1}{2}\sqrt{2(n+1)/n}$. This simplex is just covered by the $n+1$ unit balls centred at the vertices of the simplex. The quantity τ_n is defined as the ratio of the total volume of the sectors of the balls, lying in the simplex, to the volume of the simplex. It is surprising how close this is to the bound of Rogers in the opposite direction.

The situation is not so satisfactory in the case of packing. The story about upper bounds for $\delta(B_n)$ is as follows.

Using a very elegant method Bichfeldt (1929) proved that

$$\delta(B_n) \leq \frac{n+2}{2} 2^{-n/2}.$$

His result was subsequently improved by Rankin (1947) and Rogers (1958). Rogers proved that $\delta(B_n) \leq \sigma_n$, where σ_n is a quantity defined in a similar way as the bound τ_n. It is the ratio of the total volume of the sectors of $n+1$ unit balls centred at the vertices of a regular simplex of sidelengths 2, lying in the simplex, to the volume of the whole simplex. The inequality $\delta(B_n) \leq \sigma_n$ was proved independently of Rogers also by Baranovskiĭ (1964). Since the asymptotic behaviour of σ_n is $\sigma_n \sim (n/e)2^{-n/2}$, the improvement on Bichfeldt's bound is only a constant factor e/2 when n is large.

A significant improvement of Blichfeldt's bound was obtained when some mathematicians working on information theory adopted some algebraic and analytic methods from coding theory. The first who achieved an improvement of

Blichfeldt's bound in the exponential order was Sidel'nikov (1973), (1974) who proved that

$$\delta(B_n) \leq 2^{-(0\cdot 509 + o(1))n} \quad (n \to \infty).$$

Later Levenšteĭn (1975) proved that

$$\delta(B_n) \leq 2^{-(0\cdot 523 + o(1))n} \quad (n \to \infty)$$

and Kabatjanskiĭ and Levenšteĭn (1978) proved that

$$\delta(B_n) \leq 2^{-(0\cdot 599 + o(1))n} \quad (n \to \infty).$$

This is the best known upper bound for $\delta(B_n)$ up to now.

L. Fejes Tóth (1975a) raised the problem of finding the minimum of the volume of the convex hull of k non-overlapping unit balls in n-dimensional Euclidean space and formulated the following conjecture: for $n \geq 5$ the volume of the convex hull of k non-overlapping unit balls attains its minimum if the centres of the balls are equally spaced on a line with distance 2, so that the convex hull of the balls becomes a "sausage". Partial results about this conjecture are contained in the papers of Betke and Gritzmann (1983), Betke, Gritzmann and Wills (1983), Kleinschmidt, Pachner and Wills (1983) and Wills (1983). It was conjectured by Wills (1982) that under certain restrictions the surface area of the convex hull of the balls behaves similarly as the volume. The dual counterpart of the above "sausage conjecture" is also due to Wills (1983): In n-dimensional Euclidean space with $n \geq 5$ the volume of a convex body which can be covered by k unit balls attains its maximum if the centres of the balls are equally spaced on a line with distance approximately equal to $2/\sqrt{d}$.

Special results about packings and coverings in low dimensional spaces are contained in the papers of Csóka (1977), Dumir and Khassa (1973a), Golser (1977), Hollai (1974), (1975a), (1975b), Horváth (1972a), (1974), (1975) and Molnár (1978).

1.2 *Packing of balls in spherical space*

Let $\delta(n,\varphi)$ be the maximum density of balls (spherical caps) of angular diameter φ forming a packing in the n-dimensional spherical space S^n, i.e. on the surface of the $(n + 1)$-dimensional unit ball. Since, in the small, spherical geometry is the same as Euclidean geometry, we have

$$\lim_{\varphi \to 0} \delta(n,\varphi) = \delta(B_n).$$

Thus any bound for $\delta(n,\varphi)$ implies an upper bound for $\delta(B_n)$. The results of Sidel'nikov, Levenšteĭn and Kabatjanskiĭ and Levenšteĭn mentioned in the previous section can be obtained in this way as consequences of results concerning packings of balls in spherical space.

Let $M(n,\varphi)$ be the maximum number of points of S^n such that the distance between any two of them is at least φ. Denoting by $\Omega_n(\varphi)$ the volume of a ball of diameter φ in S^n, we have

$$\delta(n,\varphi) = M(n,\varphi)\Omega_n(\varphi)/\Omega_n(\pi).$$

Thus the problems of determining $\delta(n,\varphi)$ or $M(n,\varphi)$ are equivalent, however most of the results are easier to formulate in terms of $M(n,\varphi)$.

The problem of finding $M(2,\varphi)$ and the corresponding extremal configuration of points was raised by the biologist Tammes (1930). A detailed account about this case is given in the book of L. Fejes Tóth (1972a). For some recent results about the case $n = 2$ we refer to the papers of Karabinta and Székely (1973) and Székely (1974).

The first result about $M(n,\varphi)$ for general n is due to Davenport and Hajós (1951). They proved that

$$M(n,\varphi) = n + 2 \quad \text{for} \quad \pi/2 < \varphi < \pi/2 + \arcsin\frac{1}{n+1}$$

and

$$M(n,\pi/2) = 2(n + 1).$$

This implies that if $n + 3$ balls can be packed into S^n then so can $2(n + 1)$. The theorem of Davenport and Hajós was rediscovered by Rankin (1955), who also determined the exact values of $M(n,\varphi)$ for all $\varphi \geq \pi/2$ and proved that

$$M(n,\varphi) \leq \frac{\sqrt{\pi\cos\varphi}\, n^{3/2}}{2^{(n+1)/2}\sin^n\varphi/2}(1 + o(1)) \quad (n \to \infty).$$

(See also Bloh (1956).) In view of

$$\frac{\Omega_n(\varphi)}{\Omega_n(\pi)} \sim \frac{\sin^{n-1}\varphi/2}{\sqrt{2\pi n}\cos\pi/2} \quad (n \to \infty)$$

this implies

$$\delta(B_n) = \lim_{\varphi\to 0}\delta(n,\varphi) \leq (1 + o(1))\frac{n}{2^{n/2}} \quad (n \to \infty),$$

a bound for $\delta(B_n)$ which is asymptotically equivalent with the bound of Blichfeldt. A lower bound for $M(n,\varphi)$ is

$$\left(\frac{1}{\sin\varphi} + o(1)\right)^n \leq M(n,\varphi) \quad (n \to \infty)$$

proved by Shannon (1959) and Wyner (1965).

A smaller upper bound for $M(n,\varphi)$ was proposed by Coxeter (1954), (1963) and L. Fejes Tóth (1953) and proved recently by Böröczky (1978). It is more convenient to phrase this result as an upper bound for $\delta(n,\varphi)$. Böröczky's upper bound for $\delta(n,\varphi)$ is the spherical analogue of Rogers "simplex bound" for $\delta(B_n)$: the volume of the part of $n + 1$ balls of diameter φ centred at the vertices of a regular simplex of sidelength φ, lying in the simplex, divided by the volume of the simplex. This bound can be attained if S^n admits a regular tiling by simplexes of sidelength φ. These tilings of S^n are $\{3,3,\ldots,3\}$ with $\varphi = \pi - \arccos 1/(n + 1)$, $\{4,3,\ldots,3\}$ with $\varphi = \pi/2$ and $\{5,3,3\}$ for $n = 3$ with $\varphi = \pi/5$ (see Coxeter (1948)). For $\varphi = \pi - \arccos 1/(n + 1)$ and $\varphi = \pi/2$ Böröczky's simplex bound implies the bounds for $M(n,\varphi)$ anticipated by Davenport and Hajós, in the last case it yields $M(3,\pi/5) = 120$. For large values of n Coxeter (1963) examined the asymptotic behaviour of Böröczky's bound and found that it improves by a factor 2/e the bound of Rankin.

Another upper bound for $M(n,\varphi)$ can be obtained using a surprising connection between $M(n,\varphi)$ and the coefficients of the expansion of a real polynomial in terms of certain Jacobi polynomials. The Jacobi polynomials $P_i^{(\alpha,\beta)}(X)$, $i = 0, 1, \ldots$, $\alpha > -1$, $\beta > 1$, form a complete system of orthogonal polynomials on $[-1.1]$ with respect to the weight function $(1-x)^\alpha(1+x)^\beta$. Information about these polynomials can be found in the books of Abramowitz and Stegun (1972), Askey (1975) and Szegö (1975). The following theorem, often referred to as the "linear programming bound", is due to Delsarte, Goethals and Seidel (1977).

Let

$$f(x) = \sum_{i=0}^{k} f_i P_i^{((n-3)/2,(n-3)/2)}(x)$$

be a real polynomial for which

$$f(x) \le 0 \quad \text{for} \quad -1 \le x \le \cos\varphi \quad \text{and} \quad f_0 > 0, f_1 \ge 0, \ldots, f_k \ge 0.$$

Then

$$M(n,\varphi) \le f(1)/f_o.$$

Alternative proofs of this theorem were given by Kabatjanskiĭ and Levenšteĭn (1978), Lloyd (1980) and Odlyzko and Sloane (1979). A survey of the linear programming method can be found in the paper of Sloane (1982).

With the aid of polynomials which have been used previously by McElice, Rodemich, Rumsey and Welch (1977) for the study of sphere packings in Hamming space, Kabatjanskiĭ and Levenšteĭn were able to prove that for $\varphi \le \varphi^* < 63°$

$$M(n,\varphi) \le (2^{0\cdot 599 + 0(1)} \sin\varphi/2)^n \quad \text{when} \quad n \to \infty.$$

This implies that

$$\delta(B_n) = \lim_{\varphi \to 0} M(n,\varphi) \frac{\Omega_n(\varphi)}{\Omega_n(\pi)} \le 2^{-(0\cdot 599 + o(1))n} \quad \text{when} \quad n \to \infty.$$

The bound in the above theorem has been proved to be exact for some special values of n and φ. In particular, we have

$M(2,\arccos 1/\sqrt{5}) = 12$, $M(4,\arccos 1/5) = 16$,

$M(5,\arccos 1/4) = 27$, $M6,\arccos 1/3) = 56$,

$M(7,\pi/3) = 240$, $M(20,\arccos 1/9) = 112$,

$M(20,\arccos 1/7) = 162$, $M(21,\arccos 1/11) = 100$,

$M(21,\arccos 1/6) = 275$, $M(21,\arccos 1/4) = 891$,

$M(22,\arccos 1/5) = 552$, $M(22,\arccos 1/3) = 4600$,

$M(23,\pi/3) = 196560$.

The optimal arrangements of points for the above values of n and φ are described in the paper of Delsarte, Goethals and Seidel (1977). The polynomials yielding the desired upper bounds were found by Levenšteĭn (1979).

Special attention is due to the case $\varphi = \pi/3$. For, $M(n-1, \pi/3)$ is nothing else but the maximal number $N(B_n)$ of disjoint unit balls that can touch the unit ball B_n in n-dimensional Euclidean space. The problem of finding $N(B_n)$ has a long and interesting history about which a fascinating account is given in Coxeter's paper (1963). We shall discuss this and some related problems in section 2.4. Here we only mention that the above results imply that $N(B_8) = 240$ and $N(B_{24}) = 196560$. These numbers of neighbours are realised by the densest lattice packing of balls in 8 dimensions and by the Leech lattice of balls in 24 dimensions.

The bounds $N(B_8) \leq 240$ and $N(B_{24}) \leq 196560$ were proved also by Odlyzko and Sloane (1979). They used the polynomials

$$f(x) = \frac{320}{3}(x+1)(x+1/2)^2 x^2 (x-1/2)$$

and

$$g(x) = \frac{1490944}{15}(x+1)(x+1/2)^2(x+1/4)^2 x^2 (x-1/4)^2(x-1/2)$$

to deduce these bounds. These polynomials have the expansions

$$f(x) = \sum_{i=0}^{6} f_i p_i^{(2\cdot5, 2\cdot5)}(x),$$

$$f_0 = 1, \quad f_1 = \frac{16}{7}, \quad f_2 = \frac{200}{63}, \quad f_3 = \frac{832}{231},$$

$$f_4 = \frac{1216}{429}, \quad f_5 = \frac{5120}{3003}, \quad f_6 = \frac{2560}{4641}$$

and

$$g(x) = \sum_{i=0}^{10} g_i p_i^{(10\cdot5, 10\cdot5)}(x),$$

$$g_0 = 1, \quad g_1 = \frac{48}{23}, \quad g_2 = \frac{1144}{425},$$

$$g_3 = \frac{12992}{3825}, \quad g_4 = \frac{73888}{22185},$$

$$g_5 = \frac{2169856}{687735}, \quad g_6 = \frac{59062016}{25365285},$$

$$g_7 = \frac{4472832}{2753575}, \quad g_8 = \frac{23855104}{28956015},$$

$$g_9 = \frac{7340032}{20376455}, \quad g_{10} = \frac{7340032}{80848515}.$$

The polynomials f(x) and g(x), as well as some other polynomials establishing good upper bounds for $N(B_n)$ when $n \leq 24$, were found by solving a linear programming problem set up for the coefficients of polynomials which meet some weaker conditions than the conditions of the theorem of Delsarte, Goethals and Seidel. Bannai and Sloane (1981) have shown that the configurations attaining the bounds $N(B_8) = 240$ and $N(B_{24}) = 196560$ are essentially unique.

We do not know what is the exact relation of the bound of Böröczky and the bound of Delsarte, Goethals and Seidel. The asymptotic behaviour of these bounds and the numerical results for $\varphi = \pi/3$, $n \leq 23$ support the conjecture that the bound $M(n,\varphi) \leq f(1)/f_0$ is better than Böröczky's simplex bound for all $n > 2$ and $\varphi < \pi/2$.

About problems concerning coverings in spherical spaces of high dimensions much less is known. The only result in this direction we have to mention is that of Rogers (1963) establishing the existence of reasonably thin coverings by equal balls.

An interesting problem is to find a set of k points in S^n such that the biggest equal balls about these points which do not overlap form a densest packing and the smallest equal balls about the points which cover the space form a thinnest covering. Such a set of points, as well as the corresponding the number k is called *favourable*. For $n = 2$ G. Fejes Tóth and L. Fejes Tóth (1980) proved that there are only finitely many favourable numbers. It is conjectured that there are only finitely many favourable numbers in any dimensions and for $n = 2$ the only favourable numbers are 2, 3, 4, 5, 6 and 12.

1.3 *Problems in hyperbolic geometry*

After dealing with problems in Euclidean and spherical space it is natural to ask: What can we say about packing and covering in hyperbolic space? Here a new difficulty presents itself: How can we define the density of a system of bodies scattered in the whole hyperbolic space? We can try to define the density of a system of sets in hyperbolic space as we did in Euclidean space, i.e. by the limiting value of the density with respect to a sphere C(r) of radius r with a fixed centre 0. But since for a fixed value of h the volume of the spherical shell $C(r + h) - C(r)$ is of the same order of magnitude as the volume of C(r), the argument used in Euclidean space to prove that the limiting value is independent of the choice of 0 does not work in hyperbolic space. Indeed, examples constructed by Böröczky (1974) show that in hyperbolic space the density cannot be defined in this way.

However, Böröczky (1978) succeeded in extending Rogers' simplex bound to the hyperbolic space: If in the n-dimensional hyperbolic space we have a packing of balls of equal radii r, then the density of each ball with respect to its Dirichlet cell is at most $\sigma_n(r)$, where $\sigma_n(r)$ is defined analogously as Rogers' σ_n, namely as the volume of the part of $n + 1$ balls of radius r centred at the vertices of a regular simplex of sidelength 2r, lying in the simplex, divided by the volume of the simplex.

One is inclined to interpret this result as an upper bound for the density of a packing of equal balls in hyperbolic space. Unfortunately, such an interpretation is not justified. Böröczky constructed a packing $\mathscr{P}$ of equal balls and associated with each ball $B \in \mathscr{P}$ two polytopes P_B^1 and P_B^2 with the following properties: Each ball $B \in \mathscr{P}$ is contained in both of the polytopes P_B^1 and P_B^1. For fixed i, $i = 1,2$, the polytopes P_B^i, $B \in \mathscr{P}$ are congruent to each other, and they form a tiling of the space. However the polytopes P_B^1 and P_B^2 have different volumes.

These and a set of further ingenious examples of Böröczky show that there is no reasonable way to define the density of a system of sets with respect to the whole hyperbolic space. In order to avoid this difficulty and to extend some results about densest packing and thinnest covering in spherical and Euclidean spaces to hyperbolic space L. Fejes Tóth (1968b) introduced the notion of solidity. A packing $\mathscr{P}$ is said to be *solid* if no finite subsets of $\mathscr{P}$ can be rearranged so as to form, together with the rest of the sets, a packing not congruent to $\mathscr{P}$. Analogously, a covering $\mathscr{C}$ is said to be *solid* if no finite subset of $\mathscr{C}$ can be rearranged so as to form, together with the rest of the sets, a covering not congruent with $\mathscr{C}$. These definitions apply in any space of constant curvature. In Euclidean and spherical space the solidity of a packing of congruent sets implies that it has maximal density among all packings of congruent copies of these sets. Analogously, the solidity of a covering of congruent sets implies that it has minimal density among all coverings of congruent replicas of these sets.

L. Fejes Tóth (1968b) proved that for any integer $p \geq 2$ the face-incircles of the regular tiling $\{p,3\}$ form a solid packing and for $p \geq 3$ the face-circumcircles of $\{p,3\}$ form a solid covering. He conjectured that also the face-incircles of any trihedral Archimedean tiling form a solid packing and the face-circumcircles of any trihedral Archimedean tiling form a solid covering. With the aid of a general theorem L. Fejes Tóth (1968b) and G. Fejes Tóth (1974) were able to confirm this conjecture for many spherical and hyperbolic tilings. Some of these have three different types of tiles.

A result of Danzer (1975) (see also Böröczky (1983)) concerning the densest packing of eleven spherical caps implies that the packing consisting of the face-incircles of the tiling $\{5,3\}$ remains solid if we remove one circle. L. Fejes Tóth conjectured that this remains true for any $\{p,3\}$ with $p \geq 6$ instead of $\{5,3\}$. Recently A. Bezdek (1979) proved this conjecture for $p \geq 8$. The question, whether it is true for $p = 7$ and for the especially interesting case when $p = 6$ is still undecided.

The result of Böröczky (1978) referred to above implies that the packing consisting of the inspheres of facets of the regular hyperbolic tiling $\{5,3,3,3\}$ is solid.

The notion of the solidity of a packing and a covering has been generalized in various directions by G. Fejes Tóth (1974) and L. Fejes Tóth (1980d). Vermes (1979) examined packings of hypercircle-domains in the plane. Further results about packings in hyperbolic space are contained in the papers of K. Bezdek (1982a), (1983a), (1983b), Csóka (1975a), (1975b) and Molnár (1975). We conclude this section by quoting the following nice theorem of K. Bezdek (1982a): If in the hyperbolic plane a circle C is packed with at least two equal circles then the density of the circles with respect to C is at most $\pi/\sqrt{12}$. The odd thing about this theorem is the fact that the face-incircles of any tiling $\{p,3\}$ with $p > 6$ form a packing such that the density of each circle in its cell is greater than $\pi/\sqrt{12}$. This fact provides further evidence for the impossibility of the definition of density.

1.4 *Lattice packing and lattice covering*

Let $\mathbf{a}_1,\ldots,\mathbf{a}_n$ be n linearly independent vectors in n-dimensional Euclidean space. The set $\Lambda = \Lambda(\mathbf{a}_1,\ldots,\mathbf{a}_n)$ of all vectors of the form

$$\mathbf{a} = u_1\mathbf{a}_1 + \cdots + u_n\mathbf{a}_n,$$

where $u_1\ldots,u_n$ are arbitrary integers, is called a *lattice*. A system $\mathscr{S}$ consisting of the translates $\mathscr{S} + \mathbf{a}$, $\mathbf{a} \in \Lambda$ of a given set $\mathscr{S}$ is called a lattice arrangement of $\mathscr{S}$. If each point of the space belongs to at most one of the sets of $\mathscr{S}$ then $\mathscr{S}$ is called a *lattice packing* and if each point belongs to at least one of the sets of $\mathscr{S}$ then $\mathscr{S}$ is called a *lattice covering*.

Special interest was taken in the investigation of lattice packings and lattice coverings because of their connection with number theory. This connection was first observed by Gauß (1831), but it remained for Minkowski to undertake a systematic study of geometric methods in number theory. Minkowski's work laid the foundations of a flourishing branch of mathematics called by him the geometry of numbers. We are interested here in the geometric aspects of lattice packings and lattice coverings. For results in the geometry of numbers we refer to the books of Cassels (1959) and Lekkerkerker (1969). Recent developments in the geometry of numbers are reviewed in the survey paper of Gruber (1979).

The *lattice packing density* $\delta_L(S)$ and the *lattice covering density* $\vartheta_L(S)$ of the set S are defined by

$$\delta_L(S) = \sup d(S)$$

and

$$\vartheta_L(S) = \inf d(S),$$

where S ranges over all lattice packings and lattice coverings of S, respectively. (It is easy to see that for a lattice arrangement S we have $d_-(S) = d_+(S)$, so that the density d(S) exists.) First we recall some results about $\delta_L(S)$ and $\vartheta_L(S)$.

The theorem of Minkowski-Hlawka and its refinements imply the existence of dense lattice packings of convex bodies. In particular, the theorem of Schmidt yields

$$\delta_L(S) \geq (\log 2 + o(1))n/2^n \quad (n \to \infty)$$

for symmetrical convex n-dimensional sets and

$$\delta_L(S) \geq (\pi \log 2 + o(1))n^{3/2}/4^n \quad (n \to \infty)$$

for all convex n-dimensional sets.

Rogers (1959) proved that

$$\vartheta_L(S) \leq n^{\log_2 \log_e n + C}$$

for some constant c and all n-dimensional convex sets. He proved an even stronger result for balls, namely

$$\vartheta_L(B_n) \leq cn(\log_2 n)^{\frac{1}{2}\log_2 2\pi e}$$

for some constant c.

The problem of the densest lattice packing of n-dimensional balls is solved for $n \leq 8$. $\delta_L(B_n)$ was determined for $n = 2$ implicitely by Lagrange (1773), for $n = 3$ by Gauß (1831), for $n = 4$ and 5 by Korkine and Zolotareff (1872), (1877) and for $n = 6$, 7 and 8 by Blichfeldt (1934). We have

$$\delta_L(B_2) = \frac{\pi}{\sqrt{12}} = 0{\cdot}9060\ldots, \quad \delta_L(B_3) = \frac{\pi}{\sqrt{18}} = 0{\cdot}7404\ldots,$$

$$\delta_L(B_4) = \frac{\pi^2}{16} = 0{\cdot}6168\ldots, \quad \delta_L(B_5) = \frac{\pi^2}{15\sqrt{2}} = 0{\cdot}4652\ldots,$$

$$\delta_L(B_6) = \frac{\pi^3}{48\sqrt{3}} = 0{\cdot}3729\ldots, \quad \delta_L(B_7) = \frac{\pi^3}{105} = 0{\cdot}2952\ldots,$$

$$\delta_L(B_8) = \frac{\pi^4}{384} = 0{\cdot}2536\ldots.$$

Up to a similarity the densest lattice packing of n-dimensional balls with $n \leq 8$ is uniquely determined. For $n = 6$, 7 and 8 the uniqueness was recently proved by Vetcinkin (1980).

Less is known about the thinnest lattice covering with balls. We have

$$\vartheta_L(B_2) = \frac{2\pi}{\sqrt{27}} = 1{\cdot}2091\ldots, \quad \vartheta_L(B_3) = \frac{5\sqrt{5\pi}}{24} = 1{\cdot}4635\ldots,$$

$$\vartheta_L(B_4) = \frac{2\sqrt{5}\pi^2}{25} = 1{\cdot}7655\ldots, \quad \vartheta_L(B_5) = \frac{5^{3/2}7^{1/2}\pi^2}{2^4 3^{11/2}} = 2{\cdot}1242\ldots.$$

These values were found by Kershner (1939), Bambah (1954), Delone and Ryškov (1963) and Ryškov and Baranovskiĭ (1975), (1976), respectively.

The papers of Barnes and Sloane (1983), Best (1980), Ignat'ev (1964), (1966), Leech (1964), (1967a), (1967b), (1969), Leech and Sloane (1970a), (1970b), (1971), Sloane (1972), (1977), (1981) and Vetcinkin (1974) are concerned with the construction of dense lattice and non-lattice packings of balls in n-dimensional space with moderately large values of n. Most of these constructions make use of error correcting codes. The main idea of these constructions can be explained as follows:

A binary code is a set of n-dimensional binary vectors. The Hamming distance of two vectors (or codewords) of the code is the number of digits where they differ. The greater the distance of two codewords is, the easier it is to distinguish one from the other. Thus reliable storage and transmission of information requires the construction of codes having a large number of codewords with possibly great minimial distance between them. The theory of error-correcting codes has elaborated a great variety of methods to construct such codes.

From a geometric point of view a binary code is a subset of the vertices of the n-dimensional unit cube. Denoting with d the minimum Hamming distance between the codewords, the balls centred at the codewords with radii $\frac{1}{2}\sqrt{d}$ form a packing. The constructions mentioned above give different methods in order to extend such a packing of balls to a packing in the whole space.

An especially interesting packing constructed by Leech from error correcting codes is a very dense lattice packing of balls in 24-dimensional space. It is conjectured to be the densest lattice packing of balls in 24 dimensions. In this packing each ball is touched by exactly 196560 other balls.

The papers of Leech and Sloane (1971) and Sloane (1977), (1981) give excellent surveys on sphere packings constructed from error-correcting codes, as well as references about coding theory.

It is still an open problem whether the density of an arbitrary packing of equal balls can exceed the density of the densest lattice packing of balls. Rogers (1964) conjectured that $\delta_L(B_n) \leq \delta(B_n)$ for all sufficiently large n. For n = 10, 11 and 13 non-lattice packings of equal balls have been constructed whose density exceeds the conjectured values of $\delta_L(B_n)$.

Bambah and Sloane (1982) and Conway, Parkner and Sloan (1982) examined the Leech lattice and some lattices derived from it and showed that by blowing up the balls they provide very economical lattice coverings.

Obviously we have

$$\delta_L(S) \leq \delta(S) \quad \text{and} \quad \vartheta_L(S) \geq \vartheta(S).$$

Combining these relations with the results of section 1.1, we have

$$\delta_L(B_n) \leq 2^{-(0{\cdot}599 + o(1))n} \quad (n \to \infty)$$

and

$$\vartheta_L(B_n) \geq \frac{(1 + o(1))n}{e\sqrt{e}} \quad (n \to \infty)$$

Some interesting results were initiated by the remark of Heppes (1960) that in three-dimensional space any lattice packing of balls can be transilluminated in three linearly independent directions. Hortobágyi (1971) generalized the result of Heppes by proving that to any lattice packing of open unit balls we can find three infinite circular cylinders of radius $3\sqrt{2}/4 - 1 = 0{\cdot}06065\ldots$ with linearly independent axes which do not intersect any of the balls. The example of the densest lattice packing of balls shows that the constant $3\sqrt{2/4} - 1$ cannot be replaced by a smaller one. Another theorem of Hortobágyi (1977), (1983a) states that any lattice packing of balls can be transilluminated in six non-parallel directions. It is, again, the example of the densest lattice packing of balls which shows that the number six cannot be replaced by a greater one. Hortobágyi (1983b) also showed that the theorems above remain true for lattice packings of bodies of constant width 2.

The above results of Heppes and Hortobágyi have been extended for spaces of higher dimensions by Horváth (1970), (1972) and Ryškov and Horváth (1975). In particular, Horváth showed that in n-dimensional space to any lattice packing of balls there is an (n − 2)-dimensional subspace which does not intersect any of the balls.

Inspired by this result G. Fejes Tóth (1976b) raised the following problem: Find the thinnest n-dimensional lattice of closed balls such that any k-dimensional subspace ($0 \leq k \leq n - 1$) has a point in common with at least one ball. For $k = 0$ we have the problem of the thinnest lattice covering of balls. Special attention is due also to the case when $k = n - 1$. A system of n-dimensional bodies is said to be *separable* if there is an $(n - 1)$-dimensional plane not intersecting any of the bodies but containing on both sides a body. Using this terminology the case $k = n - 1$ of the above problem can be formulated as follows: Find the thinnest non-separable lattice of n-dimensional balls.

According to a surprising theorem of Makai (1978) the problem of finding the thinnest non-separable lattice of a convex body S is equivalent with the problem of finding the densest lattice packing of the polar of the difference body of S. In particular the problem of finding the thinnest non-separable lattice of balls is equivalent with the problem of finding the densest lattice packing of balls. Thus the solution of the above problem of G. Fejes Tóth is known for $k = 0, n \leq 5$ as well as for $k = n - 1 \leq 7$, but it is not known in any other cases. Special attention is due to the case when $n = 3$ and $k = 1$. It is conjectured that the density of a lattice packing of three-dimensional balls such that each line intersects at least one of the balls is at least $\pi\sqrt{18}/4$ and equality is attained only if the centres of the balls form a face-centred cubic lattice.

Makai conjectured that the density of a non-separable lattice of any n-dimensional convex set S is at least $(n + 1)/2^n n!$ and at least $1/n!$ if S is centrally symmetric, and equality holds only for certain lattices of simplexes and cross polytopes, respectively. The case when $n = 2$ was settled by L. Fejes Tóth and Makai (1974) and Makai (1978).

A system of convex bodies is said to be *connected* if the point-set union of the bodies is connected. Groemer (1966), (1968b) examined connected lattices of convex bodies and proved among others the following nice theorem: In n-dimensional space the density of a connected lattice of convex bodies is at least $1/n!$. This theorem was rediscovered by L. Fejes Tóth (1973a) who also described the case of equality.

Let V be the volume and F the surface area of an n-dimensional closed convex set S with interior points. Generalizing theorems of Bender (1962) $(n = 2)$ and Wills (1968) $(n = 3)$ Hadwiger (1970) showed that if $V/F > 1/2$ then S contains a point with integer coordinates. This means that the lattice of points with integer coordinates generates a lattice covering with translates of S. According to a more general result of Bokowski, Hadwiger and Wills (1972) S contains at least $V - F/2$ points with integer coordinates. These results were starting point of a new brand of the geometry of numbers, a survey about which is given by Betke and Wills (1979).

Some further results on lattice packings and lattice coverings are contained in the papers of Bleicher (1975), Butler (1972), Conway and Sloane (1982), Delone, Dolbilin, Riškov and Štogrin (1970), Dolbilin (1974), Holyman (1970) and Sloane (1978), (1979).

1.5 *Multiple packing and covering*

A system of sets is said to form a k-*fold packing* if each point of the space belongs to at most k sets of the system. Analogously, a system of sets is said to form a k-*fold covering* if each point of the space belongs to at least k sets of the system. It seems that the notion of multiple packing and multiple covering was introduced independently by Davenport and L. Fejes Tóth (see the notes in the first papers on the subject by Few (1953) and Heppes (1955)).

The theory of multiple packing and covering has an extensive literature. Many of the results about simple packings and coverings have been extended to multiple packings and coverings. In this section we give an account on the results about the quantities $\delta^{(k)}(S)$, $\vartheta^{(k)}(S)$, $\delta_L^{(k)}(S)$ and $\vartheta_L^{(k)}(S)$, which are defined analogously to the simple packing and covering densities $\delta(S) = \delta^{(1)}(S)$, $\vartheta(S) = \vartheta^{(1)}(S)$, $\delta_L(S) = \delta_L^{(1)}(S)$ and $\vartheta_L(S) = \vartheta_L^{(1)}(S)$.

One can obtain lower bounds for $\delta^{(k)}(S)$ and upper bounds for $\vartheta^{(k)}(S)$ using the known bounds for $\delta^{(1)}(S)$ and $\vartheta^{(1)}(S)$ and the trivial relations

$$\delta^{(k)}(S) \geq k\delta^{(1)}(S) \quad \text{and} \quad \vartheta^{(k)}(S) \leq k\vartheta^{(1)}(S).$$

Similar inequalities hold for lattices:

$$\delta_L^{(k)}(S) \geq k\delta_L^{(1)}(S) \quad \text{and} \quad \vartheta_L^{(k)}(S) \leq k\vartheta_L^{(k)}(S).$$

When S is an n-dimensional ball we can obtain a stronger result for k-fold packings using the following observation of Few (1964): If we replace the balls of a packing of unit balls by concentric balls of radius $2k/(k+1)$, we obtain a k-fold packing. This yields

$$\delta^{(k)}(B_n) \geq [2k/(k+1)]^{n/2}\delta^{(1)}(B_n)$$

and

$$\delta_L^{(k)}(B_n) \geq [2k/(k+1)]^{n/2}\delta_L^{(1)}(B_n).$$

These results are very poor if k is large. For large values of k a better lower bound for $\delta^{(k)}(S)$ can be obtained by a result of Erdös and Rogers (1962). They proved that for large n the n-dimensional space can be covered by translates of any convex set S with density $n \log n + n \log\log n + 4n$ so that no point is covered more than $e(n \log n + n \log\log n + 4n)$ times. This implies that there is a constant c such that for any convex n-dimensional set S

$$\delta^{(k)}(S) \geq ck$$

when $k \geq e(n \log n + n \log\log n + 4n)$.

At a colloquium on discrete geometry held in Salzburg, 1975 Few announced a result for multiple coverings analogous to the one above but he has not published it so far.

Cohn (1976) modified the method which Rogers used to prove the inequality $\vartheta_L(S) \leq n^{\log_2 \log_e n + c}$, referred to in section 1.1, in order to prove the existence of thin k-fold lattice coverings. In particular, he proved that

$$\vartheta_L^{(k)}(S) \leq [(k+1)^{1/n} + 8n]^n$$

for all convex n-dimensional sets. This implies that for fixed n

$$\vartheta_L^{(k)}(S) \sim k \quad \text{as} \quad k \to \infty.$$

We continue to discuss upper bounds for $\delta^{(k)}(B_n)$ and lower bounds for $\vartheta^{(k)}(B_n)$. Few (1964) proved that

$$\delta^{(k)}(B_n) \le (1 + n^{-1})[(n+1)^k - 1][k/(k+1)]^{n/2}$$

Unfortunately this bound is worse than the trivial bound

$$\delta^{(k)}(B_k) \le k$$

when $k > \sqrt{n}$. For some values of k it is also worse than the trivial in low dimensions, for example, it fails to prove that $\delta^{(2)}(B_n) < 2$ when $n \le 30$. For $k = 2$ Few (1968) established the stronger bound

$$\delta^{(2)}(B_n) \le \tfrac{4}{3}(n+2)(\tfrac{2}{3})^{n/2}$$

but even this is worse than the trivial bound when $n \le 10$.

Modifying a method of Schmidt (used in the case when $k = 1$) Florian (1978a) proved that $\delta^{(k)}(S) < k$ and $\vartheta^{(k)}(S) > k$ for any smooth convex set, but this method does not provide explicit bounds even in the case when S is a ball. G. Fejes Tóth (1976a), (1979) established a non-trivial upper bound for $\delta^{(k)}(B_n)$ and a non-trivial lower bound for $\vartheta^{(k)}(B_n)$ for all values of k and n. These bounds are explicit but for $n \ge 3$ too complicated to be quoted here. For fixed k the upper bound for $\delta^{(k)}(B_n)$ tends exponentially to zero when $n \to \infty$, but at a slower rate than the bound of Few and the lower bound for $\vartheta^{(k)}(B_n)$ tends to a constant depending on k when $n \to \infty$. On the other hand, for fixed n the difference of k and these bounds tends to zero as $k \to \infty$ which shows that they are rather weak when k is large compared with n.

For $k > 1$ the problem of the densest k-fold lattice packing of balls is solved for $n = 2$, $k \le 7$ and $n = 3$, $k = 2$ and the problem of the thinnest k-fold lattice covering with balls is solved for $n = 2$, $k \le 5$ and $n = 3$, $k = 2$. We have

$$\left.\begin{aligned} \delta_L^{(2)}(B_2) &= 2\delta_L^{(1)}(B_2) = \frac{\pi}{\sqrt{3}} = 1{\cdot}813\ldots \\ \delta_L^{(3)}(B_2) &= 3\delta_L^{(1)}(B_2) = \frac{\sqrt{3}\pi}{2} = 2{\cdot}720\ldots \\ \delta_L^{(4)}(B_2) &= 4\delta_L^{(1)}(B_2) = \frac{2\pi}{\sqrt{3}} = 3{\cdot}627\ldots \end{aligned}\right\} \quad \text{Heppes (1959)}$$

$$\left.\begin{aligned} \delta_L^{(5)}(B_2) &= \frac{4\pi}{\sqrt{7}} = 4{\cdot}749\ldots \\ \delta_L^{(6)}(B_2) &= \frac{35\pi}{8\sqrt{6}} = 5{\cdot}611\ldots \end{aligned}\right\} \quad \text{Blundon (1963a)}$$

$$\delta_L^{(7)}(B_2) = \frac{8\pi}{\sqrt{15}} = 6{\cdot}489\ldots \qquad \text{Bolle (1976)}$$

$$\delta_L^{(2)}(B_3) = \frac{8\pi}{9\sqrt{3}} = 1{\cdot}612\ldots \qquad \text{Few and Kanagasabapathy (1969)}$$

$$\left.\begin{aligned} \vartheta_L^{(2)}(B_2) &= 2\vartheta_L^{(1)}(B_2) = \frac{4\pi}{\sqrt{27}} = 2{\cdot}418\ldots \\ \vartheta_L^{(3)}(B_2) &= \frac{\pi\sqrt{27138 + 2910\sqrt{97}}}{216} = 3{\cdot}435\ldots \\ \vartheta_L^{(4)}(B_2) &= \frac{25\pi}{18} = 4{\cdot}363\ldots \end{aligned}\right\} \quad \text{Blundon (1957)}$$

$$\vartheta_L^{(5)}(B_2) = \frac{32}{7\sqrt{7}} = 5.428\ldots \quad \text{Subak (1960), Temesvári (1983a)}$$

$$\vartheta_L^{(6)}(B_2) = \frac{98}{27\sqrt{3}} = 6.583\ldots \quad \text{Subak (1960)}$$

$$\vartheta_L^{(7)}(B_2) = 7.672\ldots \quad \text{Haas (1976)}$$

$$\vartheta_L^{(2)}(B_3) = \frac{8\pi}{\sqrt{3}\sqrt{76\sqrt{6} - 159}} = 2.784\ldots \quad \text{Few (1967b).}$$

Constructions of dense k-fold lattice packings and thin k-fold lattice coverings of circles and balls for some special values of k can be found in Blundon (1963b), (1964), (1978), Bolle (1976), G. Fejes Tóth and Florian (1975), Linhart (1983), Purdy (1973) and Yang (1980). Linhart (1983) gave an algorithm, which determines $\delta_L^{(k)}(B_2)$ and $\vartheta_L^{(k)}(B_2)$ with any prescribed accuracy. With the aid of this algorithm he calculated the values of $\vartheta_L^{(k)}(B_2)$ and $\vartheta_L^{(k)}(B_2)$ when $k \leq 20$ with an accuracy of at least two decimals.

Bolle (1976) proved that there are positive constants c_1, c_2, c_3 and c_4 such that

$$k - c_1 k^{2/5} \leq \delta_L^{(k)}(B_2) \leq k - c_2 k^{1/4}$$

and

$$k + c_3 k^{1/4} \leq \vartheta_L^{(k)}(B_2) \leq k + c_4 k^{2/5}.$$

In a recent paper Bolle (1983) showed that the exponent 1/4 in the upper bound for $\delta_L^{(k)}(B_2)$ and in the lower bound for $\vartheta_L^{(k)}(B_2)$ is best possible in the sense that there are positive constants c_5 and c_6 and infinite sequences N and N′ of real numbers so that

$$\delta_L^{(k)}(B_2) \geq k - c_5 k^{1/4} \quad \text{for all} \quad k \in N,$$
$$\vartheta_L^{(k)}(B_2) \leq k + c_6 k^{1/4} \quad \text{for all} \quad k \in N'.$$

Another result of Bolle (1979, 1982) says that there are positive constants c(n) and C(n) depending only on n such that for $n \not\equiv 1 \bmod (4)$

$$\delta_L^{(k)}(B_n) \leq k - c(n) k^{(n-1)/2n},$$
$$\vartheta_L^{(k)}(B_n) \geq k + C(n) k^{(n-1)/2n}$$

and for $n \equiv 1 \bmod (4)$

$$\delta_L^{(k)}(B_n) \leq k - c(n)k^{(n-3)/2n},$$

$$\vartheta_L^{(k)}(B_n) \geq k + C(n)k^{(n-3)/2n}.$$

Dumir and Hans-Gill (1972a), (1972b) showed that

$$\delta_L^{(2)}(S) = 2\delta_L^{(1)}(S)$$

and

$$\vartheta_L^{(2)}(S) = 2\vartheta_L^{(1)}(S)$$

for any centro-symmetric convex two-dimensional set. These are generalizations of the above mentioned results of Heppes and Blundon concerning lattice double packings and lattice double coverings of circles. A generalization of these results in another direction was given by G. Fejes Tóth (1977b).

The papers of L. Fejes Tóth (1966a) and Blachnan and Few (1963) are concerned with multiple packings in spherical space. Groemer (1979a), Marley (1973, 1974) and Robinson (1979) investigated multiple tilings. Further results about multiple packings and multiple coverings can be found in the papers of Blundon (1977), Horváth (1982), Horváth and Temesvári (1982), Pach (1980), Temesvári (1983b) and in the survey paper of Few (1967a).

2. Problems in the plane

2.1 *Circle-packings and circle-coverings*

From the great variety of recent results about circle-packings and circle-coverings we first mention some results concerning packings and coverings with congruent circles. It is known that the density of a packing of equal circles is at most $\pi/\sqrt{12}$ and the density of a covering with equal circles is at least $2\pi/\sqrt{27}$. These results may be interpreted as follows: If we want to pack a "great" but given number of equal circles into a possibly small square (or another domain of given shape) or cover with the circles a possibly big square, then we have to arrange the centres of the circles in the vertices of a regular triangular lattice. However, if the number of circles is small the optimal arrangement will generally be very different from this distribution. Thus it is an interesting problem, stimulated also by practical applications, to determine the densest packing of n equal circles into a domain of given shape and the thinnest covering of the domain with n equal circles for some special kinds of domains and some special values of n.

The densest packing of n equal circles into a square was determined by Schaer and Meir (1965) for $n \leq 8$ by Schaer (1965) for $n = 9$ and by Wengerodt (1983a), (1983b) for $n = 16$ and $n = 25$. For some values of n greater than 9 Goldberg (1970) and Schlüter (1979) and Schmitz and Kirchmer (1982) constructed economical packings of n equal circles into a square. Ruda (1969) investigated packings of equal circles in rectangles. The problem of densest packing of n equal circles into a circle was solved by Pirl (1967) for $n \leq 10$, who also gave constructions for some other values of n. Further examples of dense packings of equal circles in a circle are contained in the papers of Kravitz (1967) and Goldberg (1971).

It is rather easy to determine the thinnest covering of a circle with $n = 3, 4$ and 7 equal circles (see e.g. problem 109, pp. 105–107 in the book of Škljarskiĭ, Čencov and Jaglom (1974)). Neville (1915) constructed a covering of a circle with five equal circles and stated that it is optimal, but his proof is not exact. A rigorous proof was recently given by K. Bezdek (1979), (1983c), who also determined the thinnest covering of a circle with six equal circles.

An interesting problem is to find the densest packing of equal circles into a parallel strip. Molnár conjectured that the density d of a packing of unit circles in a parallel strip of width w satisfies the inequality

$$d \leq \frac{(n+1)(n+2)\pi}{2w\{(n+1) + 2\sqrt{1-(w-\sqrt{3})^2}\}}, \quad n = \left[\frac{w-2}{\sqrt{3}}\right].$$

The following construction shows that this bound cannot be improved for any value of w. Let T be a set of $(n+1)(n+2)/2$ circles containing $1, 2, \ldots, n+1$ circles of $n+1$ consecutive rows of the densest lattice packing of unit circles, such that the convex hull of the centres of the circles of T is a regular triangle of sidelength 2n. We place triangular blocks of circles congruent to T touching alternately the two bordering lines of the strip, such that each block touches its two neighbours. Molnár observed that in the case when $w = 2 + n\sqrt{3}$, $n = 1, 2, \ldots$, the validity of his conjecture can be derived from an inequality of Groemer (1960). Recently Kertész (1982) proved Molnár's conjecture also in the case when $2 + \sqrt{3} \leq w < 2 + \sqrt{8}$.

A circle C is said to be a *supporting circle* of the circle-packing $\mathscr{P}$ if it has no common interior point with any of the circles of $\mathscr{P}$ and touches at least three circles of $\mathscr{P}$. The packing $\mathscr{P}$ is called a *ρ-system* if the greatest lower bound of the radii of all supporting circles of $\mathscr{P}$ is equal to the positive number ρ. Molnár (1977), who introduced this notion determined the densest ρ-system of unit circles for all positive numbers ρ.

We say that a packing of circles is *blockingfree* if each circle of the packing can continuously be moved arbitrarily far from its original position without overlapping during the motion any other circle of the packing. What is the maximal density of a blockingfree packing of equal circles? The author, who raised this problem, conjectured that the density of such a circle-packing is at most $\pi\sqrt{3}(\sqrt{5}-1)/12 = 0{\cdot}5604\ldots$ A blockingfree packing of unit circles with this density can be constructed as follows: Let $\mathscr{P}_0$ be the lattice packing of unit circles generated by the vectors $(2,0)$ and $(0, \sqrt{3} + \sqrt{15})$. Let $\mathscr{P}_1$ be the packing obtained by translating each circle of $\mathscr{P}_0$ by the vector $(1, \sqrt{3})$. The union of $\mathscr{P}_0$ and $\mathscr{P}_1$ consists of double rows of circles, so that each circle is touched by four others, and the double rows are separated by corridors in which a unit circle can be moved without disturbing the circles of the packing.

This conjecture was confirmed by Blind (1969). Earlier Heppes (1967) came very close to the solution by proving that the density of a blockingfree packing of equal circles is at most $0{\cdot}5651\ldots$ Heppes also generalized the problem of G. Fejes Tóth. A packing of unit circles is r-*approachable* if a circular "vehicle" of radius r can be brought into contact with each circle of the packing by a continuous motion without disturbing the circles of the packing. Heppes conjectured that the density d of an r-approachable packing of unit circles satisfies the inequality

$$d \leq \frac{\pi}{\sqrt{4(1+r)^2-1}+\sqrt{3}}$$

for any positive number r. Modifying slightly the above construction it is easy to obtain an r-approachable packing of unit circles whose density attains this upper bound.

The conjecture of Heppes was proved in a series of papers by G. Blind (1972), (1976), (1977). The original proof of G. Blind is rather complicated; a much simpler proof can be found in the joint paper of G. Blind and R. Blind (1979). Some partial results about r-approachable packings in n-dimensional Euclidean space and in two-dimensional spherical space are contained in the papers of G. Blind and R. Blind (1978) and G. Blind (1981).

Consider a system of equal circles with density d and let s denote the density of the part of the plane which is covered by the circles. Let h(d) be the supremum of s taken over all systems of equal circles with density d. It is easily seen that $\delta(B_2) = \sup\{d : d = h(d)\}$ and $\vartheta(B_2) = \inf\{d : h(d) = 1\}$. Thus the problem of determining h(d) is a generalization of the problems of densest packing and thinnest covering of equal circles. This problem was raised and solved by L. Fejes Tóth (1972a, p. 80). If f(x) denotes the area of the intersection of a circle of unit area and a concentric regular hexagon of area x, then we have

$$h(d) = df(1/d).$$

It turned out that this result has interesting applications in the theory of location (Matérn and Persson (1965), Bollobás (1972), (1973)) and its spherical analogue is an important tool in the theory of convex polyhedra (see Linhart (1977a), (1979) and Florian (1979a) for recent results in this direction). The paper of Landau and Slepian (1966) treats a generalization of the above problem in n-dimensional spherical space, as well as some applications in information theory. A rough upper bound for the quantity analogous to h(d) in high-dimensional Euclidean space was given in the paper of Erdös, Few and Rogers (1964).

Different proofs and various generalizations of the relation $h(d) = df(1/d)$ are contained in the papers of Bollobás (1973), G. Fejes Tóth (1972), (1973) and Florian (1975b). G. Fejes Tóth (1972) proved that we cannot cover a greater part of the plane with a system of circles of given density if instead of congruent circles we are allowed to use circles of not too different sizes. To make this statement precise we define the function w(x) as follows:

$$w(x) = \frac{6 \arctan q \tan \frac{\pi}{6} - 7 \arctan q \tan \frac{\pi}{7}}{5 \arctan q \tan \frac{\pi}{5} - 6 \arctan q \tan \frac{\pi}{6}},$$

where

$$q = \sqrt{3\left(\frac{\sqrt{12}}{x} - 1\right)},$$

when $\sqrt{27}/2\pi < x < \sqrt{12}/\pi$, and $w(x) = 0$ otherwise. Then we have the following theorem:

If $\mathscr{S}$ is a system of circles with density d such that the quotient of the areas of any two circles of $\mathscr{S}$ is at least $w(1/d)$, then the density of the part of the plane covered by the circles of $\mathscr{S}$ is at most $df(1/d)$.

This is a generalization of a result of G. Blind (1969) concerning packings and coverings with circles of not too different sizes.

Another result about packing of incongruent circles reads as follows (G. Fejes Tóth (1980)):

The density of a packing of circles arising by intersecting a packing of equal balls by a plane is at most $\pi/\sqrt{12}$.

Though the circle-packing under consideration may contain circles of very different sizes, its density cannot exceed the density of the densest packing of equal circles. An analogous result for covering does not hold: Intersecting a covering with equal balls by a plane we can obtain circle-coverings with density arbitrarily close to 1. But the problem of finding the thinnest covering with circles arising by intersecting a lattice-covering of balls seems to be interesting.

Florian, Hárs and Molnár (1979) proved that the density d of a ρ-system of circles whose radii are contained in the interval $[\varepsilon,1]$, $\varepsilon > 0$, satisfies the inequality

$$d \leq \frac{\arccos \dfrac{1}{1+\rho}}{\sqrt{2\rho + \rho^2}}.$$

This upper bound cannot be improved in the cases when

$$\rho = \frac{2\sqrt{3}}{3} - 1, \quad \rho = \sqrt{2} - 1 \quad \text{and} \quad \rho = 1.$$

Dorninger (1973) investigated coverings with circles with the constraint that the angles of any triangle spanned by the centres of three circles of the covering with nonempty intersection are not less than a certain constant ω, $0 < \omega \leq \pi/3$. He showed that the density d of such a covering satisfies

$$d \geq \frac{\omega(\pi - 2\omega)}{\pi} \tan\omega + \omega\cot\omega.$$

Equality is attained if $\omega = \pi/6$, $\omega = \pi/4$ or $\omega = \pi/3$ and the covering consists of the face-circumcircles of the Archimedean tiling (3,12,12), (4,8,8) and (6,6,6), respectively.

Let H be a convex polygon with at most six sides, $C_1,\ldots,C_n$ a finite set of circles and d the density of the circles with respect to H. Let $M_\varepsilon(x_1,\ldots,x_n)$ denote the power mean of exponent ε of the numbers $x_1,\ldots,x_n$. L. Fejes Tóth (1980a), (1980c) proved that

$$\frac{1}{d} \leq 1 + \frac{\sqrt{12} - \pi}{\pi} \frac{M_{1/3}(|C_1|,\ldots,|C_n|)}{M_1(|C_1|,\ldots,|C_n|)}$$

if the circles are packed, into H and

$$\frac{1}{d} \leq 1 - \frac{1}{4}\left(1 - \frac{\sqrt{27}}{4\pi}\right)\frac{M_{1/3}(|C_1|,\ldots,|C_n|)}{M_1(|C_1|,\ldots,|C_n|)}$$

if the circles cover H. The question, if in the second inequality the constant $\frac{1}{4}(1-(\sqrt{27}/4\pi))$ can be replaced by $(\sqrt{27}-2\pi)/2\pi$, is still open. In some cases the following inequalities are stronger than the ones above:

$$d \leq \frac{\pi}{\sqrt{12}} \frac{M_1(|C_1|,\ldots,|C_n|)}{M_\alpha(|C_1|,\ldots,|C_n|)}, \quad \alpha = 0{\cdot}912$$

if the circles are packed into H and

$$d \geq \frac{2\pi}{\sqrt{27}} \frac{M_1(|C_1|,\ldots,|C_n|)}{M_\infty(|C_1|,\ldots,|C_n|)}, \quad \alpha = 1{\cdot}255$$

if the circles cover H. These inequalities, due to G. Fejes Tóth (1982), are sharpenings of inequalities by L. Fejes Tóth (1972a p. 77) who proved them only for $\alpha \leq 0{\cdot}77$ and $\alpha \geq 2{\cdot}11$, respectively.

Let C be a unit circle and $\mathscr{P} = \{C_i\}$ a packing of circles into C. $\mathscr{P}$ is called *almost perfect* (alternately a solid or complete packing), if the residual set of uncovered points $R(\mathscr{P}) = C - \bigcup_{i=1}^{\infty} C_i$ has plane Lebesgue measure 0. Let r_i be the radius of C_i. The starting point of several investigations was the observation of Mergelyan (1952) that while for an almost perfect packing $\sum_{i=1}^{\infty} r_i^2 = 1$, the sum of the radii diverges: $\sum_{i=1}^{\infty} r_i = \infty$. A simple proof of Mergelyan's theorem was given by Wesler (1960). Inspired by Mergelyan's theorem Melzak (1966) considered the sum $M(\alpha,\mathscr{P}) = \sum_{i=1}^{\infty} r_i^\alpha$ and defined the *exponent* $e(\mathscr{P})$ of $\mathscr{P}$ as

$$e(\mathscr{P}) = \inf\{\alpha : M(\alpha,\mathscr{P}) < \infty\}.$$

It is a longstanding unsolved problem to determine the set of real numbers which occur as the exponent of an almost perfect packing of circles into a circle. In connection with this problem special interest is due to a certain class of packings introduced by Melzak (1966) and called by him osculatory packings. To obtain an *osculatory packing* into C we pack a finite number of circles into C and then proceed by induction inserting at each stage one of the largest circles that fit into the part of C left uncovered by the previous circle. Melzak proved that osculatory packings are almost perfect and the exponent of any osculatory packing $\mathscr{P}$ satisfies $1{\cdot}035 < e(\mathscr{P}) < 1{\cdot}999971$. He conjectured that all osculatory packings of C have the same exponent S and the set of real numbers which are the exponent of some almost perfect packing into a circle is the closed interval [S,2]. The first part of this conjecture was confirmed by Wilker (1967). As to the second part Boyd (1972) proved that the set of possible exponents contains the interval [S,2] and Wilker (1973) proved that the set of possible exponents is an interval (E,2] or [E,2]. The question whether S = E is still undecided.

The bounds $1{\cdot}035 < S < 1{\cdot}999971$ of Melzak were subsequently improved by Wilker (1967) and Boyd (1970a), (1971a), (1971b), (1973a). Boyd (1971b) gave an algorithm which, in principle, allows us to determine S to arbitrary accuracy, however, the convergence to S is rather slow and the amount of computation needed increases rapidly with the desired degree of precision. Using this algorithm Boyd (1973a) could show that

$$1{\cdot}300197 < S < 1{\cdot}314534.$$

This lends considerable weight to the heuristic estimate $S \simeq 1{\cdot}306951$, obtained by Melzak (1969).

Osculatory packings of spheres are treated in the papers of Boyd (1970b), (1973b), (1973c), (1974) and Larman (1966a). A closely related problem is to determine the possible values of the Hausdorff dimension of the residual set of an almost perfect packing. Results in this direction are contained in the papers of Boyd (1973d), Hirst (1967), Larman (1966b), (1967), (1968) and Wilker (1977). Possible applications of almost perfect packings are mentioned in the papers of Davis (1965) and Gilbert (1964).

For further results on circle-packings and circle-coverings we refer to the papers of A. Bezdek (1983a), (1983b), (1983c), K. Bezdek (1982b), G. Blind (1974), (1975), Dumir and Khassa (1973b), Erdös and Pach (1980), L. Fejes Tóth (1983a), (1983b), L. Fejes Tóth and Heppes (1980), Krötenheerdt (1974), Krötenheerdt, Mammitzsch and Richter (1974), Lampert and Csóka (1973), Molnár (1979–80) and Wegner (1983).

2.2 *Packing and covering by convex sets*

Let S be a convex set in the Euclidean plane. Let H(S) denote the area of the hexagon of minimal area circumscribed about S. According to a well-known theorem of L. Fejes Tóth (1972a p. 86) we have

$$\delta(S) \leq \frac{|S|}{H(S)}.$$

If S is centro-symmetric then a theorem of Dowker (1944) implies that there is a centro-symmetric hexagon of area H(S) circumscribed about S. Since a centro-symmetric hexagon admits a lattice tiling of the plane, we have

$$\delta(S) = \frac{|S|}{H(S)} = \delta_L(S)$$

for all centrally symmetric convex sets S. Another consequence of the inequality above is the fact, first observed by Rogers (1951), that the density of a packing by translates of any convex set S is at most $\sigma_L(s)$. A simple alternative proof of this result was recently given by L. Fejes Tóth (1983d).

As an application of the inequality $\delta(S) \leq |S|/H(S)$, L. Fejes Tóth (1967a), (1971b) determined $\delta(S)$ in the cases when S is the parallel domain of a rectangle and when S is the intersection of two congruent circles. Another application of the above inequality is the following theorem of Groemer and Heppes (1975): If S is a centrally symmetric convex set that is not a quadrangle or a hexagon then there is a line through the centre of S which cuts S into two convex sets, say S_1 and S_2, so that $\delta(S_1) = \delta(S_2) > \delta(S)$.

The papers of Chakerian and Lange (1971) and Kuperberg (1982) contain lower bounds for $\delta(S)$. In particular, Kuperberg showed that $\delta(S) > 3/4$ for all convex sets in the plane. This can be compared with results of Fáry (see L. Fejes Tóth (1972a p. 100) and Tammela (1970) which state that $\delta_L(S) \geq 2/3$ for all convex sets and $\delta(S) > 0{\cdot}892656$ for all centrally symmetric convex sets.

Turning to coverings, it is conjectured that for any convex set S we have $\vartheta(S) \geq |S|/h(S)$, where h(S) denotes the area of the hexagon of maximal area inscribed into S. This has been proved only under a restriction. We say that the sets A and B *cross* each other if none of the sets $A - B$ and $B - A$ is connected. According to L. Fejes Tóth (1972a p. 86) the density of a covering with congruent copies of a convex set S, such that no two sets of the covering cross each other is at least $|S|/h(S)$. Though it seems to be intuitively obvious that a covering with crossing sets is not economical, the difficulty in proving the above conjecture lies in getting rid of the restriction of non-crossing sets. The difficulty is illustrated by a covering of a convex domain with congruent convex sets, constructed by Wegner (1980), which cannot be rearranged so as to cover the domain without crossing each other. When S is centrally symmetric then the bound $|S|/h(S)$ is attained by the density of the thinnest lattice covering of S. Since two translated copies of a convex set cannot cross each other, it follows that the density of a covering with translates of a centro-symmetric convex set S is at least $\vartheta_L(S)$. This statement involves two conditions: central symmetry and parallel orientation. It is a longstanding open problem whether the statement remains true if we drop one of these conditions.

Bambah, Rogers and Zassenhaus (1964) proved that the density of a covering with translates of a convex set S is at least $|S|/2t(S)$, where t(S) is the area of a triangle of maximal area inscribed into S. If S is centro-symmetric then we have $h(S) = 2t(S)$, thus this bound is equivalent with the bound $|S|/h(S)$. If S has no centre of symmetry neither of the above bounds can be attained, and now this now that bound is better; for instance the bound $|S|/h(S)$ is stronger if S is a triangle and the bound $|S|/2t(S)$ is stronger if S is a regular pentagon.

For a given convex set S of unit area let $f(x) = f(x,S)$ be the maximum of the area of the intersection of S and a convex hexagon of area x. Let $\bar{f}(x)$ be the least concave function not less than f(x). G. Fejes Tóth (1972) proved the following theorem:

If $\mathscr{S}$ is a system of sets congruent to S with density d, such that no two sets of $\mathscr{S}$ cross each other, then the density s of the part of the plane covered by the sets of $\mathscr{S}$ satisfies the inequality

$$s \leq d\bar{f}(1/d).$$

If S is a circle then the hexagon of area x having a maximal area of intersection with S is a regular hexagon concentric with S. It is also easy to check that in this case f(x) is concave, thus we obtain the theorem of L. Fejes Tóth mentioned in section 2.1 as a special case of the theorem above. This theorem also implies the bound $\delta(S) \leq |S|/H(S)$, as well as the lower bound $|S|/h(S)$ for the density of a covering with not crossing congruent copies of S.

If S is centrally symmetric then the bound $d\bar{f}(1/d)$ is best possible for any value of d. This rests on the fact that if S is centrally symmetric and $f(x) = \bar{f}(x)$, then there is a centro-symmetric convex hexagon of area x, such that $|S \cap H| = f(x)$.

(See G. Fejes Tóth (1977a) Lemma 1, which has a weaker statement but whose proof applies also to this case.) Thus, if $f(1/d) = \bar{f}(1/d)$, then we can choose a centro-symmetric convex hexagon of area $1/d$ which covers a part of area $f(1/d)$ of S. This hexagon generates a lattice arrangement of S with density d covering a part of the plane of density $d\bar{f}(1/d)$. If $f(1/d) \neq \bar{f}(1/d)$, then we can proceed as follows: We choose two positive numbers, d_1 and d_2, such that $f(1/d_i) = \bar{f}(1/d_i)$, $i = 1,2$; $1/d = \lambda/d_1 + (1-\lambda)/d_2$ and $\bar{f}(1/d) = \lambda f(1/d_1) + (1-\lambda)\, f(1/d_2)$ with some number λ satisfying $0 < \lambda < 1$. We construct two lattice arrangements of S, say $\mathscr{S}_1$ and $\mathscr{S}_2$, with density d_1 and d_2, respectively, such that $\mathscr{S}_i$ covers a part of the plane of density $d_i f(1/d_i)$, $i = 1,2$. We divide the plane into two sets, E_1 and E_2 having densities $d\lambda/d_1$ and $d(1-\lambda)/d_2$. E_1 and E_2 may be taken to be complementary angular regions of angles $2\pi d\lambda/d_1$ and $2\pi d(1-\lambda)/d_2$. Then the arrangement $\mathscr{S}$ consisting of the sets of $\mathscr{S}_i$ which are contained in E_i, $i = 1,2$, has density d, no two sets of $\mathscr{S}$ cross each other and the density of the part of the plane covered by the sets of $\mathscr{S}$ is $d\bar{f}(1/d)$.

A centro-symmetric convex set for which the function $f(x)$ is not concave was constructed by G. Fejes Tóth (1977a). For such a set the bound $d\bar{f}(1/d)$ cannot be attained by a lattice arrangement if $f(1/d) \neq \bar{f}(1/d)$. This fact lends special weight to the theorem claiming that a densest packing of congruent centro-symmetric convex sets can always be realised by a lattice packing, as well as to the analogous theorem concerning the covering with translates of a centro-symmetric convex set.

Let $\mathscr{P}$ be a packing and r the supremum of the radii of the circles disjoint to the sets of $\mathscr{P}$. We call $1/r$ the *closeness of the packing* $\mathscr{P}$. In a covering $\mathscr{C}$ we consider the supremum R of the radii of the circles contained in the intersection of two sets of $\mathscr{C}$ and call $1/R$ the *looseness* of $\mathscr{C}$ (cf. L. Fejes Tóth (1976)). L. Fejes Tóth (1978) proved that the closeness of an arbitrary packing of translates of a convex set cannot exceed the closeness of the closest lattice packing of the set, and pointed out that this statement remains valid if we measure the closeness of a packing instead of circles by means of an arbitrary figure. Special interest is due to the case, investigated by Linhart (1978), when we measure the closeness of a packing or the looseness of a covering of translates of a convex set S by homothetic replicas of S or of $-S$. L. Fejes Tóth (1978a) and A. Bezdek (1980) constructed examples showing that the closeness of an arbitrary packing of congruent centro-symmetric sets can exceed the closeness of the closest lattice packing of the set.

The closeness of a packing and the looseness of a covering can be defined also in more general spaces. The problem of the closest packing of equal balls and the problem of the loosest covering with equal balls are equivalent and related to a problem of Ryskov (1974). Böröczky (1982) showed that if in Euclidean 3-space equal balls form a closest packing or a loosest covering, then the centres of the balls form a body-centred cubic-lattice. Horváth (1977), (1980) investigated the closest lattice packing of balls in higher-dimensional spaces. L. Fejes Tóth (1976) gave an upper bound for the closeness of a packing of equal circles in spherical geometry.

Let $\{S_i\}$ be a countable collection of convex sets in the plane. We say that *the plane can be covered by* $\{S_i\}$, or that $\{S_i\}$ *permits a covering*, if there are rigid motions τ_i, $i = 1,2,\ldots$, such that the sets $\{\tau_i S_i\}$ for a covering. Chakerian and Groemer (1974) gave necessary and sufficient conditions that $\{S_i\}$ permits a covering of the plane. If d_i,p_i and w_i denote the diameter, perimeter and minimal width of S_i, respectively, and $\delta_i = \min\{1,d_i\}$, then any of the following conditions is necessary and sufficient that $\{S_i\}$ permits a covering:

$$\sum w_i\delta_i = \infty, \quad \sum \frac{|S_i|}{1 + d_i} = \infty, \quad \sum \frac{|S_i|}{1 + p_i} = \infty.$$

Analogous results for higher-dimensional spaces are contained in the papers of A. Bezdek and K. Bezdek (1983), Chakerian and Groemer (1978) Groemer (1976a), (1976b), (1979b), (1980), (1981a), (1981b), (1982), (1983a), (1983b) and Makai and Pach (1983).

A packing is said to be *totally separable* if any two sets of the packing can be separated by a straight line avoiding all sets of the packing. G. Fejes Tóth and L. Fejes Tóth (1973) proved that the density of a totally separable packing consisting of congruent copies of a convex set S is at most $|S|/Q(S)$, where Q(S) denotes the area of a quadrangle of minimal area circumscribed about S. This bound is exact for centrally symmetric sets.

We say that a set A is *illuminated* by the set B if to each boundary-point P of A there is a point Q of B such that no point of A lies on the open segment PQ. This definition, due to L. Fejes Tóth (1977b), is less restrictive than the definition occuring in the book of Boltjanskii and Gohberg (1965). L. Fejes Tóth (1977b) proved that the union of m circles can be illuminated by 2m points, and the union of m convex sets not crossing each other can be illuminated by $\max\{2m,4m - 7\}$ points. It is conjectured that the minimal number of points needed to illuminate the union of m convex sets in an arbitrary position is of order m^2. Linhart (1981) gave an upper bound for the number of points needed to illuminate the union of m balls in n-dimensional space.

The papers of Beck and Bleicher (1971), Bérczi and Nagy (1980), G. Blind and R. Blind (1982), G. Fejes Tóth and L. Fejes Tóth (1982), L. Fejes Tóth (1971a), (1980b), (1980c), (1982a), (1982b), L. Fejes Tóth and Florian (1982), Makai (1972) and Vásárhelyi (1983) contain further results about packings and coverings by convex sets.

2.3 *Permeability of layers of convex sets*

A system of disjoint open convex sets contained in a parallel strip is said to form a *layer*. We shall always consider the smallest parallel strip containing the layer; its width w is called the *width of the layer*. We shall call a continuous rectifiable curve which connects the two boundary lines of the strip without intersecting any of the sets, a *path crossing the layer*. The *permeability* p of the layer is defined by

$$p = \frac{w}{\inf l},$$

where l denotes the length of a path crossing the layer.

These definitions were introduced by L. Fejes Tóth (1966b), who is responsible also for the first results on the subject. He proved that the permeability of a layer of congruent circles is greater than $\sqrt{27}/2\pi$. He proved even a little more, namely that the permeability of a layer of unit circles of width w satisfies

$$p \geq w\left\{\frac{2\pi}{\sqrt{27}}(w-2)+2\right\}^{-1}.$$

This inequality is exact when w is of the form $k\sqrt{3}+2$ ($k = 0,1,\ldots$) and the layer consists of $k+1$ consecutive rows of the densest lattice packing of unit circles. Recently Florian (1978b) determined the infimum of the permeabilities of all layers of width w consisting of unit circles for all values $w \geq 2$. The extremal configurations are identical with the packings described in section 2.1 in connection with Molnár's conjecture concerning the densest packing of unit circles in a parallel strip of width w.

For a given convex set S we define the quantity $\pi(S)$ as the infimum of the permeabilities of all layers consisting of translates of S. Hortobágyi (1976) showed that $\pi(S) = \sqrt{27}/2\pi$ for any convex set S of constant width, and Florian (1979b) proved that $\pi(S) = 3/4$ when S is a regular triangle or a regular hexagon.

The difference set $DS = S + (-S)$ of S is a circle if S is a set of constant width and DS is a regular hexagon if S is a regular triangle. Thus, the above results imply that $\pi(S) = \pi(DS)$ if S is either a set of constant width or a regular triangle. Inspired by this observation Gruber asked, whether it is true that $\pi(S) = \pi(DS)$ for any convex set S. Recently Florian (1980d) gave a method for determining $\pi(S)$ for all convex set S which enabled him to answer the problem of Gruber in the affirmative.

It is by no means trivial that the permeability of a layer of incongruent circles can be less than $(\sqrt{27}/2\pi) = 0{\cdot}826993\ldots$ L. Fejes Tóth (1966b) constructed a layer of incongruent circles with permeability 0·823225.... Danzer (private communication) improved this result by showing the existence of a layer of circles with permeability 0·8224.

It is interesting to observe that for layers of parallelograms the situation is quite different than for circle-layers. For, we have the following nice theorem of L. Fejes Tóth (1968a) (see also Bollobás (1968)):

If P is a parallelogram and p the permeability of a layer of parallelograms similar to P then there is a layer of translates of P with permeability not greater than p. An interesting unsolved problem is to find all convex sets which share this property with the parallelogram.

The above theorem enables us to determine the infimum of the permeabilities of all layers consisting of parallelograms similar to P. We emphasize the following special case: The permeability of a layer of squares is at least 2/3.

Confirming two conjectures of L. Fejes Tóth (1978b), G. Fejes Tóth (1978) proved the following theorems:

Given a packing of open unit circles, any two points of the plane lying outside the circles at distance d from one another can be connected by a path evading the circles and having length at most $(2\pi/\sqrt{27})(d-2)+\pi$.

Given a packing of open squares with sidelength not exceeding 1, any two points of the plane lying outside the squares can be connected by a path evading the squares and having length at most $(3d+1)/2$.

Using a mean value argument Pach (1977) proved a statement similar to the second theorem with the weaker bound $\frac{3}{2}d + 4\sqrt{d} + 1$, but his method did not work for packings of circles. The method used in the proof of the theorems above applies also in more general cases. In particular, combining this method with ideas used by Florian to determine $\pi(S)$ one can prove the following theorem:

Let S be a convex set with perimeter u. Let $\mathcal{P}$ be a packing of translates of S. Then any two points of the plane lying outside the sets of $\mathcal{P}$ at distance d from one another can be connected by a path evading the sets of $\mathcal{P}$ and having length at most $d/\pi(S) + u$.

2.4 *Problems connected with the number of neighbours in a packing*

Two sets of a packing are called *neighbours* if their boundaries have a point in common. Let N(S) be the maximum number of neighbours of S in all packings of congruent copies of S. The problem of finding $N(B_3)$ was a point of controversy between Newton and D. Gregory: Can a rigid material ball be brought into contact with 13 other balls of the same kind and size? Gregory said "yes" and Newton said "no", but it was only 180 years later that the problem, known as the problem of thirteen spheres, was settled by Hoppe in favor of Newton (see Bender (1874)). For this reason L. Fejes Tóth (1969a) suggested that we call N(S) the *Newton number* of S.

According to Hoppe we have $N(B_3) = 12$. Further, as mentioned in section 1.2, we have $N(B_8) = 240$ and $N(B_{24}) = 196560$. This exhausts our knowledge of Newton numbers in greater than two dimensions. On the other hand, we know the Newton number of some special convex sets in the plane. It is obvious, that the Newton number of a circle is 6. The conjecture of L. Fejes Tóth (1969b) that the Newton number of a regular n-gon is 12 for $n = 3$, 8 for $n = 4$ and 6 for $n > 4$ was confirmed by Böröczky (1971) in the case when $n \neq 5$ and by Linhart (1973b) in the case when $n = 5$. Schopp (1970) proved that the Newton number of a set of constant width is at most 7. The Newton number of a Reuleaux-triangle is 7. According to a theorem of L. Fejes Tóth (1967b) the Newton number of a convex set S of diameter d and width w satisfies the inequality

$$N(S) < (4 + 2\pi)\frac{d}{w} + 2 + \frac{w}{d}.$$

This inequality is sharp in many cases. For instance, it implies that the Newton number of an isoceles triangle with $w/d = \sin \pi/9$ is 64. Another general bound for the Newton number of convex sets involving other parameters was given by Hortobágyi (1972, 1975).

The maximal number H(S) of neighbours of S in all packings of translates of S is called the *Hadwiger number* of S. This name was introduced by L. Fejes Tóth (1970) because of the nice theorem of Hadwiger (1957) that

$$H(S) \leq 3^n - 1$$

for all n-dimensional convex sets. The case of equality characterizes the parallelotopes (see Groemer (1961) and Grünbaum (1964)). Swinnerton-Dyer

(1953) showed that

$$H(S) \geq n^2 + n$$

for all n-dimensional convex sets. In particular, we have $6 \leq H(S) \leq 8$ for any two-dimensional convex set S. Grünbaum (1964) rediscovered the inequality above and also showed that the number 7 cannot occur as the Hadwiger number of a convex set in the plane. Smith (1975) proved the inequality $H(S) \geq n^2 + n$ for compact n-dimensional sets with interior points.

Further problems arise if we consider the neighbours of higher order of a set. Two sets of a packing are called *neighbours of order* k *or* k-th neighbours if there are $k - 1$ sets $A_1,\ldots,A_{k-1}$ of the packing, but not less than $k - 1$, such that in the sequence $A = A_0, A_1,\ldots,A_{k-1}, A_k = B$ A_{i-1} is a neighbour of A_i for $i = 1,\ldots,k$. The k-*th Newton number* $N_k(S)$ and the k-*th Hadwiger number* $H_k(S)$ of a convex set S are defined as the maximal number of neighbours of order $\leq k$ of a set in a packing consisting of congruent copies of S or translates of S, respectively. The second Newton number of a circle is 18. (L. Fejes Tóth and Heppes (1967)). For $k > 2$ the exact value of $N_k(B_2)$ is not known. On the other hand, a result of L. Fejes Tóth (1969) gives fairly good information about the asymptotic behaviour of $N_k(B_2)$ when k is large. We have $\lim_{k\to\infty} N_k(B_2)/k^2 = 2\pi/\sqrt{3}$.

The k-th Hadwiger number of an arbitrary two-dimensional convex set S satisfies the inequalities

$$3k(k + 1) \leq H_k(S) \leq 4k(k + 1).$$

The inequality on the right hand side is a special case of the more general inequality $H_k(S) \leq (2k + 1)^n - 1$ valid for the k-th Hadwiger number of an n-dimensional convex set S. The proof is similar to that of the special case $k = 1$ given by Hadwiger (1957), Groemer (1961) and Grünbaum (1964). The case of equality characterizes again the parallelotopes. The inequality $3k(k + 1) \leq H_k(S)$ follows from the fact that in a lattice packing of S in which each set has exactly six neighbours, each set has 6i neighbours of order i, and thus $3k(k + 1)$ neighbours of order less than k. There are other packings in which the number of neighbours of order less than k of a set is $3k(k + 1)$. For a strictly convex set S these packings can be obtained by the "greedy" construction, starting from S, adding to it the maximal number of first neighbours, then the maximal number of second neighbours and so on. In each step of this construction we can choose the position of one new neighbour of a certain order arbitrarily but then the positions of the rest of the neighbours of this order are uniquely determined (L. Fejes Tóth (1972b)). However a lower bound for $\liminf_{k\to\infty} H_k(S)/k^2$ given by L. Fejes Tóth (1975b) shows that for each set S this greedy construction is only economical up to a certain value $k_0(S)$. L. Fejes Tóth and Heppes (1977) have shown that for $k > 24$ $H_k(S) > 3k(k + 1)$ for any convex set.

A packing of congruent convex sets is called *maximal* if each set of the packing has as many neighbours as its Newton number. It is an interesting problem to characterize the numbers which can occur as the number of neighbours of a set in a maximal packing of convex sets. Gács (1972) proved that there are only finitely many such numbers; the conjecture of L. Fejes Tóth that such a number cannot be greater than 21 was recently confirmed by Linhart (1977b). The faces of the tiling dual to the Archimedean tiling (3,12,12) are isosceles triangles with angles 30°, 30° and 120°. Each face of this tiling has exactly 21 neighbours, but it is still undecided whether the Newton number of such a triangle is equal to 21 or not.

According to another conjecture of L. Fejes Tóth the faces of all spherical and Euclidean regular tilings $\{p,q\}$ form a maximal packing, but among the regular hyperbolic tilings there are only finitely many with this property. The first part of this conjecture was confirmed by Böröczky (1971), the second part was established in a series of papers by A. Florian (1975c) and A. Florian and H. Florian (1975a,b). Linhart (1975) showed that the regular hyperbolic tiling $\{7,3\}$ is maximal.

Let K = K(n) be the minimal number with the property that there exists a tiling of the plane consisting of K sorts of convex tiles such that each face of the tiling has exactly n neighbours. The function K(n) exists and we have $K(n) \leq (n+1)/2$ (see G. Fejes Tóth (1971)). It is conjectured that $\lim_{n\to\infty} K(n) = \infty$. The odd thing about this conjecture is that in spaces with dimension more than 2 there are tilings with congruent bricks such that each brick has exactly n neighbours when n is sufficiently large. This was observed independently by D. Nagy and A. Bezdek and it can be seen by a slight modification of a construction of Zamorzaev (1965) which he used for a quite different purpose. Another interesting problem is to find the natural numbers n for which K(n) = 1. It is conjectured that the only numbers for which K(n) = 1 are 6,7,8,9,10,12,14,16,21 (L. Fejes Tóth (1969a)). A similar statement for the class of edge-to-edge tilings was proved by K. Bezdek (1977–80).

A packing in which each set has at least n neighbours is called an n-*neighbour packing*. It is obvious that there are 5-neighbour packings of translates of a parallelogram with density zero. On the other hand, the existence of such a packing characterizes the parallelograms:

Any five-neighbour packing of translates of a convex set other than a parallelogram has a density greater than or equal to 3/7. This result is due to Makai (1977). He proved an analogous result for centro-symmetric sets: Any five-neighbour packing of translates of a centro-symmetric convex set other than a parallelogram has a density greater than or equal to 9/14. Equality can be attained for certain arrangements of triangles and affin regular hexagons, respectively.

Concerning six-neighbour packings L. Fejes Tóth (1973b) proved the following theorem: The density of a six-neighbour packing of translates of a convex set is at least 1/2. Equality holds only for a packing of triangles meeting by three at each vertex.

It follows from the result of Grünbaum mentioned above, that 7 cannot occur as the Hadwiger number of a convex set in the plane, that the only convex sets which permit a seven-neighbour packing by translates are the parallelograms. The only eight-neighbour packing of convex sets is a lattice of parallelograms.

Confirming a conjecture of L. Fejes Tóth (1977c) Bárány, Füred; and Pach (1983) proved that in a six-neighbour packing of circles the circles are either congruent or circles of arbitrarily small radius occur.

Further results about problems connected with the number of neighbours in a packing of convex sets in the plane are contained in the papers of Chvátal (1975), Linhart (1973a), (1974), Österreicher and Linhart (1981) and Wegner (1971).

Results on the subject in higher-dimensional spaces are found in the papers of Banai and Sloane (1981), G. Fejes Tóth (1981) and L. Fejes Tóth and Sauer (1977). From among these we emphasize the following nice result of Banai and Sloane:

If in 8-dimensional or 24-dimensional Euclidean space congruent balls form a maximal packing, then they form a densest lattice packing and a Leech-lattice, respectively.

Acknowledgement

I should like to thank professors H. Groemer, J. Linhart and J. Wilker for valuable comments on the first draft of this manuscript.

References

Abramowitz, M. and Stegun, I.A.
(1972) Handbook of mathematical functions, National Bureau of Standards Applied Math. Series 55, U.S. Dept. Commerce, Washington, D.C.
Askey, R.
(1975) Orthogonal polynomials and special functions, Regional Conference Lectures in Applied Mathematics, SIAM 21.
Bambah, R.P.
(1954) On lattice coverings by spheres, Proc. Nat Inst. Sci. India **20**, 25–52.
(1970) Packing and covering, Math. Student **38**, 133–138.
(1971) Geometry of numbers, packing and covering and discrete goemetry, Math. Student **39**, 117–129.
Bambah, R.P., Rogers, C.A. and Zassenhaus, H.
(1964) On coverings with convex domains, Acta Arith. **9**, 191–207.
Bambah, R.P. and Solane, N.J.
(1982) On a problem of Ryškov concerning lattice coverings, Acta Arith. **42**, 107–109.
Bannai, E. Solane, N.J.A.
(1981) Uniqueness of certain spherical codes, Canad. J. Math. **33**, 437–449.
Baranovskiĭ, E.P.
(1964) On packing of n-dimensional Euclidean spaces by equal spheres (Russian), Izv. Vysš. Učebn. Zved. Matematika **39**, 14–24.
Bárány, I., Füredi, Z. and Pach, J.
(1983) Discrete convex functions and proof of the six circle conjecture of Fejes Tóth (to appear).
Barnes, E.S. and Sloane, N.J.A.
(1983) New lattice packings of spheres, Canad. J. Math. **35**, 117–130.
Beck, A. and Blecker, M.N.
(1971) Packing convex sets into a similar set, Acta Math. Acad. Sci. Hungar. **22**, 283–303.
Bender, C.
(1874) Bestimmung der grössten Anzahl gleich grosser Kugeln, welche sich auf eine Kugel von demselben Radius, wie die übrigen, auflegen lassen, Arch. Math. Phys. **56**, 302–312.
Bender, E.A.
(1962) Area-perimeter relations for two-dimensional lattices, Amer. Math. Monthly **69**, 742–744.
Bérczi, Sz. and Nagy, D.
(1980) Periodicity of extremal goemetric arrangements (densest packing, thinnest covering, tessellations), Acta Geol. Sci. Hungar. **23**, 173–200.
Best, M.R.
(1980) Binary codes with a minimum distance of four, IEEE Trans. Inform. Theory **IT-26**, 738–742.
Betke, U. and Wills, J.M.
(1979) Stetige und diskrete Funktionale konvexer Körper, Contributions to geometry, Proc. Symp. Siegen 1978, 226–237.
Betke, U. and Gritzmann, P.
(1983) Über L. Fejes Tóths Wurstvermutung in kleinen Dimensionen.

Betke, U., Gritzmann, P. and Wills, J.M.
(1983) Slices of Fejes Tóth's sausage conjecture, Mathematika (to appear).
Bezdek, A.
(1979) Solid packing of circles in the hyperbolic plane, Studia Sci. Math. Hungar. **14**, 203–207.
(1980) Remark on the closest packing of convex discs, Studia Sci. Math. Hungar. **15**, 283–285.
(1983a) Ausfüllung und Überdeckung der Ebene durch Kreise, Ann. Univ. Sci. Budapest. Eötvös Sect. Math. (to appear).
(1983b) Locally separable circle packings, Studia Sci. Math. Hungar. **18** (to appear).
(1983c) On the thinnest double saturated packing of equal circles, Studia Sci. Math. Hungar. **18** (to appear).
Bezdek, A. and Bezdek, K.
(1983) Eine hinreichende Bedingung für die Überdeckung des Einheitswürfels durch homothetische Exemplare in dem n-dimensionalen euklidischen Raum, Beiträge zur Algebra und Geometrie (to appear).
Bezdek, K.
(1977–80) Mosaics with each face having the same number of neighbours, Mat. Lapok **28**, 297–303.
(1979) Optimal coverings of circles (Hungarian) Thesis, Budapest.
(1982a) Ausfüllung eines Kreises durch kongruente Kreise in der hyperbolischen Ebene, Studia Sci. Math. Hungar. **17** (to appear).
(1982b) Über Ionenpackungen, Studia Sci. Math. Hungar. **17** (to appear).
(1983a) Ausfüllungen in der hyperbolischen Ebene durch endliche Anzahl kongruenter Kreise, Ann. Univ. Sci. Budapest. Eötvös Sect. Math. (to appear).
(1983b) Über einige optimale Konfigurationen von Kreisen, Ann. Univ. Sci. Budapest. Eötvös Sect. Math. (to appear).
(1983c) Über einige Kreisüberdeckungen, Beiträge zur Algebra und Geometrie (to appear).
Blachman, N.M. and Few, L.
(1963) Multiple packing of spherical caps, Mathematika **10**, 84–88.
Bleicher, M.N.
(1975) The thinnest three dimensional point lattice trapping a sphere, Studia Sci. Math. Hungar. **10**, 157–170.
Blichfeldt, H.F.
(1929) The minimum value of quadratic forms and the closest packing of spheres, Math. Ann. **101**, 605–608.
(1934) The minimum values of positive quadratic forms in six, seven and eight variables, Math. Z. **39**, 1–15.
Blind, G.
(1969) Über Unterdeckungen der Ebene durch Kreise, J. reine angew. Math. **236**, 145–173.
(1972) Zugängliche Unterdeckungen der Ebene durch kongruente Kreise, J. Reine Angew. Math. **257**, 29–46.
(1974) Überdeckung der Ebene durch inkongruente Kreise, Math. Z. **140**, 179–194.
(1975) Unterdeckung der Ebene durch inkongruente Kreise, Arch. Math. (Basel) **26**, 441–448.
(1976) r-zugängliche Unterdeckungen der Ebene durch kongruente Kreise I, J. Reine Angew. Math. **288**, 1–23.
(1977) r-zugängliche Unterdeckungen der Ebene durch kongruente Kreise II, J. Reine Angew. Math. **289**, 1–29.
(1981) ρ-zugängliche Unterdeckungen der Sphäre durch kongruente Kreise, Resultate Math. **4**, 141–154.
Blind, G. and Blind, R.
(1978) Zugänglichkeit von Kugelpackungen im R^n, Arch. Math. (Basel) **30**, 438–439.
(1979) r-zugängliche Unterdeckungen der Ebene durch kongruente Kreise, Studia Sci. Math. Hungar. **14**,
(1982) Eine Abschätzung für die Dichte der dichtesten Packung mit Reuleaux-Dreiecken, Studia Sci. Math. Hunger. **17** (to appear).
Bloh, E.L.
(1956) On the most dense arrangement of spherical segments on a hypersphere (Russian), Izv. Akad. Nauk SSSR Ser. Mat. **20**, 707–712.
Blundon, W.J.
(1957) Multiple covering of the plane by circles, Mathematika **4**, 7–16.

(1963a) Multiple packing of circles in the plane, J. London Math. Soc. **38**, 176–182.
(1963b) Note on a paper of A. Heppes, Acta Math. Acad. Sci. Hungar. **14**, 317.
(1964) Some lower bounds for density of multiple packing, Canad. Math. Bull. **7**, 565–572.
(1977) A three-fold non-lattice covering, Canad. Math. Bull. **20**, 29–31.
(1978) A nine-fold packing, Acta Math. Acad Sci. Hungar. **32**, 293–294.
Bokowski, J.H., Hadwiger, H. and Wills, J.M.
(1972) Eine Ungleichung zwischen Volumen, Oberfläche und Gitterpunktanzahl konvexer Körper im n-dimensionalen euklidischen Raum, Math. Z. **127**, 363–364.
Bolle, U.
(1976) Mehrfache Kreisanordnungen in der euklidischen Ebene, Dissertation, Universität Dortmund.
(1979) Dichteabschätzungen für mehrfache gitterförmige Kugelanordnungen im R^m, Studia Sci. Math. Hungar. **14**, 51–68.
(1982) Dichteabschätzungen für mehrfache gitterförmige Kugelanordnungen im R^m II, Studia Sci. Math. Hungar. **18** (to appear).
(1983) Über die Dichte mehrfacher, gitterförmigen Kreisanordnungen in der Ebene, Studia Sci. Math. Hungar. **18** (to appear).
Bollobás, B.
(1968) Remarks to a paper of L. Fejes Tóth, Studia Sci. Math. Hungar. **3**, 373–379.
(1972) The optimal structure of market areas, J. Economic Theory **4**, 174–179.
(1973) The optimal arrangement of producers, J. London Math. Soc. (2) **6**, 605–613.
Boltjanski, W.G. and Gockberg, I.Z.
(1965) Theorems and problems in combinatorial goemetry (Russian), Nauka, Moscow; German translation: Sätze und Probleme der kombinatorischen Geometrie, Berlin (1972).
Boyd, D.W.
(1970a) Lower bounds for the disk-packing constant, Math. Comp. **24**, 697–704.
(1970b) Osculatory packings by spheres, Canad. Math. Bull. **13**, 59–64.
(1971a) On the exponent of an osculatory packing, Canad. J. Math. **23**, 355–363.
(1971b) The disc-packing constant, Aequationes Math. **7**, 182–193.
(1972) Disk packings which have non-extreme exponents, Canad. Math. Bull. **15**, 341–344.
(1973a) Improved bounds for the disk-packing constant, Aequationes Math. **9**, 99–106.
(1973b) An algorithm for generating the sphere coordinates in a three-dimensional osculatory packing, Math. Comp. **27**, 369–377.
(1973c) The osculatory packing of a three dimensional sphere, Canad. J. Math. **25**, 303–322.
(1973d) The residual set dimension of the Apollonian packing, Mathematika **20**, 170–174.
(1974) A new class of infinite sphere packings, Pacific J. Math. **50**, 383–398.
Böröczky, K.
(1971) Über die Newtonsche Zahl regulärer Vielecke, Periodica Math. Hungar. **1**, 113–119.
(1974) Sphere packing in spaces of constant curvature I (Hungarian), Mat. Lapok **25**, 265–306.
(1978) Packing of spheres in spaces of constant curvature, Acta Math. Acad. Sci. Hungar. **32**, 243–261.
(1982) Closest packing and loosest covering of the space with balls, Studia Sci. Math. Hungar. **17** (to appear).
(1983) The problem of Tammes for n = 11, Studia Sci. Math. Hungar. (to appear).
Butler, G.J.
(1972) Simultaneous packing and covering in Euclidean space, Proc. London Math. Soc. (3) **25**, 721–735.
Cassels, J.W.S.
(1959) An introduction to the geometry of numbers, Grundl. Math. Wiss. 99 Springer 1959, 1972. Berlin.
Chakerian, G.D. and Groemer, H.
(1974) On classes of convex sets that permit plane coverings, Israel J. Math. **19**, 305–311.
(1978) On coverings of euclidean space by convex sets, Pacific J. Math. **75**, 77–80.
Chakerian, G.D. and Lange, L.H.
(1971) Geometric extremum problems, Math. Mag. **44**, 57–69.
Chvátal, V.
(1975) On a conjecture of Fejes Tóth, Periodica Math. Hungar. **6**, 357–362.
Cohn, M.J.
(1976) Multiple lattice covering of space, Proc. London Math. Soc. III. **32**, 117–132.

Conway, J.H., Parker, F.R.S. and Sloane, N.J.A.
(1982) The covering radius of the Leech lattice, Proc. R. Soc. London A **380**, 261–290.
Conway, J.H. and Sloane, N.J.A.
(1982) Voronoi regions of lattices, 2nd moments of polytopes, and quantization, IEEE Transactions on Information Theory **IT-28**, 211–226.
Coxeter, H.S.M.
(1948) Regular polytopes, Methuen, London.
(1954) Arrangements of equal spheres in non-Euclidean spaces, Acta Math. Acad. Sci. Hungar. **4**, 263–274.
(1963) An upper bound for the number of equal nonoverlapping spheres that can touch another of the same size, Proceedings of Symposia in Pure Mathematics, Amer. Math. Soc. **7**, Convexity, 53–71.
Coxeter, H.S.M., Few, L. and Rogers, C.A.
(1959) Covering space by equal spheres, Mathematika **6**, 147–157
Csóka, G.
(1975a) On the number of congruent spheres hiding a given sphere (Russian), Ann. Univ. Sci. Budapest. Eötvös Sect. Math. **18**, 55–60.
(1975b) On the densest packing of spheres in a cylinder in hyperbolic space (Russian), Ann. Univ. Sci. Budapest. Eötvös Sect. Math. **18**, 60–67.
(1977) The number of congruent spheres that cover a given sphere of three-dimensional space, not less than 30 (Russian), Studia Sci. Math. Hungar. **12**, 323–334.
Danzer, L.
(1975) Endliche Punktmengen auf der S^2 mit maximalem Mindestabstand. Kolloquium über Diskrete Geometrie, Salzburg 1975.
Davenport, H. and Hajós, G.
(1951) Problem 35, Mat. Lapok **2**, 68.
Davis, P.J.
(1965) Simple quadratures in the complex plane, Pacific J. Math. **15**, 813–824.
Delone, B.N., Dolbilin, N.P., Ryškov, S.S. and Štogrin, M.I.
(1970) A new construction in the theory of lattice coverings of an n-dimensional space by equal spheres, Izv. Akad. Nauk SSSR Ser. Mat. **34**, 289–298 = Math. USSR Izvestija **4**, 293–302.
Delone, B.N. and Ryškov, S.S.
(1963) Solution of the problem on the least dense lattice covering of a 4-dimensional space by equal spheres (Russian), Dokl. Akad. Nauk SSSR **152**, 523–524 = Soviet Math. Dokl. **14**, 1333–1334.
Delsarte, P., Goethals, J.M. and Seidel, J.J.
(1977) Spherical codes and designs, Geometriae Dedicata **6**, 363–388.
Dolbilin, N.P.
(1974) On the covering radius of hyperbolically equivalent n-dimensional lattices and the boundedness of the index of a divided parallelepiped (Russian), Dokl. Akad. Nauk SSSR **214**, 1002–1004 = Soviet Math. Dokl. **15**, 277–279.
Dorninger, D.
(1973) Überdeckung der Ebene durch inkongruente Kreise, Elem. Math. **28**, 105–107.
Dowker, C.H.
(1944) On minimum circumscribed polygons, Bull. Amer. Math. Soc. **50**, 120–122.
Dumir, V.C. and Hans-Gill, R.J.
(1972a) Lattice double coverings in the plane, Indian J. Pure Appl. Math. **3**, 466–480.
(1972b) Lattice double packings in the plane, Indian J. Pure Appl. Math. **3**, 481–487.
Dumir, V.C. and Khassa. D.S.
(1973a) Saturated systems of symmetric convex domains; results of Eggleston, Bambah and Woods, Proc. Camb. Phil. Soc. **74**, 107–116.
(1973b) A conjecture of Fejes Tóth on saturated systems of circles, Proc. Camb. Phil. Soc. **74**, 453–460.
Erdös, P. Few, L. and Rogers, C.A.
(1964) The amount of overlapping in partial coverings of space by equal spheres, Mathematika **12**, 171–184.
Erdös, P. and Pach, J.
(1980) On a problem of L. Fejes Tóth, Discrete Math. **30**, 103–109.

Erdös, P. and Rogers, C.A.
(1962) Covering space with convex bodies, Acta Arith. **7**, 281–285.
Fejes Tóth, G.
(1971) Über Parkettierungen konstanter Nachbarnzahl, Studia Sci. Math. Hungar. **6**, 133–135.
(1972) Covering the plane by convex discs, Acta Math. Acad. Sci. Hungar. **23**, 263–270.
(1973) Sum of moments of convex polygons, Acta Math. Acad. Sci. Hungar. **24**, 417–421.
(1974) Solid sets of circles, Studia Sci. Math. Hungar. **9**, 101–109.
(1976a) Multiple packing and covering of the plane with circles, Acta Math. Acad. Sci. Hungar. **27**, 135–140.
(1976b) Research problem 18, Period. Math. Hungar. **7**, 89–90.
(1977a) On the intersection of a convex disc and a polygon, Acta Math. Acad. Sci. Hungar. **29**, 149–153.
(1977b) A problem connected with multiple circle-packings and circle coverings, Studia Sci. Math. Hungar. **12**, 447–456.
(1978) Evading convex discs, Studia Sci. Math. Hungar. **13**, 453–461.
(1979) Multiple packing and covering of spheres, Acta Math. Acad. Sci. Hungar. **34**, 165–176.
(1980) On the section of a packing of equal balls, Studia Sci. Math. Hungar. **15**.
(1981) Ten-neighbour packing of equal balls, Period. Math. Hungar. **12**, 125–127.
(1982) Packing and covering with incongruent circles, Studia Sci. Math. Hungar. **17** (to appear).
Fejes Tóth, G. and Fejes Tóth, L.
(1973) On totally separable domains, Acta Math. Acad. Sci. Hungar. **24**, 229–232.
(1980) Dictators on a planet, Studia Sci. Math. Hungar. **15**, 313–316.
(1982) Packing the plane with Minkowskian sums of convex sets, Studia Math. Hungar. **17** (to appear).
Fejes Tóth, G. and Florian, A.
(1975) Mehrfache gitterförmige Kreis- und Kugelanordnungen, Monatsh. Math. **79**, 13–20.
Fejes Tóth, L.
(1953) On close-packings of spheres in spaces of constant curvature, Publ. Math. Debrecen **3**, 158–167.
(1966a) Mehrfache Kreisunterdeckungen und Kreisüberdeckungen auf der Kugel, Elem. Math. **21**, 34–35.
(1966b) On the permeability of a circle-layer, Studia Sci. Math. Hungar. **1**, 5–10.
(1967a) On the arrangement of houses in a housing estate, Studia Sci. Math. Hungar. **2**, 37–42.
(1967b) On the number of equal discs that can touch another of the same kind, Studia Sci. Math. Hungar. **2**, 363–367.
(1968a) On the permeability of a layer of parallelograms, Studia Sci. Math. Hungar. **3**, 195–200.
(1968b) Solid circle-packings and circle-coverings, Studia Sci. Math. Hungar. **3**, 401–409.
(1969a) Remarks on a theorem of R.M. Robinson, Studia Sci. Math. Hungar. **4**, 441–445.
(1969b) Scheibenpackungen konstanter Nachbarnzahl, Acta Math. Acad. Sci. Hungar. **20**, 375–381.
(1970) Über eine affininvariante Masszahl bei Eipolyedern, Studia Sci. Math. Hungar. **5**, 173–180.
(1971a) Über Scheiben mit richtungsinvarianter Packungsdichte, Elem. Math. **26**, 58–59.
(1971b) The densest packing of lenses in the plane (Hungarian), Mat. Lapok **22**, 209–213.
(1972a) Lagerungen in der Ebene, auf der Kugel und im Raum, zweite Auflage, Grundl. Math. Wiss. 65, Springer-Verlag, Berlin-Heidelberg-New York.
(1972b) A closing theorem (Hungarian), Mat. Lapok **23**, 9–12.
(1973a) On the density of a connected lattice of convex bodies, Acta Sci. Math. Hungar. **24**, 373–376.
(1973b) Five-neighbour packing of convex discs, Periodica Math. Hungar. **4**, 221–229.
(1975a) Research problem 13, Periodica Math. Hungar. **6**, 197–199.
(1975b) On Hadwiger numbers and Newton numbers of a convex body, Studia Sci. Math. Hungar. **10**, 111–115.
(1976) Close packing and loose covering with balls, Publ. Math. Debrecen **23**, 323–326.
(1977a) Dichteste Kugelpackung. Eine Idee von Gauß, Abh. Braunschweigischen Wiss. Gesell. **27**, 311–321.
(1977b) Illumination of convex discs, Acta Math. Acad. Sci. Hungar. **29**, 355–360.
(1977c) Research problem 21, Periodica Math. Hungar. 8, 103–104.
(1978a) Remarks on the closest packing of convex discs, Comment. Math. Helvetici **33**, 536–541.
(1978b) Research problem 24, Periodica Math. Hungar. **9**, 173–174.
(1980a) Some density-bounds for packing and covering with incongruent circles, Studia Sci. Math. Hungar. **15**, 63–70.

(1980b) Approximation of convex domains by polygons, Studia Sci. Math. Hungar. **15**, 133–138.
(1980c) Packing and covering with convex discs, Studia Sci. Math. Hungar. **15**, 93–100.
(1980d) Solid packing of circles in the hyperbolic plane, Studia Sci. Math. Hungar. **15**, 299–302.
(1982a) Packing of r-convex discs, Studia Sci. Math. Hungar. **17** (to appear).
(1982b) Packing and covering with r-convex discs, Studia Sci. Math. Hungar. **17** (to appear).
(1983a) Compact packing of circles, Studia Sci. Math. Hungar. (to appear).
(1983b) Hiding a circle with circles (Hungarian), Mat. Lapok (to appear).
(1983c) Density bounds for packing and covering with convex discs, Expositiones Mathematicae (to appear).
(1983d) On the densest packing of convex discs, Mathematika (to appear).

Fejes Tóth, L. and Florian, A.
(1982) Packing and covering with convex discs, Mathematika 29.

Fejes Tóth, L. and Heppes, A.
(1967) A variant of the problem of the thirteen spheres, Canad. J. Math. **19**, 1092–1100
(1977) A remark on the Hadwiger numbers of a convex disc, Studia Sci. Math. Hungar. **12**, 409–412.
(1980) Multi-saturated packings of circles, Studia Sci. Math. Hungar. **15**, 303–307.

Fejes Tóth, L. and Makai, E. Jr.
(1974) On the thinnest non-separable lattice of convex plates, Studia Sci. Math. Hungar. **9**, 191–193.

Fejes Tóth, L. and Sauer, N.
(1977) Thinnest packing of cubes with a given number of neighbours, Canad. Math. Bull. **20**, 501–507.

Few, L.
(1953) The double packing of spheres, J. London Math. Soc. **28**, 297–304.
(1964) Multiple packing of spheres, J. London Math. Soc. **39**, 51–54.
(1967a) Multiple packing of spheres: a survey, Proc Colloquium Convexity (Copenhagen 1965), Københavns Univ. Mat. Inst., Kopenhagen, 88–93.
(1967b) Double covering with spheres, Mathematika **14**, 207–214.
(1968) Double packing of spheres: A new upper bound, Mathematika **15**, 88–92.

Few, L. and Kanagasabapathy, P.
(1969) The double packing of spheres, J. London Math. Soc. **44**, 141–146.

Florian, A.
(1975a) Neuere Ergebnisse in der diskreten Geometrie, Berichte Math. Stat. Sekt. Forschungszentrum Graz, Bericht Nr. **45** (1975).
(1975b) Integrale auf konvexen Mosaiken, Periodica Math. Hungar. **6**, 23–38.
(1975c) Reguläre hyperbolische Mosaike und Newtonshce Zahlen, Periodica Math. Hungar. **6**, 97–101.
(1978a) Mehrfache Packung konvexer Körper, Österreich. Akad. Wiss. Math.-Natur. Kl. Sitzungsber. II, **186**, 373–384.
(1978b) On the permeability of layers of discs, Studia Sci. Math. Hungar. **13**, 123–132.
(1979a) Neuere Entwicklung über reguläre Polyeder, Contributions to geometry, Proc. Symp. Siegen 1978, 238–247.
(1979b) Über die Durchlassigkeit gewisser Scheibenschichten, Österreich. Akad. Wiss. Math.-Natur. Kl. Sitzungsber. II, **188**, 417–427.
(1980a) Kugelwolken, Berichte Math.-Stat. Sekt. Forschungszentrum Graz, Bericht Nr. **143** (1980).
(1980b) Durchlässigkeit einer Scheibenschicht, Berichte Math.-Stat. Sekt. Forschungszentrum Graz, Bericht Nr. **144** (1980).
(1980c) Newtonsche und Hadwigersche Zahlen, Berichte Math. Stat. Sekt. Forschungszentrum Graz, Bericht Nr. **145** (1980).
(1980d) Über die Durchlässigkeit einer Schicht konvexer Scheiben, Studia Sci. Math. Hungar. **15**, 201–213.

Florian, A. and Florian, H.
(1975a) Reguläre hyperbolische Mosaike und Newtonsche Zahlen II, Periodica Math. Hungar. **6**, 179–183.
(1975b) Reguläre hyperbolische Mosaike und Newtonsche Zahlen III, Österreich. Akad. Wiss. Math.-Natur. Kl. Sitzungber. II, **184**, 29–40.

Florian, A., Hárs, L. and Molnár, J.
(1979) On the ρ-systems of circles, Acta Math. Acad. Sci. Hungar. **34**, 205–221.

Gács, P.
(1972) Packing of convex sets in the plane with a great number of neighbours, Acta Math. Acad. Sci. Hungar. **23**, 383–388.
Gauß, C.F.
(1831) Untersuchungen über die Eigenschaften der positiven ternären quadratischen Formen von Ludwig August Seeber, Göttingische gelehrte Anzeigen, 1831 Juli 9 = J. Reine Angew. Math. **20** (1840), 312–320 = Werke II, 188–196.
Gilbert, E.N.
(1964) Randomly packed and solidly packed spheres, Canad. J. Math. **16**, 286–298.
Goldberg, M.
(1970) The packing of equal circles in a square, Math. Magazine, **43**, 24–30.
(1971) Packing of 14, 16, 17 and 20 circles in a circle, Math. Magazine **44**, 134–139.
Golser, G.
(1977) Dichteste Kugelpackungen im Oktaeder, Studia Sci. Math. Hungar. **12**, 337–343.
Groemer, H.
(1960) Über die Einlagerung von Kreisen in einen konvexen Bereich, Math. Zeitschr. **73**, 285–294.
(1961) Abschätzungen für die Anzahl der konvexen Körper, die einen konvexen Körper berühren, Monatsh. Math. **65**, 74–81.
(1963) Existenzsätze für Lagerungen im Euklidischen Raum, Math. Zeitschr. **81**, 260–278.
(1966) Zusammenhängende Lagerungen konvexer Körper, Math. Zeitschr. **94**, 66–78.
(1968a) Existenzsätze für Lagerungen in metrischen Räumen, Monatsh. Math. **72**, 325–334.
(1968b) Einige Bemerkungen über zusammenhängende Lagerungen, Monatsh. Math. **72**, 212–216.
(1976a) On a covering property of convex sets, Proc. Amer. Math. Soc. **59**, 346–352.
(1976b) Some packing and covering problems, Amer. Math. Monthly **83**, 726–727.
(1979a) On multiple space subdivisions by zonotopes, Monatsh. Math. **86**, 185–188.
(1979b) Space coverings by translates of convex sets, Pacific J. Math. **82**, 379–386.
(1980) On finite classes of convex sets that permit space coverings, Amer. Math. Monthly **87**, 286–288.
(1981a) On coverings of plane convex sets by translates of strips, Aequationes Math. **22**, 215–222.
(1981b) On coverings of convex sets by translates of slabs, Proc. Amer. Math. Soc. **82**, 261–266.
(1982) Covering and packing properties of bounded sequences of convex sets, Mathematika **29**, 18–31.
(1983a) On space coverings by unbounded convex sets, J. Combinatorial Theory, Ser. A.
(1983b) On coverings of spheres by convex sets (to appear).
Groemer, H. and Heppes, A.
(1975) Packing and covering properties of split disks, Studia Sci. Math. Hungar. **10**, 185–189.
Gruber, P.M.
(1979) Geometry of numbers, Contributions to geometry, Proc. Symp. Siegen 1978, 186–225.
Grünbaum, B.
(1964) On a conjecture of Hadwiger, Pacific J. Math. **11**, 215–219.
Grünbaum, B. and Shephard, G.C.
(1983) Tilings and patterns.
Haas, A.
(1976) Die dünnste siebenfache gitterförmige Überdeckung der Ebene durch kongruente Kreise, Dissertation, Wien.
Hadwiger, H.
(1957) Über Treffanzahlen bei translationsgleichen Eikörpern, Archiv Math. **8**, 212–213.
(1970) Volumen und Oberfläche eines Eikörpers, der keine Gitterpunkte überdeckt, Math. Z. **116**, 191–196.
Heppes, A.
(1955) Über mehrfache Kreislagerungen, Elem. Math. **10**, 125–127.
(1959) Mehrfache gitterförmige Kreislagerungen in der Ebene, Acta Math. Acad. Sci. Hungar. **10**, 141–148.
(1960) Ein Satz über gitterförmige Kugelpackungen, Ann. Univ. Sci. Budapest. Eötvös Sect. Math. **3–4**. 89–90.
(1967) On the densest packing of circles not blocking each other, Studia Sci. Math. Hungar. **2**, 257–263.
Hirst, K.E.
(1967) The apollonian packing of circles, J. London Math. Soc. **42**, 281–291.

Hlawka, E.
(1944) Zur Geometrie der Zahlen, Math. Z. **49**, 285–312.
Hollai, M.
(1974) Lagerungen inkongruente Kugeln I, Ann. Univ. Sci. Budapest. Eötvös Sect. Math. **17**, 73–90.
(1975a) Doppelgitterförmige Lagerungen inkongruenter Kreise und Kugeln, Ann. Univ. Sci. Budapest. Eötvös Sect. Math. **18**, 75–86.
(1975b) Lagerungen inkongruenter Kugeln II, Ann. Univ. Sci. Budapest. Eötvös Sect. Math. **18**, 101–116.
Hortobágyi, I.
(1971) Durchleutung gitterförmiger Kugelpackungen mit Lichtbündeln, Studia Sci. Math. Hungar. **6**, 147–150.
(1972) The Newton number of convex plane regions (Hungarian), Mat. Lapok **23**, 313–317.
(1975) Über die auf Scheibenklassen bezügliche Newtonsche Zahl der konvexen Scheiben, Ann. Univ. Sci. Budapest. Eötvös Sect. Math. **18**, 123–127.
(1976) Über die Durchlässigkeit einer aus Scheiben konstanter Breite bestehenden Schicht, Studia Sci. Math. Hung. **11**, 383–387.
(1977) Extreme problems connected with packings of sets of constant width, Thesis, Budapest (Hungarian).
(1983a) Über die Anzahl derjenigen Richtungen in denen sich die gitterförmige Kugelpackungen durchleuchten lassen. Studia Sci. Math. Hungar. (to appear).
(1983b) Durchleuchtung gitterförmiger Packungen von Körpern konstanter Breite mit Lichtbündeln, Studia Sci. Math. Hungar. (to appear).
Horváth, J.
(1970) Über die Durchsichtigkeit gitterförmigen Kugelpackungen, Studia Sci. Math. Hungar. **5**, 421–426.
(1972a) On the densest packing of an n-dimensional cylinder with unit balls (Russian), Ann. Univ. Sci. Budapest. Eötvös Sect. Math. **15**, 139–143.
(1972b) Remarks on the theory of clouds of spheres (Russian), Ann. Univ. Sci. Budapest. Eötvös Sect. Math. **15**, 145–150.
(1974) Die Dichte einer Kugelpackung in einer 4-dimensionalen Schicht, Studia Sci. Math. Hungar. **5**, 195–199.
(1975) On the densest packing of unit balls in 3- and 4-dimensional slabs (Russian), Ann. Univ. Sci. Budapest. Eötvös Sect. Math. **18**, 171–176.
(1977) Close lattice packings of balls in lattices of the first type (Russian), Ann. Univ. Sci. Budapest. Eötvös Sect. Math. **20**, 191–194.
(1980) Close lattice packing of unit balls in the space E^n (Russian), Geometry of positive quadratic forms. Trudy Mat. Inst. Steklov. **152**, 216–231.
(1982) Über die Enge der gitterförmigen k-fachen Packungen, die Lockerheit der gitterförmigen k-fachen Überdeckungen und die k-Enge der gitterförmigen Punktmengen, Beiträge zur Algebra und Geometrie **16** (to appear)
Horváth, J. and Temesvári, A.H.
(1982) Über Dichte und Enge von doppelgitterförmigen zweifachen Kreispackungen, Studia Sci. Math. Hungar. **17** (to appear).
Hoylman, D.J.
(1970) The densest lattice packing of tetrahedra, Bull. Amer. Math. Soc. **76**, 135–137.
Ignat'ev, N.K.
(1964) On a practical method of finding dense packings of n-spheres (Russian), Sibirsk. Mat. Ž. **5**, 815–819.
(1966) On finding dense packings of n-dimensional spheres (Russian), Sibirsk. Mat. Ž. **7**, 820–825.
Kabatjanskiĭ, G.A. and Levenšteĭn, V.I.
(1978) Bounds for packings on the sphere and in space (Russian), Problemy Peredači Informacii **14**, 3–25 = Problems of information transmission **14**, 1–17.
Karabinta, A. and Székely, E.
(1973) Sur les empilements optimaux de circles congruents sur la sphère, Ann. Univ. Sci. Budapest. Eötvös Sect. Math. **16**, 143–154.
Kershner, R.
(1939) The number of circles covering a set, Amer. J. Math. **61**, 665–671.
Kleinschmidt, P., Pachner, U. and Wills, J.M.

(1983) On L. Fejes Tóth's sausage conjecture (to appear).
Kertész, G.
(1982) On the problem of parasites (Hungarian), Dissertation, Budapest.
Korkine, A. and Zolotareff, G.
(1872) Sur les formes quadratiques positives quaternaires, Math. Annalen **5**, 581–583, and (1877) Sur les formes quadratiques, Math. Annalen **11**, 242–249.
Kravitz, S.
(1967) Packing cylinders into cylindrical containers, Math. Mag. **40**, 65–71.
Krötenheerdt, O.
(1974) Die flächenkleinsten Dreiecke, die zwei gegebene, sich von außen berührende Kreise enthalten, Beiträge zur Algebra und Geometrie **2**, 145–154.
Krötenheerdt, O.
(1975) Probleme der Lagerung von Figuren, Mitteilungen der Math. Gesselschaft der DDR **1**, 65–74.
Krötenheerdt, O., Mammitzsch, C. and Richter, P.
(1974) Die flächenkleinsten Vierecke, die zwei gegebene, sich von außen berührende Kreise enthalten, Beiträge zur Algebra und Geometrie **3**, 167–183.
Kuperberg, W.
(1982) Packings convex bodies in the plane with density greater than 3/4, Geometriae Dedicata **13**, 149–155.
Lagrange, J.L.
(1773) Recherches d'arithmetique, Nouv. Mem. Acad. Roy. Sc. Belle Lettres, Berlin 1773, 265–312 = Oeuvres III, 693–758.
Lampert, D. and Csóka, G.
(1973) On the density of regular system of circles (Russian), Ann. Univ. Sci. Budapest. Eötvös Sect. Math. **16**, 69–85.
Landau, H.J. and Slepain, O.
(1966) On the optimality of the regular simplex code, Bell System Techn. J. **45**, 1247–1272.
Larman, D.G.
(1966a) On the exponent of the convergence of a packing of spheres, Mathematika **13**, 57–59.
(1966b) A note on the Besicovich dimension of the closest packing of spheres in R_n, Proc. Camb. Phil. Soc. **62**, 193–195.
(1967) On the Besicovitch dimension of the residual set of arbitrary packed discs in the plane, J. London Math. Soc. **42**, 292–302.
(1968) On packings of unequal spheres in R_n, Canad. J. Math. **20**, 967–969.
Leech, J.
(1964) Some sphere packings in higher space, Canad. J. Math. **16**, 657–682.
(1967a) Nones on sphere packings, Canad. J. Math. **19**, 251–267.
(1967b) Five-dimensional nonlattice sphere packings, Canad. Math. Bull. **10**, 387–393.
(1969) Six and seven dimensional nonlattice sphere packings, Canad. Math. Bull. **12**, 151–155.
Leech, J. and Sloane, N.J.A.
(1970a) New sphere packings in dimensions 9–15. Bull. Amer. Math. Soc. **76**, 1006–1010.
(1970b) New sphere packings in more than thirty-two dimensions, Proc. Second Chapel Hill Conf. on Combin. Math. and Applic., Chapel Hill, N.C., 1970, 345–355.
(1971) Sphere packings and error-correcting codes, Canad. J. Math. **23**, 718–745.
Lekkerkerker, C.G.
(1969) Geometry of numbers, Bibl. Math. 8. Groningen: Wolters-Noordhoff et Amsterdam: North-Holland.
Levenšteĭn, V.I.
(1975) The maximal density of filling an n-dimensional Euclidean space with equal balls, Mat. Zametki **18**, 301–311 = Math. Notes **18**, 765–771.
(1979) Bounds for packing in n-dimensional Euclidean space (Russian), Dokl. Akad. Nauk SSSR **245**, 1299–1303 = Soviet Math. Dokl. **20**, 417–421.
Linhart, J.
(1973a) Über einige Vermutungen von L. Fejes Tóth, Acta Math. Acad. Sci. Hungar. **24**, 199–201.
(1973b) Die Newtonsche Zahl von regelmäßigen Fünfecken, Periodica Math. Hungar. **4**, 315–328.
(1974) Endliche n-Nachbarnpackungen in der Ebene und auf der Kugel, Periodica Math. Hungar. **5**, 301–306.

(1975) Ein reguläres hyperbolisches Maximalmosaik, Österreich. Akad. Wiss. Math.-Natur. Kl. Sitzungsber. II, **184**, 473–485.
(1977a) Über die Kantenlängensumme von Dreieckspolyedern, Monatsh. Math. **83**, 25–36.
(1977b) Scheibenpackungen mit nach unten beschränkter Nachbarnzahl, Studia Sci. Math. Hungar. **12**, 281–293.
(1978) Closest packing and closest coverings by translates of a convex disc, Studia Sci. Math. Hungar. **13**, 157–162.
(1979) Kantenkrümmung und Umkugelradius konvexer Polyeder, Acta Math. Acad. Sci. Hungar. **34**, 1–2.
(1981) Die Beleuchtung von Kugeln, Geometriae Dedicata **10**, 145–154.
(1983) Eine Methode zur Berechnung der Dichte einer dichtesten gitterförmigen k-fachen Kreispackung, Arbeitsber. Math. Inst. Univ. Salzburg.

Lloyd, S.P.
(1980) Hamming association schemes and codes on spheres, SIAM J. Math. Anal. **11**, 488–505.

Makai, E. Jr.
(1972) On centrosymmetric convex domains with a packing density independent of the direction, Studia Sci. Math. Hungar. **7**, 423–424.
(1977) Packing problems in the Euclidean plane, Thesis, Budapest (Hungarian).
(1978) On the thinnest non-separable lattice of convex bodies, Studia Sci. Math. Hungar. **13**, 19–27.

Makai, E. Jr. and Pach, J.
(1983) Controlling of function classes and coverings of Euclidean space, Studia Sci. Math. Hungar. **18** (to appear).

Marley, G.C.
(1973) Multiple subdivisions of E^n, Rocky Mountain J. Math. **3**, 583–589.
(1974) Multiple subdivisions of the plane, Math. Magazine **47**, 202–206.

Matérn, B. and Person, O.
(1965) On the extremum properties of the equilateral triangular lattice and the regular hexagonal network, Förtechning Över, Rapporter ack Uppsatser Stockholm, Nr. 7.

Mc Elice, P.J., Rodemick, E.R., Rumsey, H. Jr. and Welch, L.R.
(1977) New upper bounds on the rate of a code via the Delsarte—MC Williams inequalities, IEEE Trans. Inform. Theory, **23**, 157–166.

Melzak, Z.A.
(1966) Infinite packings of disks, Canad. J. Math. **18**, 838–852.
(1969) On the solid packing constant for circles, Math. Comp. **23**, 169–172.

Mergelian, S.N.
(1952) Uniform approximations to functions of a complex variable, Uspehi Mat. Nauk **7**, 31–122.

Molnár, J.
(1975) Sur les empilements optimeaux des sphères dans une sphère de l'espace à courbure constant à n-dimensions, Ann. Univ. Sci. Budapest. Eötvös Sect. Math. **18**, 87–99.
(1977) On the ρ-system of unit circles, Ann. Univ. Sci. Budapest. Eötvös Sect. Math. **20**, 195–203.
(1978) Packing of congruent spheres in a strip, Acta Math. Acad. Sci. Hungar. **31**, 173–183.
(1979–80) On the packing of unit circles in a convex domain, Ann. Univ. Sci. Budapest, Eötvös Sect. Math. **22–23**. 113–123.

Neville, E.H.
(1915) On the solution of numerical functional equations, Proc. London Math. Soc. (2) **14**, 308–326.

Odlyzko, A.M. and Sloane, N.J.A.
(1979) New bounds on the number of unit spheres that can touch a unit sphere in n-dimensions, J. Combin. Theory Ser. A **26**, 210–214.

Österreicher, F. and Linhart, J.
(1981) Packungen kongruenter Stäbchen mit konstanter Nachbarnzahl, Elem. Math, **37**, 5–16.

Pach, J.
(1977) On the permeability problem, Studia Sci. Math. Hungar. **12**, 419–424.
(1980) Decomposition of multiple packing and covering Diskrete Geometrie, 2. Kolloq., Inst. Math. Univ. Salzburg 1980, 169–178.

Pirl, U.
(1967) Der Mindesabstand von n in Einheitskreisscheibe gelegenen Punkten, Math. Nachr. **40**, 111–124.

Purdy, G.B.
(1973) The lattice triple packing of spheres in Euclidean space, Trans. Amer. Math. Soc. **181**, 457–470.
Rankin, R.A.
(1947) On the closest packing of spheres in n-dimensions, Ann. Math. **48**, 1062–1081.
(1955) The closest packing of spherical caps in n-dimensions, Proc. Glasgow Math. Assoc. **2**, 139–144.
Robinson, R.M.
(1979) Multiple tilings of n-dimensional space by unit cubes, Math. Z. **166**, 225–264.
Rogers, C.A.
(1951) The closest packing of convex two-dimensional domains, Acta Math. **86**, 309–321.
(1957) A note on coverings, Mathematika **4**, 1–6.
(1958) The packing of equal spheres, Proc. London Math. Soc. (3) **8**, 447–456.
(1959) Lattice coverings of space, Mathematika **6**, 33–39.
(1963) Covering a sphere with spheres, Mathematika **10**, 157–164.
(1964) Packing and covering, Cambridge tracts **54**, Cambridge Univ. Press, Cambridge.
Rogers, C.A. and Shephard, G.C.
(1957) The difference body of a convex body, Arch. Math. **8**, 220–233.
Ruda, M.
(1969) Packing of squares in rectangles (Hungarian), MTA III. Oszt. Közleményei **19**, 73–87.
Ryškov, S.S.
(1974) Density of an (r,R)-system (Russian), Mat. Zametki **16**, 474–454 = Math. Notes **16**, 855–858.
(1977) The geometry of positive quadratic forms (Russian), Proc. Intern. Congress of Math. (Vancouver 1974) I. 501–506 = Amer. Math. Soc. Transl. II Ser. **109**, 27–32.
Riškov, S.S. and Bavanouskiĭ, E.P
(1975) Solution of the problem of the least dense lattice covering of five-dimensional space by equal spheres (Russian), Dokl. Akad. Nauk SSSR **222**, 39–42 = Soviet Math. Dokl. **6**, 586–590.
(1976) 29 c-types of n-dimensional primitive parallelohedra (with applications to the theory of coverings), Trudy Mat. Inst. Steklov **127**, (Russian).
Ryškov, S.S. and Horvath, J. (1975)
Estimation of the radius of a cylinder that can be imbedded in every lattice packing of n-dimensional unit balls (Russian), Mat. Zametki **17**, 123–128 = Math. Notes **17**, 72–75.
Saaty, T.L. and Alexander, J.M.
(1975) Optimization and the geometry of numbers: packing and covering, SIAM Review **17**, 475–519.
Schaer, J.
(1965) The densest packing of 9 circles in a square, Canad. Math. Bull. **8**, 273–277.
Schaer, J. and Meir, A.
(1965) On a geometric extremum problem, Canad. Math. Bull. **8**, 21–27.
Schlüter, K.
(1979) Kreispaçkungen in Quadraten, Elem. Math. **34**, 12–14.
Schmidt, W.M.
(1961) Zur Lagerung kongruenter Körper im Raum, Monatsh. Math. **65**, 154–158.
(1963) On the Minkowski-Hlawka theorem, Illinois J. Math. **7**, 18–23.
Schmitz, M. and Kirchner, K.
(1982) Eine Verteilung von 13 Punkten auf einem Quadrat, Wiss. Zeitschrift Pädagogischen Hochschule Erfurt-Mühlhausen 18, 113–115.
Schopp, J.
(1970) Über die Newtonsche Zahl einer Scheibe konstanter Breite, Studia Sci. Math. Hungar. **5**, 475–478.
Shannon, C.E.
(1959) Probability of error for optimal codes in a Gaussian channel, Bell System Tech. J. **38**, 611–656.
Sidel'nikov, V.M.
(1973) The densest packing of balls on the surface of the n-dimensional Euclidean sphere, and the number of vectors of a binary code with prescribed code distance (Russian), Dokl. Akad. Nauk SSSR 213, 1029–1032 = Soviet Math. Dokl. **14**, 1851–1855.
(1974) New estimates for the closest packing of spheres in n-dimensional Euclidean space (Russian), Mat. Sb. **95**, 148–158 = USSR Sb. **24**, 147–157.

Škljarskiĭ, D.O., Čencov, N.N. and Jaglom, I.M.
(1974) Geometrical estimates and problems from combinatorial geometry (Russian), Library of the Mathematica Circle, No. 13, Izdat. "Nauka", Moscow.
Sloane, N.J.A.
(1972) Sphere packings constructed from BCH and Justesen codes, Mathematika **19**, 183–190.
(1977) Binary codes, lattices, and sphere packings, Combinatorial Surveys, P.J. Cameron Ed., Academic Press, London—New York, 117–164.
(1978) Codes over GF(4) and complex lattices, J. Algebra **52**, 168–181.
(1979) Self-dual codes and lattices, Proceedings of Symposia in Pure Mathematics, Amer. Math. Soc. **34**, Relation between combinatorics and other parts of mathematics, 273–308.
(1981) Tables of sphere packings and spherical codes, IEEE Trans. Inform. Theory, IT27, 327–338.
(1982) Recent bounds for codes, sphere packings and related problems obtained by linear programming and other methods, Contemporary Mathematics **9**, 153–185.
Smith, M.J.
(1975) Packing translates of a compact set in Euclidean space, Bull. London Math. Soc. **7**, 129–131.
Subak, H.
(1960) Mehrfache gitterförmige Überdeckungen der Ebene durch Kreise, Dissetrtation, Wien.
Szegö, G.
(1975) Orthogonal polynomials, Amer. Math. Soc. Colloquium Publications 29, Amer. Math. Soc. Providence, Rhode Island, Fourth edition.
Székely, E.
(1974) Sur le problème de Tammes, Ann. Univ. Sci. Budapest. Eötvös Sect. Math. **17**, 157–175.
Swinnerton-Dyer, H.P.F.
(1953) Extremal lattices of convex bodies, Proc. Camb. Phil. Soc. **49**, 161–162.
Tammela, P.
(1970) An estimate of the critical determinant of a two-dimensional convex symmetric domain (Russian), Izv. Vysš. Učebn. Zaved. Mat. **12** (103), 103–107.
Tammes, R.M.L.
(1930) On the origine of number and arrangement of the places of exit on the surface of pollen grains, Rec. Trav. Bot. Neerl. **27**, 1–84.
Temesvári, A.H.
(1983a) Die dünnste gitterförmige 5-fache Kreisüberdeckung der Ebene, Studia Sci. Math. Hungar. **18** (to appear).
(1983b) Die dünnste doppelgitterförmige 2-fache Kreisüberdeckung in der Ebene, Studia Sci. Math. Hungar. **18** (to appear).
Vásárhelyi, E.
(1983) Über eine Überdeckung mit kongruenten Dreiecken, Beiträge zur Algebra und Geometrie (to appear).
Vermes, I.
(1979) Ausfüllungen der hyperbolischen Ebene durch kongruente Hyperzykelbereiche, Periodica Math. Hungar. **10**, 217–229.
Vetčinkin, N.M.
(1974) The packing of equal n-dimensional balls constructed from error-correcting codes (Russian), Studies in the geometry of positive quadratic forms, Ivanov Gos. Univ. Učen. Zap. **89**, 87–91.
(1980) Uniqueness of classes of positive quadratic forms, on which values of Hermite constants are reached for $6 \leq n \leq 8$ (Russian), Geometry of positive quadratic forms. Trudy Mat. Inst. Steklov **152**, 34–86.
Wegner, G.
(1971) Bewegungsstabile Packungen konstanter Nachbarnzahl, Studia Sci. Math. Hungar. **6**, 431–438.
(1980) Zu einem ebenen Überdeckungsproblem, Studia Sci. Math. Hungar. 15.
(1983) Über endliche Kreispackungen in der Ebene, Studia Sci. Math. Hungar. **18** (to appear).
Wengerodt, G.
(1983a) Die dichteste Packung von sechzehn Kreisen in einem Quadrat, Beiträge zur Algebra and Geometrie (to appear).
(1983b) Die dichteste Packung von 25 Kreisen in einem Quadrat, Ann. Univ. Sci. Budapest. Eötvös Sect. Math. (to appear).

Wesler, O.
(1960) An infinite packing theorem for spheres, Proc. Amer. Math. Soc. **11**, 324–326.
Wilker, J.B.
(1967) Open disk packings of a disk, Canad. Math. Bull. **10**, 395–415.
(1973) The interval of disk packing exponents, Proc. Amer. Math. Soc. **41**, 255–260.
(1977) Sizing up a solid packing, Period. Math. Hungar. **8**, 117–134.
Wills, J.M.
(1968) Ein Satz über konvexe Mengen und Gitterpunkte, Monatsh. Math. **72**, 451–463.
(1982) Research problem 30, Periodica Math. Hungar. **13**, 75–76.
(1983) Research problem 31, Periodica, Math. Hungar.
Wyner, A.D.
(1965) Capabilities of bounded discrepancy decoding, Bell System Tech. J. **44**, 1061–1122.
Yang, L.J.
(1980) Multiple lattice packings and coverings of spheres, Mh. Math. **89**, 69–76.
Zamorzaev, A.M.
(1965) On non-normal regular partitions of Euclidean space (Russian), Dokl. Akad. Nauk SSSR **161**, 30–32.

G. Fejes Tóth
Mathematical Institute
Hungarian Academy of Sciences
Realtanoda u. 13–15
P. Box 428
H–1395 Budapest (Hungary)

Wolfgang Weil

Stereology: A Survey for Geometers

1. Introduction

1A. *Foreword*

The determination of characteristic geometric properties of (usually 3-dimensional) objects by investigations of sections, projections, intersections with test sets, or other transformed images is a problem inherent to most of the experimental sciences. Surprisingly enough, formulas for such purposes have been established by people working in quite different fields of life, materials and earth sciences for over a hundred years until, some twenty years ago, the common theoretical (and in fact mathematical) background was realized. The new discipline dealing with problems of this type was called *stereology*. It was only then that mathematicians pointed out that most of the stereological results stem from classical formulas in integral geometry and that the validity of the formulas presupposes certain random structures either of the underlying sets or of the images taken of them. The development of stereological models therefore was greatly influenced by the growth of stochastic geometry. On the other hand, parts of stochastic geometry (random sets, point processes of convex bodies) have their roots in stereological questions. It seems, however, that stereology as an important field of applications of (convex) geometry is not as well-known to geometers as it should be. This was the motivation for the following survey which, although it is of an introductory nature, will present some of the recent developments with detailed, yet not exhaustive references. Since there are several geometric models which can serve as a basis of stereology, a lot of cross connections with other mathematical fields, and a variety of stereological problems of especial type, we had to restrict ourselves to a part of stereology which allows a unified treatment within a justifiable extent. The model, we present here, is based on convex geometry (methods from differential geometry and geometric measure theory are mentioned only occasionally) and the integral geometry of convex bodies (and unions thereof) is a crucial part of our considerations. The theory of random sets and point processes is investigated mainly in view of random versions of the kinematic integral formulas. We have, however, tried to be more complete with the references and give short comments to further developments and the corresponding literature at the end of each section.

Since the emphasis will be laid on the theoretical aspect, it is consistent to work in the setting of d-dimensional space. Moreover, we will use the term ‘stereology’ in a wider sense including not only methods and problems for sections (intersections by flats) but also for intersections by test sets and for projections.

The section headings are as follows.

1B. *Examples*

The earliest question of geometric type in probability theory, Buffon's famous needle problem (published in 1777), can be interpreted as the first stereological example since it can be used as a procedure to estimate the length of a needle or the distance of a grid. However, the first practical problem of stereological type was studied by Delesse in 1847:

(a) *Volume portion of minerals in rocks.* Let a piece of rock be given in which a certain amount of a mineral is embedded. In order to determine the relative proportion by volume, Delesse studied a planar polished section of the rock and used the relative areas of the two phases as an estimator. Refinements of this method were obtained by Rosiwal in 1898 and Glagoleff in 1933. Rosiwal used line segments and Glagoleff lattice points, both placed on the planar section. By measuring the lengths of intersections or by simply counting points, they describe procedures to 'determine' the volume portion.

(b) *Lung diffusion capacity.* Problems of measurement by sections for other parameters have not such a long history. In 1945 Tomkeieff considered the determination of surface area which is of great interest in biology and medicine. The gas exchange of the human (or animal) lung, for instance, depends on the alveolar surface of the lung, i.e. the area of the surface between the tissue (alveoli) and the air. To measure this area, only a part of lung tissue can be examined and usually only a thin slice can be observed under the microscope. Thus, in this case, a three-dimensional quantity (surface area) has to be determined, using measurements in a plane image obtained by sectioning with a thickened plane (cylinder) and projection of the intersection on to that plane.

More examples are found in the literature which will be referred to in more detail in 1C. From examples (a) and (b) the structure of stereological problems appears as follows (and this will also be the structure of this survey article). First, the class of sets which are of interest has to be fixed and the functionals which are to be measured must be defined. Then, a connection between functionals in sections and projections and functionals of the original set must be found. This is done by mean value formulas (integral geometry). Since, in practice, not the mean values of all sections (projections) are measured but only values of some sections (projections), the unknown parameter can only be estimated. To do this, a random structure, either of the sections (projections) or of the underlying set has to be assumed. Finally, since the results are estimations, a statistical analysis is necessary.

1C. *Literature*

A survey as our's which describes a certain part of stereology under a certain aspect cannot cover the whole range of sterological literature nor can it touch all the fields which are related to stereology. Therefore, the following bibliographical remarks are more concerned with stuff which is not treated in the sequel. Even for those parts of the theory which are presented in the following, the references are not complete. For instance, we have not quoted a lot of material in integral geometry since there is an extensive bibliography in the book of Santaló [1976a].

There is no mathematical textbook on stereology. The books of Saltykov [1974], Underwood [1970], and Weibel [1980] are written for practical purposes and do not give a rigorous mathematical treatment. Mathematical surveys on stereological problems from a statistical point of view are given by Clarke [1982], Coleman [1979], Davy [1978a], Moran [1972], and Stoyan [1979a], although with quite different particularity and aims. The most extensive one is the work of Davy [1978a]. Books on integral geometry, geometric probability, and random set theory containing stereological applications are those of Kendall-Moran [1963], Matheron [1967], [1975], and Santaló [1976a]. An introduction to stochastic geometry is announced by Stoyan-Mecke [1982]. Hints to stereological literature are also contained in the survey articles on research in geometrical probability by Baddeley [1977a], [1982b], Little [1974], and Moran [1966], [1969]. The following collections of mathematical articles are also related to stereology: Ambartzumian [1980], Harding-Kendall [1974], Miles-Serra [1978], Nicholson [1972]. Shorter surveys on special topics which touch stereology are written by Ambartzumian [1975], [1978], Cowan [1978], Cruz-Orive-Weibel [1981], Davy [1978b], König-Stoyan [1980], Mecke [1982], Miles [1981b], Monari [1978], Nicholson [1978], Ripley [1976b], Serra [1972], [1978], [1981], Stoyan [1982b], Stoyan-Koschitzki [1980], Streit [1976], [1980], and Watson [1978].

As we have pointed out we will not describe the practical aspects of stereology. These, together with numerous examples, methods and generalizations can be found in the books of Saltykov [1974], Underwood [1970], and Weibel [1980] mentioned above and in the following collections, reports of symposia and the proceedings of the international congresses of the ISS (International Society for Stereology): Adam et al. [1980], Chermant [1978], De Hoff-Rhines [1968], Elias [1967], Exner [1975], ISS [1963], Weibel-Elias [1967], Weibel et al. [1972]. Further papers on stereology are published frequently in the journals 'Acta Stereologica' and 'Journal of Miscroscopy'.

Next, we give some comments on related topics which are not, or only occasionally, mentioned further on.

Combinatorial integral geometry is concerned with solutions and generalizations of the classical Buffon-Sylvester problem. For a survey and interrelations with stereology see Ambartzumian [1977], [1982] and Baddeley [1981], [1982a].

The projection methods which we describe in this article are based on the geometric properties of the projection of a set. Analytical versions of projections of functions which are of great importance in medicine (*X-ray methods*) do not fit in this context. They involve Radon transforms and belong to the analytic branch of integral geometry. For surveys, see Shepp-Kruskal [1978], Smith et al. [1977], and Zalcman [1980].

Of course, X-ray problems as well as stereological questions can be viewed as parts of the general concept of reconstructing shapes (of sets or graphs of functions) from transformed images. A corresponding theory (*pattern synthesis*) is presented in Grenander [1976]. The methods, however, are quite different and are based on assumptions (of the random structure of objects) which are in most cases not relevant for random set theory and stereology.

The main part of this article will deal with the presentation of models for stereological problems which are based on convex geometry. We said already that there exist (partially similar) models using methods from differential geometry or geometric measure theory. It is an important question for applications which of the models fits best for the practical problem. For example, the value of the inner surface area of the lung (example (b) in section 1B.) will depend on the magnification used for the measurements. This is of course due to the systematic errors inherent to image performing and analysing equipments but it may also support the thesis that the alveolar surface does not have a specific (finite) area at all and should better be described by *fractals* (sets of fractal dimensions). We will not discuss this further and refer the interested reader to the book of Mandelbrot [1977].

Modern digital image analysers allow not only the automatic measurement of geometrical quantities but also the adaptation of models to the practical problem. For that purpose, set transformations are implemented (outer and inner parallel sets as well as combinations and generalizations). The corresponding theory (*mathematical morphology*) was developed for the construction of an image analyser and is strongly connected with random set theory; see Haas et al. [1967], Matheron [1967], [1975], Serra [1969] and especially Serra [1982] for further references and interrelations with image analysis.

The theory of random sets and point processes of convex bodies, as it will be presented in this article, contains as special cases the patterns built by (processes of) points in d-dimensional space. While obviously the geometrical theory of shape and form then collapses, other problems occur and different methods have to be used. For references, see the articles in Lewis [1972] and the books on statistical analysis of *spatial patterns* by Bartlett [1975] and Ripley [1981]. The latter also contains a section on stereology. Ordinary point patterns can be viewed also as special cases of patterns built by (point processes of) flats. Although *flat networks* are of an immense stereological significance, either by themselves or by the polytopes built from them, the results are too numerous and would require a separate survey article. Here we refer to the books of Matheron [1975], Ripley [1981], Santaló [1976] and remark that many of the results are due to Miles [1964], [1969], [1970a], [1970b], [1971a], [1971b], [1973b].

Finally we mention shortly the problem of overlap of particles which frequently arises in stereology. Here, a statistical theory has been developed for the line and the plane which is presented in Roach [1968]. For more recent results on problems of *clumping*, see Hadwiger-Streit [1970], Pitts [1981], Rehder [1980], Scheaffer [1973], and the literature cited there.

It is clear that these topics are not represented adequately in the literature list. On the other hand, some references (especially to recent results) are included, although they do not fit into the setup of the paper and are therefore not mentioned in the text.

1D. *Notations and definitions*

The setting of this article is the d-dimensional Euclidean space $\mathbb{R}^d$ with scalar product $\langle\cdot,\cdot\rangle$, norm $\|\cdot\|$, and Lebesgue measure $\lambda = \lambda_d$. S^{d-1} denotes the unit sphere and ω the spherical Lebesgue measure. We use $\varkappa_d$ for the volume of the d-dimensional unit ball B. Thus, we have $\varkappa_d = \lambda(B) = d^{-1} \cdot \omega(S^{d-1})$.

In $\mathbb{R}^d$, the following classes of sets are used throughout: $\mathscr{C}^d$, the class of compact sets; $\mathscr{K}^d$, the class of convex bodies (convex sets in $\mathscr{C}^d$); $\mathscr{R}^d$, the convex ring (finite unions of convex bodies); $\mathscr{S}^d$, the class of sets $A \subset \mathbb{R}^d$ such that $A \cap K \in \mathscr{R}^d$ for all $K \in \mathscr{K}^d$. We equip $\mathscr{C}^d \setminus \{\emptyset\}$ and the subclasses $\mathscr{R}^d \setminus \{\emptyset\}$ and

$\mathscr{K}^d \setminus \{\emptyset\}$ with the Hausdorff metric d. By a lower index k we denote the subclass of those sets having dimension k.

For the integral geometric formulas, we need the following measure spaces: the rotation group SO_d with invariant measure ν, the group of rigid motions G_d with invariant measure μ, and the homogeneous space $\mathscr{E}^d_q$ of q-flats (affine subspaces) with invariant measure μ_q, $q = 0,\ldots,d-1$. Moreover, let $\mathscr{L}^d_q$ denote the set of q-dimensional (linear) subspaces of $\mathbb{R}^d$. All these sets are equipped with their natural topologies, the measures are defined on the corresponding Borel sets, and the normalizations are as follows. We put $\nu(SO_d) = 1$ and

$$\mu = \gamma \circ (\lambda \otimes \nu)$$

where $\gamma: \mathbb{R}^d \times SO_d \to G_d$ is the map defined by $\gamma(t,\delta)x = \delta x + t$, $x \in \mathbb{R}^d$, $\otimes$ denotes the product measure and $\circ$ the image measure. In the same way, let

$$\mu_q = \gamma_q \circ (\lambda_{d-q} \otimes \nu)$$

where λ_{d-q} is the Lebesgue measure on the orthogonal complement $L^\perp$ of a fixed subspace $L \in \mathscr{L}^d_q$, and $\gamma_q: L^\perp \times SO_d \to \mathscr{E}^d_q$ is given by $\gamma_q(t,\delta) = \delta(L + t)$. If necessary, we generalize these concepts slightly by assuming the measures to be complete.

For a set $A \subset \mathbb{R}^d$ and a subspace $L \in \mathscr{L}^d_q$ let $A|L$ denote the orthogonal projection of A on to L. Further, let $p(A,\cdot)$ be the nearest point map (metric projection), i.e. for $x \in \mathbb{R}^d$ let $p(A,x)$ be the point $y \in A$ with $\|x - z\| > \|x - y\|$ for all $z \in A$, $z \neq y$, provided such a (unique) point y exists. Note that for $K \in \mathscr{K}^d$, $p(K,\cdot)$ is defined on all of $\mathbb{R}^d$ and is continuous.

2. Basic model assumptions

2A. *The convex ring*

It is clear that different stereological problems may lead to models of different generality. For instance, volume measurements require only measurability conditions, area measurements are possible for sets with piecewise smooth boundary, etc. We will, however, lay stress on a universal model for all, or nearly all, stereological problems to be considered. Such a model should be built up by sets simple enough to allow a rich variety of global geometric parameters like volume and surface area but also local notions like curvatures. On the other hand, the model should be rich enough to include the sets occurring in practice, at least after some simplifications of the latter. Two such models (for compact sets) have been presented in the literature, the convex ring $\mathscr{R}^d$ (Hadwiger [1956], [1957]) and the sets of positive reach (Federer [1959]). While the convex ring is based on convex geometry, the theory of sets of positive reach may be viewed as a generalization of classical differential geometry. It should be mentioned, however, that the concept of sets of positive reach includes both, the class $\mathscr{K}^d$ of convex bodies and the sets in $\mathbb{R}^d$ with smooth boundary, whereas the convex ring does not contain sets with smooth boundaries others than convex ones or their disjoint unions. More precisely the opposite role of the two concepts is made evident by the following result which, although it has a simple proof, is not found in the literature.

(2.1) **Theorem.** *A set* $K \in \mathscr{R}^d$ *is of positive reach if and only if* K *is the disjoint union of convex bodies* $K_1, \ldots, K_n$.

The definition of fundamental, global and local geometric functionals for sets of positive reach is given in Federer [1959], for the convex ring we will do this in some detail in the following. Both models are 'rich enough', since the finite point sets which are obviously both, of positive reach and elements of $\mathscr{R}^d$, are dense in $\mathscr{C}^d$. This simple fact strikingly exhibits a difficulty with which we shall be faced in the sequel. Any suitable non-trivial functional on $\mathscr{R}^d$ will be discontinuous. Extensions of functionals defined on $\mathscr{R}^d$ to larger classes by continuity properties are therefore not possible. Moreover, a passage from sets in $\mathscr{R}^d$ to sets of positive reach and a comparison of the corresponding functionals is not practicable in the general setting of Hausdorff metric. The lacking notion is that of an appropriate convergence of sets in $\mathscr{R}^d$ to sets of positive reach and vice versa. For the applications, this means that all geometrical parameters which are measured in practice will behave quite sensitive to slight transformations of the material or its images.

We would like to point out that, as a common generalization of the sets of positive reach and the convex ring $\mathscr{R}^d$, the class of Hausdorff rectifiable sets can be considered (see Federer [1969]). But, although there are some integral geometric formulas for Hausdorff measures, this class does not, at the moment, admit the range of functionals and integral formulas which we have for the convex ring. A 'minimal' common generalization is obviously given by the class of sets which are finite unions of sets of positive reach. Extensions of fundamental functionals and of kinematic integral formulas to this class are announced in Zähle [1982a].

From the remarks above it is clear that, for functionals T on $\mathscr{R}^d$, we can only claim continuity on $\mathscr{K}^d$. As further conditions on $\mathscr{K}^d$ we will pose compatibility with motions and additivity in the following elementary sense:

$$T(K \cup K') + T(K \cap K') = T(K) + T(K')$$

for $K, K' \in \mathscr{K}^d$ with $K \cup K' \in \mathscr{K}^d$. These restrictions can be motivated from stereological practice, and this is done by several authors. From the theoretical point of view, their importance lies in the fact that they guarantee a sufficiently rich integral geometric theory.

We will now study three types of functionals with these properties on $\mathscr{K}^d$, global, local and vector functionals. We give characterization theorems and extensions to the convex ring $\mathscr{R}^d$. It should be remarked that (under the conditions mentioned above) additive extensions to $\mathscr{R}^d$ are always possible by a general theorem of Groemer [1978]. Further, since additive functionals are valuations, the following considerations partially fit into the general theory of valuations, for which we refer to the survey article of McMullen-Schneider [1983].

2B. *Intrinsic volumes*

A scalar (or global) functional T on $\mathscr{K}^d$ is a map $T: \mathscr{K}^d \to \mathbb{R}$. As we have announced, we impose on T the following conditions:

(i) $T(gK) = T(K)$ for all $g \in G_d$ and all $K \in \mathcal{K}^d$ (motion invariance),
(ii) T is continuous on $\mathcal{K}^d$,
(iii) T is additive on $\mathcal{K}^d$.

Familiar functionals of this type are the quermassintegrals $W_i, i = 0,\ldots,d$. Instead of the W_i, we will use the *intrinsic volumes* $V_j, j = 0,\ldots,d$, in the following. These functionals, introduced by McMullen [1975], are connected with the classical quermassintegrals by the following equation

$$\varkappa_{d-j}V_j(K) = \binom{d}{j} W_{d-j}(K), \quad j = 0,\ldots,d, \quad K \in \mathcal{K}^d.$$

As a definition, we can use the Steiner formula which for the intrinsic volumes has the following form

(2.2) $$\lambda_d(K + \alpha B) = \sum_{j=0}^{d} \alpha^{d-j}\varkappa_{d-j}V_j(K), \quad \alpha > 0, \quad K \in \mathcal{K}^d.$$

Note that $V_d = \lambda_d$ is the volume, V_{d-1} is half the surface area,..., V_1 is $d\varkappa_d/2\varkappa_{d-1}$ times the mean width, and $V_0 \equiv 1$ on $\mathcal{K}^d \setminus \{\emptyset\}$

The central role of the intrinsic volumes in convex geometry is emphasized by the following characterization theorem of Hadwiger [1951], [1952], [1957].

(2.3) **Theorem.** *Let* $T: \mathcal{K}^d \to \mathbb{R}$ *be motion invariant, continuous, and additive. Then, there are numbers* $\beta_0,\ldots,\beta_d \in \mathbb{R}$ *with*

$$T = \sum_{i=0}^{d} \beta_i V_i.$$

As further properties of the intrinsic volumes we mention positivity, monotony, the representation as integrals of curvature functions in the smooth case, and the existence of kinematic integral formulas, especially the representation as projection integrals (see next section). The latter is very often used as definition by induction on the dimension and is obviously responsible for the name 'quermassintegral.' We remarked already that extensions of the intrinsic volumes to $\mathcal{R}^d$ will necessarily loose some of their properties.

An additive extension was studied by Hadwiger [1956], [1957]. It can be defined in the obvious way by

(2.4) $$V_j(K) = \sum_{v \in S(n)} (-1)^{|v|-1} V_j(K_v)$$

for $K = \bigcup_{i=1}^{n} K_i \in \mathcal{R}^d, K_i \in \mathcal{K}^d, j \in \{0,\ldots,d\}$.

Here we have used the following abbreviations: S(n) denotes the family of all nonvoid subsets of $\{1,\ldots,n\}$, $|v|$ is the cardinality of v, and K_v equals $K_{i_1} \cap \cdots \cap K_{i_k}$ if $v = \{i_1,\ldots,i_k\}$. If we extend the characteristic function of parallel bodies

$$c_\alpha(K,x) = \begin{cases} 1 & \text{if } d(K,\{x\}) \leq \alpha \\ 0 & \text{otherwise} \end{cases}$$

$K \in \mathscr{K}^d$, $\alpha > 0$, $x \in \mathbb{R}^d$, additively to $\mathscr{R}^d$

$$c_\alpha(K,\cdot) = \sum_{v \in S(n)} (-1)^{|v|-1} c_\alpha(K_v,\cdot),$$

for $K = \bigcup_{i=1}^n K_i$, $K_i \in \mathscr{K}^d$, then the following extension of the Steiner formula (2.2) holds (see Matheron [1975])

$$\int_{\mathbb{R}^d} c_\alpha(K,x)\, d\lambda(x) = \sum_{j=0}^{d} \alpha^{d-j} \varkappa_{d-j} V_j(K), \tag{2.5}$$

$\alpha > 0$, $K \in \mathscr{R}^d$. The additive extension V_j still represents important geometric functionals. On $\mathscr{R}^d$, V_d is the volume, V_{d-1} is half the surface area, and V_0 is the Euler-Poincaré characteristic. Of course, for $j < d-2$, V_j is no longer positive.

A positive extension was first considered by Matheron [1975] (see also Davy [1976b]) and established in full detail by Schneider [1980a]. It is based on the following extension of $c_\alpha(K,\cdot)$ to $\mathscr{R}^d$. For $x \in \mathbb{R}^d$ and $K \in \mathscr{R}^d$ call a point $y \in K$ a projection of x on to K if there is a neighbourhood U of y such that $\|x - z\| > \|x - y\|$ for all $z \in U \cap K$, $z \neq y$. For $\alpha > 0$, let $\bar{c}_\alpha(K,x)$ be the number of projections y of x on to K with $\|y - x\| \leq \alpha$. This number is finite and we have $\bar{c}_\alpha(K,\cdot) = c_\alpha(K,\cdot)$ for $K \in \mathscr{K}^d$. For $K \in \mathscr{R}^d$ the following Steiner-type formula holds

$$\int_{\mathbb{R}^d} \bar{c}_\alpha(K,x)\, d\lambda(x) = \sum_{j=0}^{d} \alpha^{d-j} \varkappa_{d-j} \bar{V}_j(K), \tag{2.6}$$

$\alpha > 0$, $K \in \mathscr{R}^d$, by which positive functionals $\bar{V}_j$ are defined.

A different positive extension mentioned by several authors can be easily constructed using the interpretation of $V_i(K)$ as quermassintegral (see section 3). Let

$$\bar{\bar{V}}_j(A) = c_j \int_{SO_d} \lambda_j(A|\delta L)\, d\nu(\delta), \quad L \in \mathscr{L}_j^d,$$

with

$$c_j = \frac{\binom{d}{j}\varkappa_d}{\varkappa_j \varkappa_{d-j}}, \quad j = 0,\ldots,d.$$

These functionals $\bar{\bar{V}}_0,\ldots,\bar{\bar{V}}_d$ are defined even for more general sets A than those of the convex ring (see e.g. Ohmann [1952], [1954], [1955], [1956]).

2C. *Curvature measures*

By a local functional T on $\mathscr{K}^d$ we mean a map T which to each $K \in \mathscr{K}^d$ attaches a finite (nonnegative) Borel measure $T(K,\cdot)$ on $\mathbb{R}^d$. Since $T(K,\cdot)$ should describe the local behaviour of K, we state as a first condition:

(i) If $\beta \subset \mathbb{R}^d$ is open and if $K_1, K_2 \in \mathscr{K}^d$ are convex bodies with $\beta \cap K_1 = \beta \cap K_2$, then $T(K_1,\beta') = T(K_2,\beta')$ for all Borel sets $\beta' \subset \beta$.

As in the case of scalar functionals we demand moreover:

(ii) $T(gK,g\beta) = T(K,\beta)$ for all $g \in G_d$, all $K \in \mathscr{K}^d$, and all Borel sets $\beta \subset \mathbb{R}^d$ (motion invariance),
(iii) T is continuous on $\mathscr{K}^d$ (w.r.t. weak convergence of measures),
(iv) T is additive on $\mathscr{K}^d$.

Examples of local functionals are the *curvature measures* $\psi_0(K,\cdot),\ldots,\psi_d(K,\cdot)$ of Federer [1959], investigated further by Matheron [1975] and Schneider [1978a] (see Schneider [1979a] for a survey and references). They can be introduced by a local version of the Steiner formula (2.2). For $K \in \mathscr{K}^d$, $\alpha > 0$, and a Borel set $\beta \subset \mathbb{R}^d$ let

$$A_\alpha(K,\beta) = \{x \in \mathbb{R}^d : 0 \leq \|x - p(K,x)\| \leq \alpha, \quad p(K,x) \in \beta\}.$$

Then we have

$$(2.7) \quad \lambda(A_\alpha(K,\beta)) = \sum_{j=0}^{d} \alpha^{d-j}\varkappa_{d-j}\psi_j(K,\beta).$$

The curvature measures are local analogues of the intrinsic volumes since

$$V_j(K) = \psi_j(K,\mathbb{R}^d), \quad j = 0,\ldots,d.$$

$\psi_d(K,\cdot)$ is just the Lebesgue measure λ restricted to K whereas $\psi_0(K,\cdot),\ldots,\psi_{d-1}(K,\cdot)$ are supported by bdK. $\psi_{d-1}(K,\cdot)$ is half the $(d-1)$-dimensional Hausdorff measure on bdK and $\psi_0(K,\beta)$ is the ω-measure of the spherical image of $\beta \subset$ bdK. In the smooth case $\psi_0(K,\cdot),\ldots, \psi_{d-1}(K,\cdot)$ are integrals of functions of the principal curvatures (see Schneider [1979a] for details).

The following characterization theorem of Schneider [1978a] is the analogue of (2.3) for curvature measures.

(2.8) **Theorem.** *Let* T *be a local functional on* $\mathscr{K}^d$, *attaching to each* $K \in \mathscr{K}^d$ *a finite Borel measure* $T(K,\cdot)$ *on* $\mathbb{R}^d$ *which is locally defined in the sense of* (i). *Assume moreover that* T *is motion invariant, continuous, and additive. Then, there are numbers* $\beta_0,\ldots,\beta_d \in \mathbb{R}$ *with*

$$T(K,\cdot) = \sum_{i=0}^{d} \beta_i\psi_i(K,\cdot), \quad K \in \mathscr{K}^d.$$

Extensions of curvature measures to the convex ring $\mathscr{R}^d$ have been studied by Schneider [1977b], [1980a]. They rely on the obvious local versions of the functions c_α and $\bar{c}_\alpha$. For $K \in \mathscr{K}^d$, $\alpha > 0$, $\beta \subset \mathbb{R}^d$ Borel, and $x \in \mathbb{R}^d$, let

$$c_\alpha(K,\beta,x) = \begin{cases} 1 & \text{if} \quad 0 \leq d(K,\{x\}) \leq \alpha \quad \text{and} \quad p(K,x) \in \beta \\ 0 & \text{otherwise} \end{cases}$$

For $K = \bigcup_{i=1}^{n} K_i$ define

$$c_\alpha(K,\beta,\cdot) = \sum_{v \in S(n)} (-1)^{|v|-1} c_\alpha(K_v,\beta,\cdot)$$

and let $\bar{c}_\alpha(K,\beta,x)$ be the number of projections $y \in K$ with $\|y - x\| \leq \alpha$ and $y \in \beta$.

Then the following local versions of (2.4) and (2.5) define additive extensions $\psi_0(K,\cdot),\ldots,\psi_d(K,\cdot)$ resp. positive extensions $\bar{\psi}_0(K,\cdot),\ldots,\bar{\psi}_d(K,\cdot)$:

$$(2.9)\quad \int_{\mathbb{R}^d} c_\alpha(K,\beta,x)\,d\lambda(x) = \sum_{j=0}^{d} \alpha^{d-j}\varkappa_{d-j}\psi_j(K,\beta),$$

$$(2.10)\quad \int_{\mathbb{R}^d} \bar{c}_\alpha(K,\beta,x)\,d\lambda(x) = \sum_{j=0}^{d} \alpha^{d-j}\varkappa_{d-j}\bar{\psi}_j(K,\beta),$$

$K \in \mathscr{R}^d$, $\alpha > 0$, $\beta \subset \mathbb{R}^d$ a Borel set. The method for the positive extension stems from Matheron [1975]. In the following, we will call $\psi_i(K,\cdot)$ the *signed curvature measures* and $\bar{\psi}_i(K,\cdot)$ the *absolute curvature measures* of K.

2D. *Curvature centroids*

The last group of functionals we want to study are the vector functionals T, i.e. the maps $T:\mathscr{K}^d \to \mathbb{R}^d$. We look for functionals which (when extended to the convex ring) describe the location as well as the dispersion of a set. In order to get a vector space of such functionals, we interpret compatibility with motions not by the motion equivariance

$$T(gK) = gT(K), \quad g \in G_d,$$

but in the following way:

(i) $T(\delta K) = \delta T(K)$ for all $\delta \in SO_d$,
and $T(K + x) - T(K)$ is parallel to
x for all $x \in \mathbb{R}^d$, $K \in \mathscr{K}^d$.

The further conditions are the same as in the previous sections:

(ii) T is continuous on $\mathscr{K}^d$,
(iii) T is additive on $\mathscr{K}^d$.

An obvious example of a set of vector functionals on $\mathscr{K}^d$ is given by the *curvature centroids* $s_0(K),\ldots,s_d(K)$. Here, $s_i(K)$ is defined by

$$s_i(K) = \frac{1}{V_i(K)} \int_{\mathbb{R}^d} x\psi_i(K,dx), \quad i = 0,\ldots,d.$$

Of course, in order that $s_i(K)$ be defined, $V_i(K)$ must not vanish. $s_d(K)$ is the ordinary centroid of K, and $s_0(K)$ is the well-known Steiner point. In view of the weighting factor $V_i(K)$, the curvature centroids do not satisfy the conditions we have imposed on a vector functional T, in particular, they are not additive. Hence, let us look at the ordinary *moment vectors*

$$r_i(K) = V_i(K)s_i(K) = \int_{\mathbb{R}^d} x\psi_i(K,dx), \quad i = 0,\ldots,d.$$

The r_i are defined for all $K \in \mathscr{K}^d$, they obey (i), and are continuous and additive. They, as well as the curvature centroids, were introduced and studied independently by Schneider [1972a], [1972b] and Hadwiger (see Hadwiger-Schneider [1971]). Notice that we have changed their notations slightly in order to make them fit into the concept of curvature measures and intrinsic volumes. For the r_i, the following Steiner formula holds:

(2.11) $\int_{K+\alpha B} x \, d\lambda(x) = \sum_{j=0}^{d} \alpha^{d-j}\varkappa_{d-j} r_j(x),$

$\alpha > 0, K \in \mathscr{K}^d.$

Again, there is a characterization theorem (Hadwiger-Schneider [1971], Schneider [1972b]).

(2.12) **Theorem.** *Let* $T: \mathscr{K}^d \to \mathbb{R}^d$ *be a vector functional which is continuous, additive, and compatible with motions in the sense of* (i). *Then, there are numbers* $\beta_0, \ldots, \beta_d \in \mathbb{R}$ *with*

$$T(K) = \sum_{i=0}^{d} \beta_i r_i(K), \quad K \in \mathscr{K}^d,$$

or (if K *is* d*-dimensional)*

$$T(K) = \sum_{i=0}^{d} \beta_i V_i(K) s_i(K), \quad K \in \mathscr{K}^d_d.$$

Obviously, there are extensions r_i and $\bar{r}_i$ of the moment vectors to the convex ring. These can be defined by means of the signed and absolute curvature measures. For the additive extension also the direct method works (see Hadwiger-Schneider [1971], Schneider [1972b]). For sets $K \in \mathscr{R}^d$ with $V_i(K) \neq 0$ resp. $\bar{V}_i(K) \neq 0$, curvature centroids then can be defined by

$$s_i(K) = \frac{r_i(K)}{V_i(K)} \text{ resp. } \bar{s}_i(K) = \frac{\bar{r}_i(K)}{\bar{V}_i(K)}.$$

$s_d(K) = \bar{s}_d(K)$ is the (ordinary) centroid of K.

Although further results can be obtained (e.g. Steiner formulas) we will not follow this line since the situation is quite similar to that of the intrinsic volumes and curvature measures.

2E. *Cylinders*

For a $q \in \{0, \ldots, d-1\}$ let $L \in \mathscr{L}^d_q$ be fixed and let $K \subset L^\perp$ be a compact set. We call $K + L$ a cylinder with basis K and consider the set $\mathscr{Z}(q,K)$ of all cylinders (properly) congruent to $K + L$. In the obvious way, $\mathscr{Z}(q,K)$ can be topologized and becomes a homogeneous space (w.r.t the motion group). We define the invariant measure $\mu_{q,K}$ on $\mathscr{Z}(q,K)$ by

$$\mu_{q,K} = \gamma_{q,K} \circ (\lambda_{d-q} \otimes \nu),$$

where λ_{d-q} is the Lebesgue measure on $L^\perp$ and $\gamma_{q,K}: L^\perp \times SO_d \to \mathscr{Z}(q,K)$ is defined by $\gamma_{d,K}(t,\zeta) = \zeta(K + L + t)$.

2F. *Fields of bodies*

By definition, a set $A \in \mathscr{S}^d$ is closed and has the property that $A \cap K \in \mathscr{R}^d$ for all $K \in \mathscr{K}^d$. Obviously this implies that $A = \bigcup_{i=1}^{\infty} K_i$ for a sequence of convex bodies $K_i \in \mathscr{K}^d$. The sequence $\{K_i\}$ is sometimes called a *field* of convex bodies. For a field $\{K_i\}$ the density D_j of the intrinsic volume V_j can be defined by

$$D_j = \lim_{\alpha \to \infty} \frac{1}{V_d(\alpha B)} \sum_{K_i \subset \alpha B} V_j(K_i),$$

provided the limit exists. In the same way, mean values M_j can be introduced by

$$M_j = \lim_{\alpha \to \infty} \frac{1}{N_\alpha} \sum_{K_i \subset \alpha B} V_j(K_i)$$

where N_α is the number of bodies of the field contained in αB. As a next step, one can look for existence results for D_j and M_j, interrelations between D_j and M_j, and study problems of integral geometric or probabilistic type for fields of bodies. Such investigations were started by Fejes Toth-Hadwiger [1947], [1948] for d = 2 and for general d by Giger-Hadwiger [1968], Hadwiger [1974], and Santaló [1970b].

For a while, fields of bodies seemed to be an important tool also in stereology and formulas of stereological type were investigated (Giger [1967a], [1967b], [1970], [1972b]). The development of general point process theory, however, gave a much more powerful tool for the theoretical investigation of such fields by viewing them as realizations of a (Poisson) point process on $\mathscr{K}^d$ (see section 6). For practical purposes, it seems to be equally realistic or unrealistic to assume that a field $\{K_i\}$ is the realization of a (Poisson) point process (with certain additional properties) or to assume the existence of densities D_j or mean values M_j.

The investigations described above therefore seem to be of no great importance in modern stereology and we have mentioned them for historical reasons only.

2G. *Notes and comments*

(a) The curvature measures which Federer [1959] introduced for sets of positive reach are signed measures which are additive. Hence they may be viewed as analogues of the additive extension $\psi_j(K,\cdot)$, $K \in \mathscr{R}^d$. Similarly, for compact sets of positive reach, the total curvature measures correspond to the functionals V_j on $\mathscr{R}^d$. By the results announced in Zähle [1982a] both concepts are unified. However, it is an open question whether continuity results hold, for instance, when sets $K \in \mathscr{R}^d$ are approximated from outside by sets of positive reach in a suitable way (a partial result was obtained by Brehm-Kühnel [1982]). Analogues of $\bar{V}_j$ and $\bar{\psi}_j$ have also been studied for smooth sets or sets of positive reach. Such (integrals of) absolute curvatures are due to Baddeley [1980a], Federer [1959], Santalo [1970]. Baddeley [1980a] even generalized the integrals of absolute curvatures to certain Hausdorff rectifiable sets, but it seems that his notion does not coincide on $\mathscr{R}^d$ with $\bar{V}_j$.

(b) As we have already indicated, the convex ring is not the most general concept in which intrinsic volumes V_j (quermassintegrals) can be considered, but perhaps the simplest one. Hadwiger [1959] introduced the class of normal bodies (which contains $\mathscr{R}^d$) and extends the functionals $V_0,\ldots,V_d$ to normal bodies K. For V_0, the characteristic, this is done directly, for $V_1,\ldots,V_d$ by the Crofton formulas (see section 3). The definition of normality is however more intricate and the functionals are defined just by those integral formulas which are important for stereological applications. An external geometric characterization of the V_j, e.g. via a Steiner formula, is not known in this general setting.

(c) For a convex body K one can easily find a sequence $K_k \in \mathscr{R}^d$ with $K_k \to K$ in the Hausdorff metric and such that $\psi_j(K_k,\cdot)$ or $\bar{\psi}_j(K_k,\cdot)$ converges weakly for $j = 0,\dots,d-1$, but with $\lim \psi_j(K_k,\cdot) \neq \psi_j(K,\cdot)$ resp. $\lim \bar{\psi}_j(K_k,\cdot) \neq \bar{\psi}_j(K,\cdot)$ for at least one j. This shows that an extension by continuity arguments on larger classes of sets is not possible, although this is sometimes suggested in the literature (see e.g. Serra [1982]).

(d) On the screen of an image analyser sets are represented by a collection of lattice points. For the measurement of functionals it would therefore be necessary first to define them appropriately for point rasters and then to think of approximation properties. Although this is an important question for applications, we will not discuss it further and refer to Bernroider [1978], Giger [1972a], [1975], and especially Serra [1982].

(e) There is another set of measures which are well-known as local analogues of the intrinsic volumes and used for a long time in the theory of convex bodies. These are the surface area measures of Aleksandrov-Fenchel-Jessen (see Schneider [1979a] for a survey) and they share a lot of properties with the curvature measures. Especially, they fulfill a local Steiner formula, there is a characterization theorem (Schneider [1975a]), and extensions to $\mathscr{R}^d$ are possible (Schneider [1980a]). However, there is no Crofton-type integral formula for them and this fact makes them less interesting for stereological applications.

(f) The characterization theorems (2.3), (2.8), (2.12) are an important tool to prove integral geometric formulas. In this connection, a general theorem of Schneider [1978a] is of interest which characterizes external angles of polytopes and which implies many of the formulas given in section 3 (see Schneider [1979b], [1980b]). Hadwiger [1975b] considered certain general functionals of convex bodies and proved a characterization theorem for them from which he derived some integral formulas (see 3G).

3. Integral geometric formulas

3A. *Introduction*

In the following we will use the setup of measure theory to present a variety of integral geometric formulas. This is not self-evident since frequently the literature concerned with integral geometry, geometric probability, stochastic geometry, and stereology is based on differential forms or on the somewhat vague concept of integral geometric (or kinematic) densities. Exceptions are Matheron [1975] and Schneider [1979b], the latter giving a thorough introduction of the invariant measures used on the spaces $G_d, S0_d, \mathscr{E}^d_q, \mathscr{L}^d_q$.

Besides these measure spaces, we use in 3E spaces of cylinders (as they were introduced in 2E). For the translation formulas in 3F we identify the translation group with $\mathbb{R}^d$.

The formulas which are directly relevant for stereological purposes are presented in 3B, 3C, 3D and 3E. Most of them can be proved by the characterization theorems (2.3), (2.8), (2.12). 3G contains a collection of formulas for which not in all cases a stereological use is known. Here we have only surveyed recent results, for earlier contributions see Santaló [1976a]. Some of the formulas are connected with contact problems for convex bodies, see Weil [1979b] for a survey.

3B. *Principal kinematic formulas*

The following principal kinematic formula is classical for convex bodies (see e.g. Hadwiger [1957], Santaló [1976a]). The extension to $\mathscr{R}^d$ is immediate from (2.4).

(3.1) **Theorem.** *Let* $K, K' \in \mathscr{R}^d$ *and* $j \in \{0,\ldots,d\}$. *Then*

$$\int_{G_d} V_j(K \cap gK')\, d\mu(g) = \sum_{k=j}^{d} \alpha_{djk} V_k(K) V_{d+j-k}(K')$$

with

$$\alpha_{djk} = \frac{\binom{k}{j} \varkappa_k \varkappa_{d+j-k}}{\binom{d}{k-j} \varkappa_j \varkappa_d}.$$

(3.1) also holds for compact sets K, K' of positive reach (Federer [1959]). Although then, for specific $g \in G_d$, $K \cap gK'$ need not be of positive reach, this is true for μ-almost all g.

For applications, (3.1) can be interpreted as to give information on a set K when a test body K′ is moved into a random position and the intersection with K is observed (see 4A). In some cases, however, one may only be able to observe the parts on the boundary $K \cap g\,\mathrm{bd}K'$. However, bdK′ needs neither be an element of $\mathscr{R}^d$ nor of positive reach. Hence, $V_j(K \cap g\,\mathrm{bd}K')$ is in general not defined and a formula analogously to (3.1) is not known. Therefore, formulas of this type occurring in the stereological literature are valid only in special cases, e.g. if $K \in \mathscr{R}^d$ and K′ is a polytope (or a union $\bigcup_{i=1}^n P_i$ of polytopes) for then bdK′ is in $\mathscr{R}^d$, or if $K \in \mathscr{K}^d$ and K′ is a parallel body $K' = K'' + \alpha B$, $\alpha > 0$, for then bdK′ is of positive reach. A conjecture for general convex bodies K, K' is set up by Firey [1979b].

Such kinematic formulas for boundaries (of convex bodies) are connected with another problem. In case $K, K' \in \mathscr{K}^d$ and $j = 0$, the left hand side of (3.1) equals

$$\mu(\{g : K \cap gK' \neq \emptyset\}).$$

This value plays an important role as a normalizing constant (see 4A), e.g. in the practical case where K represents a sampling window in which randomly positioned sets K′ are observed. In some cases, due to edge effects, only those positions gK′ of K′ are observable (or observed) which fulfill $gK' \subset K$. A general formula for

$$\mu(\{g : gK' \subset K\}),$$

$K, K' \in \mathscr{K}^d$, is however not available. Solutions are known in some special cases (see Santaló [1976a] for a survey, and Pohl [1981] for a recent result). The situation is simple if K' can roll freely in K (see Firey [1979a], Weil [1982]). The freely rolling condition can be expressed for instance as $K = L + \alpha B$ and $K' + L' = \alpha B$ for some $\alpha > 0$ and bodies $L, L' \in \mathscr{K}^d$. In that case one gets

$$(3.2) \quad \mu(\{g : gK' \subset K\}) = \sum_{k=0}^{d} (-1)^k \alpha_{d0k} V_k(K) V_{d-k}(K').$$

If K' is not a summand of αB but small enough, (3.2) is still valid with the intrinsic volumes of the freely rolling cover of K' (Miles [1974b], Weil [1982]). An interesting conjecture connecting freely rolling bodies with a kinematic integral formula is due to Matheron [1978].

The principal kinematic formula (3.1) can be easily iterated to express integrals of the form

$$\int_{G_d} \cdots \int_{G_d} V_j(K \cap g_1 K_1 \cap \cdots \cap g_n K_n)\, d\mu(g_1) \cdots d\mu(g_n)$$

(see Streit [1970] for explicit formulas).

For curvature measures the principal kinematic formula was proved by Federer [1959] (for sets of positive reach). A proof using the characterization theorem (2.8) as well as the extension to $\mathscr{R}^d$ are due to Schneider [1978a], [1980a]. The result is as follows.

(3.3) **Theorem.** *Let* $K, K' \in \mathscr{R}^d, \beta, \beta' \subset \mathbb{R}^d$ *be Borel sets, and* $j \in \{0, \ldots, d\}$. *Then*

$$\int_{G_d} \psi_j(K \cap gK', \beta \cap g\beta')\, d\mu(g) = \sum_{k=j}^{d} \alpha_{djk} \psi_k(K, \beta) \psi_{d+j-k}(K', \beta').$$

Obviously, the coefficients are the same as in (3.1).

Finally, there is a similar formula for the curvature centroids obtained by Hadwiger-Schneider [1971] and Schneider [1972b]. For simplicity, we formulate the result for the moment vectors r_j.

(3.4) **Theorem.** *Let* $K, K' \in \mathscr{R}^d$ *and* $j \in \{0, \ldots, d\}$. *Then*

$$\int_{G_d} r_j(K \cap gK')\, d\mu(g) = \sum_{k=j}^{d} \alpha_{djk} V_{d+j-k}(K') r_k(K).$$

Obviously, by taking $\beta = \beta' = \mathbb{R}^d$ in (3.3), we get (3.1) and with $\beta' = \mathbb{R}^d$ and Fubini's theorem, we get (3.4). Thus, (3.3) is the most general of these formulas. Schneider [1980b] even improved this by giving a principal kinematic formula for external angles of polytopes from which he derived (3.3).

3C. *Intersections by flats*

In the following, we describe the Crofton formulas and their extensions to $\mathscr{R}^d$, to curvature measures, and to curvature centroids. Since the situation is quite similar to that of the foregoing subsection, we will be brief.

We start with the Crofton formulas for the intrinsic volumes (Hadwiger [1957], Santaló [1976a]).

(3.5) **Theorem.** *Let* $K \in \mathscr{R}^d$, $q \in \{0,\ldots,d-1\}$, *and* $j \in \{0,\ldots,q\}$. *Then*

$$\int_{\mathscr{E}_q^d} V_j(K \cap E)\, d\mu_q(E) = \alpha_{djq} V_{d+j-q}(K).$$

Notice that α_{djq} is the coefficient occurring also in (3.1) (with $k = q$).

For compact sets of positive reach, the formula is due to Federer [1959].

For later use, we mention that in case $K \in \mathscr{K}^d$ and $j = 0$, (3.5) reads

(3.6) $\mu_q(\{E : K \cap E \neq \emptyset\}) = \alpha_{d0q} V_{d-q}(K)$.

Davy-Miles [1977] showed that (3.5) and (3.6) hold true for arbitrary compact sets K with the intrinsic volumes V_j replaced by the projection integrals $\bar{V}_j$.

As a local analogue of (3.5), the following formula was proved by Federer [1959] (for sets of positive reach) and Schneider [1978a], [1980a] (for $K \in \mathscr{K}^d$ and $K \in \mathscr{R}^d$).

(3.6) **Theorem.** *Let* $K \in \mathscr{R}^d$, $\beta \subset \mathbb{R}^d$ *be a Borel set,* $q \in \{0,\ldots,d-1\}$, *and* $j \in \{0,\ldots,q\}$. *Then*

$$\int_{\mathscr{E}_q^d} \psi_j(K \cap E, \beta)\, d\mu_q(E) = \alpha_{djq} \psi_{d+j-q}(K, \beta).$$

From (3.6), one immediately deduces the corresponding result for curvature centroids due to Hadwiger-Schneider [1971], Schneider [1972b] (again formulated for moment vectors).

(3.7) **Theorem.** *Let* $K \in \mathscr{R}^d$, $q \in \{0,\ldots,d-1\}$, *and* $j \in \{0,\ldots,q\}$. *Then*

$$\int_{\mathscr{E}_q^d} r_j(K \cap E)\, d\mu_q(E) = \alpha_{djq} r_{d+j-q}(K).$$

3D. *Projection formulas*

As we have remarked, the intrinsic volumes of convex bodies can be defined in different ways. One possibility is by the Steiner formula, a second one which extends to $\mathscr{R}^d$ uses the Crofton formulas. A third and probably most familiar procedure is to define $V_j(K)$ as the mean of the appropriate content (quermass) of $K|L$ over all j-flats L. The generalization to projections on flats of other dimensions is given by the following Cauchy-Kubota formulas (see e.g. Hadwiger [1957], Santaló [1976a]).

(3.8) **Theorem.** *Let* $K \in \mathscr{K}^d$, $q \in \{0,\ldots,d\}$, $L \in \mathscr{L}_q^d$, *and* $j \in \{0,\ldots,q\}$. *Then*

$$\int_{SO_d} V_j(K|\zeta L)\, d\nu(\zeta) = \beta_{djq} V_j(K)$$

with

$$\beta_{djq} = \frac{\binom{q}{j} \varkappa_q \varkappa_{d-j}}{\binom{d}{j} \varkappa_d \varkappa_{q-j}}.$$

Of course, the integral can be easily transformed into an integration on the space $\mathscr{L}^d_q$.

As we have indicated, (3.8) can be proved with the characterization theorem (2.3). However, this method is quite indirect and there are a lot of more elementary proofs according to the definition used for V_j. Miles [1975] surveys some methods.

It is easily seen that (3.8) is no longer true on $\mathscr{R}^d$ as far as the usual notion of the projection of a set is used. However, for $j = q$, an extension to $\mathscr{R}^d$ can be obtained in the following way (due to Hadwiger [1956]). For $j = q$, we can write the left side of (3.8) as a double integral

$$\int_{SO_d} \int_{\zeta L} V_0(K \cap (x + \zeta L^{\perp}))\, d\lambda_q(x)\, d\nu(\zeta).$$

Here λ_q is the Lebegue measure on ζL (we suppress the dependence on the rotation ζ). In this form, (3.8) also holds for sets in $\mathscr{R}^d$. More generally, from the definition of μ_q and (3.5) one immediately gets the following result (which has analogues for the curvature measures and the moment vectors).

(3.9) **Theorem.** *Let* $K \in \mathscr{R}^d$, $q \in \{0,\dots,d\}$, $j \in \{0,\dots,d-q\}$ *and* $L \in \mathscr{L}^d_q$. *Then*

$$\int_{SO_d} \int_{\zeta L} V_j(K \cap (x + \zeta L^{\perp}))\, d\lambda_q(x)\, d\nu(\zeta) = \alpha_{djd-q} V_{j+q}(K).$$

(3.9) is implicitly contained in the work of Groemer [1972] on quermass-integrals and projections of functions (and non-convex sets).

For the projection integrals $\bar{\bar{V}}_j$, projection formulas which generalize (3.8) are simple consequences of their definition. They hold e.g. for all compact sets.

(3.10) **Theorem.** *Let* $K \subset \mathbb{R}^d$ *be compact,* $q \in \{0,\dots,d\}$, $L \in \mathscr{L}^d_q$, *and* $j \in \{0,\dots,q\}$. *Then*

$$\int_{SO_d} \bar{\bar{V}}_j(K|\zeta L)\, d\nu(\zeta) = \beta_{djq} \bar{\bar{V}}_j(K).$$

Let us turn now to the curvature measures. The following generalization of (3.8) was proved in Weil [1979b].

(3.11) **Theorem.** *Let* $K \in \mathscr{K}^d$, $\beta \subset \mathrm{bd}K$ *be a Borel set,* $q \in \{1,\dots,d\}$, $L \in \mathscr{L}^d_q$, *and* $j \in \{0,\dots,q-1\}$. *Then*

$$\int_{SO_d} \psi_j(K|\zeta L, \beta|\zeta L)\, d\nu(\zeta) = \beta_{djq} \psi_j(K,\beta).$$

The corresponding statement for $j = q$ is false. Although, in general $\beta|\zeta L$ need not be a Borel set this is true for the intersection of $\beta|\zeta L$ with the (relative) boundary of $K|\zeta L$ for at least ν-almost all ζ by a result of Schneider [1978b].

For the moment vectors r_i of a convex body K, projection formulas are known only in the case $i = 0$ (Hadwiger-Schneider [1971]).

(3.12) **Theorem.** *Let* $K \in \mathscr{K}^d$, $L \in \mathscr{L}^d_q$, *and* $q \in \{0,\dots,d\}$. *Then*

$$\int_{SO_d} r_0(K|\zeta L)\, d\nu(\zeta) = \beta_{d0q} r_0(K).$$

In (3.11), only those boundary points of K (in β) contribute to the integrand which have a tangent (d − q)-flat orthogonal to ζL. This indicates how the notion of projection of a non-convex set can be modified. For example, for a compact set A with smooth boundary bd A (or more generally for a smooth surface) one may take the projection (on to a subspace $L \in \mathscr{L}_q^d$) of all points $x \in \mathrm{bd}A$ such that there is a tangent (d − q)-flat at x, orthogonal to L. With this definition of the projection, integral formulas for smooth surfaces have been considered by Baddeley [1980a], Langevin-Shifrin [1982], Miles [1980a], and Slavskii [1975] (stereological applications are discussed in De Hoff [1978], [1980]). Corresponding formulas for the convex ring and the (additive or positive) extensions of the intrinsic volumes or curvature measures have not been investigated.

3E. *Integral formulas for cylinders*

The set $\mathscr{Z}(q,K')$ of cylinders congruent to $K' + L, K' \subset L^{\perp}$, reduces to $\mathscr{E}_q^d$ if K′ consists of one point only. In the other extreme case, namely when $q = 0$ (and thus $K' \in \mathscr{K}^d$) $\mathscr{Z}(q,K')$ coincides with $\{gK' : g \in G_d\}$. Therefore, a common generalization of the principal kinematic formulas and of the Crofton formulas to a result valid for cylinders may be expected. For the intrinsic volumes and convex bodies the following result can be found in Santaló [1976a], the extension to $\mathscr{R}^d$ is immediate.

(3.13) **Theorem.** *Let* $K \in \mathscr{R}^d, q \in \{0,\dots,d-1\}, j \in \{0,\dots,d\}, L \in \mathscr{L}_q^d$, *and* $K' \in \mathscr{R}^d$ *with* $K' \subset L^{\perp}$. *Then*

$$\int_{\mathscr{Z}(q,K')} V_j(K \cap Z)\, d\mu_{q,K'}(Z) = \sum_{k=j}^{d} \alpha_{djk} V_k(K) V_{d+j-k-q}(K').$$

Notice that the coefficients α_{djk} are the same as in (3.1) and that functionals V_m with index $m < 0$ are defined to be zero.

(3.13) was generalized to curvature measures by Schneider [1980b]. The formulation must be slightly different from (3.13) since we have to pay attention to boundary sets of a basis of a cylinder.

(3.14) **Theorem.** *Let* $K \in \mathscr{R}^d, q \in \{0,\dots,d-1\}, j \in \{0,\dots,d\}, L \in \mathscr{L}_q^d$, *and* $K' \in \mathscr{R}^d$ *with* $K' \subset L^{\perp}$. *Let* $\beta \subset \mathbb{R}^d$ *and* $\beta' \subset L^{\perp}$ *be Borel sets. Then*

$$\int_{SO_d} \int_{L^{\perp}} \psi_j(K \cap \zeta(K' + L + t), \beta \cap \zeta(\beta' + L + t))\, d\lambda_{d-q}(t)\, d\nu(\zeta)$$

$$= \sum_{k=j}^{d} \alpha_{djk} \psi_k(K,\beta) \psi_{d+j-k-q}(K',\beta').$$

From (3.14), we easily get a cylinder formula for moment measures (and curvature centroids) as a common generalization of (3.4) and (3.7).

(3.15) **Theorem.** *Let* $K \in \mathscr{R}^d$, $q \in \{0,\dots,d-1\}$, $j \in \{0,\dots,d\}$, $L \in \mathscr{L}_q^d$, *and* $K' \in \mathscr{R}^d$ *with* $K' \subset L^{\perp}$. *Then*

$$\int_{\mathscr{Z}(q,K')} r_j(K \cap Z)\, d\mu_{q,K'}(Z) = \sum_{k=j}^{d} \alpha_{djk} V_{d+j-k-q}(K') r_k(K).$$

A problem that specifically arises with cylinders is that of projected thick sections. A convex body K is intersected by a randomly placed cylinder $K' + L$ and the intersection $K \cap (K' + L)$ is projected on to the generating subspace L of $K' + L$. A mean value formula for the intrinsic volumes of such projections was proved by Schneider [1981a]. The case where K′ is a ball (of suitable dimension) was settled earlier by Matheron [1976] and rediscovered for dimension three by Voß [1980]. For a cylinder $Z = K' + L$ and $K \in \mathscr{K}^d$ we denote by $(K \cap Z)|Z$ the projection of $K \cap Z$ on to L.

(3.16) **Theorem.** *Let* $K \in \mathscr{K}^d$, $q \in \{0,\ldots,d-1\}$, $j \in \{0,\ldots,d\}$, $L \in \mathscr{L}^d_q$, *and* $K' \subset L^\perp$. *Then*

$$\int_{Z(q,K')} V_j((K \cap Z)|Z)\, d\mu_{q,K'}(Z) = \sum_{k=j}^{d} \gamma_{djkq} V_k(K) V_{d+j-q-k}(K')$$

with

$$\gamma_{djkq} = \frac{\binom{q}{j} \varkappa_q \varkappa_k \varkappa_{d-k}}{\binom{d}{d-k} \varkappa_{q-j} \varkappa_d \varkappa_j}.$$

Different formulas occur when 'intersecting' of cylinders is interpreted in another way. Davy [1978a] gives corresponding formulas.

3F. *Translation formulas*

The integral formulas we have discussed so far concern measure spaces which obey an invariance property w.r.t. rigid motions. This is correlated to the requirement of compatibility with motions which we have posed on the functionals in section 2 and the link is given by the characterization theorems. In practice however, there are a lot of problems where the assumption of motion invariance for the structure is not realistic (e.g. if certain directions are preferred) and should be replaced by the weaker condition of translation invariance. But, first of all no description of all translation invariant, continuous, additive (real) functionals is known (see McMullen-Schneider [1983] for the problem of a characterization theorem for translation invariant, continuous valuations) and then, consequently, there is much less information on integrals with respect to the translation group even for the classical functionals. The known results are presented in the following.

Most of them are concerned with translative versions of the principal kinematic formula. Starting with the volume, we have the following formula

(3.17) $$\int_{\mathbb{R}^d} V_d(K \cap (K' + x))\, d\lambda(x) = V_d(K) V_d(K').$$

This formula is true not only for $K, K \in \mathscr{R}^d$ but for all measurable sets in $\mathbb{R}^d$ (with V_d replaced by λ). Also for the surface area, the principal kinematic formula holds already for the translation group.

(3.18) $$\int_{\mathbb{R}^d} V_{d-1}(K \cap (K' + x))\, d\lambda(x) = V_{d-1}(K) V_d(K') + V_d(K) V_{d-1}(K')$$

Here $K,K' \in \mathscr{R}^d$, but the formula is valid for all closed sets with boundaries of σ-finite Hausdorff measure $\mathscr{H}^{d-1}$ (and of course V_{d-1} and V_d replaced by $\mathscr{H}^{d-1}$ and λ). (3.17) and (3.18) follow from a single formula which holds in the more general setting of groups and homogeneous spaces. This was settled, together with proofs and further references, by Groemer [1980b] and Schneider [1981b]. From their results also follows that corresponding formulas are true for the curvature measures.

$$\int_{\mathbb{R}^d} \psi_d(K \cap (K' + x), \beta \cap (\beta' + x))\, d\lambda(x) = \psi_d(K,\beta)\psi_d(K',\beta'), \tag{3.19}$$

$$\int_{\mathbb{R}^d} \psi_{d-1}(K \cap (K' + x), \beta \cap (\beta' + x))\, d\lambda(x) = \psi_{d-1}(K,\beta)\psi_d(K',\beta') + \psi_d(K,\beta)\psi_{d-1}(K',\beta'), \tag{3.20}$$

$K,K' \in \mathscr{R}^d$, $\beta,\beta' \subset \mathbb{R}^d$ Borel sets. For the other intrinsic volumes V_j a formula of translative integral geometry is known only if $j = 0$. Since, for convex bodies $K,K' \in \mathscr{K}^d$, $K \cap (K' + x) \neq \emptyset$ if and only if $x \in K + (-K')$, where $-K' = \{-y : y \in K'\}$, we have

$$\int_{\mathbb{R}^d} V_0(K \cap (K' + x))\, d\lambda(x) = V_d(K + (-K')).$$

Using mixed volumes we therefore have the following result.

$$\int_{\mathbb{R}^d} V_0(K \cap (K' + x))\, d\lambda(x) = \sum_{i=0}^{d} \binom{d}{i} V(\underbrace{K,\ldots,K}_{i}, \underbrace{-K',\ldots,-K'}_{d-i}), \tag{3.21}$$

$K,K' \in \mathscr{K}^d \setminus \{\emptyset\}$. (3.21) is found in Groemer [1977] and Streit [1975], the two-dimensional case was also treated by Miles [1974a] who rediscovered a result of Blaschke (see also Firey [1977]). In the plane, the formula can be extended to more general sets with an appropriate definition of mixed areas. Iterations of these formulas as well as slight generalizations to measures with densities (instead of λ) are given in Streit [1973], [1975].

Concerning translation formulas for intersections of sets with flats, there is only one simple but important result. By Fubini's theorem we have

$$\int_{L^\perp} V_q(K \cap (L + x))\, d\lambda_{d-q}(x) = V_d(K) \tag{3.22}$$

for all sets $K \in \mathscr{R}^d$, $q \in \{0,\ldots,d-1\}$, and a fixed $L \in \mathscr{L}_q^d$ (of course, the result holds for all measurable sets K with V_q and V_d the corresponding Lebesgue measures).

3G. *Notes and comments*

(a) A fourth group of classical formulas concerns mean values for bodies $K + \zeta K'$, $K,K' \in \mathscr{K}^d$, over all rotations ζ. For the intrinsic volumes the result is as follows (Hadwiger [1957], Santaló [1976a]).

$$\int_{SO_d} V_j(K + \zeta K')\, d\nu(\zeta) = \sum_{k=0}^{j} \delta_{djk} V_k(K) V_{j-k}(K') \tag{3.23}$$

with

$$\delta_{djk} = \frac{\binom{d-k}{j-k} \varkappa_{d-k}\varkappa_{d+k-j}}{\binom{d}{j-k} \varkappa_{d-j}\varkappa_d}, \quad j \in \{0,\ldots,d\}.$$

A corresponding local result was proved in Weil [1979b].

$$(3.24)\quad \int_{SO_d} \psi_j(K + \zeta K', \beta + \zeta\beta')\, d\nu(\zeta) = \sum_{k=0}^{j} \delta_{djk}\psi_k(K,\beta)\psi_{j-k}(K',\beta')$$

Here $j \in \{0,\ldots,d\}$, $K,K' \in \mathscr{K}^d$, and $\beta \subset \mathrm{bd}K$, $\beta' \subset \mathrm{bd}K'$ are Borel sets.

(b) Integral formulas for non-intersecting convex bodies have been treated by several authors. A survey is found in Weil [1979b]. The most general results are as follows. For bodies $K,K' \in \mathscr{K}^d$ with $K \cap K' = \varnothing$ let $r[K,K']$ be the usual distance of the point sets K,K' (minimal distance of points) and let $x[K,K']$, $y[K,K']$ be the points of K and K' with $r[K,K'] = \|x[K,K'] - y[K,K']\|$ (provided these points exist uniquely). Consider a function f on $(0,\infty) \times \mathrm{bd}K \times \mathrm{bd}K'$ such that the integrals in the following sum are finite. Then

$$\int_{K \cap gK' = \varnothing} f(r[K,gK'],x[K,gK'],g^{-1}y[K,gK'])\, d\mu(g)$$

$$(3.25)\quad = \sum_{k=0}^{d-1} \sum_{j=0}^{d-k-1} \varepsilon_{djk} \int_0^\infty \int_{\mathrm{bd}K} \int_{\mathrm{bd}K'} r^{d-k-j-1} f(r,x,y)\psi_k(K,dx)\psi_j(K',dy)\, dr$$

where

$$\varepsilon_{djk} = \frac{\binom{d-k-1}{j}(d-k)\varkappa_{d-j}\varkappa_{d-k}}{\binom{d}{j}\varkappa_d}.$$

If, for $E \in \mathscr{E}^d_q$, we similarly define $r[K,E]$ and $x[K,E]$ and if h is a function on $(0,\infty) \times \mathrm{bd}K$, then

$$\int_{K \cap E = \varnothing} h(r[K,E],x[K,E])\, d\mu_q(E)$$

$$(3.26)\quad = \sum_{j=0}^{d-q-1} \varepsilon_{djq} \int_0^\infty \int_{\mathrm{bd}K} r^{d-k-q-1} h(r,x)\psi_j(K,dx)\, dr.$$

Special cases of (3.25) and (3.26) are due to Bokowski et al. [1976] and Hadwiger [1975a] (see also Groemer [1980a]).

Translation formulas for nonintersecting convex bodies (in the spirit of (3.25)) have also been considered. Results involving mixed volumes and mixed surface area measures are due to Groemer [1980a] and Weil [1979b].

(c) For the surface area measures (see 2E(e)) there are a lot of local integral formulas which are in correspondence with formulas for curvature measures. There are analogues for (3.11), (3.24), (3.25) and (3.26). The proofs are due to Hadwiger [1975b], Schneider [1975b], [1977a], Weil [1979a], [1981] (see also the survey articles of Schneider [1979a] and Weil [1979b]).

(d) The integral geometric formulas presented so far can all be derived in a measure theoretic setup. For a lot of other integral formulas such a treatment is not known. We mention especially the formulas for multiple integrals of flats (or subspaces) which 'change the order of integration'; i.e., an integration of q-flats $E_q^{(1)},\ldots,E_q^{(k)}$ contained in a p-flat E_p, $kq < p$, which is then integrated over $\mathscr{E}_p^d$, is expressed by an integral over the p-flats $\tilde{E}_p$ containing $\tilde{E}_q^{(1)},\ldots,\tilde{E}_q^{(k)}$ when these are integrated over $\mathscr{E}_q^d$. Formulas of this type are discussed in Santaló [1976a]; special cases are the Blaschke-Petkantschin formula and 'Crofton's Second Theorem'. More recent applications and generalizations have been given by Davy [1978a], Miles [1979], Piefke [1978a], [1978b], Ruben-Miles [1980].

4. Stochastic interpretation

4A. *Random intersections and projections*

The integral formulas derived in the last section can be easily interpreted stochastically and transformed into mean value (expectation) formulas for random (inter-) sections and random projections (such interpretations are classical in integral geometry, see e.g. Santaló [1936]). For this purpose, we only have to normalize the underlying measures. In case of ν this is immediate, in the other cases (μ,λ and μ_q) we first have to restrict our considerations to a set of finite measure. The spaces $G_d, SO_d, \mathscr{E}_q^d, \mathbb{R}^d$ with their Borel σ-algebras then turn into a probability space $(\Omega',\mathscr{A}',P')$. As a typical example, let us consider $\Omega' = \{E \in \mathscr{E}_q^d : E \cap K_0 \neq \emptyset\}$, $\mathscr{A}' = \{A \subset \Omega' : A \text{ Borel set in } \mathscr{E}_q^d\}$, and $P' = \mu_q|_{\Omega'}/\mu_q(\Omega')$ where $\mu_q|_{\Omega'}$ denotes the restriction of μ_q to the set Ω'. Here $K_0 \subset \mathbb{R}^d$ is arbitrary except that we have to assume $0 < \mu_q(\Omega') < \infty$. A random q-flat X_q intersecting K_0 (with distribution P') then is a random element of Ω' (more precisely, X_q is the identity map $X_q:\Omega' \to \Omega'$). As usual in probability theory, a more abstract point of view is sometimes preferable. Therefore we will use the term 'random q-flat X_q intersecting K_0' for any measurable map of an (abstract) probability space $(\Omega,\mathscr{A},P)$ into $(\Omega',\mathscr{A}')$ which has image measure $P_{X_q} = P'$. We call P_{X_q} the distribution of X_q and denote by $\mathbb{E}f(X_q)$ the expectation of $f(X_q)$ for a real or vector valued function f on Ω^1. Then we can transform, for example, (3.5) into the following form

$$\mathbb{E}V_j(K \cap X_q) = \frac{1}{\mu_q(\Omega')} \int_{\mathscr{E}_q^d} V_j(K \cap E)\, d\mu_q(E)$$

$$= \frac{\alpha_{djq} V_{d+j-q}(K)}{\mu_q(\{E : E \cap K_0 \neq \emptyset\})}.$$

Here $K \in \mathscr{R}^d$, but such that

$$\int_{\Omega'} V_j(K \cap E)\, d\mu_q(E) = \int_{\mathscr{E}_q^d} V_j(K \cap E)\, d\mu_q(E). \tag{4.1}$$

Since we want to use such formulas in order to get information on K, and since we are assuming K_0 to be observable, we are interested in explicit expressions of $\mu_q(\{E : E \cap K_0 \neq \emptyset\})$ in terms of geometric functionals of K_0. This is possible (and given by (3.5)) if $K_0 \in \mathscr{K}^d$. Moreover, in this case, (4.1) is fulfilled if and only if $K \subset K_0$. In order to guarantee that $\mu_q(\Omega') > 0$ we suppose moreover that K_0 has inner points. Thus we will use the following assumptions. K_0 is a d-dimensional convex body with known (or observable) intrinsic volumes and $K \subset K_0$ is a set in $\mathscr{R}^d$ which we want to investigate. Since K_0 mostly serves as a reference set for the generation of the random q-flat X_q these assumptions are in accordance with practice where one may choose a suitable convex body K_0 (ball, box, cube, etc.), large enough to contain the material of interest, K. In those cases where K_0 is of interest too but not convex one gets (4.3) below with $V_{d-q}(K_0)$ replaced by $\bar{V}_{d-q}(K_0)$.

In the case of the principal kinematic formula (3.1) we are, of course, in a quite similar situation. We consider a set K to be explored which is contained in a reference set K_0. We have a test set K' moving around and concentrate on those $g \in G_d$ where $K_0 \cap gK' \neq \emptyset$. The normalizing factor for μ then is $\mu(\{g: K_0 \cap gK' \neq \emptyset\})$. This is explicitly known from (3.1) if K_0 and K' are convex bodies. (3.1) then implies a formula for the expectation

$$\mathbb{E}V_j(K \cap \xi K')$$

if $K \in \mathscr{R}^d$ and ξ is the random motion with distribution $P_\xi = \mu|_{\Omega'}/\mu(\Omega')$, $\Omega' = \{g: K_0 \cap gK' \neq \emptyset\}$. In order to have $\mu(\Omega') > 0$ we assume $K_0 \in \mathscr{K}^d_d$.

In the following two theorems we summarize the results for the two situations just described (random intersections by q-flats resp. by moving test sets) and the intrinsic volumes. Corresponding results follow for the curvature measures (from (3.3) and (3.6)) and the curvature centroids (from (3.4) and (3.7)).

(4.2) **Theorem.** *Let* $K_0 \in \mathscr{K}^d_d$, $K' \in \mathscr{K}^d$, $K \in \mathscr{R}^d$ *with* $K \subset K_0, j \in \{0,\dots,d\}$, *and let* ξ *be a random motion with distribution*

$$P_\xi = \frac{\mu|_{\Omega'}}{\mu(\Omega')},$$

where $\Omega' = \{g \in G_d: K_0 \cap gK' \neq \emptyset\}$. *Then*

$$\mathbb{E}V_j(K \cap \xi K') = \frac{\sum_{k=j}^{d} \alpha_{djk} V_k(K) V_{d+j-k}(K')}{\sum_{k=0}^{d} \alpha_{d0k} V_k(K_0) V_{d-k}(K')}.$$

(4.3) **Theorem.** *Let* $K_0 \in \mathscr{K}^d_d$, $K \in \mathscr{R}^d$ *with* $K \subset K_0$, $q \in \{0,\dots,d-1\}$, $j \in \{0,\dots,q\}$, *and let* X_q *be a random q-flat with distribution*

$$P_{X_q} = \frac{\mu_q|_{\Omega'}}{\mu_q(\Omega')},$$

where $\Omega' = \{E \in \mathscr{E}^d_q: K_0 \cap E \neq \emptyset\}$. *Then*

$$\mathbb{E}V_j(K \cap X_q) = \frac{\alpha_{djq} V_{d+j-q}(K)}{\alpha_{d0q} V_{d-q}(K_0)}.$$

For random projection formulas we take $\Omega' = \mathscr{L}_q^d$ and P′ as the probability measure induced by ν. Then (3.8) implies the following result.

(4.4) **Theorem.** *Let* $K \in \mathscr{K}^d$, $q \in \{0,\ldots,d\}$, $j \in \{0,\ldots,q\}$, *and let* Y_q *be a random* q-*subspace with distribution* P_{Y_q} *induced by* ν. *Then*

$$\mathbb{E}V_j(K|Y_q) = \beta_{djq}V_j(K).$$

Again there is a corresponding result for curvature measures. Moreover, most of the integral formulas of section 3 can be interpreted as expectation formulas for sections, projections, projected thick sections, etc. It is not necessary to go into the details since the method of transposition is always the same.

The statistical interpretation of such mean value formulas is the following. Consider, for example, (4.3). For a certain realization X_q the value $V_j(K \cap X_q)$ (which we assume to be measured) gives an estimate of the right side

$$\frac{\alpha_{djq}V_{d+j-q}(K)}{\alpha_{doq}V_{d-q}(K_0)}.$$

(4.3) implies that this estimator is unbiased. Equivalently, $(\alpha_{doq}/\alpha_{djq})$ $V_{d-q}(K_0)\,V_j(K \cap X_q)$ is an unbiased estimator for $V_{d+j-q}(K)$. (4.2) and (4.4) also give unbiased estimators for the intrinsic volumes of K or combinations thereof. More generally, the role of integral geometry in stereology may be described as to yield a range of unbiased estimators for geometrical functionals (see section 7 for a further discussion). Such probabilistic and statistical interpretations of integral geometric results for intrinsic volumes have been presented in more detail in Davy [1978a], Davy-Miles [1977], Giger [1967a], [1970], [1972b], Giger-Hadwiger [1968], Hadwiger [1974], Kendall-Moran [1963], Matheron [1976], Miles [1972], [1973a], [1974a], [1976] Miles-Davy [1976], [1977], Santaló [1976a], Streit [1970], [1973], [1975]. Stereological applications of the integral formulas for moment vectors and curvature centroids have been given by Davy [1980a], [1980b], [1981]. The integral formulas for curvature measures which may serve for the estimation of local shape parameters are as yet not exploited in the literature.

In dimension $d = 3$ which is the most interesting case for applications, (4.3) gives us three unbiased estimators for the volume of a set $K \in \mathscr{R}^d$, which are proportional to $V_2(K \cap X_2), V_1(K \cap X_1), V_0(K \cap X_0)$ respectively, two unbiased estimators for the surface area of K, proportional to $V_1(K \cap X_2), V_0(K \cap X_1)$, and one estimator for $V_1(K)$ (mean width if K is convex), namely a multiple of $V_0(K \cap X_2)$. If K is convex, projection methods may be used. Here, from (4.4) we get (besides trivial relations for the characteristic) two unbiased estimators for the mean width and one for the surface area.

Notice that $V_0(K)$, for $K \in \mathscr{R}^d$, cannot be estimated by flat sections and $V_d(K)$, for $K \in \mathscr{K}^d$, cannot be estimated by projections (within this concept of integral geometry). Especially the first of these two observations contains a serious problem for practical stereology. Quite often, the set $K \in \mathscr{R}^d$ is the union of disjoint (convex) particles $K = \bigcup_{i=1}^n K_i$. Their number n then coincides with

$V_0(K)$. Instead of $V_j(K)$, $j \geq 1$, one is really interested in the mean value

$$\frac{1}{n}\sum_{i=1}^{m} V_j(K_i)$$

which in view of the additivity equals $(1/n)V_j(K)$. Hence these mean values can only be estimated if the particle number $V_0(K)$ is known (or can be estimated itself). Different approaches to overcome this difficulty have been considered in the literature. The easiest case is when intersections of K with a randomly situated test set $K' \in \mathscr{K}^d$ are available and all intrinsic volumes $V_j(K \cap \xi K')$, $j = 0,\ldots,d$, can be measured. If we assume $V_d(K') > 0$, then the system of linear equations given by (4.2) can be solved and the solution gives unbiased estimators for $V_j(K)$, $j = 0,\ldots,d$, (see Santaló [1976a], pp. 282–286).

Studying the intersection of K with a randomly situated (and infinitely extended) wedge, Miles-Davy [1978] proposed an unbiased estimator for n based on observed curvatures and angles at the intersection points of the wedge edge with K. However, this method works for smooth particles and does not transfer to general convex bodies. Moreover, it seems to be a difficult method for practice.

Finally, as Matheron [1976] observed, the particle number n can be estimated from projected thick sections if (at least two) sections of different thickness are used.

So far we have described unbiased estimators based on single experiments like intersection by a random flat, projection on to a random subspace, etc. The results obviously carry over to the case of multiple replications of an experiment and an unbiased estimator is given by the sample mean of the single estimators.

Another easy generalization is to n-stage procedures of different type, as they are usual in practice. For example, if characteristic functionals (per unit volume) of tissues are to be estimated, this may be done in three steps. First, a piece of tissue is cut out randomly (intersection with test set), then a random section of this piece is observed under the microscope (or a projected thick section), and finally the characteristic parameters of the plane image are estimated by using random lines, segments, points, etc. Again, if the different steps of the procedure are chosen independently (and according to the appropriate distributions) the formulas given in this subsection or in section 3 and suitably combined will yield unbiased estimators.

4B. *Weighted sampling*

The unbiased estimators for the intrinsic volumes $V_j(K)$ obtained from (4.3), require the knowledge of intrinsic volumes of the reference set K_0. Although these can be measured in some practical cases this will cause additional expense. Moreover, in those cases where K is a piece cut out from a larger amount of material by K_0 one is frequently interested in parameters which do not depend on the magnitude of K_0, for instance in intrinsic volumes of K per unit volume of K_0. From (4.3), one is led to choose $V_j(K \cap X_q)/V_q(K_0 \cap X_q)$ as an estimator for $\alpha_{djq} V_{d+j-q}(K)/V_d(K_0)$. This is, of course, an estimator but not an unbiased one. Davy and Miles showed that by an appropriate choice of the distribution of X_q the estimator above becomes unbiased.

The procedure is based on a simple trick. Let us look, for instance, at the Crofton formulas (3.5), applied to the situation $K \subset K_0, K_0 \in \mathscr{K}^d, K \in \mathscr{R}^d$. If the function $f: \mathscr{E}_q^d \to \mathbb{R}_+$ is such that $\int_{K_0 \cap E \neq \emptyset} f(E)\, d\mu_q(E)$ exists and is finite then (3.5) implies

$$(4.5) \quad \int_{\mathscr{E}_q^d} \frac{V_j(K \cap E)}{f(E)} d\tilde{P}(E) = \frac{\alpha_{djq} V_{d+j-q}(K)}{\int_{K_0 \cap E \neq \emptyset} f(E)\, d\mu_q(E)}$$

where $\tilde{P}$ is the probability measure given by

$$\tilde{P}(A) = \frac{\int_{A \cap \{K_0 \cap E \neq \emptyset\}} f(E)\, d\mu_q(E)}{\int_{K_0 \cap E \neq \emptyset} f(E)\, d\mu_q(E)}, \quad A \subset \mathscr{E}_q^d \text{ Borel.}$$

Hence

$$\frac{V_j(K \cap X_q)}{f(X_q)}$$

is an unbiased estimator for

$$\frac{\alpha_{djq} V_{d+j-q}(K)}{\int_{K_0 \cap E \neq \emptyset} f(E)\, d\mu_q(E)},$$

if X_q has distribution $\tilde{P}$. For example, choosing

$$f(E) = V_k(K_0 \cap E), \quad k \in \{0,\ldots,q\},$$

we obtain

$$(4.6) \quad \int_{\mathscr{E}_q^d} \frac{V_j(K \cap E)}{V_k(K_0 \cap E)} d\tilde{P}(E) = \frac{\alpha_{djq} V_{d+j-q}(K)}{\alpha_{dkq} V_{d+k-q}(K_0)}.$$

In case $k = 0$, this is (4.3), in case $k = q$ we have the result announced above. Notice that (4.5) and (4.6) are true even if K_0, the reference set, is in $\mathscr{R}^d$ and not convex (provided the weighting function f is still nonnegative). Moreover, similar formulas hold for the projection integrals V_j and compact sets K, K_0, $K \subset K_0$.

With an obvious interpretation, a procedure based on (4.6) for $k > 0$ (or more generally on (4.5)) is called weighted sampling. It can be introduced, in a similar way, for intersections by test sets or projections and for other functionals (curvature measures and curvature centroids). Since there is no mathematical difficulty to do this, we will not give the details and refer to the work of Davy [1978a], [1980a], [1980b], [1981], Davy-Miles [1977], Miles [1972], [1978], [1980c], Miles-Davy [1976], [1977]. For problems of weighted sampling in connection with multi-stage procedures, see Jensen-Gundersen [1982].

4C. *Generation of probability measures*

So far, we have constructed probability measures for geometric objects from the invariant measures on SO_d and $\mathbb{R}^q$ (from which the invariant measures on $G_d, \mathscr{E}_q^d$,

$\mathscr{L}^d_q$, etc. are derived) either directly or by a weighting procedure. From two different points of view, it is an important question how random objects distributed according to these probability measures can be generated geometrically. First, the existence of such a generating procedure is a necessary prerequisite for practical applications. On the other hand, random flats intersecting a convex body, for example, can be generated in a lot of elementary and intuitive ways (remember the classical example of Bertrand). It is therefore a natural question to which distributions these geometrical procedures lead.

We will only consider the case of random q-flats which is the most interesting one both theoretically and for applications. A construction for a random q-flat X_q hitting a set K_0 with distribution P′ as in 4A can be easily obtained from the definition of μ_q (in the literature, such a flat X_q is sometimes called a *uniform isotropic* random q-flat). Choose a ball C with $K_0 \subset C$ (we assume C to be centered around the origin) and a random q-subspace Y_q(this can be easily managed if $d = 3$ and $q = 1$ or 2). Next, choose a uniformly distributed random point x in $Y_q^\perp \cap C$ and consider $x + Y_q$. If $x + Y_q$ does not intersect K_0 repeat this procedure independently until a flat $X_q = x + Y_q$ hits K_0. Then, X_q is a uniform isotropic random q-flat. Alternatively, if one can choose a point x uniformly distributed in $K_0|Y_q^\perp$ then $X_q = x + Y_q$ intersects K_0 and has distribution P′.

Davy-Miles [1977] (see also Davy [1978a], Miles-Davy [1976]) gave procedures to obtain weighted random flats. We mention only the case (important for applications) of an area-weighted random 2-flat X_2 in R^3. This can be obtained by first choosing a uniformly distributed random point $x \in K_0$ and then (independently) a plane X_2 through x with uniformly distributed direction.

As far as the intuitive geometrical methods to generate random flats hitting a convex body $K \in \mathscr{K}^d$ are concerned, we only mention some references. Random q-flats generated by $q + 1$ independent and uniformly distributed points in K are studied in Kingman [1969]. Their distribution is found and compared with the uniform isotropic case. An extensive literature exists for random secants of convex boxies. Here, various models of generation have been considered and for $d = 2$ or 3 and simple shaped bodies (circles, squares, balls, cubes, boxes) distributions and moments have been given explicitely (see e.g. Coleman [1969], [1981], Ehlers-Enns 1981, Enns-Ehlers [1978], [1980], Horowitz [1965], Kellerer [1971], Kendall-Moran [1963], Kingman [1965], Matérn [1960], Piefke [1979], Solomon [1978]. Voß [1982]).

A uniformly isotropic random line through a three-dimensional convex body K can be obtained also as the line determined by two independent points on bd K distributed according to the normalized measure $\psi_{d-1}(K,\cdot)$ (surface randomness), but this is not true for other dimensions (see Hadwiger [1968], Santaló [1976a]).

4D. *Notes and comments*

(a) In connection with projection formulas for surfaces, Miles [1980a] remarked that the different ways to generate random q-flats intersecting a convex body $K \in \mathscr{K}^d_d$ lead, by a limit process, to different 'natural' measures for q-flats touching K (in boundary points). However, also in this case there is one measure distinguished by its connection with μ_q via a local Steiner-type formula (see Weil [1979b], [1981]).

(b) Instead of uniformly distributed flats, segments, etc. in the sense of probability theory, a fixed sequence of geometric objects (flats, segments) may be chosen to estimate functionals of sets provided this sequence is 'uniformly distributed' in a non-probabilistic sense. Kulle [1971] and Maak [1970] studied a concept of this type where the 'uniform distribution' of the sequence is related to the underlying set. For segments, Kulle-Reich [1973] showed the existence of universal sequences which are uniformly distributed in the sense of number theory and by which the surface area of any smooth surface in a fixed cube of $\mathbb{R}^d$ can be measured by counting intersections. Since the first k elements of such sequences can be taken quite arbitrary, randomly chosen segments seem to be preferable in practice.

5. Random sets

5A. *Definition of random sets*

The methods of integral geometry and their stochastic interpretation, though they result from a simple and powerful mathematical theory, are in many cases not transferable to practical situations. For instance, independent replications of random plane sections of a 3-dimensional solid body are difficult to generate. More practicable are several parallel cuts of a material with prescribed fixed orientation. Methods of this type may be justified if the structure under examination can be viewed as the outcome of a random process such that different (parallel) sections contain (stochastically) the same information and, moreover, the distribution of the structure is independent of orientation and location.

A theory of random sets was developed by Kendall [1974] and Matheron [1967], [1969], [1972a], [1972b], [1975], based on earlier results of Choquet [1955]. Especially, Matheron's notion of a random closed set X in $\mathbb{R}^d$ proved to be a useful tool in the further development of stochastic geometry. To define a random closed set in $\mathbb{R}^d$, nothing but a suitable topology has to be chosen on the space $\mathscr{F}^d$ of closed subsets of $\mathbb{R}^d$. This topology is derived from the following base of open sets in $\mathscr{F}^d$

$$\{A \in \mathscr{F}^d : A \cap K = \emptyset, \quad A \cap G_i \neq \emptyset, \quad i = 1,\ldots,n\},$$

where $K \in \mathscr{C}^d$, $G_i \subset \mathbb{R}^d$ open, and $n = 0,1,2,\ldots$. On compact subsets of $\mathscr{C}^d \setminus \{\emptyset\}$ this topology coincides with the one given by the Hausdorff metric. Further topological properties of $\mathscr{F}^d$ are fully discussed in Matheron [1975]. Consider, now, the σ-algebra $\mathbb{B}$ of Borel subsets of $\mathscr{F}^d$. Then, a *random closed set* X (in $\mathbb{R}^d$) simply is a measurable mapping

$$X : (\Omega,\mathscr{A},P) \to (\mathscr{F}^d,\mathbb{B}),$$

where $(\Omega,\mathscr{A},P)$ is an abstract probability space. The induced probability P_X on $(\mathscr{F}^d,\mathbb{B})$ is called the distribution of X. Actually, $\mathbb{B}$ is generated by the class of sets

$$\{A \in \mathscr{F}^d : A \cap K = \emptyset\}, \quad K \in \mathscr{C}^d,$$

and P_X is uniquely determined by the functional $K \mapsto P_X(\{A \in \mathscr{F}^d : A \cap K = \emptyset\})$ on $\mathscr{C}^d$. The class of these functionals can be characterized by a theorem of Choquet [1955]. Since the rotation group and the translation group operate in an obvious way on $\mathscr{F}^d$, we can call X isotropic if P_X is invariant under rotations, and stationary if P_X is translation invariant. We speak of a *random compact set* X if $P_X(\mathscr{C}^d) = 1$, and of a *random convex body* if $P_X(\mathscr{K}^d) = 1$.

In view of the role played by the invariant functionals in integral geometry, and in order to transfer some of the integral geometric results to random sets, we will lay emphasis on random closed sets X which are stationary or even stationary and isotropic. Stationary random closed sets X are almost surely (a.s.) unbounded. This follows from the fact that the closed convex hull coX of a stationary random closed set X is a stationary random closed set, too, and therefore coX is a.s. trivial, i.e. has values only in $\{\emptyset, \mathbb{R}^d\}$ (Matheron [1975], p. 84). In order to work with geometrical functionals in the following, we will therefore concentrate on random sets X obeying $P(X \in \mathscr{S}^d) = 1$. We then speak of a random set in $\mathscr{S}^d$.

By the choice of the topology on $\mathscr{F}^d$, the usual operations on random sets lead to random sets again. For instance, for a random set X in $\mathscr{S}^d$, $E \in \mathscr{E}^d_q$, and $K \in \mathscr{K}^d$, $X \cap E$ is a random set in $\mathscr{S}^d$ and $X \cap K$ is a random set with values in $\mathscr{R}^d$.

As we mentioned, the intrinsic volumes V_i, the curvature measures ψ_i, and the moment vectors r_i are continuous on $\mathscr{K}^d \setminus \{\emptyset\}$, but on $\mathscr{R}^d$ and for general i neither the additive nor the positive extensions are continuous (or not even upper (lower) semicontinuous). Therefore, the question of measurability occurs. Apparently there is no proof of it in the literature but according to private communications of R. Schneider and J.A. Wieacker, the measurability of the curvature measures ψ_i (and thus of the other functionals) can be proved. For $\bar{V}_i$ the measurability is given in Matheron [1975], more generally it is shown that the map

$$K \mapsto \int_{\mathbb{R}^d} f(x)\bar{\psi}_j(K,dx)$$

is measurable on $\mathscr{R}^d$ for every lower semicontinuous real function $f \geq 0$ on $\mathbb{R}^d$. These results especially imply the measurability of the volume and the surface area, and of their local analogues. An extension of such measurability properties to Hausdorff measures is due to Zähle [1982b].

5B. *Densities*

In view of our motivation to introduce random sets we would now expect to get formulas of integral geometric type for the intrinsic volumes (and other quantities), analogous to those of section 3 but with fixed test set K′ and fixed q-flat E, and with the set K replaced by a random set X with motion invariant distribution. However, as we have seen, such an X will a.s. be unbounded and, therefore, in order to make use of intrinsic volumes and expectations, we have to consider intersections $X \cap C$ (for instance by balls C) and study limits as C increases. The following considerations of this type are due to Davy [1978a]. Yet it should be remarked that we have to base our considerations on the assumption that $\mathbb{E}V_i(X \cap K)$ is finite and depends continuously on $K \in \mathscr{K}^d \setminus \{\emptyset\}$.

For a stationary and isotropic random set X in $\mathscr{S}^d$ fulfilling these conditions, we consider the functional

$$T_j : K \mapsto \mathbb{E}V_j(X \cap K), \quad K \in \mathscr{R}^d.$$

Since T_j is additive, motion invariant, and continuous on $\mathscr{K}^d$, (2.3) implies that $T_j(K)$ is a combination of $V_k(K)$, $k = 0,\ldots,d$. This is made precise by the following result due to Davy [1978a].

(5.1) **Theorem.** *Let* X *be a stationary isotropic random set in* $\mathscr{S}^d$ *fulfilling the continuity condition mentioned above. Let* $j \in \{0,\ldots,d\}$ *and* $K \in \mathscr{R}^d$. *Then*

$$\mathbb{E}V_j(X \cap K) = \sum_{k=j}^{d} \alpha_{djk} V_k(K) D_{d+j-k}(X)$$

with

$$D_{d+j-k}(X) = \lim_{r \to \infty} \left(V_d(rK')\right)^{-1} \mathbb{E}V_{d+j-k}(X \cap rK')$$

for all d-*dimensional convex bodies* $K' \subset \mathbb{R}^d$.

The quantity $D_j(X)$ may be called the *density* of V_j for the random set X. Notice that (5.1) may be viewed as an analogue of (3.1).

Under our assumptions, for a flat $E \in \mathscr{E}_q^d$, the density $D_j(X \cap E_q)$ exists and does not depend on the choice of E. The following result of Davy [1978a] gives the analogue of Crofton's formula.

(5.2) **Theorem.** *Let* X *be a stationary, isotropic random set in* $\mathscr{S}^d$ *fulfilling the continuity condition mentioned above. Let* $q \in \{0,\ldots,d\}$, $E \in \mathscr{E}_q^d$, *and* $j \in \{0,\ldots,q\}$. *Then*

$$D_j(X \cap E) = \alpha_{djq} D_{d+j-q}(X).$$

Both results can be interpreted as to give natural estimators for the right sides.

If two independent random sets X,X′ in $\mathscr{S}^d$ are considered, then $X \cap X'$ and $X \cup X'$ are random sets in $\mathscr{S}^d$. If X,X′, and $X \cap X'$ fulfill the conditions of this subsection, the following version of the kinematic formula holds.

$$(5.3) \quad D_j(X \cap X') = \sum_{k=j}^{d} \alpha_{djk} D_k(X) D_{d+j-k}(X'), \quad j = 0,\ldots,d.$$

Since D_j is additive by definition, this gives us a formula for $D_j(X \cup X')$, too (see Davy [1978a] for further details).

The representation of $D_j(X)$ as a limit can be easily improved with the use of (5.1). Let C_0 be the cube of unit edge length given by

$$C_0 = \{x \in \mathbb{R}^d : -\tfrac{1}{2} \le x_i \le \tfrac{1}{2}, \quad i = 1,\ldots,d\}$$

and let ∂C_0 be the 'upper right' boundary

$$\partial C_0 = \left\{x \in C_0 : \max_{i=1,\ldots,d} x_i = \frac{1}{2}\right\}.$$

Since ∂C_0 is in $\mathscr{R}^d$ we may consider

$$\mathbb{E}V_j(X \cap C_0) - \mathbb{E}V_j(X \cap \partial C_0).$$

Because of

$$V_j(C_0) - V_j(\partial C_0) = \begin{cases} 1 & \text{for } j = d \\ 0 & \phantom{\text{for }} j < d \end{cases},$$

(5.1) implies the following.

(5.4) **Corollary.** *Let* X *be a random set as in* (5.1) *and let* $j \in \{0,\ldots,d\}$. *Then*

$$D_j(X) = \mathbb{E}[V_j(X) \cap C_0) - V_j(X \cap \partial C_0)].$$

(5.4) gives an unbiased estimator for $D_j(X)$.
The Crofton formula (5.2) now reads as follows.

(5.5) **Corollary.** *Let* X *be a random set as in* (5.2), *let* $q \in \{0,\ldots,d\}$ *and* $j \in \{0,\ldots,q\}$. *Let* $E \in \mathscr{E}_q^d$ *be a supporting* q*-flat of* C_0 *such that* $E \cap C_0$ *is a* q*-face of* C_0. *Then*

$$\mathbb{E}[V_j(X \cap E \cap C_0) - V_j(X \cap E \cap \partial C_0)] = \alpha_{djq} D_{d+j-q}(X).$$

5C. *Ergodicity*

Although the results of 5B generate natural or even unbiased estimators for the densities D_i of a random set X, the model has a serious disadvantage. If we look at a tissue as the realization of a random set we cannot work with independent replications to lower the mean square error. The question is whether (or for which random sets) we can obtain densities (i.e. expectations) like $D_j(X)$ or $D_j(X \cap E)$ by observations of a single realization $X(\omega)$ or $X(\omega) \cap E$ of our random set X or the section $X \cap E$. This would be possible if we have (for $K' \in \mathscr{K}_d^d$)

$$(5.6)\qquad \frac{V_j(X \cap rK')}{V_d(rK')} \xrightarrow[r\to\infty]{} D_j(X) \quad \text{a.s.}$$

Results of this type (or with convergence in the mean) are called *ergodic*.

Ergodic theorems for scalar and vector valued functionals on set valued stochastic processes are presented in a unified approach by Nguyen [1979] and Nguyen-Zessin [1979]. An ergodic theorem for random sets themselves is due to Hess [1979]. The results of Nguyen and Zessin can be formulated also for Banach space valued functionals (e.g. support functions) and, therefore, used to give an ergodic theorem for certain random sets, too (see Wieacker [1982]).

Nguyen and Zessin apply their theorems to a special class of random sets (Boolean models, see next section) but one can also derive similar results for stationary random sets X in $\mathscr{S}^d$. For that purpose, two conditions must be imposed on X. First an integrability condition for $|V_j(X \cap K)|$, $K \in \mathscr{K}^d$, which is fulfilled, e.g. if $2^{m(X)}$ is an integrable random variable, where $m(X(\omega))$ denotes the minimal number n of bodies $K_i \in \mathscr{K}^d$ such that

$$X(\omega) \cap C_0 = \bigcup_{i=1}^{n} K_i.$$

Then, a condition of ergodic or mixing type is needed. X is mixing if

(5.7) $\quad P(X \in A_1, X + nx \in A_2) \underset{n\to\infty}{\longrightarrow} P(X \in A_1)\cdot P(X \in A_2),$

for all $A_1, A_2 \in \mathbb{B}$, $x \in \mathbb{R}^d \setminus \{0\}$. Intuitively, (5.7) expresses the fact that X behaves asymptotically independent at places in $\mathbb{R}^d$ which are at distance nd, $d > 0$ with $n \to \infty$.

For a stationary random set X in $\mathscr{S}^d$ fulfilling these two conditions, the ergodic theorems of Nguyen-Zessin [1979] imply that (5.6) holds for a.s. as well as mean convergence; where $D_j(X)$ is, in this case, defined by

$$D_j(X) = \mathbb{E}[V_j(X \cap C_0) - V_j(X \cap \partial C_0)].$$

5D. *Notes and comments*

(a) The measurability of occurring variables can be forced obviously by a suitable choice of the σ-algebra on $\mathscr{S}^d$. This is done in similar situations by different authors (e.g. Mecke [1980], Pohlmann [1980], Pohlmann et al. [1981], Mecke-Stoyan [1980b]).

(b) It seems to be an open problem whether for a stationary random set X in $\mathscr{S}^d$ the expectation $\mathbb{E}V_j(X \cap K)$ depends continuously on $K \in \mathscr{K}^d \setminus \{\emptyset\}$ (finiteness of the expectations being assumed).

(c) The cases $j = d$ and $j = d - 1$ are of a special type. Since $V_d = \bar{V}_d$ and $V_{d-1} = \bar{V}_{d-1}$, both are positive. Thus, for any stationary random set X in $\mathscr{S}^d$ with suitable existence assumptions on the following expectations we have that

$$\mathbb{E}V_d(X \cap \cdot)$$

extends to a translation invariant measure on $\mathbb{R}^d$, from which we get

$$\mathbb{E}V_d(X \cap K) = D_d(X)V_d(K),$$

where

$$D_d(X) = \mathbb{E}V_d(X \cap C_0).$$

This follows from results of Matheron [1975], from which one may also derive a version of (5.1) for stationary X and $j = d - 1$.

For the volume there is also a 'stationary' version of (5.2).

(d) Matheron [1975] proved the existence of densities $\bar{D}_j(X)$ for the positive extensions $\bar{V}_j$ and stationary X. He makes use of the fact that the curvature measures ψ_j are locally defined and may, therefore, be extended to sets $K \in \mathscr{S}^d$ as Radon measures. $\bar{D}_j(X)$ then stems from the expectation measure of the random measure $\bar{\psi}_j(X), \cdot)$. It is not clear whether (or under which assumptions) a similar treatment of the additive extension ψ_j is possible.

(e) It seems likely that the densities $D_j(X)$ can be defined without the assumption of isotropy but corresponding results are not present in the literature.

(f) Mecke [1981] proved formulas corresponding to section 5B for stationary and isotropic random sets which have piecewise smooth boundary and for densities of Hausdorff measures.

For stationary random sets built by curves in the plane (fibre processes) results are obtained, without isotropy, for the density of the length in Mecke [1980a], Mecke-Stoyan [1980b], Stoyan et al [1980]. Especially there are formulas of Crofton type involving an angle distribution. In the isotropic case the result corresponds to (5.2) (with $d = 2$, $q = 1$, and $j = 0$).

Fibre processes in $\mathbb{R}^3$ are considered in a similar way by Mecke-Nagel [1980], the corresponding Crofton-formula for stationary fibre processes involves a directional distribution. Stereological applications are given also in Ohser [1981]. A corresponding work for surface processes in R^3 is done in Pohlmann [1980] and Pohlmann et al. [1981]. Since the surface processes considered there are just random sets in $\mathcal{S}^3$, the results are direct generalizations of some of the formulas in 5B. Especially there is a version of (5.2) for $d = 3$ and $j = q - 1$ for stationary random sets X which rests on certain distributions of normal directions.

These results for dimensions two and three are generalized to arbitrary dimensions, Hausdorff rectifiable sets, and Hausdorff measures by Zähle [1982b] (there is also a generalization to Riemannian spaces, see Zähle [1982c]).

6. Point processes

6A. *Introduction*

The difficulties we had in the last section with measurability, continuity, and ergodicity for random sets X in $\mathcal{S}^d$ are connected with the fact that for a set $K \in \mathcal{S}^d$ the representation $K = \bigcup_{i=1}^{\infty} K_i, K_i \in \mathcal{K}^d$, is not unique and there is no 'natural' representation of this type, in general. The random set model will be much easier to deal with if we start with a random process generating as an outcome a collection $\{K_i\}$ of convex bodies K_i and then, as a secondary notion, consider the random set X in $\mathcal{S}^d$ given by the union

$$X(\omega) = \bigcup_{i=1}^{\infty} K_i.$$

This model is in accordance with many practical situations when sets $K \in \mathcal{S}^d$ are present which are the union of recognizable and distinguishable bodies $K_i \in \mathcal{K}^d$ (grains, cells, etc.). Moreover, in such cases one frequently is not only interested in functionals of K but also of the individual sets K_i and mean values thereof. Random processes Y which have as realization collections of points $\{s_i\}$ in a suitable space are called point processes. Point processes can be investigated on quite general topological spaces. The corresponding theory is found e.g. in Neveu [1977] or Snyder [1975]. The important spaces for geometric point processes are, of course, $\mathscr{K}^d, \mathscr{R}^d, \mathscr{C}^d, \mathscr{E}^d_q$, and spaces of cylinders. A great deal of the geometric theory of point processes is developed in Matheron [1975]. As we already announced in 1C, we will not speak on point processes on $\mathscr{E}^d_q$ (flat networks) or on $\mathbb{R}^d$ (point patterns). Rather we will concentrate on point processes on $\mathscr{K}^d$. It should, however, be remarked that parts of the general considerations are true for point processes on $\mathscr{R}^d$ or $\mathscr{C}^d$ and some of the formulas in 6B allow generalizations to point processes of cylinders.

A point process Y on $\mathscr{K}^d$ can be defined as a measurable map from a probability space $(\Omega,\mathscr{A},P)$ in the set of collections $\{K_i\}$ of bodies $K_i \in \mathscr{K}^d$ such that $\bigcup_{i=1}^{\infty} K_i \in \mathscr{S}^d$ when a suitable σ-algebra is defined on this set. For technical reasons and in order to delete labellings of bodies, we represent collections $\{K_i\}$ by point measures η,

$$\eta = \sum_{i=1}^{\infty} \varepsilon_{K_i}$$

where ε_K is the one-point measure at $K \in \mathscr{K}^d$. In order to simplify the considerations, we concentrate on simple point processes, i.e. that only point measures η with atoms of measure 1 are allowed (no multiple points). For each such point measure η on $\mathscr{K}^d$ let K_η be the union of the bodies which support η and let $\mathscr{M}$ be the set of point measure η such that

$$\eta(\{K \in \mathscr{K}^d : K \cap C \neq \emptyset\}) < \infty$$

for all $C \in \mathscr{C}^d$ (we say, η is locally finite). On $\mathscr{M}$ we consider the σ-algebra $\mathbb{M}$ generated by the mappings

$$\eta \mapsto \eta(A)$$

where A runs through the Borel subsets of $\mathscr{K}^d$. A *point process* Y on $\mathscr{K}^d$ is then a measurable map of an abstract probability space $(\Omega,\mathscr{A},P)$ into $(\mathscr{M},\mathbb{M})$. Point processes can be viewed as special random closed subsets of $\mathscr{M}$ (see Ripley [1976a], [1976b]). Point processes Y, Y′ having the same distribution $P_Y = P_{Y'}$ on $(\mathscr{M},\mathbb{M})$ are called equivalent. Stationarity and isotropy of a point process Y is defined by the invariance of P_Y, as usual. If, for each $\omega \in \Omega$, we define $X_Y(\omega) = K_{Y(\omega)}$, then X_Y is a random set in $\mathscr{S}^d$ and stationarity or isotropy of Y implies the same property for X_Y, but the converse is false. An important notion for point processes is the intensity measure ϑ. For a Borel set $A \subset \mathscr{K}^d$, $\vartheta(A)$ is the mean number of bodies in A, i.e.

$$\vartheta(A) = \int_\Omega Y(\omega)(A)\, dP(\omega).$$

ϑ is a Borel measure on $\mathscr{K}^d$ which we assume to be locally finite.

Looking at the translation classes of convex bodies we can give another interpretation of a point process on $\mathscr{K}^d$. Let

$${}^0\mathscr{K}^d = \{K \in \mathscr{K}^d : s_0(K) = 0\}$$

be a system representing these translation classes (there are of course other systems possible). Then ${}^0\mathscr{K}^d \times \mathbb{R}^d$ is homeomorphic to $\mathscr{K}^d$, and a homeomorphism is given by

$$f:(K,t) \mapsto K + t.$$

${}^0\mathscr{K}^d$ may be called the shape space of the point process Y on $\mathscr{K}^d$. Moreover, if π_2 denotes the projection of ${}^0\mathscr{K}^d \times \mathbb{R}^d$ on to $\mathbb{R}^d$, and if η is a point measure on $\mathscr{K}^d$, then $\pi_2(f^{-1}(\eta))$ is obviously an ordinary point measure on $\mathbb{R}^d$. Thus, we can interpret a point process Y of convex bodies as a process $\hat{Y}$ of points in $\mathbb{R}^d$ such that to each point t is attached a random body K from the shape space ${}^0\mathscr{K}^d$. Point processes on $\mathscr{K}^d$ therefore fit into the general concept of marked point processes (see e.g. Kerstan et. al. [1974] or Neveu [1977]).

For a stationary point process Y, the intensity measure ϑ is translation invariant. Hence, from a standard argument it follows that there exist uniquely a probability measure P_0 on ${}^0\mathscr{K}^d$ and a $c > 0$ such that

$$\vartheta = f \circ (P_0 \otimes c\lambda). \tag{6.1}$$

P_0 may be called the shape distribution of Y.

The most important example of a point process is the Poisson process. It plays the role in general point process theory and in stochastic geometry that Gaussian random variables play in classical statistics. To give the definition, a point process Y on $\mathscr{K}^d$ with intensity measure ϑ is a Poisson process if (i) the extended real random variable $N_A: \omega \mapsto Y(\omega)(A)$ has a Poisson distribution with parameter $\vartheta(A)$, for all Borel sets $A \subset \mathscr{K}^d$, and (ii) the random variables $N_{A_1},\ldots,N_{A_k}$ are independent for disjoint sets $A_1,\ldots,A_k$, $k = 1,2,\ldots$. For a given locally finite measure ϑ on $\mathscr{K}^d$ there always exists a unique (up to equivalence) Poisson process Y on $\mathscr{K}^d$ with intensity ϑ. Y is stationary (isotropic) if and only if ϑ is translation invariant (rotation invariant). Moreover, for a Poisson process Y on $\mathscr{K}^d$, $\hat{Y}$ is a Poisson process on $\mathbb{R}^d$ and if Y is stationary (isotropic) then $\hat{Y}$ has the same property. From (6.1), we therefore can interpret a stationary Poisson process in the following way. Y is generated by a stationary point process in $\mathbb{R}^d$ with a mean number of c points per unit volume. To each point independently a shape $K \in {}^0\mathscr{K}^d$ is attached with shape distribution P_0. We also get the following property of a stationary Poisson process Y on $\mathscr{K}^d$. Take a $K_0 \in \mathscr{K}^d_d$ and suppose that n bodies of Y have the property that their Steiner point falls into K_0. Then, under this condition, the n Steiner points are independently and uniformly distributioned in K_0 (this is a statement on conditional distributions!). Conversely, a stationary Poisson process Y on $\mathscr{K}^d$ can be constructed by approximation in the following way. Given a $c > 0$ and a distribution P_0 on ${}^0\mathscr{K}^d$. Take a sequence C_n, $n = 1,2,\ldots$, of balls (or other convex bodies) such that $c \cdot V_d(C_n) = n$. For each n, consider n independent and uniformly distributed points $\xi_1,\ldots,\xi_n$ in C_n and attach to each ξ_i (independently) a random body $Z_i \in {}^0\mathscr{K}^d$ with distribution P_0. Then, by the set $\{\xi_1 + Z_1,\ldots,\xi_n + Z_n\}$ a point process Y_n on $\mathscr{K}^d$ is defined which, for $n \to \infty$, 'tends' to a stationary Poisson process Y on $\mathscr{K}^d$ with intensity measure

$$\vartheta = f \circ (P_0 \otimes c\lambda).$$

A precise formulation of this statement as well as a proof is found in Wieacker [1982]. Very often, such an approximation property of Poisson processes is used without proof (e.g. Davy [1976a], [1978a], Miles [1970a]) and frequently there is made no difference between fields in the sense of 2F and stationary Poisson processes (e.g. Giger [1968]).

6B. *Densities and mean values*

In this subsection we consider a stationary Poisson process Y on $\mathscr{K}^d$ and the corresponding stationary random set $X = X_Y$ in $\mathscr{S}^d$ (some authors call X a stationary Boolean model with compact, convex grains). For X, the difficulties of section 5 do not occur, especially for each $K \in \mathscr{K}^d$, $V_j(X \cap K)$ is a random variable with finite expectation $\mathbb{E}V_j(X \cap K)$ (which, moreover, depends continuously on $K \in \mathscr{K}^d$), the densities $D_j(X)$ exist, and the limits are ergodic. The precise statements are as follows (with $C_0, \partial C_0$ as in (5.4)).

(6.2) **Theorem.** *Let* Y *be a stationary Poisson process on* $\mathscr{K}^d$ *and* X *the corresponding random set. Let* $j \in \{0,\dots,d\}$ *and* $K' \in \mathscr{K}^d_d$.
Then the limit

$$\lim_{r\to\infty} \frac{V_j(X \cap rK')}{V_d(rK')} = : D_j(X)$$

exists in the mean and a.s. Moreover we have

$$D_j(X) = \mathbb{E}[V_j(X \cap C_0) - V_j(X \cap \partial C_0)].$$

Especially, this means that $D_j(X)$ is a constant independent of K'. Nguyen-Zessin [1979] showed that the limit exists and equals a conditional expectation. Wieacker [1982] proved that Y (or X) is mixing and thus the conditional expectations are constant a.s.

Since the random set X is the union of the grains of the point process Y one could expect, in view of the stationarity, a connection between the densities D_j and mean values of the grains. The following result is given in Wieacker [1982].

(6.3) **Theorem.** *Let* Y *be a stationary Poisson process on* $\mathscr{K}^d$ *with intensity measure* ϑ, *and* X *the corresponding random set. Let* $j \in \{0,\dots,d\}$ *and* $K' \in \mathscr{R}^d$.
Then

$$\mathbb{E}V_j(X \cap K')$$

$$= \sum_{k=1}^{\infty} \frac{(-1)^{k+1}}{k!} \int_{\mathscr{K}^d} \cdots \int_{\mathscr{K}^d} V_j(K_1 \cap \cdots \cap K_k \cap K')\, d\vartheta(K_1) \cdots d\vartheta(K_k).$$

If Y is also isotropic, then (5.1) holds for $X = X_Y$. Moreover, in this case, the integrals in (6.3) correspond to integrals over ${}^0\mathscr{K}^d$ and G_d and can be simplified according to the formulas of Streit [1970]. Explicit expressions are due to Davy [1976a], [1978a] and (with a different proof) to Wieacker [1982]. Davy solved the problem more generally for Poisson processes of cylinders. Since the formulas are intricate we mention only the cases $j = d$ and $j = d - 1$. We use the abbreviation

$$M_j(Y) = \int_{{}^0\mathscr{K}^d} V_j(K)\, dP_0(K), \quad j = 0,\dots,d.$$

Then,

$$\begin{aligned}\mathbb{E}V_d(X \cap K') &= V_d(K')(1 - e^{-cM_d(Y)}) \\ &= V_d(K')P(0 \in X), \\ \mathbb{E}V_{d-1}(X \cap K') &= V_{d-1}(K')(1 - e^{-cM_d(Y)}) + V_d(K')e^{-cM_d(Y)}cM_{d-1}(Y) \\ &= V_{d-1}(K')P(0 \in X) + V_d(K')\frac{1}{2}\lim_{\varepsilon \downarrow 0} \varepsilon^{-1}[P(0 \in X + \varepsilon B) \\ &\quad - P(0 \in X)].\end{aligned}$$

Therefore

$$\begin{aligned}D_d(X) &= 1 - e^{-cM_d(Y)} = P(0 \in X), \\ D_{d-1}(X) &= e^{-cM_d(Y)}cM_{d-1}(Y) \\ &= \frac{1}{2}\lim_{\varepsilon \downarrow 0} \varepsilon^{-1}[P(0 \in X + \varepsilon B) - P(0 \in X)].\end{aligned}$$

In the isotropic case, the Crofton formula (5.2) holds for the densities of $X = X_Y$. Matheron [1975] also found a Crofton formula for the mean values $M_j(Y)$. Consider a q-flat E. Then, the intersection Y_q of Y with E is again a stationary and isotropic Poisson process with corresponding random set $X \cap E$. Let c_q be the density of the underlying point process in E, P_0^q the shape distribution, and $M_j(Y_q)$ the mean value of V_j w.r.t. P_0^q (since Y is stationary and isotropic, these notions depend only on q and not on E). Then

$$c_q = c\,\alpha_{d0q}\, M_{d-q}(Y).$$

(6.4)

$$M_j(Y_q) = \frac{\alpha_{djq}}{\alpha_{d0q}} \frac{M_{d+j-q}(Y)}{M_{d-q}(Y)}.$$

If instead of the intersection of a stationary and isotropic Poisson process Y with a q-flat a projected thick section is considered, i.e. the intersection with a fixed cylinder $K + L$, $L \in \mathscr{L}_q^d$, $K \subset L^\perp$, and a subsequent projection on to L, the result is again a stationary and isotropic point process Y' in L. Its mean values $M_j'(Y')$ are connected with the mean values $M_i(Y)$ of Y by a formula analogous to (3.14) (see Davy [1976a], [1978a], Miles [1976a], [1976b], for the case of a ball K). If $K = \{0\}$, then the formula reduces to (6.4). From the M_j', a formula for the densities D_j' of Y' can be derived.

Notice that the projection Y_L of a Poisson process Y on to a subspace $L \in \mathscr{L}_q^d$ is no longer a process of this type since every q-dimensional ball in L is a.s. hit by infinitely many 'grains' of Y_L.

Mean value formulas of a different type for point processes Y on $\mathscr{K}^d$ can be easily obtained in the following way. Consider a measurable real function f on $\mathscr{K}^d$, and let ξ be the sum of f(K) over all $K \in Y(\omega)$. If the expectation of the random variable ξ exists, we have

$$(6.5)\quad \mathbb{E}\xi = \int_{\mathscr{K}^d} f(K)\,d\vartheta(K).$$

This follows from the definition of ϑ. If X is stationary, then from (6.1) follows

$$\mathbb{E}\xi = c \int_{\mathbb{R}^d} \int_{0\mathscr{K}^d} f(K + x)\,dP_0(K)\,d\lambda(x).$$

Especially for a $K' \in \mathscr{K}^d$,

$$f(K) = V_j(K \cap K'),$$

and rotation invariant P_0, this implies

$$\mathbb{E}\xi = c \int_{0\mathscr{K}^d} \int_{G_d} V_j(gK \cap K')\,d\mu(g)\,dP_0(K).$$

The principal kinematic formula (3.1) leads now to the following

(6.6) **Theorem.** *Let* Y *be a stationary and isotropic point process on* $\mathscr{K}^d$, $j \in \{0,\ldots,d\}$, $K' \in \mathscr{K}^d$, *and* ξ *the sum of* $V_j(K \cap K')$ *where* K *runs through the elements of* Y. *Then*

$$\mathbb{E}\xi = c \sum_{k=j}^{d} \alpha_{djk} M_j(Y) V_{d+j-k}(K').$$

This result is due to Fava-Santaló [1978], [1979] (see also Parker-Cowan [1976] for processes of segments). Notice that no Poisson type assumptions are necessary. Moreover, stationarity and isotropy of Y can be replaced by corresponding properties of ϑ.

6C. *Notes and comments*

(a) By definition, a random set X in $\mathscr{S}^d$ is built by a locally finite 'random collection' of convex bodies. Thus, this concept appears as an immediate generalization of ordinary point processes in $\mathbb{R}^d$. But, due to the phenomenon of overlap, it makes a difference whether one considers the collection of bodies or their union set. This seems to be a technical distinction, important only for measurability questions. In the first case, measurability can be easily proved, in the second case it is more intricate, as we have seen. A positive answer to the following (apparently) open question would strikingly clarify the interrelations between geometric point processes and random sets. Does there exist to every random set X in $\mathscr{S}^d$, a point process Y on $\mathscr{K}^d$ with $X = X_Y$?

(b) Results similar to those in 6B may be obtained for some other random sets in $\mathscr{S}^d$ with stationary and isotropy conditions, based on point process theory. For example, Stoyan [1979e] studied point processes with non-overlapping grains and obtained Crofton-type formulas.

(c) For stationary Poisson processes a variety of additional results of geometrical type can be obtained. Among these are results on distance distributions, mean clumping numbers, mean visible distance and volume, etc.

(Berman [1977], Coleman [1972], [1974], Fava-Santaló [1978], [1979], Holgate [1967], Parker-Cowan [1976], Santaló [1976b], [1977], [1978], Stoyan [1979b], [1982], Wieacker [1982]).

(d) The stereological formulas for projected thick sections depend on the convexity of the particles and are no longer true if point processes on $\mathscr{R}^d$ are considered. However, generalizations to processes of non-convex particles may be possible if the notion of projection is altered (compare section 3) or the particles have lower dimensions. For point processes of fibres in $\mathbb{R}^d$, corresponding results are settled in Zähle [1982d].

(e) Frequently, in practice, fields of distinguishable (convex) particles are observed which cannot interpenetrate (e.g. bases of trees in a forest). In those cases, a (stationary) Poisson process on $\mathscr{K}^d$ can only be an approximate model, since it allows intersections. The approximation will be good if the density c of the underlying point process in $\mathbb{R}^d$ is small, relative to the shapes of the particles. In the other cases, models of (stationary) point processes on $\mathscr{K}^d$ other than the Poisson process have to be considered. Such models for non-overlapping grains have been described by many authors. A familiar method is to use thinning or stopping procedures for the underlying point process in $\mathbb{R}^d$ (see Davy [1978a], Hanisch [1980], [1981], Mürmann [1978], Stoyan [1979c]).

(f) The formulas of subsection 6B connecting the means of a stationary Poisson process on $\mathscr{K}^d$ with those of the process obtained in a section or a projected thick section can be viewed as to give first order information on the shape distribution P_0 by observations of corresponding shape distributions P_0^q in the flat $L \in \mathscr{L}_q^d$, or the projected thick section w.r.t. $L + K$, $K \subset L^\perp$. It is an old and important problem, whether (for suitable processes) the whole distribution P_0 can be determined (estimated) from the shape distribution P_0^q in the section (or projected thick section). The problem has been posed first by Wicksell [1925] who treated balls with a radius distribution having a density w.r.t. λ_1 and gave an integral equation of Abel type connecting the densities. The spherical case then was studied by a number of authors under varying assumptions. For instance, in the case of thick sections, one can distinguish the cases that the balls are opaque and the medium transparent or vice versa, one can exlude the observations of very large or very small intersection radii, one can admit cases where the true radii of intersected balls can be observed, etc. (see Bach [1965], [1976], Coleman [1978], [1980], [1982a], [1982b], Giger-Riedwyl [1970], Goldsmith [1967], Horálek [1980], Kendall-Moran [1963], Likeš [1981], Piefke [1976], Reid [1955], Santaló [1976a], Šidák [1968], Tallis [1970]).

The occurring integral equations are studied in detail by Jakeman-Anderssen [1975] and Anderssen-Jakeman [1975].

Distributions of non-spherical particles (ellipsoids, particles of the same shape, etc.) are studied by Cruz-Orive [1976], [1978], De Hoff [1962], Grimaldi-Tanasi [1981], Nicholson-Merx [1969], Santaló [1943], [1976a], Tallis [1970], Tanasi [1979], Voß [1978], Wicksell [1926].

Koschitzki [1980] and Mecke-Stoyan [1980a], [1980c] showed that Poisson type assumptions (or uniformly distributed independent particles) are not necessary.

7. Statistical applications

7A. *Unbiased estimators*

As we have already remarked, integral geometry mainly provides one with unbiased estimators for stereological situations, if the right distributional assumptions are made. Unbiased estimators yield the exact parameter 'in the mean', if they are used over a long period. Hence, for a single application to a given material, the estimator may be as bad as possible. Even, if n independent replications for the same material are available, which will lower the error by factor 1/n, we have no knowledge about the quality of an unbiased estimator as far as its mean square error is not known. In view of the unbiasedness the latter is equal to the variance of the estimator (see 7B for further discussion).

These remarks indicate the difficulties with a statistical analysis of stereological problems. Because of their highly non-parametric nature (e.g. for random sections of a deterministic material) they do not fit well into the classical statistical theory. Nevertheless, results exist (e.g. for Poisson point processes) but they lie outside the frame of integral geometry, and therefore also outside the scope of this article. We will thus mainly restrict ourselves to give some references in this final section.

7B. *Variances*

To give an expression for the variance of the unbiased estimators in section 4, integrals of the following type have to be considered:

$$\int_{G_d} V_j^2(K \cap gK')\, d\mu(g),$$

$$\int_{\mathscr{E}_q^d} V_j^2(K \cap E)\, d\mu_q(E),$$

$$\int_{SO_d} V_j^2(K|\zeta L)\, d\nu(\zeta),$$

and similar expressions for the curvature measures, the moment vectors, projected thick sections, etc. In the case of weighted sampling even more complicated integrals occur. Despite of some special cases no general formulas for such integrals are known.

Since we have given several unbiased estimators for the same quantity in section 4 one would like to compare them. Sectioning by q-flats intuitively seems to be more accurate than by r-flats with $r < q$. Davy [1978a] showed this to be true for weighted sampling but to be false for isotropic uniform flats in special situations.

Concerning a comparison of the mean square error of weighted against unweighted sampling, many authors use heuristic arguments but a general result does not exist.

7C. *Notes and comments*

(a) As we have mentioned, the generation of several independent random sections (projections) is impossible or at least difficult in many practical situations. Therefore, the use of systematic procedures is preferred. By empirical investigations, different systematic methods have been compared and results on their accuracy were obtained (see Cruz-Orive-Weibel [1981] for a survey). Systematic sampling is also discussed in Davy [1978a] and Miles [1977].

(b) If k independent random (inter-) sections or projections are available one may regard estimators other than the sample mean. For instance, weighted and unweighted samples can be mixed, ratio estimators which are biased can be used and corrected by a term estimating the bias, etc. (Davy [1978a], Miles [1978a]). Another possibility is to replace the sample mean by a mean where the results of different sections are combined with different weighting factors (not to be confused with weighted sectioning!). With the use of a linear regression model, this procedure is studied and (unbiased) estimators of minimum variance are obtained in Cruz-Orive [1980a], [1982], Cruz-Orive-Weibel [1981].

(c) It is obviously a difficult geometrical problem to find bounds for the maximal deviation when k (random or systematic) sections or projections of a convex body $K \in \mathscr{K}^d$ are examined and used to estimate $V_j(K), j \in \{0,\ldots,d\}$. For the case of projections on to q-subspaces, this problem is studied by Betke-McMullen [1982]. They obtain a solution for $q = d - 1$, a partial result for $q = 1$, and set up conjectures in the intermediate cases.

(d) From 5B, an unbiased estimator of the volume density $D_d(X)$ of a stationary random set X is given by

$$\hat{D}_d^{(K)} = \alpha_{dqq}^{-1} \frac{V_q(X \cap E \cap K)}{V_q(K)}$$

where $E \in \mathscr{E}_q^d$ and $K \in \mathscr{K}^d, K \subset E$, has dimension q. Under special assumptions on X, further information on such estimators has been obtained. First, if X (or $X \cap E$) is ergodic, then $\hat{D}_d^{(K)}$ is strongly consistent, i.e.

$$\hat{D}_d^{(rK)} \xrightarrow[r \to \infty]{} D_d(X) \quad \text{a.s.}$$

This follows from 5C and was settled also by Mase [1982]. Mase showed moreover that under stronger assumptions of mixing type (which are fulfilled for stationary Boolean models with suitably bounded compact grains), for $d = 3$ and $q = 2$.

$$\frac{1}{V_q(rK)} [\hat{D}_d^{(rK)} - D_d(X)]$$

has asymptotically a normal distribution. He also gave an estimator for the variance of $\hat{D}_d^{(rK)}$ which is asymptotically unbiased and consistent. Stoyan [1979d] studied the case $d = 3, q = 1$, and gave formulas for the variance of $\hat{D}_d^{(rK)}$ for three special random set models.

(e) For stationary point processes Y, statistical results concerning second order properties (second moment measures) are obtained by Hanisch-Stoyan [1979], [1981], Ohser [1980], Ripley [1976c], [1981]. A survey on statistical methods for general point processes is given by Krickeberg [1980].

(f) A lot of literature of statistical nature exists for the Wicksell problem and its generalizations. Here, classical statistical procedures (e.g. maximum likelihood estimation) apply if the shape distributions are assumed to be suitably parametrized (see Cruz-Orive [1978], Dupač [1980], Evans-Clarke [1975], Jakeman-Scheaffer [1978], Keiding et al. [1972], Mase [1980], Nicholson [1970], [1976], [1978], Nicholson-Merckx [1969], Watson [1971]).

Acknowledgement

The author is greatful to the many collegues who helped to fill gaps in the list of references and sent preprints. Special thanks go to R. Schneider and J.A. Wieacker for their critical comments on an earlier version of this article.

References

Adam, H., G. Bernroider, H. Haug (Ed.)
1980 Fifth international congress for stereology, proceedings. G. Fromme, Wien-München 1980. Reprinted from Mikroskopie (Wien) **37** (Suppl.) (1980).
Ambartzumian, R.V.
1975 The solution to the Buffon-Sylvester problem and stereology. Proc. Intern. Congr. Math., Vancouver 1974, Canad. Math. Congress 1975, 137–141.
1977 Stochastic geometry from the standpoint of integral geometry. Adv. Appl. Prob. **9** (1977), 792–823.
1978 Needles and wedges as tools for integration. In Miles-Serra (1978), 271–278.
1982 Combinatorial integral geometry. Wiley, Chichester-New York-Brisbane-Toronto-Singapore 1982.
Ambartzumian, R.V. (Ed.)
1980 Combinatorial principles in stochastic geometry. (Russian). Akad. Nauk Armjan. SSR, Erewan 1980.
Anderssen, R.S., A.J. Jakeman
1975 Abel type integral equations in stereology. II. Computational methods of solution and the random spheres approximation. J. Microscopy **105** (1975), 135–154.
Bach, G.
1965 Über die Bestimmung von charakteristischen Größen einer Kugelverteilung aus der unvollständigen Verteilung der Schnittkreise. Metrika **9** (1965), 228–233.
1976 Über die Auswertung von Schnittflächenverteilungen. Biometrical J. **18** (1976), 407–412.
Baddeley, A.
1977a A fourth note on recent research in geometrical probability. Adv. Appl. Prob **9** (1977), 824–860.
1977b Integrals on a moving manifold and geometrical probability. Adv. Appl. Prob. **9** (1977), 588–603.
1980a Absolute curvatures in integral geometry. Math. Proc. Camb. Phil. Soc. **88** (1980), 45–58.
1980b A limit theorem for statistics of spatial data. Adv. Appl. Prob. **12** (1980), 447–461.
1981 Combinatorial foundations of stochastic geometry. Proc. London Math. Soc. (3) **42** (1981), 151–177.
1982a Appendix A to Ambartzumian (1982), 194–214.
1982b Stochastic geometry: an introduction and reading-list. Int. Statist. Rev. **50** (1982), 179–193.
Bartlett, M.S.
1975 The statistical analysis of spatial pattern. Chapman & Hall, London 1975.
Berman, M.
1977 Distance distributions associated with Poisson processes of geometric figures. J. Appl. Prob. **14** (1977), 195–199.
Bernroider, G.
1978 The foundation of computational geometry: Theory and application of the point-lattice-concept within modern structure analysis. In Miles-Serra (1978), 153–170.

Betke, U., P. McMullen
1982 Estimating the sizes of convex bodies from projections. (to appear).
Bokowski, J., H. Hadwiger, J.M. Wills
1976 Eine Erweiterung der Croftonschen Formeln für konvexe Körper. Mathematika **23** (1976), 212–219.
Brehm, U., W. Kühnel
1982 Smooth approximation of polyhedral surfaces regarding curvatures. Geometriae Dedicata **12** (1982), 435–461.
Chermant J.L. (Ed.)
1978 Quantitative analysis of microstructures in biology, material sciences and medicine. Riederer, Stuttgart 1978.
Choquet, G.
1955 Theory of capacities. Ann. Inst. Fourier **5** (1955), 131–295.
Clarke, K.R.
1982 Statistical techniques in stereology. Commun. Statist.-Theor. Meth. A**10** (1982), 1459–1478.
Coleman, R.
1969 Random paths through convex bodies. J. Appl. Prob. **6** (1969), 430–441.
1972 Sampling procedures for the lengths of random straight lines. Biometrika **59** (1972), 415–426.
1974 The distance from a given point to the nearest end of one member of a random process of linear segments. In Harding-Kendall (1974), 192–201.
1978 The stereological analysis of two-phase particles. In Miles-Serra (1978), 37–48.
1979 An introduction to mathematical stereology. Memoir Series, Aarhus 1979.
1980 The distribution of the sizes of spheres from observations through a thin slice. In Adam et al. (1980), 68–73.
1981 Intercept lengths of random probes through boxes. J. Appl. Prob. **18** (1981), 276–282.
1982a The sizes of spheres from profiles in a thin slice. I. Opaque spheres. Biom. J. **24** (1982), 273–286.
1982b The sizes of spheres from profiles in a thin slice. II. Transparent spheres. Biom. J. (to appear).
Corssin, S.
1955 A measure of the area of a homogeneous random surface in space. Quart. Appl. Math. **12** (1955), 404–408.
Cowan, R.
1978 The use of the ergodic theorems in random geometry. Adv. Appl. Prob. **10** (Suppl.) (1978), 47–57.
1980 Properties of ergodic random mosaic processes. Math. Nachr. **97** (1980), 89–102.
Cruz-Orive, L.-M.
1976 Particle size-shape distributions: The general spheroid problem I. J. Microscopy **107** (1976), 235–253.
1978 Particle size-shape distributions: The general spheroid problem II. J. Microscopy **112** (1978), 153–167.
1980a Best linear unbiased estimators for stereology. Biometrics **36** (1980), 595–605.
1980b On the estimation of particle number. In Adam et al. (1980), 79–85.
1980c On the estimation of particle number. J. Microscopy **120** (1980), 15–27.
1982 The use of quadrats and test systems in stereology, including magnification corrections. J. Microscopy **125** (1982), 89–102.
Cruz-Orive, L.-M., A.O. Myking
1979 A rapid method for estimating volume ratios. J. Microscopy **115** (1979), 127–136.
1981 Stereological estimation of volume ratios by systematic sections. J. Microscopy **122** (1981), 143–157.
Cruz-Orive, L.-M., E.R. Weibel
1981 Sampling designs for stereology. J. Microscopy **122** (1981), 235–257.
Davy, P.
1976a Projected thick sections through multi-dimensional particle aggregates. J. Appl. Prob. **13** (1976), 714–722. Correction: J. Appl. Prob. **15** (1978), 456.
1976b Review of Matheron (1975). MR **52** # 6828.
1978a Stereology—a statistical viewpoint. Thesis, Australian National Univ. 1978.
1978b Aspects of random set theory. Adv. Appl. Prob. **10** (Suppl.) (1978), 28–35.

1980a The estimation of centroids in stereology. In Adam et al. (1980), 94–100.
1980b The stereology of location. J. Appl. Prob. **17** (1980), 860–864.
1981 Interspersion of phases in a material. J. Microscopy **121** (1981), 3–12.
Davy, P., R.E. Miles
1977 Sampling theory for opaque spatial specimens. J. Royal Statist. Soc. B **39** (1977), 56–65.
De Hoff, R.T.
1962 The determination of the size distribution of ellipsoidal particles from measurements made on random plane sections. Trans Metall. Soc. AIME **224** (1962), 474–481.
1978 Stereological uses of the tangent count. In Miles-Serra (1978), 99–113.
1980 Stereological information contained in the integral mean curvature. In Adam et al. (1980), 32–36.
1981 Stereological meaning of the inflection point count. J. Microscopy **121** (1981), 13–19.
De Hoff, R.T., F.N. Rhines (Ed.)
1968 Quantitative microscopy. McGraw-Hill, New York 1968.
Dupač, V.
1980 Parameter estimation in the Poisson field of discs. Biometrica **67** (1980), 187–190.
Ehlers, P.F., E.G. Enns
1981 Random secants of a convex body generated by surface randomness. J. Appl. Prob. **18** (1981), 157–166.
Elias, H. (Ed.)
1967 Stereology. Proceedings of the second international congress for stereology, Chicago 1967. Springer, Berlin-Heidelberg-New York 1967.
Enns, E.G., P.F. Ehlers
1978 Random paths through a convex region. J. Appl. Prob. **15** (1978), 144–152.
1980 Random paths originating within a convex region and terminating on its surface. Austral. J. Statist. **22** (1980), 60–68.
Evans, D.A., K.R. Clarke
1975 Estimation of embedded particle properties from plane section intercepts. Adv. Appl. Prob. **7** (1975), 542–560.
Exner, H.E., (Ed.)
1975 Quantitative analysis of microstructures in medicine, biology and material development. Riederer, Stuttgart 1975.
Fava, N.A., L.A. Santaló
1978 Plate and line segment processes. J. Appl. Prob. **15** (1978), 494–501.
1979 Random processes of manifolds in $\mathbb{R}^n$. Z. Wahrscheinlichkeitstheorie verw. Gebiete **50** (1979), 85–96.
Federer, H.
1959 Curvature measures. Trans. Amer. Math. Soc. **93** (1959), 418–491.
1969 Geometric measure theory. Springer, Berlin-Heidelberg-New York 1969.
Fejes Tóth, L., H. Hadwiger
1947 Mittlere Trefferzahlen und geometrische Wahrscheinlichkeiten. Experientia **3** (1947), 366–369.
1948 Über Mittelwerte in einem Bereichsystem. Bull. Inst. polytechn. Jassy **3** (1948), 29–35.
Firey, W.J.
1977 Addendum to R.E. Miles' paper on the fundamental formula of Blaschke in integral geometry. Austral. J. Statist. **19** (1977), 155–156.
1979a Inner contact measures. Mathematika **26** (1979), 106–112.
1979b Problem 18. In Contributions to Geometry, Proc. Geom. Symp. Siegen 1978, Ed. J. Tölke and J.M. Wills. Birkhäuser, Basel-Boston-Stuttgart 1979, 259.
Giger, H.
1967a Ermittlung der mittleren Masszahlen von Partikeln eines Körpersystems durch Messungen auf dem Rand eines Schnittbereichs. Z angew. Math. Phys. **18** (1967), 883–888.
1967b A system of basic stereologic formulae. In Elias (1967), 257–258.
1968 Zufallsmoirés. Optica Acta **15** (1968), 511–519.
1970 Grundgleichungen der Stereologie I. Metrika **16** (1970), 43–57.
1972a Rasterable pointsets. In Weibel et al. (1972), 197–202.
1972b Grundgleichungen der Stereologie II. Metrika **18** (1972), 84–93.
1975 Rasterbare Texturen. Z. Angew. Math. Phys. **26** (1975), 521–536.
1977 Zufallsmoirés. Z. Angew. Math. Phys. **28** (1977), 205–212.

Giger, H., H. Hadwiger
1968 Über Treffzahlwahrscheinlichkeiten im Eikörperfeld. Z. Wahrscheinlichkeitstheorie verw. Gebiete **10** (1968), 329–334.
Giger, H., H. Riedwyl
1970 Bestimmung der Grössenverteilung von Kugeln aus Schnittkreisradien. Biometrische Z. **12** (1970), 156–162.
Goldsmith, P.L.
1967 The calculation of true particle size distributions from the sizes observed in a thin slice. Brit. J. Appl. Phys. **18** (1967), 813–830.
Grenander, U.
1976 Pattern synthesis. Lectures in pattern theory I. Springer, New York 1976.
Grimaldi, V., C. Tanasi
1981 Statistical distribution of convex nonoverlapping particles in the Euclidean n-dimensional space. Revista Un. Mat. Argentina **30** (1981), 11–18.
Groemer, H.
1972 Eulersche Charakteristik, Projektionen und Quermaßintegrale. Math. Ann. **198** (1972), 23–56.
1977 On translative integral geometry. Arch. Math. **39** (1977), 324–330.
1978 On the extension of additive functionals on classes of convex sets. Pacific J. Math. **75** (1978), 397–410.
1980a The average distance between two convex sets. J. Appl. Prob. **17** (1980), 415–422.
1980b The average measure of the intersection of two sets. Z. Wahrscheinlichkeitstheorie verw. Gebiete **54** (1980), 15–20.
Gundersen, H.J.G.
1977 Notes on the estimation of the numerical density of arbitrary profiles: the edge effect. J. Microscopy **111** (1977), 219–223.
Haas, A., G. Matheron, J. Serra
1967 Morphologie mathématique et granulométries en place. Ann. Mines **11** (1967), 736–753, **12**, 767–782.
Hadwiger, H.
1951 Beweis eines Funktionalsatzes für konvexe Körper. Abh. Math. Sem. Univ. Hamburg **17** (1951), 69–76.
1952 Additive Funktionale k-dimensionaler Eikörper, I. Arch. Math. **3** (1952), 470–478.
1956 Integralsätze im Konvexring. Abh. Math. Sem. Univ. Hamburg **20** (1956), 136–154.
1957 Vorlesungen über Inhalt, Oberfläche und Isoperimetrie. Springer, Berlin 1957.
1959 Normale Körper im euklidischen Raum und ihre topologischen und metrischen Eigenschaften. Math. Z. **71** (1959), 124–140.
1968 Geometrische Wahrscheinlichkeiten bei Durchstichen von Geraden durch Kugelflachen. Mitteil. Vereinig. schweiz. Versicherungsmath. **68** (1968), 27–35.
1974 Räumlich-geometrische Wahrscheinlichkeiten und Mittelwerte. Z. angew. Mech. **54** (1974), 664–667.
1975a Eine Erweiterung der kinematischen Hauptformel der Intergralgeometrie. Abh. Math. Sem. Univ. Hamburg **44** (1975), 84–90.
1975b Eikörperrichtungsfunktionale und kinematische Integralformeln. Lecture Notes, Bern 1975.
Hadwiger, H., R. Schneider
1971 Vektorielle Integralgeometrie. Elemente Math. **26** (1971), 49–57.
Hadwiger, H., F. Streit
1970 Über Wahrscheinlichkeiten räumlicher Bündelungserscheinungen. Monatsh. Math. **74** (1970), 30–40.
Hanisch, K.-H.
1980 On classes of random sets and point process models. Elektron. Informationsverarb. Kybernet. **16** (1980), 498–502.
1981 On classes of random sets and point process models. Serdica **7** (1981), 160–166.
Hanisch, K.-H., D. Stoyan
1979 Formulas for the second-order analysis of marked point processes. Math. Operationsforsch. Statist., Ser. Statistics **10** (1979), 555–560.
1981 Stereological estimation of the numerical density of the radial distribution function of centres of spheres. J. Microscopy **122** (1981), 131–141.

Harding, E.F., D.G. Kendall (Ed.)
1974 Stochastic Geometry. Wiley, London-New York-Sydney 1974.
Hess, C.
1979 Théorème ergodique et loi forte des grands nombres pour des ensembles aléatoires. C.R. Acad. Sci. Paris **288** (1979), 519–522.
Holgate, P.
1967 The angle-count method. Biometrika **54** (1967), 615–623.
Horálek, V.
1980 On evaluating exposures containing both extracted and sectioned particles. In Adam et al. (1980), 90–93.
1981 On the decomposition of particle size distribution in the extraction replica method. Aplikace matematiky **26** (1981), 401–417.
Horowitz, M.
1965 Probability of random paths across elementary geometrical shapes. J. App. Prob. **2** (1965), 169–177.
International Society for Stereology (Ed.)
1963 Proceedings of the first international congress for stereology. Vienna 1963.
Jakeman, A.J., R.S. Anderssen
1975 Abel type integral equations in stereology. I. General discussion. J. Microscopy **105** (1975), 121–134.
Jakeman, A.J., R.L. Scheaffer
1978 On the properties of product integration estimators for linear functionals of particle size distribution. Utilitas Math. **14** (1978), 117–128.
Janson, S., O. Kallenberg
1981 Maximizing the intersection density of fibre processes. J. Appl. Prob. **18** (1981), 820–828.
Jensen, E.B., H.J.G. Gundersen
1982 Stereological ratio estimation based on counts from integral test systems. J. Microscopy **125** (1982), 51–66.
Jensen, E.B., H.J.G. Gundersen, R. Osterby
1979 Determination of membrane thickness distribution from orthogonal intercepts. J. Microscopy **115** (1979), 19–33.
Keiding, N., S.T. Jensen, L. Ranek
1972 Maximum likelihood estimation of the size distribution of liver cell nuclei from the observed distribution in a plane section. Biometrics **28** (1972), 813–829.
Kellerer, A.M.
1971 Considerations on the random transversal of convex bodies and solutions for general cylinders. Radiation Res. **47** (1971), 359–376.
Kendall, D.G.
1974 Foundations of a theory of random sets. In Harding-Kendall (1974), 322–376.
Kendall, M.G., P.A.P. Moran
1963 Geometrical probability. Griffin, New York 1963.
Kerstan, J., K. Matthes, J. Mecke
1974 Unbegrenzt teilbare Punktprozesse. Akademie-Verlag, Berlin 1974.
Kingman, J.F.C.
1965 Mean free paths in a convex reflecting region. J. Appl. Prob. **2** (1965), 162–168.
1969 Random secants of a convex body. J. Appl. Prob. **6** (1969), 660–672.
König, D., D. Stoyan
1980 Stereological formulas through random marked point processes. In Adam et al. (1980), 46–49.
Koschitzki, S.
1980 Some stereological problems for random discs in $\mathbb{R}^3$. Math. Operationsforsch. Statist., Ser. Statistics **11** (1980), 75–83.
Koschitzki, S., D. Stoyan
1977 Intersections of a random process of linear segments with a given straight line. Serdica **3** (1977), 344–346.
Krickeberg, K.
1980 Statistical problems on point processes. In Mathematical Statistics. Banach Center Publ. **6**, Warsaw 1980, 197–222.

Kulle, R.-D.
1971 Über die Messung des spezifischen Flächeninhalts. Nachr. Akad. Wiss. Göttingen. II. Math.-phys. Kl. 1971, 209–215.
Kulle, R.-D., A. Reich
1973 Flächenmessung mit gleichverteilten Folgen. Nachr. Akad. Wiss. Göttingen. II. Math.-phys. Kl. 1973, 217–225.
Langevin, R., T. Shifrin
1982 Polar varieties and integral geometry. Amer. J. Math. **104** (1982), 553–605.
Lewis, P.A.W. (Ed.)
1972 Stochastic point processes. Statistical analysis, theory and applications. Wiley, NewYork 1972.
Likeš, J.
1981 Determination of number and size of spherical particles from extraction replicas. Biom. J. **23** (1981), 795–810.
Little, D.V.
1974 A third note on recent research in geometrical probability. Adv. Appl. Prob. **6** (1974), 103–130.
Maak, W.
1970 Über Längen- und Inhaltsmessung. Nachr. Akad. Wiss. Göttingen. II. Math.-phys. Kl. 1970. 57–66.
Mandelbrot, B.B.
1977 Fractals: Form, chance, dimension. Freeman, San Francisco-London 1977.
Mase, S.
1980 Maximum likelihood estimation of size distribution for a Poisson aggregate of balls. Technical Report **22,** Hiroshima University 1980.
1982 Asymptotic properties of stereological estimators of volume fraction for stationary random sets. J. Appl. Prob. **19** (1982), 111–126.
Matern,B.
1960 Spatial variation. Medd. statens. skogsforskningsinst. **49** (1960), 5.
Matheron, G.
1967 Éléments pour une théorie des milieux poreux. Masson, Paris 1967.
1969 Théorie des ensembles aléatoires. École de Mines, Paris 1969.
1972a Ensembles fermés aléatoires, ensembles semimarkoviens et polyèdres poissoniens. Adv. Appl. Prob. **4** (1972), 508–541.
1972b Random sets theory and its applications to stereology. In Weibel et al. (1972), 15–23.
1975 Random sets and integral geometry. Wiley, New York-London-Sydney-Toronto 1975.
1976 La formule de Crofton pour les sections épaisses. J. Appl. Prob. **13** (1976), 707–713.
1978 La formule de Steiner pour les érosions. J. Appl. Prob. **15** (1978), 126–135.
McMullen, P.
1975 Non-linear angle-sum relations for polyhedral cones and polytopes. Math. Proc. Camb. Phil. Soc. **78** (1975), 247–261.
McMullen, P., R. Schneider
1983 Valuations on convex bodies. (to appear).
Mecke, J.
1980a Formulas for stationary planar fibre processes III-Intersections with fibre systems. Math. Operationsforsch. Statist., Ser. Statistics **12** (1980), 201–210.
1980b Palm methods for stationary random mosaics. In Ambartzumian (1980), 124–132.
1981 Stereological formulas for manifold processes. Probab. Math. Statist. **2** (1981), 31–35.
1982 Modelle der stochastischen Geometrie. In Geobild '82, Friedrich-Schiller-Universität, Jena 1982, 9–17.
Mecke, J., W. Nagel
1980 Stationäre räumliche Faserprozesse und ihre Schnittzahlrosen. Elektron. Informationsverarb. Kybernet. **16** (1980), 475–483.
Mecke, J., D. Stoyan
1980a A general approach to Buffon's needle and Wicksell's corpuscle problem. In Ambartzumian (1980), 164–171.
1980b Formulas for stationary planar fibre processes I-General theory. Math. Operationsforsch. Statist., Ser. Statistics **11** (1980), 267–279.
1980c Stereological problems for spherical particles. Math. Nachr. **96** (1980), 311–317.

Miles, R.E.
1964 Random polygons determined by random lines in a plane I, II. Proc. Nat. Acad. Sci. USA **52** (1964), 901–907, 1157–1160.
1969 Poisson flats in Euclidean spaces. I: A finite number of random uniform flats. Adv. Appl. Prob. **1** (1969), 211–237.
1970a On the homogeneous planar Poisson process. Mathematical Biosciences **6** (1970), 85–127.
1970b A symposis of 'Poisson flats in Euclidean spaces'. Izv. Akad. NaukArmjan. SSR Ser. Mat. **5** (1970), 263–285. Reprinted in Harding-Kendall (1974), 202–227.
1971a Poisson flats in Euclidean spaces. II: Homogeneous Poisson flats and the complementary theorem. Adv. Appl. Prob. **3** (1971), 1–43.
1971b Isotropic random simplices. Adv. Appl. Prob. **3** (1971), 353–382.
1972 Multi-dimensional perspectives on stereology. In Weibel et al. (1972), 181–195.
1973a On the information derivable from random plane and line sections of an aggregate of convex particles embedded in an opaque medium. Proc. 4th Conf. Prob. Theory, Brasov 1971, Acad. Rep. Soc. Romania 1973, 305–317.
1973b The various aggregates of random polygons determined by random lines in a plane. Advances in Math. **10** (1973), 256–290.
1974a The fundamental formula of Blaschke in integral geometry and geometrical probability, and its iteration, for domains with fixed orientations. Austral. J. Statist. **16** (1974), 111–118.
1974b On the elimination of edge effects in planar sampling. In Harding-Kendall (1974), 228–247.
1975 Direct derivations of certain surface integral formulae for the mean projections of a convex set. Adv. Appl. Prob. **7** (1975), 818–829.
1976a On estimating aggregate and overall characteristics from thick sections by transmission microscopy. In Unterwood et al. (1976), 3–12.
1976b Estimating aggregate and overall characteristics from thick sections by transmission microscopy. J. Microscopy **107** (1976), 227–233.
1978a The importance of proper model specification in stereology. In Miles-Serra (1978), 115–136.
1978b The sampling, by quadrats, of planar aggregates J. Microscopy **113** (1978), 257–267.
1979 Some new integral geometric formulae, with stochastic applications. J. Appl. Prob. **16** (1979), 592–606.
1980a The random tangential projection of a surface. Adv. Appl. Prob. **12** (1980), 425–446.
1980b The stereological application of integrals of powers of curvature and absolute curvature for planar curve data. In Adam et al. (1980), 27–31.
1980c On the underlying relationship of plane section stereology to statistical sampling theory. In Adam et al. (1980), 13–18.
1981a Stereological formulae based upon planar curve sections of surfaces in space. J. Microscopy **121** (1981), 21–27.
1981b A survey of geometrical probability in the plane, with emphasis on stochastic image modeling. In Image Modeling, Ed. A. Rosenfeld, Academic Press, New York-London-Toronto-Sydney-San Francisco 1981, 277–300.

Miles, R.E., P. Davy
1976 Precise and general conditions for the validity of a comprehensive set of stereological fundamental formulae. J. Microscopy **107** (1976), 211–226.
1977 On the choice of quadrats in stereology. J. Microscopy **110** (1977), 27–44.
1978 Particle number or density can be stereologically estimated by wedge sections. J. Microscopy **113** (1978), 45–51.

Miles, R.E., J. Serra (Ed.)
1978 Geometrical probability and biological structures. Lect. Notes Biomath. **23**, Springer, Berlin-Heidelberg-New York 1978.

Monari, P.
1978 I fondamenti statistici della stereometria nella ricerca biologica. Statistica, Bologna **38** (1978), 507–515.

Moran, P.A.P.
1966 A note on recent research in geometrical probability. J. Appl. Prob. **3** (1966), 453–463.
1968 Statistical theory of a high-speed photoelectric planimeter. Biometrika **55** (1968), 419–422.
1969 A second note on recent research in geometrical probability. Adv. Appl. Prob. **1** (1969), 73–89.
1972 The probabilistic basis of stereology. In Nicholson (1972), 69–91.

Mürmann, M.G.
1978 Poisson point processes with exclusion. Z. Wahrscheinlichkeitstheorie verw. Gebiete **43** (1978), 23–38.
Neveu, J.
1977 Processus ponctuels. Lecture Notes in Mathematics 598. Springer, Berlin-Heidelberg-New York 1977.
Nicholson. W.L.
1970 Estimation of linear properties of particle size distributions. Biometrika **57** (1970), 273–297.
1976a Estimation of linear functionals by maximum likelihood. In Underwood et al. (1976), 19–24.
1976b Estimation of linear functionals by maximum likelihood. J. Microscopy **107** (1976), 323–334.
1978 Application of statistical methods in quantitative microscopy. J. Microscopy **113** (1978), 223–239.
Nicholson, W.L. (Ed.)
1972 Proc. Symp. Statist. Prob. Problems in Metallurgy, Seattle 1971. Suppl. Adv. Appl. Prob. 1972.
Nicholson, W.L., K.R. Merckx
1969 Unfolding particle size distributions. Technometrics **11** (1969), 707–723.
Nguyen, X.X.
1979 Ergodic theorems for subadditive spatial processes. Z. Wahrscheinlichkeitstheorie verw. Gebiete **48** (1979), 159–176.
Nguyen, X.X., H. Zessin
1979 Ergodic theorems for spatial processes. Z. Wahrscheinlichkeitstheorie verw. Gebiete **48** (1979), 133–158.
Ohmann, D.
1952 Ungleichungen zwischen den Quermassintegralen beschränkter Punktmengen I. Math. Annalen **124** (1952), 265–276.
1954 Ungleichungen zwischen den Quermassintegralen beschränkter Punktmengen II. Math. Annalen **127** (1954), 1–7.
1955 Eine Verallgemeinerung der Steinerschen Formel. Math. Annalen **129** (1955), 209–212.
1956 Ungleichungen zwischen den Quermassintegralen beschränkter Punktmengen III. Math. Annalen **130** (1956), 386–393.
Ohser, J.
1980 On statistical analysis of the Boolean model. Elektron. Informationsverarb. Kybernet. **16** (1980), 651–653.
1981 A remark on the estimation of the rose of directions of fibre processes. Math. Operationsforsch. Statist., Ser. Statistics **12** (1981), 581–585.
Parker, P., R. Cowan
1976 Some properties of line segment processes. J. Appl. Prob. **13** (1976), 96–107.
Piefke, F.
1976 Estimation of linear properties of spherical bodies in thin foils from their projections. In Underwood et al. (1976), 497–498.
1978a Beziehungen zwischen der Sehnenlängenverteilung und der Verteilung des Abstandes zweier zufälliger Punkte im Eikörper. Z. Wahrscheinlichkeitstheorie verw. Gebiete **43** (1978), 129–134.
1978b Zwei integralgeometrische Formeln für Paare konvexer Körper. Z. angew. Math. Phys. **29** (1978), 664–669.
1979 The chord length distribution of the ellipse. Litovsk. Mat. Sb. **19** (1979), 45–54.
Pitts, E.
1981 The overlap of random particles and similar problems: Expressions for variance of coverage and its analogue. SIAM J. Appl. Math. **41** (1981), 493–498.
Pohl, W.F.
1981 The probability of linking of random closed curves. Lect. Notes Math. 894, Springer, Berlin-Heidelberg-New-York 1981, 113–126.
Pohlmann, S.
1980 Stationäre Flächenstückprozesse im $\mathbb{R}^3$. Elektron. Informationsverarb. Kybernet. **16** (1980), 495–497.
Pohlmann, S., J. Mecke, D. Stoyan
1981 Stereological formulas for stationary surface processes. Math. Operationsforsch. Statist., Ser. Statistics **12** (1981), 429–440.

Rehder, W.
1980 The asymptotic distribution of random molecules. Adv. Appl. Prob. **12** (1980), 640–654.
Reid, W.P.
1955 Distribution of sizes of spheres in a solid from a study of slices of the solid. J. Math. Phys. **34** (1955), 95–102.
Ripley, B.D.
1976a Locally finite random sets: Foundations for point process theory. Ann. Prob. **4** (1976), 983–994.
1976b The foundations of stochastic geometry. Ann. Prob. **4** (1976), 995–998.
1976c The second-order analysis of stationary point processes. J. Appl. Prob. **13** (1976), 255–266.
1981 Spatial statistics. Wiley, New York-Chichester-Brisbane-Toronto 1981.
Roach, S.A.
1968 The theory of random clumping. Methuen, London 1968.
Ruben, H., R.E. Miles
1980 A canonical decomposition of the probability measure of sets of isotropic random points in $\mathbb{R}^n$. J. Multivariate Anal. **10** (1980), 1–18.
Russel, A.M., N.S. Josephson
1965 Measurement of area by counting. J. Appl. Prob. **2** (1965), 339–351.
Saltykov, S.A.
1974 Stereometrische Metallographie. Deutscher Verlag f. Grundstoffind., Leipzig 1974.
Santaló, L.A.
1936 Integralgeometrie 5. Über das kinematische Maß im Raum. Hermann, Paris 1936.
1943 On the probable distribution of corpuscles in a body, deduced from the distribution of its sections, and analogous problems. Revista Un. Mat. Argentina **9** (1943), 145–164.
1970a Mean values and curvatures. Izv. Akad. Nauk Armjan. SSR Ser. Mat. **5**, 286–295. Reprinted in Harding-Kendall (1974), 165–175.
1970b Probabilities on convex bodies and cylinders. (Spanish). Revista Un. Mat. Argentina **25** (1970), 95–104.
1976a Integral geometry and geometric probability. Addison-Wesley, Reading 1976.
1976b On random segments in E_n. (Spanish). Rev. Univ. Nac. Tucumán A **26** (1976), 229–238.
1977 Sets of segments on surfaces. (Spanish). Math. Notae **26** (1977), 63–72.
1978 Random processes of linear segments and graphs. In Miles-Serra (1978), 279–294.
Scheaffer, R.L.
1973 Tests for uniform clustering and randomness. Comm. Stat. **2** (1973), 479–492.
Schneider, R.
1972a Krümmungsschwerpunkte konvexer Körper I. Abh. Math. Sem. Univ. Hamburg **37** (1972), 112–132.
1972b Krümmungsschwerpunkte konvexer Körper II. Abh. Math. Sem. Univ. Hamburg **37** (1972), 204–217.
1975a Kinematische Berührmaße für konvexe Körper. Abh. Math. Sem. Univ. Hamburg **44** (1975), 12–23.
1975b Kinematische Berührmaße für konvexe Körper und Integralrelationen für Oberflächenmaße. Math. Ann. **218** (1975), 253–267.
1977a Eine kinematische Integralformel für konvexe Körper. Arch. Math. **28** (1977), 217–220.
1977b Kritische Punkte und Krümmung für die Mengen des Konvexrings. L'Enseignement Math. **23** (1977), 1–6.
1978a Curvature measures of convex bodies. Ann. Mat. Pura Appl. **116** (1978), 101–134.
1978b Kinematic measures for sets of colliding convex bodies. Mathematika **25** (1978), 1–12.
1979a Boundary structure and curvature of convex bodies. In Contributions to Geometry, Proc. Geom. Symp. Siegen 1978, Ed. J. Tölke and J.M. Wills. Birkhäuser, Basel-Boston-Stuttgart 1979, 13–59.
1979b Integralgeometrie. Lecture Notes, Freiburg 1979.
1980a Parallelmengen mit Vielfachheit und Steiner-Formeln. Geom. Ded. **9** (1980), 111–127.
1980b Curvature measures and integral geometry of convex bodies. Rend. Sem. Mat. Univers. Politecn. Torino **38** (1980), 79–98.
1981a Crofton's formula generalized to projected thick sections. Rend. Circ. Mat. Palermo, Ser. II, **30** (1981), 157–160.
1981b A local formula of translative integral geometry. Arch. Math. **36** (1981), 466–469.

Serra, J.
1969 Introduction à la morphologie mathématique. Ecole de Mines, Paris 1969.
1972 Stereology and structuring elements. In Weibel et al. (1972), 93–103.
1978 One, two, three, ... infinity. In Miles-Serra (1978), 137–152.
1981 The Boolean model and random sets. In Image Modeling, Ed. A. Rosenfeld, Academic Press, New York-London-Toronto-Sydney-San Francisco 1981, 343–370.
1982 Image analysis and mathematical morphology. Academic Press, London et al. 1982.
Shepp, L.A., J.B. Kruskal
1978 Computerized tomography: The new medical X-ray technology. Amer. Math. Monthly **85** (1978), 420–439.
Šidák, Z.
1968 On the mean number and size of opaque particles in transparent bodies. In Studies in Mathematical Statistics, Ed. Sarkadi and Vincze, Akadémiai Kiadó, Budapest 1968, 161–168.
Slaviskiĭ, V.V.
1975 Integral-geometric relations with an orthogonal projection for surfaces. Siberian Math. J. **16** (1975), 275–284.
Smith, K.T., D.C. Solmon, S.L. Wagner
1977 Practical and mathematical aspects of the problem of reconstructing objects from radiographs. Bull. Amer. Math. Soc. **83** (1977), 1227–1270.
Snyder, D.L.
1975 Random point processes. Wiley, New York 1975.
Solomon, H.
1978 Geometric probability. Soc. Ind. Appl. Math., Philadelphia 1978.
Solomon, H., M.A. Stephens
1980 Approximations to densities in geometric probability. J. Appl. Prob. **17** (1980), 145–153.
Stoyan, D.
1979a Applied stochastic geometry: A survey. Biom. J. **21** (1979), 693–715.
1979b On some qualitative properties of the Boolean model of stochastic geometry. Z. angew. Math. Mech. **59** (1979), 447–454.
1979c Interrupted point processes. Biom. J. **21** (1979), 607–610.
1979d On the accuracy of lineal analysis. Biom. J. **21** (1979), 439–449.
1979e Proofs of some fundamental formulas of stereology for non-Poisson grain models. Math. Operationsforsch. Statist., Ser. Optimization **10** (1979), 575–583.
1982a A remark on the line transect method. Biom. J. **24** (1982), 191–195.
1982b Probleme und Methoden der Statistik für geometrische Strukturen. In Geobild '82, Friedrich-Schiller-Universität, Jena 1982, 18–37.
Stoyan, D., S. Koschitzki
1980 Some models and problems of stochastic geometry. (Russian). In Ambartzumian (1980), 133–156.
Stoyan, D., J. Mecke
1982 Stochastische Geometrie- Eine Einführung. (to appear).
Stoyan, D., J. Mecke, S. Pohlmann
1980 Formulas for stationary planar fibre processes II—Partially oriented fibre systems. Math. Operationsforsch. Statist., Ser. Statistics **11** (1980), 281–286.
Streit, F.
1970 On multiple integral geometric integrals and their applications to probability theory. Can. J. Math. **22** (1970), 151–163.
1973 Mean-value formulae for a class of random sets. J.R. Statist. Soc. **B35** (1973), 437–444.
1975 Results on the intersections of randomly located sets. J. Appl. Prob. **12** (1975), 817–823.
1976 On methods and problems of geometrical stochastics. Bull. Intern. Statist. Inst. **46** (2) (1976), 600–605.
1980 Analysis of spatially distributed objects. In Adam et al. (1980), 37–41.
Sukiasyan, G.S.
1980 On processes of chords on straight lines intersecting random fields of circles on the plane. (Russian). Dokl. Akad. Nauk Arm. SSR **70** (1980), 297–300.
Tallis, G.M.
1970 Estimating the distribution of spherical and elliptical bodies in conglomerates from plane sections. Biometrics **26** (1970), 87–103.

Tanasi, C.
1979 Statistical distribution of convex spherical domains in the unit sphere. Rend. Sem. Mat. Univers. Politecn. Torino **37** (1979), 139–144.
Underwood, E.E.
1970 Quantitative stereology. Addison-Wesley, Reading 1970.
1972 The stereology of projected images. In Weibel et al. (1972), 25–44.
Underwood, E.E., R. de Wit, G.A. Moore (Ed.)
1976 Stereology 4. Proceedings of the 4th international congress for stereology. U.S. National Bureau of Standards Special Publ. No. 431.
Voß, K.
1978 Zur numerischen Auswertung von Schnittflächenverteilungen IV. Biometrical J. **20** (1978), 425–434.
1980 Exakte stereologische Formeln und Näherungslösungen für konvexe Körper. Elektron. Informationsverarb. Kybernet. **16** (1980), 485–491.
1982 Powers of chords for convex sets. Biometrical J. **24** (1982), 513–516.
Watson, G.S.
1971 Estimating functionals of particle size distributions. Biometrika **58** (1971), 483–490.
1978 Characteristic statistical problems of stochastic geometry. In Miles-Serra (1978), 215–234.
Weibel, E.R.
1980 Stereological methods I, II. Academic Press, London-New York-Toronto 1980.
Weibel, E.R., H. Elias (Ed.)
1967 Quantitative methods in morphology. Springer, Berlin 1967.
Weibel, E.R., G. Meek, B. Ralph, P. Echlin, R. Ross (Ed.)
1972 Stereology 3. Proceedings of the third international congress for stereology. Blackwell, Oxford 1972. Reprinted from J. Microscopy **95** (1972).
Weil, W.
1979a Berührwahrscheinlichkeiten für konvexe Körper. Z. Wahrscheinlichkeitstheorie verw. Gebiete **48** (1979), 327–338.
1979b Kinematic integral formulas for convex bodies. In Contributions to Geometry, Proc. Geom. Symp. Siegen 1978, Ed. J. Tölke and J.M. Wills, Birkhäuser, Basel-Boston-Stuttgart 1979, 60–76.
1981 Zufällige Berührung konvexer Körper durch q-dimensionale Ebenen. Resultate der Mathematik **4** (1981), 84–101.
1982 Inner contact probabilities for convex bodies. Adv. Appl. Prob. **14** (1982), 582–599.
Wieacker, J.A.
1982 Translative stochastische Geometrie der konvexen Körper. Thesis, Universität Freiburg 1982.
Wicksell, S.D.
1925 The corpuscle problem I. Biometrika **17** (1925), 84–99.
1926 The corpuscle problem II. Biometrika **18** (1926), 151–172.
Zähle, M.
1982a Krümmungsmaße bei zufälligen Mengen. In Geobild '82, Friedrich-Schiller-Universität Jena 1982, 161–169.
1982b Random processes of Hausdorff rectifiable closed sets. Math. Nachr. **108** (1982), 49–72.
1982c Random set processes in homogeneous Riemannian space. (to appear).
1982d Thick section stereology for random fibres. (to appear).
Zalcman, L.
1980 Offbeat integral geometry. Amer. Math. Monthly **87** (1980), 161–175.

W. Weil
Mathematisches Institut II
Universität Karlsruhe
Englerstrasse 2
D–7500 Karlsruhe (FRG)

J.M. Wills

Semi-Platonic Manifolds

§1. Introduction

Polyhedral 2-manifolds which are locally combinatorial regular, were introduced in [10] and [11].

In this paper we investigate some of them with additional algebraic, geometric and topologic properties, which are closely related to the 5 Platonic solids.

We give a short survey, in which we confine ourselves to definitions, motivations, examples and simple results. For different and obvious reasons we are only interested in analogues of the Platonic solids in Euclidean 3-space E^3.

As the history, definitions and properties of the Platonic solids and their analogues have been excellently and exhaustively described in Coxeter's books and articles (e.g. [2], [3], [4], [5]), we only mention them as far as needed in this context.

All analogues of Platonic solids have one or more properties which differ from those of Platonic solids. So the 13 Archimedean solids have two or three types of regular faces. The 4 Kepler-Poinsot-polyhedra have regular faces with self-intersections and "false" vertices; the same holds for the 53 known uniform polyhedra [5] found by Badoureau, Brückner, Hess, Pitsch, Coxeter, Miller and Longuet-Higgins.

The compound bodies are star-polyhedra with nonconvex faces.

All these analogues have the symmetry group, resp. the rotation group of one of the Platonic solids, whereas the following two analogues have not:

(a) the regular tilings of the plane, which can be interpreted as degenerate regular polyhedra, and

(b) the apeirohedra or regular skew polyhedra by Petrie, Coxeter [4], and J. Gott [7], which are infinite (or non-compact) polyhedral 2-manifolds.

Last but not least we mention the higher dimensional regular polyhedra, which are not embeddable in E^3, but for $d = 4$ (via Schlegel-diagrams) are closely related to the Semi-Platonic manifolds.

After these different analogues (convex polyhedra in E^d, $d = 3$ and $d > 3$, star polyhedra, manifolds with self-intersections, infinite manifolds) it seems to be appropriate to introduce compact oriented 2-manifolds. There have been some attempts before in similar directions, compare e.g. [13], [14] and §3, last remark.

§2. Definitions

We use the notation of [8], [10], [11]. A polyhedral 2-manifold is a closed topological manifold in some euclidean space E^d (here always in E^3) which is the underlying point-set of a geometric 2-complex. Its facets (i.e. 2-cells) are convex polygons. So a polyhedral 2-manifold is a geometric model of an abstract 2-manifold, made up of convex polygons and, in this paper, always embedded in E^3, so necessarily oriented. We require further that adjacent faces, i.e. with a common edge, are not coplanar.

After this basic definition we call a polyhedral 2-manifold *equivelar* (i.e. with equal flags), if each of its faces has the same number p of edges and if each of its vertices belongs to the same number q of edges. A manifold of such type we write as {p,q;g}, where g denotes its genus.

If, as usual, $f_i = f_i(M)$, $i = 0,1,2$ denote the numbers of i-cells (vertices, edges, faces) of M, equivelarity implies $qf_0 = 2f_1 = pf_2$. From this and Euler's equation $f_0 - f_1 + f_2 = 2 - 2g$ follows

$$qf_0 = 2f_1 = pf_2 = 2(g-1)\left\{\frac{1}{2} - \frac{1}{p} - \frac{1}{q}\right\}^{-1} > 0. \qquad (*)$$

Equivelar manifolds have interesting extremal properties concerning the relations between the f_i, $i = 0,1,2$ and g [10], [11]. Infinitely many of them are realizable with the symmetry (resp. rotation) group of one of the Platonic solids. But if one adds a rigidity condition and a suitable topological restriction, one obtains only finitely many of them.

In the following we use the fact that a manifold M together with its interior (bounded) component, shortly $(M \cup \text{int}M)$, has the same homotopy-type as a (possibly degenerate) graph with straight lines. So this graph is a deformation-retract of $(M \cup \text{int}M)$ (compare e.g. [12]).

Definition

A polyhedral 2-manifold $M \subset E^3$ of genus g is called a *Semi-Platonic manifold* (abbrev.: SPM) if it has the following properties:

(A) Algebraic condition: M has the rotation group of one of the Platonic solids; i.e. the tetrahedral or octahedral or icosahedral group.

(B) Combinatorial condition: M is equivelar; i.e. all faces of M are convex p-gons, all vertices of M are q-valent.

(C) Geometric condition: The vertices of M lie on Platonic or Archimedean solids, and the rotation group of M acts transitively on the vertices of each solid.

(D) Topological condition: $(M \cup \text{int}M)$ has the same homotopy-type as a finite rectilinear graph R(M) in E^3 with the following properties:

(1) R(M) has the rotation group of M;
(2) All vertices of R(M) have same valence v;
(3) R(M) contains the 1-skeleton of at most two Platonic solids with the rotation group of M and at most one further vertex (i.e. a degenerate Platonic solid).

§3. Remarks

(1) For $g = 0$ (or if R(M) is a point), we obtain exactly the 5 Platonic solids.

(2) We use the word Platonic to make clear that the definition only works in E^3, whereas "regular" is used without dimension-restriction. The prefix "semi" means that some properties, which one might expect, do not hold. E.g. most of the known examples are not combinatorially regular ([9],

[11]). (L. Danzer and E. Schulte checked that at most {4,6;6}, {6,4;6}, are combinatorially regular.) Further, no example contains only regular faces.

(3) Condition (D) implies that only finitely many SPM exist. The only possible graphs are, as can easily be shown: a point, the 1-skeleton of a Platonic solid; the 1-skeleton of a 4-prism over a Platonic solid, the 1-skeleton of the 4-simplex and of the 16-cell. So the only possible genera are:

Tetrahedral group: g = 0,3,6,9,17

Octahedral group: g = 0,5,7,17,19

Icosahedral group g = 0,11,19,41,49

For $g = 49$ we do not know an SPM, for $g = 19$ one, and for all other $g > 0$ two SPMs.

With easy arguments we obtain further that the smallest positive value of $(1/2) - (1/p) - (1/q)$ is $(1/42)$. So from $g \leq 49$ and from (*) we have

$$qf_0 = 2f_1 = pf_2 \leq 42 \cdot 48 = 2016,$$

i.e the f_i, $i = 0,1,2$ are bounded, and only finitely many SPM exist. (Easy calculations give better bounds.) For an f-vector there may exist non-isomorphic SPMs, but at most finitely many, because of C).

(4) The condition D (3) needs some justification because the restriction "at most two" seems artificial. We mention, that if one omits "at most two", then surprisingly for each rotation group exactly one further graph exists, namely (see [15]).

one for the tetrahedral group with $g = 41$ and $v = 7$

one for the octahedral group with $g = 71$ and $v = 9$ and

one for the icosahedral group with $g = 253$ and $v = 10$.

(5) Condition (C) includes that in the tetrahedral case even icosahedra may occur, e.g. in {3,7;3}, which has tetrahedral rotation group and its vertices lie on icosahedra.

We remark that the icosahedron is not only a Platonic solid, but also the Archimedean snub polyhedron of the tetrahedron, which exists in a right-hand version and in a left-hand version [2].

(6) Therefore the three {3,7;g}, g = 3,5,11 have only the rotation group, but not the symmetry group of the corresponding Platonic solids.

That is because in all three cases there exists a right-hand version and a left-hand version of the snub polyhedra from which they are derived. In other words, these snub polyhedra, being chiral, occur in enantiomorphic pairs [2].

So the three {3,7;g} exist each in two dual versions. This coincides with the fact that among the Platonic solids the tetrahedron is self-dual.

(7) In all known examples there is an "inside-outside-symmetry", i.e. by an obvious inversion, which transforms the "interior" vertices outside and vice versa, the combinatorial structure of M remains unchanged.

(8) To make the SPMs rigid, one e.g. may require that the inradius of one solid coincides with the circumradii of the next solid. At the moment this seems to be of minor interest.

(9) Conditions (D) (and (C)) exclude $\{8,4;73\}$ and $\{4,8;73\}$ obtained from the 24-cell and the corresponding $\{6,4;g\}$ and $\{4,6;g\}$ obtained from the 120-cell and the 600-cell. The vertices of the corresponding retract-graphs lie on Platonic and Archimedean solids.

(10) In some cases the retract-graph R(M) is combinatorially regular, namely if it is a point or the 1-skeleton of a regular 3- or 4-polytope.

(11) For the SPMs and the apeirohedra of Coxeter, Petrie and Gott we have $(1/p) + (1/q) < (1/2)$; for the regular tilings of the plane: $(1/p) + (1/q) = (1/2)$ and for the Platonic solids, the Kepler-Poinsot-polyhedra and the uniform polyhedra (with fractional p,q) we have $(1/p) + (1/q) > (1/2)$.

(12) The condition that adjacent faces are not coplanar is essential, because there exist many additional manifolds which have all properties of SPMs except this one.

E.g. from the $\{4,6;g\}$ we obtain, by suitable triangulation (compare [10]), two $\{3,9;g\}$ with same g and same rotation group, which form an enantiomorphic pair. By a suitable triangulation of the hexagons (compare [10]), we obtain from each $\{6,4;g\}$ two $\{3,8;g\}$ with the same g which form another enantiomorphic pair. Besides these 20 manifolds there exist four $\{4,5;19\}$ and $\{3,7;g\}$, $g = 3,5,11$ two of each with coplanar adjacent faces.

(13) In [13] Stewart introduces polyhedral 2-manifolds with symmetry group of Platonic solids and with regular faces. Because different regular faces occur, and because the vertices have different valences; these manifolds could be named "Semi-Archimedean".

§4. Examples

The figures show 16 of the 19 known SPMs. It is easy to verify that (A), (B) and (D) are satisfied. To make the figures easier to understand and easier to construct, we have violated in some cases the condition in (C) that the vertices lie on Platonic or Archimedean solids.

We have also violated the condition that adjacent faces are not coplanar. The shaded parts should help to understand the shape of the SPM. To each figure of an SPM we have added its f-vector $f = (f_0,f_1,f_2) = |\chi(M)|\cdot(f'_0,f'_1,f'_2)$, where $\chi(M) = 2(1 - g)$ denotes the Euler-characteristic.

§5. Generating symmetric equivelar manifolds from the SPM

In this § we mention some simple processes to generate infinitely many manifolds using the SPM, which differ only slightly from the SPM. These examples show that the introduction of the graph R(M) and its properties is necessary.

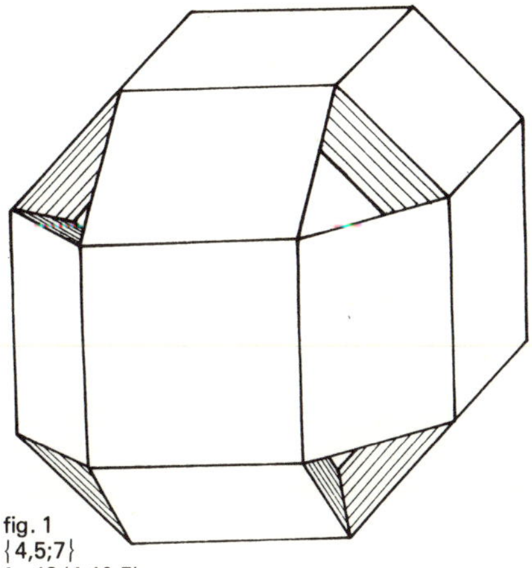

fig. 1
{4,5;7}
f = 12 (4,10,5)

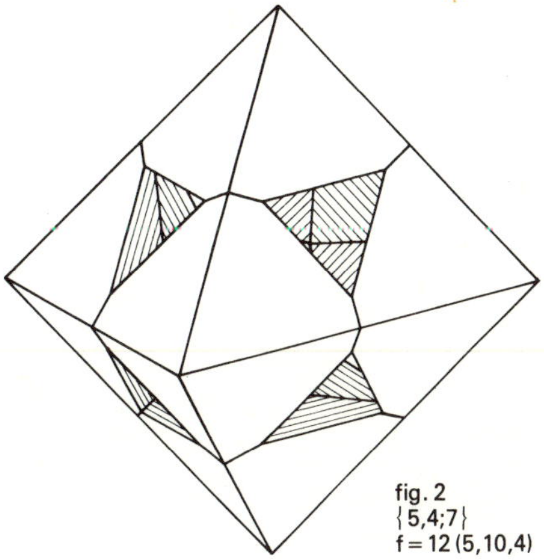

fig. 2
{5,4;7}
f = 12 (5,10,4)

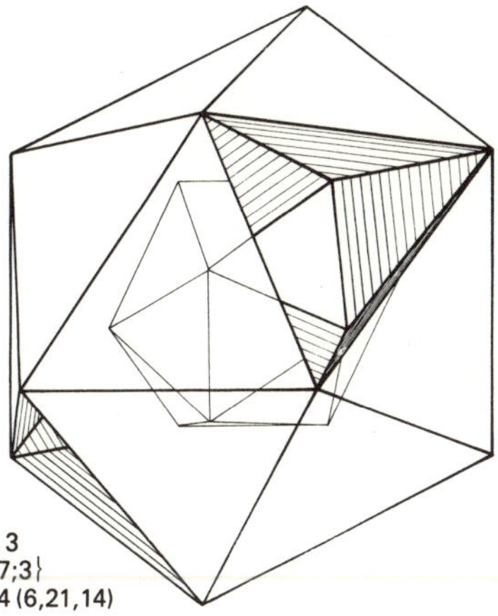

fig. 3
{3,7;3}
f = 4 (6,21,14)

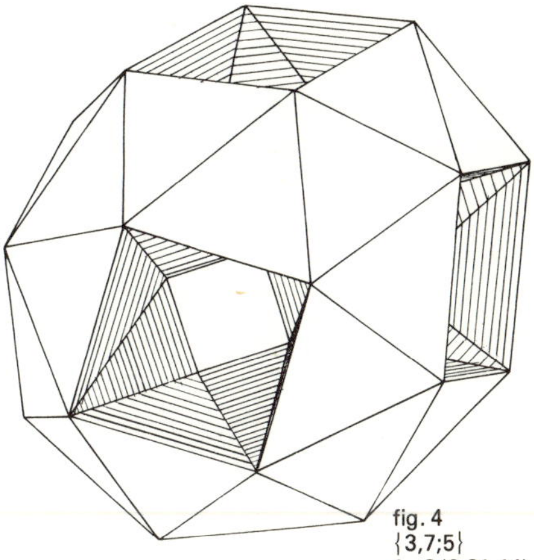

fig. 4
{3,7;5}
f = 8 (6,21,14)

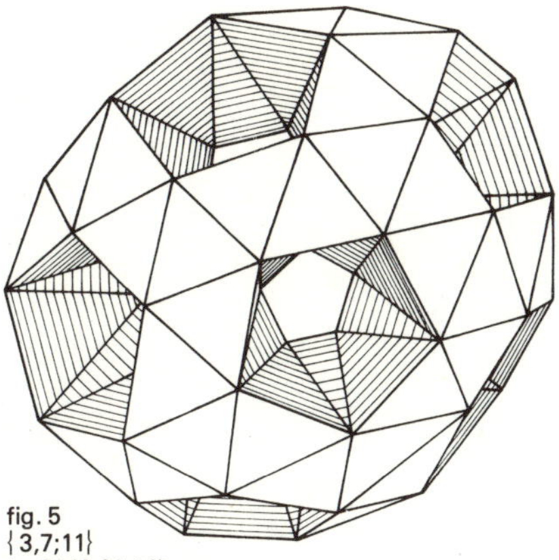

fig. 5
{3,7;11}
f = 20 (6,21,14)

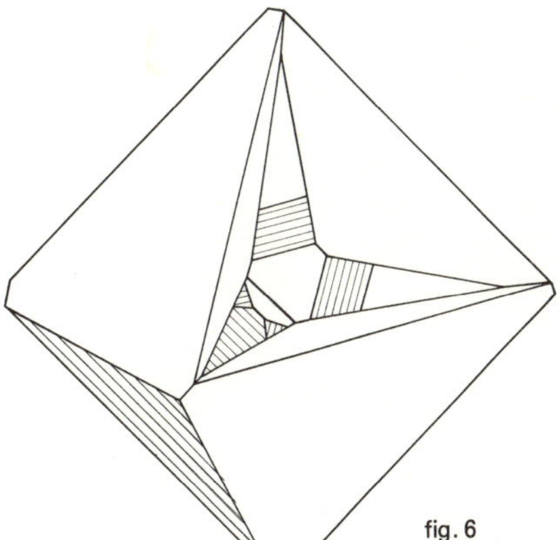

fig. 6
{6,4;6}
f = 10 (3,6,2)

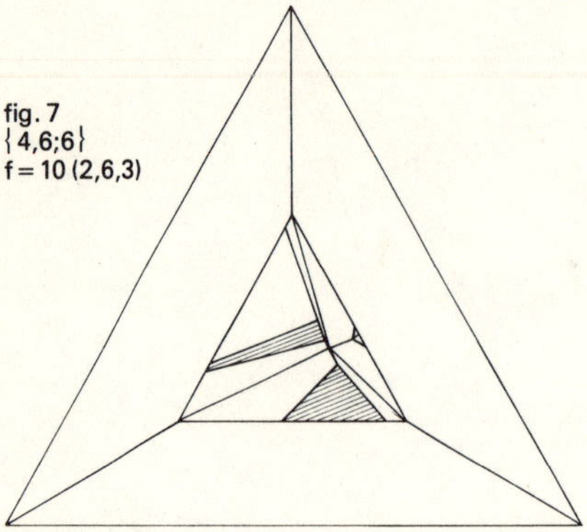

fig. 7
{4,6;6}
f = 10 (2,6,3)

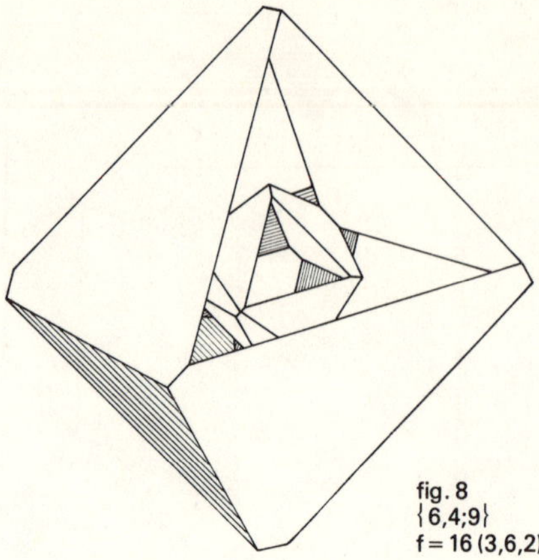

fig. 8
{6,4;9}
f = 16 (3,6,2)

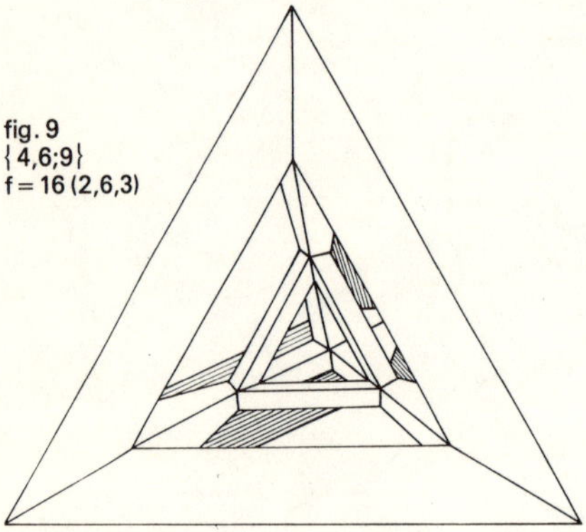

fig. 9
{4,6;9}
f = 16 (2,6,3)

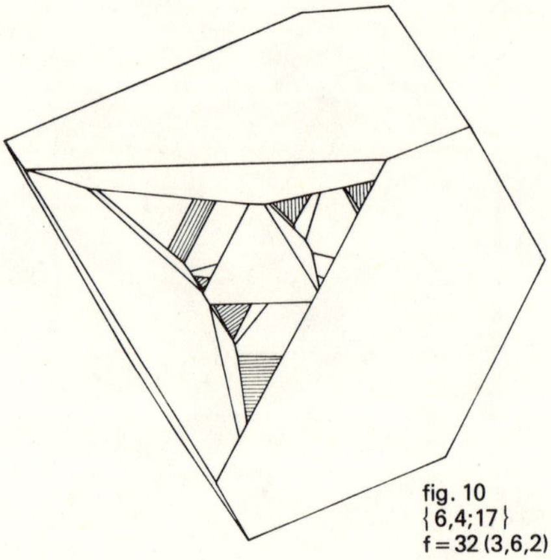

fig. 10
{6,4;17}
f = 32 (3,6,2)

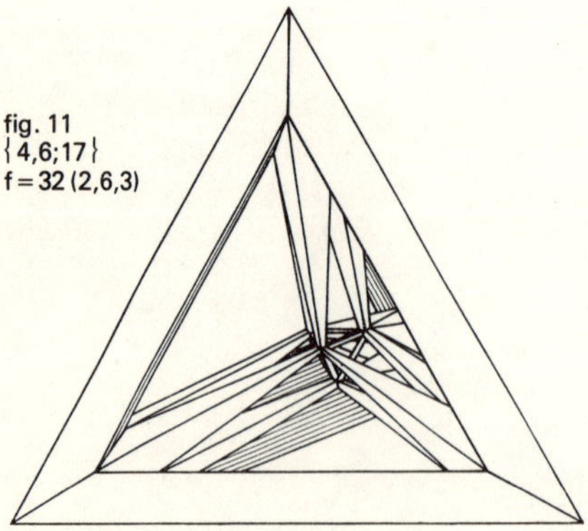

fig. 11
{4,6;17}
f = 32 (2,6,3)

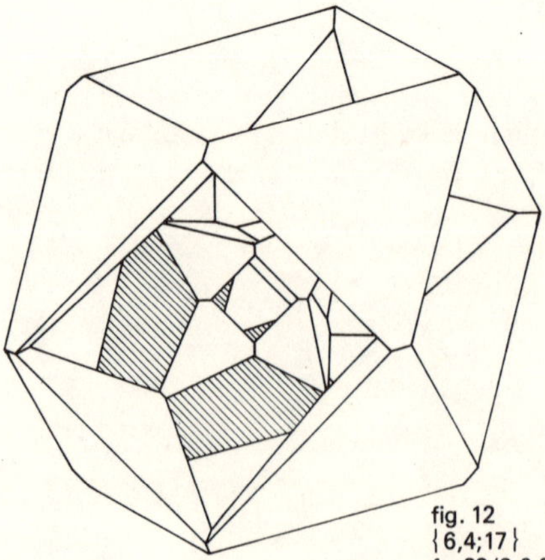

fig. 12
{6,4;17}
f = 32 (3,6,2)

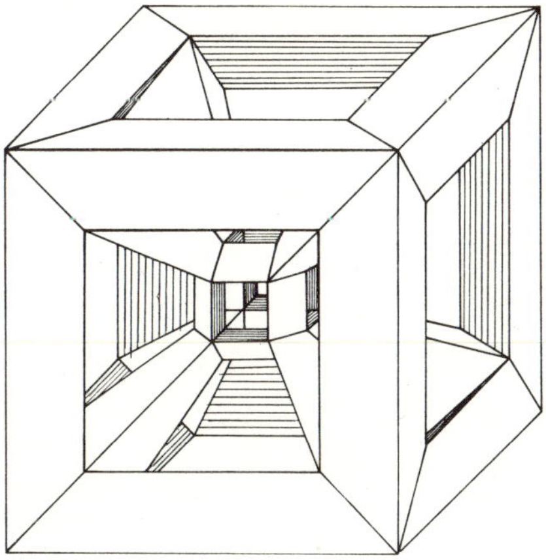

fig. 13
{4,6;17}
f = 32 (2,6,3)

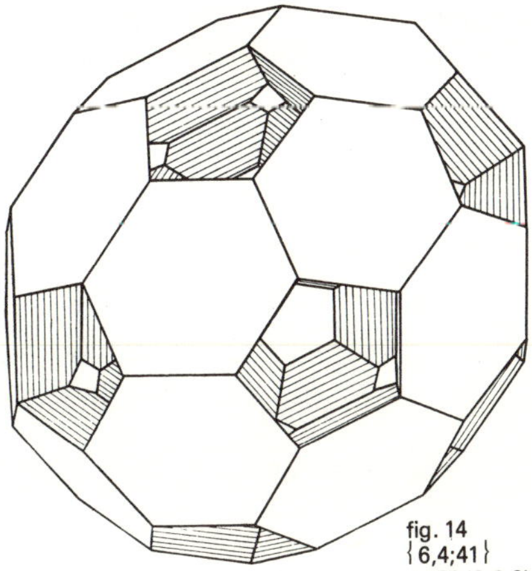

fig. 14
{6,4;41}
f = 80 (3,6,2)

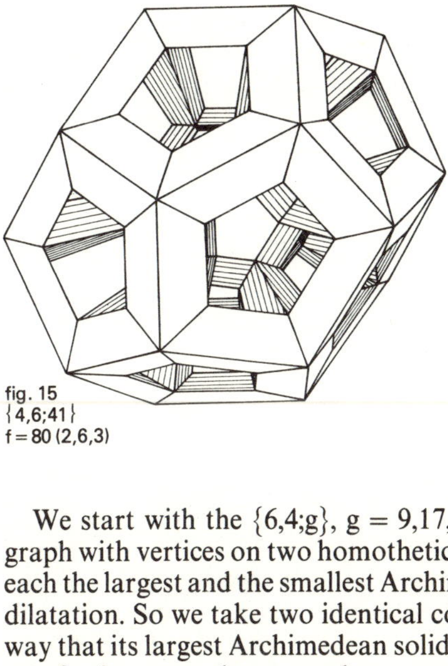

fig. 15
{4,6;41}
f = 80 (2,6,3)

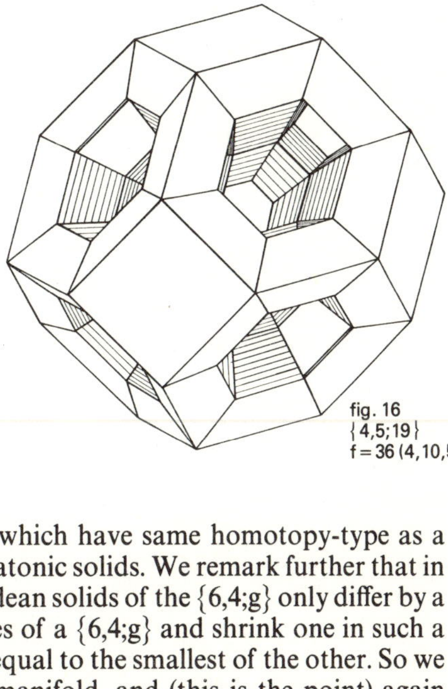

fig. 16
{4,5;19}
f = 36 (4,10,5)

We start with the {6,4;g}, g = 9,17,41 which have same homotopy-type as a graph with vertices on two homothetic Platonic solids. We remark further that in each the largest and the smallest Archimedean solids of the {6,4;g} only differ by a dilatation. So we take two identical copies of a {6,4;g} and shrink one in such a way that its largest Archimedean solid is equal to the smallest of the other. So we can fit them together to make a new 2-manifold, and (this is the point) again equivelar.

If we continue this process we obtain manifolds

$\{6,4;6(n-1)+3\}$ with tetrahedral symmetry group

$\{6,4;12(n-1)+5\}$ with octahedral symmetry group

$\{6,4;30(n-1)+11\}$ with icosahedral symmetry group,

where n = 2,3,4,... denotes the number of Platonic solids in the graph of the same homotopy type as M.

In the tetrahedral case we have the additional possibility of putting one $\{6,4;6\}$ as the interior manifold. So we obtain in addition:

$$\{6,4;6(n-1)\}, \quad n = 2,3,4,\ldots.$$

By a similar process, which we leave to the interested reader, we obtain from the

$\{4,6;g\}$, $g = 6,9,17,41$ the analogue $\{4,6;6(n-1)\}$ and
$\{4,6;6(n-1)+3\}$ with tetrahedral symmetry group
$\{4,6;12(n-1)+5\}$ with octahedral symmetry group
$\{4,6;30(n-1)+11\}$ with icosahedral symmetry group.

We obtain further manifolds by the same processes from the

$\{6,4;17\}$ and the $\{4,6;17\}$.

If we continue these processes to infinity, we obtain infinite (or non-compact) manifolds $\{4,6;\infty\}$ and $\{6,4;\infty\}$ with symmetry group of the Platonic solids.

These are rotation-symmetric analogues to the apeirohedra of Petrie and Coxeter.

We mention that we obtain other rotation-symmetric infinite manifolds by similar methods; namely

$$\{3,9;\infty\} \text{ from the } \{3,7;g\}, \quad g = 3,5,11$$

which are analogous to J. Gott's $\{3,10;\infty\}$ [7].
Further from $\{4,5;7\}$ we obtain $\{4,6;\infty\}$, and from $\{5,4;7\}$ we obtain a $\{6,4;\infty\}$, both with octahedral symmetry group; both different (and in a vague sense dual) from the above mentioned $\{4,6;\infty\}$ and $\{6,4;\infty\}$ with octahedral symmetry group.

So all of the known Semi-Platonic manifolds generate by easy constructions an infinite analogue to the Petrie-Coxeter-Gott apeirohedra with one of the rotation groups of the Platonic solids. Moreover these processes are related to the (infinite) processes of vertex-inscribing and edge-inscribing mentioned in Coxeter's [5], page 429.

At the end I want to thank U. Brehm, H.S.M. Coxeter, L. Danzer, P. McMullen, E. Schulte and G.C. Shephard for helpful discussions and suggestions.

References

[1] Brückner, M., Vielecke und Vielfläche; Teubner, Leipzig 1900.

[2] Coxeter, H.S.M., Regular polytopes, Dover (New York, 1973).

[3] Coxeter, H.S.M., Regular complex polytopes; Cambridge, England 1974.

[4] Coxeter, H.S.M., Regular skew polyhedra in three and four dimensions, and their topological analogues. Proc. London Math. Soc. (2) **43** (1937), 33–62. (Reprinted, with minor changes, in Twelve Geometric Essays, Southern Illinois University Press (Carbondale-Evansville, 1968).)

[5] Coxeter, H.S.M., M.S. Longuet-Higgins, J.C.P. Miller, Uniform Polyhedra; Phil. Trans. A. Vol. **246** (1954) 401–456.

[6] Fejes Tóth, L., Regular Figures; Pergamon (New York, 1964).

[7] Gott, J.R., Pseudopolyhedrons; Amer. Math. Monthly **74** (1967), 497–504.

[8] Grünbaum, B., Convex polytopes; Wiley, London-New York-Sydney, 1967.

[9] McMullen, P., Combinatorial regular polytopes; Matematika **14** (1967), 142–150.

[10] McMullen, P., Ch. Schulz, J.M. Wills, Equivelar polyhedral manifolds in E^3; Israel Journal of Math. **41** (1982) 331–346.

[11] McMullen, P., Ch. Schulz, J.M. Wills, Polyhedral manifolds in E^3 with unusually large genus; Israel Journal of Math.

[12] Seifert, H., W. Threlfall, Lehrbuch der Topologie; Leipzig 1934.

[13] Stewart, B.M., Michigan State University, private print, 1970.

[14] Wells, A.F., Three-dimensional Nets and Polyhedra; Wiley-Interscience (London-New York-Sydney, 1977).

[15] Wills, J. M., Finite Platonic graphs; Discrete Math., 1983.

J.M. Wills
Lehrstuhl für Mathematik II
Universität Siegen
Hölderlinstrasse 3
D–5900 Siegen (FRG)